2nd Edition

RESPONSES OF PLANTS TO ENVIRONMENTAL STRESSES

VOLUME I

Chilling, Freezing, and High Temperature Stresses

PHYSIOLOGICAL ECOLOGY

A Series of Monographs, Texts, and Treatises

EDITED BY

T. T. KOZLOWSKI

University of Wisconsin
Madison, Wisconsin

T. T. KOZLOWSKI. Growth and Development of Trees, Volumes I and II – 1971

DANIEL HILLEL. Soil and Water: Physical Principles and Processes, 1971

J. LEVITT. Responses of Plants to Environmental Stresses, 1972

V. B. YOUNGNER AND C. M. MCKELL (Eds.). The Biology and Utilization of Grasses, 1972

T. T. KOZLOWSKI (Ed.). Seed Biology, Volumes I, II, and III – 1972

YOAV WAISEL. Biology of Halophytes, 1972

G. C. MARKS AND T. T. KOZLOWSKI (Eds.). Ectomycorrhizae: Their Ecology and Physiology, 1973

T. T. KOZLOWSKI (Ed.). Shedding of Plant Parts, 1973

ELROY L. RICE. Allelopathy, 1974

T. T. KOZLOWSKI AND C. E. AHLGREN (Eds.). Fire and Ecosystems, 1974

J. BRIAN MUDD AND T. T. KOZLOWSKI (Eds.). Responses of Plants to Air Pollution, 1975

REXFORD DAUBENMIRE. Plant Geography, 1978

JOHN G. SCANDALIOS (Ed.), Physiological Genetics, 1979

BERTRAM G. MURRAY, JR. Population Dynamics: Alternative Models, 1979

J. LEVITT. Responses of Plants to Environmental Stresses, 2nd Edition. Volume I: Chilling, Freezing, and High Temperature Stresses, 1980

In Preparation

J. LEVITT. Responses of Plants to Environmental Stresses, 2nd Edition. Volume II: Water, Radiation, Salt, and Other Stresses, 1980

JAMES A. LARSEN. The Boreal Ecosystem, 1980

2nd Edition

RESPONSES OF PLANTS TO ENVIRONMENTAL STRESSES

VOLUME I

Chilling, Freezing, and High Temperature Stresses

J. LEVITT

Department of Plant Biology
Carnegie Institution of Washington
Stanford, California

1980

ACADEMIC PRESS

A Subsidiary of Harcourt Brace Jovanovich, Publishers

New York London Toronto Sydney San Francisco

ACADEMIC PRESS, INC.
111 Fifth Avenue, New York, New York 10003

United Kingdom Edition published by
ACADEMIC PRESS, INC. (LONDON) LTD.
24/28 Oval Road, London NW1 7DX

Library of Congress Cataloging in Publication Data

Levitt, Jacob, Date
Responses of plants to environmental stresses.

(Physiological ecology)
Includes bibliographies and index.
CONTENTS: v. 1. Chilling, freezing, and high temperature stresses.
1. Plants, Effect of stress on.
2. Plants--Hardiness. 3. Botany. I. Title.
QK754.L42 1979 581.2'4 79-51680
ISBN 0-12-445501-8 (v. 1)

PRINTED IN THE UNITED STATES OF AMERICA

80 81 82 83 9 8 7 6 5 4 3 2 1

Contents

IV. HIGH-TEMPERATURE STRESS

Preface

Until recently, a monograph on a specialized subject such as stress remained current and in need of only minor revisions for one or more decades. Now, the explosion of information has ended this relatively long life. Lange (1975), reviewing a mere two years of investigations in only one of the stress areas, covered 2700 publications and admitted that this did not include all the published works. On this basis, it can be estimated that the total number of publications in all areas of environmental stress *since* completion of the first edition of this monograph must be in the tens of thousands and of the same order as the total number published in the whole history of the subject *before* the first edition.

To allow this mass of publications to accumulate without attempting to incorporate them into an overall treatment of the subject would lead to a tragic loss of valuable information contributed by the intense efforts of countless able scientists. It would also leave the subject of environmental stress in an archaic and even chaotic state. Someone had to have either the courage or foolhardiness to attempt an integration of as many as possible of these newer investigations with one another and with the earlier work and to propose general principles based on this integration. Even if most of such proposals should prove to be incorrect, they will contribute to the science by suggesting the most reasonable direction of future meaningful research.

I have, therefore, taken the liberty of hypothesizing liberally throughout this second edition, which is composed of two volumes. The reader must always remember, however, that in our present state of ignorance all such hypotheses must be tested, not accepted. Some of them may even be proved incorrect on the basis of newer information appearing between completion of the manuscript and its publication.

I apologize to those scientists whose valuable work has been overlooked or inadequately treated in this first volume of the second edition.

J. Levitt

Preface to the First Edition

For many years, bits of information have been accumulating on the effects of stresses on plants. I have long felt the need to integrate these in an attempt to discover the basic principles. This need has now become more urgent due to the increasing importance of stress injuries, largely as a result of man's activities. Previously known stresses are becoming more important, and new ones are constantly arising. The practical aim is, therefore, to learn how to control the stresses, or to decrease the injuries they produce.

But the practical goal, though sufficient in itself, is not the sole reason for investigating environmental stresses. It has been said that to understand the normal cell we must study the abnormal cell. To paraphrase this statement, if we wish to understand life we must also study death. The causes of death as a result of exposure to environmental stresses are, therefore, of fundamental importance to all biology, and, for that matter, to all human activities since these are all impossible without life. An understanding of the nature of environmental stresses and of the plant's responses to them may, therefore, help to answer the age-old question: What is life?

It is, therefore, essential that we understand how stresses produce their injurious effects and how living organisms defend themselves against stresses. Why then confine our attention to plants? The simplest answer, of course, is my ignorance. But there is also another reason. The plant has succeeded in developing defenses against stresses that the animal (with few exceptions) has not developed, for instance, against freezing and drought. These also happen to be the stresses that have been most intensively studied. As a result, the research on animals has been mainly confined to responses of quite a different kind. At this stage, therefore, the resistance of plants to environmental stresses is a field in itself. This does not mean that investigations of other organisms can be completely ignored. Some of the most important aids to our understanding of the effects of stresses on plants have come from investigations of animal cells and microorganisms. Such information must, of course, be included.

I have covered four stresses in previous publications: "Frost Killing and Hardiness of Plants" (1941, Burgess, Minneapolis), "The Hardiness of Plants" (1956, Academic Press, New York), Frost, drought, and heat resistance (1958, *Protoplasmatologia* **6**), and Winter hardiness in plants [1966, *in* "Cryobiology" (H. T. Meryman, ed.), Academic Press, New York]. The first two are now out of print, and all are out-of-date. This monograph will include essentially all the environmental stresses which have been intensively investigated (with the exception of mineral deficiencies, which comprise too broad and involved a field to be incorporated with other stresses) and will attempt to bring the information on the above four stresses up-to-date. An attempt will then be made to analyze the possibilities of developing unified concepts of stress injury and resistance. The aim of this synthesis is, therefore, a comprehensive, unified, and molecular point of view. Descriptive aspects of the plant's responses have been largely excluded. For a diagnostic approach to the problem, the reader is referred to Treshow (1970, "Environment and Plant Response," McGraw-Hill, New York).

Only too often in the history of science, parallel investigations by different investigators have led to parallel but different systems of nomenclature. This has occurred in the field of stress research. Any attempt to integrate the results of such parallel investigations requires the adoption of a single, exactly defined terminology. In the case of stresses, this terminology should be applicable to all organisms, plant as well as animal. I have, therefore, attempted to introduce such a uniform terminology in this monograph. The earlier term "frost" has, for instance, been discarded in favor of "freezing," which is now used more generally by cryobiologists. Similarly, the term "tolerance" is adopted in place of the older "hardiness." It is my hope that such adoptions will clarify rather than confuse the concepts.

Unfortunately, the information explosion has prevented an all-inclusive integration. I tender my apologies to all investigators whose important contributions have not been included.

J. Levitt

I

STRESS CONCEPTS

1. Stress and Strain Terminology

The responses of plants to the severities of their environment have occupied the attention of man long before the beginnings of the science of biology (Levitt, 1941). To the farmer, plants that survive in these environments are "hardy," those that do not are "tender." The scientist, however, requires a more quantitative terminology. Therefore, in recent years, biologists have adopted the term *stress* for any environmental factor potentially unfavorable to living organisms, and *stress resistance* for the ability of the plant to survive the unfavorable factor and even to grow in its presence. Unfortunately, although stress has been exactly defined in mechanics, no such exact terminology has been developed in biology. Since the lack of exact terminology in science commonly leads to a lack of exact concepts, an attempt will first be made to apply the definitions of mechanics to biology. It must be recognized at the outset, however, that the mechanical and the biological stresses are not completely identical, and that, therefore, the terminology can be transferred only up to a point. Furthermore, the medical concept of stress (Selye,1973) is quite different from both the biological and the physical.

A. PHYSICAL STRESS AND STRAIN

According to Newton's laws of motion, a force is always accompanied by a counter force (Duff, 1937). If a body A exerts a force on body B, then body B must also exert a counter force on A. The two forces are called action and reaction and are parts of an inseparable whole, known as a stress. When subjected to a stress, a body is in a state of strain. The external force produces internal forces between contiguous parts of the body leading to a change in size or shape. The magnitude of the stress is the force per unit area. The magnitude of the strain is the change in dimension (e.g., length or volume) of the body.

Up to a point, which is specific for each body, a strain may be completely reversible. Such reversible strains are said to be *elastic*. Beyond this point, the strain will be only partially reversible, and the irreversible part is called the *permanent set* (Fig. 1.1). The permanent set is also called a *plastic strain*. The elastic strain produced in a specific body as a result of a specific stress will always be the same, and the strain is proportional to the stress. Therefore,

$$M = \text{stress/strain}$$

The constant M is known as the modulus of elasticity of the body, which differs for different bodies: the greater the modulus, the more elastic the body. The more elastic the body, the greater is its resistance to deformation (i.e., the larger the stress required to produce a unit strain). It should be noted that elasticity is *not* the same as elastic extensibility, which is a measure of the maximum possible elastic (i.e., reversible) strain. Unlike elastic strains,

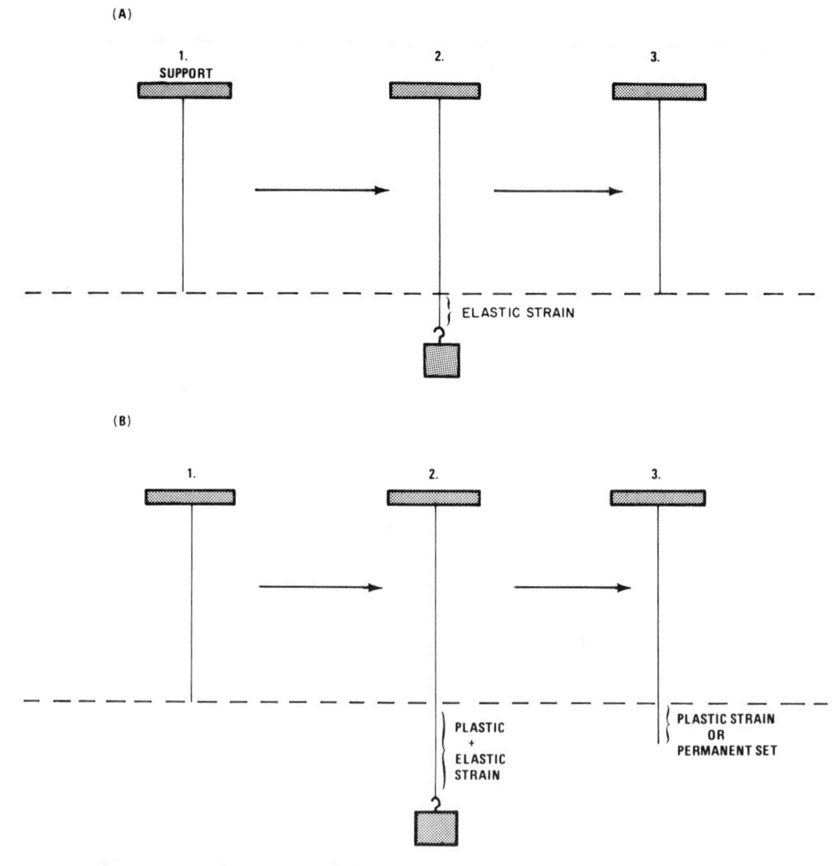

Figure 1.1. Elastic (A) and plastic (B) strains in a simple physical system.

plastic strains are not constant for specific stresses, since they may eventually lead to rupture of the body. There is, therefore, no modulus of plasticity.

B. BIOLOGICAL STRESS AND STRAIN

Biological stresses differ from mechanical stresses in two main ways. First, since the plant is able to erect barriers between its living matter and the environmental stress, the stress must be measured not in units of force but in units of energy. Second, the term stress in biology always has a connotation of possible injury—i.e., of irreversible or plastic strain. A biological stress may, therefore, be defined as any environmental factor capable of inducing a potentially injurious strain in living organisms. Since the biological stress is not necessarily a force, the biological strain is also not necessarily a change in dimension. The living organism may, however, show a physical strain or change (e.g., cessation of cytoplasmic streaming) or a chemical strain (a shift in metabolism). If either strain is sufficiently severe, the organism may suffer a permanent set, i.e., injury or death. Like the physical body, a specific organism will undergo a specific strain when subjected to a specific stress. It will, therefore, have its own modulus of elasticity, or resistance to physical or chemical change. By analogy with disease resistance, the term "elastic resistance" is more in agreement with biological terminology than modulus of elasticity. In biological systems, unlike physical systems, "plastic resistance" is more commonly measured than elastic resistance. Since plastic strains may be dependent on the time exposed to the stress, the time factor must be measured whenever the plastic resistance of biological systems is determined. The above stress terminology for the two systems is compared in Table 1.1.

The stress resistance of biological organisms is, therefore, of two main types. *Elastic resistance* is a measure of the organism's ability to prevent reversible or elastic strains (physical or chemical changes) when exposed to a specific environmental stress. *Plastic resistance* is a measure of its ability to prevent irreversible or plastic strains and, therefore, injurious physical or chemical changes.

One advantage of a precise biological terminology based on an analogy with mechanics now becomes apparent. The term resistance to environmental stresses has been mainly used for plastic resistance. The concept of an elastic resistance has not been as clearly recognized. There is, therefore, a whole field in physiology waiting to be investigated—a determination of the comparative elastic resistances of different organisms and an attempt to discover the mechanisms involved. As an example, when a corn plant is cooled from 30° to 5°C, its growth comes to a complete stop. Wheat, on the

TABLE 1.1

Stress Terminology

Term	Physical sense	Biological sense
Stress	A force acting on a body (F/A = dynes/cm^2 or bars)	An external factor acting on an organism (e.g., bars of water stress)
Strain	A change in dimension produced by a stress	Any physical or chemical change produced by a stress
Elastic strain	A reversible change in dimension	A reversible physical or chemical change
Plastic strain	An irreversible change in dimension	An irreversible physical or chemical change
Modulus of elasticity (or elastic resistance)	Stress/elastic strain	Intensity of external factor/amount of reversible physical or chemical change
Modulus of plasticity (or plastic resistance)	Not measured	Intensity of external factor producing a standard irreversible physical or chemical change[a]

[a]The organism must be exposed to the stress for a standard time.

other hand, continues to grow, though at a slower rate. In both cases, when returned to the normal growing temperature, normal growth is resumed. The strain is, therefore, reversible, i.e., elastic. Why does the corn plant suffer a greater elastic strain than the wheat plant when cooled? Or, using resistance terminology, what is the cause of the greater elastic resistance of wheat than corn when cooled?

Another advantage is that the importance of the time factor becomes obvious in the case of plastic strains. The plastic stretch of a wire may be just as dependent on the time exposed to the stress as on the stress itself. Similarly, injury to an organism is just as dependent on the time exposed to a high-temperature stress as on the high temperature used. On the other hand, this is not completely true of freezing stresses, as will be seen below.

There are two pronounced differences, however, between the responses of a nonliving body and of a living organism to stress.

1. Plastic strains in biological systems may be reparable. As in the case of the physical systems, the plastic strain will increase with the stress, producing more and more injury; the plastic strain is irreversible only in the spontaneous (thermodynamic) sense. The plant may be able to repair the strain by an active expenditure of metabolic energy. As the stress increases, the plastic strain also increases until the "rupture" point, when the strain is irreversible both thermodynamically and by metabolic repair, and the plant is killed. It is obvious, then, that stress resistance has two main components:

(a) The innate internal properties (or "forces") of the plant which oppose (i.e., resist) the production of a strain by a specific stress. (b) The repair system which reverses the strain. Only the first of these is analogous to the modulus of elasticity in physical systems.

2. Living organisms are adaptable. They are, therefore, capable of changing gradually in such a way as to decrease or prevent a strain when subjected to a stress. Both the elastic and plastic resistances of a plant to a specific stress may, therefore, increase (or decrease). This adaptation may be either stable, having arisen by evolution over a large number of generations, or unstable, depending on the developmental stage of the plant and the environmental factors to which it has been exposed. The unstable adaptation must, of course, also have arisen by evolution, but the hereditary potential is wide enough to permit large changes during the growth and development of the organism.

This adaptation is important both in the case of elastic and plastic strains. Plastic strains are by definition injurious. Therefore the adaptation leading to increased plastic resistance will obviously prevent injury by a stress which injures the unadapted organism. This kind of adaptation has been called "resistance adaptation" by Precht *et al.* (1955), since the adaptation implies a resistance to injury. Injury due to elastic strain would seem to be precluded, by analogy with nonliving systems. Although elastic strains are reversible by removal of the stress and therefore, by definition, are noninjurious, it must be realized that if they are maintained for a long enough period, they may lead to injury and even death. This may simply be due to the inability of the organism to compete with others that undergo less elastic strain when subjected to the same stress (e.g., mesophiles versus psychrophiles at low temperatures). The elastic strain may also eventually injure the plant even in the absence of competition, due to a disturbance of the metabolic balance. Thus, a low-temperature stress may simply decrease the rates of all metabolic processes reversibly, but not all may be decreased to the same degree. Therefore, if the stress is maintained for a long enough period, the strain may conceivably lead to an accumulation of toxic intermediates or to a deficiency of essential intermediates. In either case, a long enough exposure to the stress may injure or kill the organism. An adapted organism, on the other hand, may live, grow, complete its life cycle, and regenerate in the presence of the stress. This kind of adaptation has been called "capacity adaptation" by Precht (1967). Resistance adaptation may not permit growth and may merely prevent the plastic strain and therefore the injury until the stress is removed or decreased to the level permitting growth and development.

Nevertheless, both adaptations involve a *resistance* to the effects of a stress; on the one hand, a resistance to elastic strain, and, on the other, a resistance to plastic strain. It seems better, then, to include them both under the heading of stress resistance, using the terms elastic adaptation and plastic adaptation for Precht's "capacity adaptation" and "resistance adaptation," respectively. A plant showing elastic adaptation would develop elastic resistance, and one with plastic adaptation, plastic resistance (Diag. 1.1).

In the case of animals, elastic adaptation has been intensively investigated and plastic adaptation has been largely ignored (Precht, 1967). In plants, the reverse has been the case. The subject of stress resistance in plants will, therefore, be confined essentially to plastic resistance. Nevertheless, there is a dependence of plastic resistance on elastic strain. This is illustrated in Fig. 1.2, which relates the two types of strain to the stress and the time factor. The vertical graph, to the left, illustrates the instantaneous response. With increase in stress, there is an increase in elastic (reversible) strain up to the yield point, the strain being proportional to the stress. Beyond the yield point, a plastic (irreversible) strain occurs, and the strain increases more rapidly than the stress. The horizontal graph with the two vertical panels illustrates the time-dependent response. If a small stress, producing only an elastic strain is maintained for some time, two kinds of adaptation are possible (first vertical panel). (1) The strain may decrease with time to a constant low value, leading to elastic (Precht's "capacity") adaptation, or (2) the strain may remain constant. In the latter case, even though the specific strain measured is constant, secondary changes may be induced in the plant, leading to a plastic (Precht's "resistance") adaptation. As a result, if the stress is now increased to a point that produced a plastic strain in the unadapted plant, no plastic strain now occurs because of plastic adaptation (second vertical panel).

This time-dependent response of the plant to an elastic strain occurs only in certain plants. There are actually three possible responses. (1) The elastic strain may remain constant and fully reversible, without leading to other

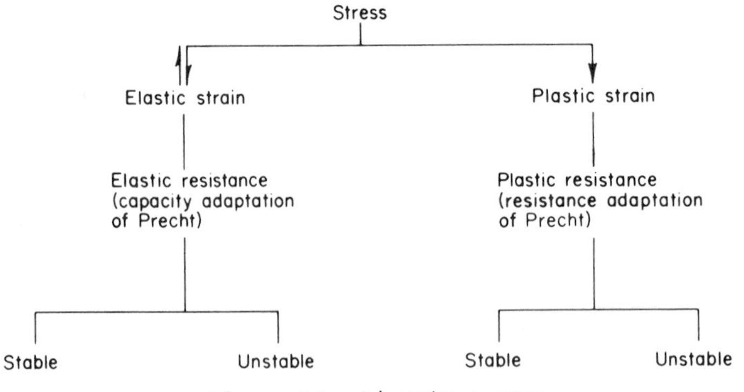

Diagram 1.1. Adaptation to stress.

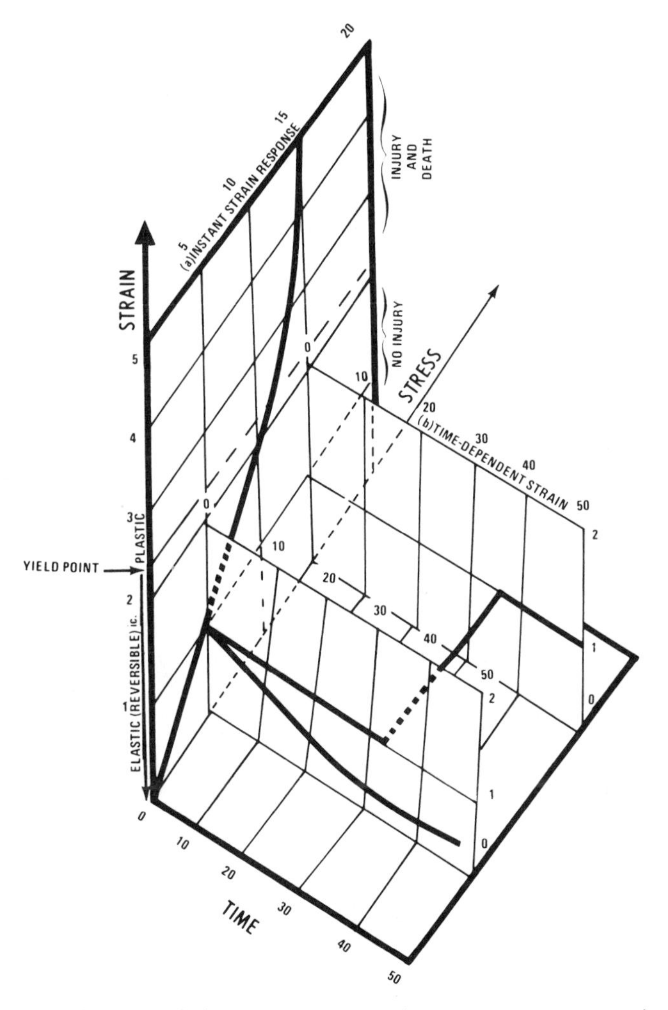

Figure 1.2. Adaptation of plants to stress. For explanation see text. Time scale may be in seconds, minutes, hours, or days, depending on the particular stress.

changes. (2) The elastic strain may be converted to an indirect plastic (injurious) strain (see Chapter 2). (3) The elastic strain may lead to secondary changes which induce either elastic or plastic adaptation, or both (Fig. 1.2). The distinction between the "tender" and "hardy" plants of the practical man now becomes clear. Those plants in which the first and second strain responses occur are, therefore, unable to adapt and belong to the "tender" group. Those in which the third effect occurs become adapted in the process and belong to the "hardy" group.

2. The Nature of Stress Injury and Resistance

A. STRESS INJURY

Although it is not possible to eliminate all of the stresses to which a plant may be exposed, it is possible to modify them or the strains that they are capable of producing. It is, therefore, essential that we understand how the stresses produce their injurious effects, and how some living organisms succeed in surviving stresses that injure others.

When a stress acts on a plant, it may produce an injury in different ways (Diag. 2.1).

1. It may induce a direct plastic strain which produces the injury. This may be called *direct stress injury* and may be recognized by the speed of its appearance. In such cases, the plant may be killed by very brief exposures to the stress (seconds or minutes). An example is the rapid freezing strain produced by a sudden low-temperature stress. If the protoplasm freezes, the ice crystals may lacerate the plasma membrane, thereby producing instant loss of semipermeability and death of the cell.

2. The stress may produce an elastic strain which is reversible and, therefore, not injurious of itself. If maintained for a long enough time, this elastic

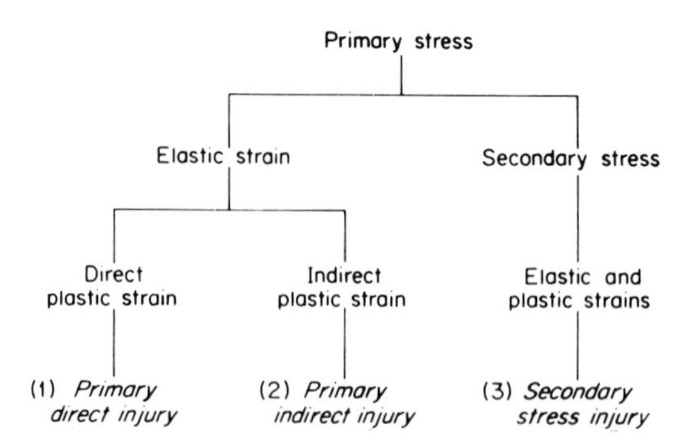

Diagram 2.1. Kinds of stress injury.

(reversible) strain may give rise to an indirect plastic (irreversible) strain which results in injury or death of the plant. This may be called *indirect stress injury*. Indirect injury may be recognized by the long exposure (hours or days) to the stress before injury is produced. An example is a chilling stress, which exposes the plant to a low temperature too high to induce freezing. In some cases, the strains may be mainly elastic—the slowdown of all the physical and chemical processes in the plant and, therefore, not injurious of themselves. In other cases, the slowdown may not be uniform for all processes and may produce a disturbance in the cell's metabolism, leading to a deficiency of a metabolic intermediate, or production of a toxic substance.

3. A stress may injure a plant, not by the strain it produces, but by giving rise to a second stress. A high temperature, for instance, may not be injurious of itself, but may produce a water deficit which may injure the plant. This may be called *secondary stress injury*. Since the secondary stress may require some time to develop, secondary stress injury may also require relatively long exposures to the primary stress. The secondary stress, in its turn, may also produce a direct or an indirect injury. Furthermore, it is conceivable that it may give rise to a tertiary stress, etc. The injury (whether direct or indirect) is, therefore, primary if caused by a primary stress, and secondary if caused by a secondary stress, etc. However, when any one stress is considered, only three kinds of injury will be discussed: (a) primary direct stress injury, (b) primary indirect stress injury, and (c) secondary stress injury (a tertiary stress may also occur).

B. STRESS RESISTANCE

1. Meaning

The modulus of elasticity (M) for physical systems is obtained from the equation:

$$M = \frac{\text{stress}}{\text{strain}} \tag{2.1}$$

As pointed out above, this modulus of elasticity measures the resistance of a body to a stress-induced elastic strain. Therefore, in the case of the plant, stress resistance (R) can be expressed in the same terms:

$$R = \frac{\text{stress}}{\text{strain}} \tag{2.2}$$

According to this equation, the stress resistance of a plant may be defined as the stress necessary to produce a specific strain. This definition may be applied to plastic as well as elastic strains. Some standard plastic strain (or injury) must be chosen, for instance, zero strain, or the stress that is just insufficient to induce a plastic strain. This is called the sublethal point, and is synonymous with the yield point in physical systems. The most commonly chosen standard strain, however, is the 50% killing point. Stress resistance then becomes the stress that is just sufficient to produce 50% killing. This value may be referred to as the K_{50} (the 50% killing point) or the LD_{50} (the 50% lethal dose).

Environmental stresses are of two main types—biotic and physicochemical ones. The former belongs to the field of pathology and ecology and will not be considered in this monograph. Among the physicochemical stresses, to which a plant may be exposed, some that have been unimportant until now may become more important due to man's explorations of outer space. Eight stresses have long been known to give rise to resistance adaptations (Diag. 2.2).

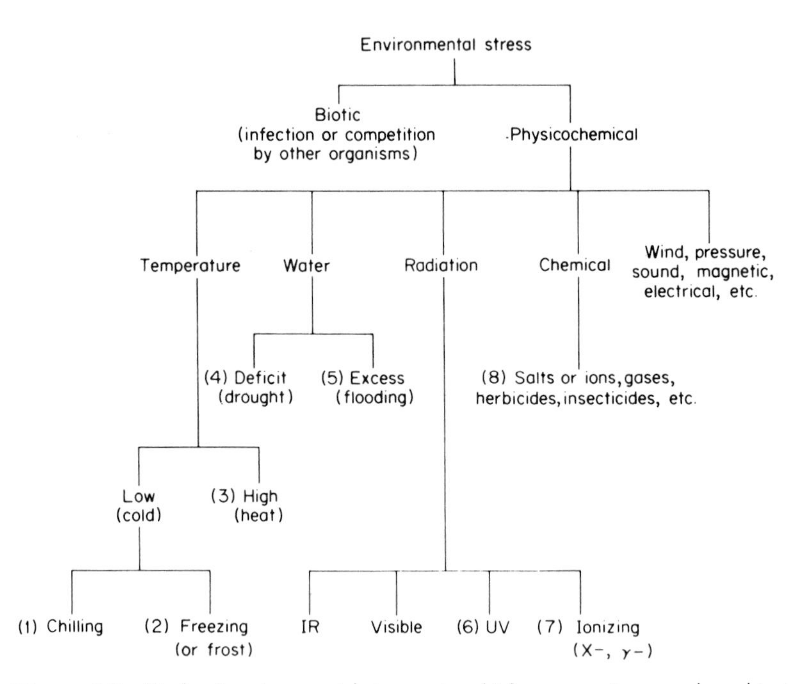

Diagram 2.2. Kinds of environmental stresses to which an organism may be subjected. Resistance against each of the numbered stresses is known.

2. Kinds of Resistance

From Eq. (2.2) above, two possible resistance methods are available to the plant: (*a*) it may increase the stress necessary to produce a specific strain, and (*b*) it may decrease the strain produced by a specific stress.

(a) Since the plant cannot alter an environmental stress external to itself, it can affect the stress only by preventing or decreasing the penetration of the stress into its tissues. This kind of resistance is called *stress avoidance*. Stress avoidance is stress resistance by avoiding thermodynamic equilibrium with the stress. The plant with avoidance is able to exclude the stress, either partially or completely, either by means of a physical barrier which insulates its living cells from the stress, or by a steady state exclusion of the stress (a chemical or metabolic barrier). By avoiding the stress, it also avoids the strain.

(b) The plant may be resistant in spite of permitting the stress to enter its tissues, provided that it decreases or eliminates the strain. This kind of resistance is called *stress tolerance*. Stress tolerance is stress resistance by an ability to come to thermodynamic equilibrium with the stress without suffering injury. The plant with stress tolerance is able to prevent, decrease, or repair the injurious *strain* induced by the stress. Though tolerating the stress, it may either avoid or tolerate the strain. Even the plastic (injurious) strain may be tolerated, provided that the injury is not irreversible and the plant possesses a repair mechanism capable of reversing it.

It should be pointed out that the term stress tolerance is sometimes used in the literature without determining whether or not the stress is tolerated internally. This is particularly true of the usage of the term "salt tolerance." In this monograph, all such uncertain cases will be called resistance, indicating that either avoidance or tolerance is involved. The meaning of the two main kinds of resistance becomes clear when they are considered in relation to specific stresses (Table 2.1).

Practical men have used the term "hardiness" as a synonym for resistance to a stress, particularly in the case of freezing, drought, and heat resistance. Horticulturists commonly use the term "acclimation" for adaptation to an environmental stress, particularly for the low-temperature stress.

The concept of hardiness is synonymous with toughness, and usually implies tolerance and not avoidance. The most common case is plants that harden (or become resistant) to freezing temperatures during the normal exposure to low temperatures in the fall. This hardening is an increase in tolerance. In at least some cases, however, hardening has been shown to involve an increase in avoidance. Soybeans that are allowed to wilt for some

TABLE 2.1

Twofold Nature of Stress Resistance

Stress	Condition of resistant plant cells exposed to the stress and surviving due to	
	Avoidance	Tolerance
1. Low (chilling) temperatures	Warm	Cold
2. High temperatures	Cool	Hot
3. Drought	High water potential	Low water potential
4. Radiation	Low absorption	High absorption
5. Salt (high conc.)	Low salt conc.	High salt conc.
6. Flooding (O_2 def.)	High O_2 content	Low O_2 content

days develop a cuticle less permeable to water than before wilting (Clark and Levitt, 1956). They are therefore able to recover from wilting without a decrease in the environmental water stress. Consequently, they increase their drought avoidance but they show no increase in drought tolerance. If hardening is taken to be strictly an increase in tolerance, such an increase in avoidance may be called a "pseudohardening."

In the case of seasonal stresses, some plants have developed an adaptation called *stress evasion*. These plants complete their life cycle before exposure to the stress by becoming dormant in the seed or spore stage. This adaptation is not a kind of resistance, but merely an ability to ensure reproduction when the stress is replaced by suitable growing conditions. *Stress avoidance* permits a plant to survive a stress though exposed to it in the vegetative state. A *stress evader* completes its vegetative stage before the stress appears.

3. Resistance to Individual Stresses

These are the theoretical possibilities, but each kind of stress must be examined individually in order to find out whether or not the plant has succeeded in developing the two kinds of resistance. The following brief survey will be discussed more fully in later chapters.

a. LOW TEMPERATURE (OR COLD)

Table 2.2 illustrates how closely the plant temperature agrees with that of its environment at low temperatures even below the freezing point. The inability of the plant to control its own temperature is further shown by the radiation cooling of plants below the air temperature at night and by the radiation heating of the cambium on the sunny side of trees to as high as

30°C above the shady side in winter. Plants are thus *poikilotherms*—they tend to assume the temperature of their environment. This means, by definition, that they cannot develop low-temperature avoidance. Exceptions are the fleshy inflorescences of the Araceae and so-called snow plants that develop temperatures above that of their environment due to rapid growth and the consequent rapid release of respiratory heat. However, this occurs only at environmental temperatures not low enough to induce injury.

The inescapable conclusion is that existing *low-temperature resistance can only be due to tolerance of low temperature.*

b. HIGH TEMPERATURE

The temperatures of leaves at high air temperatures in the full sun have long been a matter of controversy. The question has been whether or not the radiant heating by the sun can be counteracted by the transpirational cooling. At moderately high temperatures (up to 40°C) it now appears established, both from experimental evidence and theory, that the leaf temperature cannot be more than a very few degrees below the air temperature. At very high temperatures (40°–50°C) this is not true, and leaf temperatures may be considerably (as much as 15°C) below the air temperature, in this way exhibiting pronounced high-temperature avoidance. The leaves of most plants, however, survive in spite of temperatures above that of the air, and they must therefore possess high-temperature tolerance.

c. WATER DEFICIT

Aquatic plants and lower land plants are *poikilohydric* (Walter, 1955)—they rapidly come to equilibrium with the water in their environment. In some cases, they may be hydrated in the morning and air dry in the afternoon. These poikilohydric plants must therefore be drought tolerant. Higher land plants, on the other hand, may be considered *homoiohydric,* since they normally remain turgid and therefore near 100% relative humidity, although

TABLE 2.2

Temperatures (°C) of Tree Trunks (Elm and Red Fir) and of the Surrounding Air[a]

Air	Tree
−13 to −15	−12 to −14
−2.0	−1.0
−2.5	−0.5
−15.2	−14

[a]See Levitt, 1956.

daily exposed to the much lower relative humidity of the surrounding air. They, therefore, possess drought avoidance.

d. RADIATION

Ionizing radiations are highly penetrating and highly absorbed by the plant. Plants that survive such radiations must possess stress tolerance. Ultraviolet radiations may, however, be largely reflected and transmitted by the leaf as a whole or absorbed by the leaf surface. Survival of these radiations may therefore be due to avoidance.

e. SALTS

Halophytes are plants that grow in high salt soils. It has long been known that these plants contain high cell sap concentrations due to absorption of large quantities of salts. Osmotic values as high as 200 atm have been recorded. However, in at least some cases, these values are too high due to a redissolving of surface-excreted salts by the extracted juice. Even when care is taken to wash off this surface-excreted salt, high values have been found (100–130 atm) and these plants must therefore possess salt tolerance. In the case of varietal differences in mildly resistant plants (e.g., barley), it has been found that the more resistant variety excludes the salt better than the less resistant variety and therefore owes its resistance to avoidance.

The general conclusion from all the above examples, is that in the case of most stresses, the plant has succeeded in developing both stress tolerance and stress avoidance. However, even in such cases, the two kinds of stress resistance may not necessarily occur in the same plant. Therefore, the following three major classes of stress resistant plants may be found:

1. Stress tolerant but not avoiding (stress tolerant nonavoiders).
2. Stress tolerant and avoiding (stress tolerant avoiders)
3. Stress avoiding but not tolerating (stress intolerant avoiders).

Tolerance seems to be the more primitive adaptation, and avoidance more advanced (Table 2.3). This is reasonable, since tolerance involves an equilibrium state, and avoidance requires development by the plant of a mechanism to avoid equilibrium and to replace it by the steady state. Avoidance is also a more efficient adaptation; by avoiding the stress, the plant avoids both the elastic and the plastic strain. It is, therefore, able not only to survive when exposed to the stress, but also to metabolize, develop, and complete its life cycle. Tolerance (assuming plastic, but no elastic resistance), on the other hand, merely permits the plant to survive until such time that the stress is removed and the plant can recommence its normal metabolism, growth, and development.

On the same basis, drought resistance seems to be the most advanced of

TABLE 2.3

Relative Importance of Stress Tolerance and Stress Avoidance in Lower versus Higher Plants

Stress	Lower plants	Higher plants
Low temperature	Solely tolerance	Solely tolerance
Heat	Solely tolerance	Mainly tolerance
Drought	Solely tolerance	Mainly avoidance

the three types of resistance. This is also reasonable, since freezing and heat resistance no doubt had to be developed by plants before the evolution of species capable of completing their entire life cycle on land.

C. KINDS OF STRESS TOLERANCE

It is obvious that the mechanism of resistance in the case of any one stress will depend on whether the plant owes its resistance to avoidance or tolerance; the avoidance mechanism must be completely different from the tolerance mechanism. Furthermore, the avoidance mechanism need only be of one kind for any one stress, since it depends only on the nature of the stress and is independent of the kind of strain or injury. Tolerance, on the other hand, depends on the nature of the strain and the consequent injury. Since each stress may give rise to more than one kind of strain and injury (see above), there must also be more than one kind of tolerance. If both direct and indirect injury are produced by a single stress, there will be at least two tolerance mechanisms—*tolerance of plastic and elastic strains.* The stress tolerance mechanisms will further depend on whether it is due to *strain avoidance* or *tolerance*. There are therefore four possible mechanisms of stress resistance (Diag. 2.3). In each case, avoidance and tolerance are used in a quantitative sense.

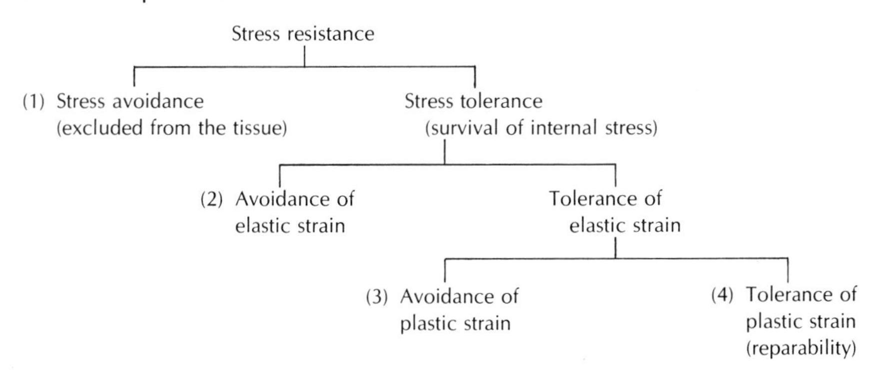

Diagram 2.3. The four possible mechanisms of stress resistance.

At first sight, the fourth resistance type appears impossible, since a plastic strain is irreversible and, therefore, cannot be tolerated. The living organism, however, may be capable of repairing, and therefore reversing a thermodynamically irreversible plastic strain. Only two of these have their counterparts in physical systems. The second mechanism—avoidance of elastic strain—would be proportional to the modulus of elasticity (stress/strain), which is really a measure of the stress that must be imposed in order to produce unit elastic strain. Obviously, the greater the stress that must be employed to produce this unit strain, the greater the elastic strain avoidance. The third mechanism—tolerance of elastic strain due to avoidance of plastic strain—is analogous to elastic extensibility. The greater the elastic extensibility of a body, the more it can be extended without suffering a plastic strain and therefore the greater its tolerance of elastic strain or avoidance of plastic strain.

The above four resistance mechanisms may themselves be subdivided. For instance, a dehydration strain resulting from a water stress is accompanied by cell contraction and solute concentration. It is, therefore, necessary to investigate all three of these elastic strains in order to understand which, if any, gives rise to injury. This problem will be discussed in detail in later chapters.

On the basis of the above concepts, the following law of stress resistance may be proposed:

Whatever the stress, a plant may achieve resistance to it by either an avoidance or a tolerance mechanism. These two mechanisms of resistance may be developed at any one of three levels—the stress level, the elastic strain level, or the plastic strain level.

3. Chilling Injury and Resistance

A. CHILLING STRESS

Although it has been recognized for centuries, the chilling stress is still very difficult to define quantitatively. As early as 1778, Bierkander reported on plants of eight species that were killed at 1° to 2°C above the freezing point (Molisch, 1896). Göppert (1830) obtained similar results. Of 56 species of tropical plants tested by Hardy in 1844, some 25 were killed at 1° to 5°C (see Molisch, 1896). Molisch (1897) suggested that low-temperature damage in the absence of freezing should be called chilling injury (Erkältung) as opposed to freezing injury (Erfrieren). On the basis of the above results, a chilling temperature can be defined as any temperature that is cool enough to produce injury but not cool enough to freeze the plant. Therefore, the theoretical definition of a chilling stress would be the number of degrees (or kelvins) that the environmental temperature is below optimum for the plant activity being measured (e.g., growth). In practice, this definition would be unworkable, since the optimum temperature for any plant process is not necessarily a constant, but may vary with other conditions. In most cases, plants do not suffer chilling injury until the temperature drops below 10°C. This may, therefore, be taken as the arbitrary zero point, and the plastic chilling stress may be defined as the number of degrees below 10°C. Unfortunately, rice (when in flower) and sugar cane may suffer chilling injury at 15°C (Adir, 1968; Tsunoda *et al.*, 1968), so that this definition would not apply to them. Similarly, the germination of cacao seeds is decreased at temperatures below 14°C (Boroughs and Hunter, 1963). Furthermore, exceptional plants (e.g., some fungi) may be killed at freezing temperatures in the absence of freezing—i.e., in the undercooled state (Lindner, 1915; Onoda, 1937). For the vast majority of plants, however, a chilling stress refers to any temperature below 10°–15°C, and down to 0°C.

As in the case of other stresses, the injury increases with the degrees of chilling. Cucumber leaves, for instance, survived 10°C for a week, showed injury at 8°C after 3 days, and at 5°C within a few hours (Minchin and Simon, 1973). In some cases, however, injury is greater at a higher than at a lower chilling temperature, for instance, at 3° than at 0°C (Lyons,1973).

B. CHILLING INJURY

The many symptoms of chilling injury are described by Lyons (1973). The symptoms which develop in the field, however, may be very different from those shown by fruit of the same plant as a result of chilling in storage (Platt-Aloia and Thomson, 1976).

Although chilling injury is primarily observed in plants from tropical or subtropical climates, certain cells of plants from temperate climates may be injured. For instance, at a foliage temperature of 0°–3°C, chilling produces sterility in wheat, when the pollen is in the stage of the first nuclear division (Toda, 1962). All three kinds of stress injury may occur: (1) direct, (2) indirect, and (3) secondary stress injury.

1. Direct Injury

Seible (1939) divides plants that suffer chilling injury into two types according to the speed of the reaction. The first type (e.g., *Episcia, Achimenes, Gloxinia*) show injured spots after hours or at the latest after a day, due to death of the protoplasm and infiltration of the intercellular spaces. The second type (e.g., *Tradescantia, Solanum, Coleus*) are more resistant. They remain perfectly normal and turgid for a full day but become soft and wilted only after chilling 5–6 days. Many fruits would fit into Seible's second type. Some apples, for instance, are injured by prolonged storage at temperatures below 36° to 40°F; but the fruit of tropical plants is much more sensitive—bananas may be injured by a few hours at temperatures below 55°F (Pentzer and Heinze, 1954).

Seible's first (rapid) type is presumably direct injury, the second (slow) type indirect injury. It is conceivable that the direct injury is due to a qualtitative, all-or-none physical change, the indirect injury to a slower quantitative, kinetic change in the chemical reactions of metabolism. However, there is no sharp line between the two. Many crop plants native to warm climates are intermediate between Seible's two types, showing injury after 24–48 hr at 0.5° to 5°C (Sellschop and Salmon, 1928). Nevertheless, at least some cases of injury appear to occur too rapidly to be explainable by metabolic disturbances, (Table 3.1) since chemical reactions occur slowly at chilling temperatures. This is particularly true of Möbius' observations of injury due to exposure to subfreezing temperatures for 1–2 min., and when a marine dinoflagellate is killed within 5 hr at 4°C (Eng-Wilmot *et al.*, 1977) or when soybean embryos are injured by imbibition at 2°C for as little as 5 min (Bramlage *et al.*, 1978). In such cases, the injury due to sudden chilling occurs so rapidly it may be called a *cold shock*. Cold shock is obviously a direct chilling injury, since it occurs too rapidly for either indirect stress effects to appear, or for development of a secondary stress.

Sachs (1864) observed a cessation of cytoplasmic streaming at 10–12°C in root hairs of cucumber and tomato, and Cohn described a *pseudoplasmolysis* in *Spirogyra* cells suddenly exposed to 0°C (see Molisch, 1897). Greeley (1901) and Livingston (1903) observed the same phenomenon. This was explained by a sudden increase in permeability resulting in leakage of cell solutes. The postulated leakage has, indeed, been confirmed by Lieberman *et al.* (1958). Chilled sweet potatoes showed five times as much leakage as the controls, almost all of it being K^+. In this case, somewhat longer chilling periods were required than for cold shock. Direct injury, therefore, may perhaps occur as a result of several hours' chilling, as well as when chilled abruptly for a few minutes. Increased leakage of K^+, for instance, was observed in as little as 3–6 hr after exposure of bean and corn root tips to 1°C (Wheaton, 1967—See Lyons, 1973), and in 5–14hr after exposure of Coleus petioles to 0.5° to 4.0°C (Katz and Reinhold, 1965). It continued to rise almost linearly for several hours, attaining a constant value after 20 hr. The damage at this time is apparently irreversible since the conductivity continued to rise if the petioles were transferred to room temperature after chilling for 22 hr. In contrast to coleus, a slight but rapid efflux occurred from leaf discs of *Cucumis sativus* floated on 1% sucrose at 2°C in the dark, but this was reversible within 4 days at 25°C (Tanczos 1977).

Two steps of leakage (mainly of K^+) were observed from leaves of *Passiflora* exposed to 0°C—a relatively slow rate, followed by a high rate during which most of the electrolyte was lost from the tissue, (Patterson *et al.,* 1976). The second step leakage occurred at about the same time as obvious lesions appeared, indicating lethal chilling injury.

These apparent increases in cell permeability as a result of somewhat longer chilling periods, seem to differ from the increase proposed by early investigators to explain chilling shock; the shock produced a rapid, spontaneous pseudoplasmolysis. In contrast, the epidermal cells of the coleus petioles plasmolyzed more slowly or not at all when chilled (Katz and Reinhold, 1965). However, the difference may be more apparent than real. The slower plasmolysis occurred in a hypertonic solution, and if there was a true increase in permeability, the plasmolyte may have penetrated the cell, slowing down or preventing plasmolysis. Casas *et al.* (1965) also suggest that membrane damage in cotyledonary tissues of cacao is a direct result of cold treatment. Restoration of viability following heat treatment could be due to a reversal of such a change. The sensitivity to low temperatures of a maize mutant (M11) also seems to be associated with membrane sensitivity (Millerd *et al.,* 1969). When grown in the light at 15°C, its plastids contain little pigment, they are deficient in ribosomes, their ultrastructure is abnormal, and the membrane shows an extreme sensitivity to light.

2. Indirect Injury

a. SOLUTE LEAKAGE

Slow chilling injury may require days or weeks of exposure to the temperature stress before its appearance (Table 3.2). Many suggestions have been made as to the cause of injury. An increase in permeability has been indicated by increased leakage, as in the case of direct injury, although the increase was only to three times the original value, after 4 weeks at 0°C, in the case of mature-green, chilling-sensitive tomato fruit (Lewis and Workman, 1964). Chilling-resistant cabbage showed no increase. Similarly, citrus leaves show chilling injury after 5 weeks at 1.7°C (Yelenosky, 1978). Light is essential for the injury to occur, the leaves becoming bleached and leaking amino acids. Since O_2 uptake is also decreased, the increase in leakage may be due to a decrease in active uptake. It has long been known that even in chilling resistant plants, such as barley, active, metabolism-dependent, ion uptake is sharply curtailed by chilling temperatures. Direct evidence of a relation between this low-temperature induced decrease in ion absorption and chilling injury has been produced by Wheaton (see Lyons, 1973). He found that chilling-sensitive roots of bean and corn, in opposition to resistant pea and wheat, not only leaked K^+, but also lost their ability to accumulate it at 1°C. One explanation for this loss is a respiratory disturbance. In the case of wheat roots (Nordin, 1977), ion influx was inhibited at chilling temperature (1° versus 25°C), but so was efflux, although to a lesser degree. The net uptake was reduced to 1/3 that of the unstressed plants, and the K^+ efflux increased to 1.5. These results were explained by an increase in per-

TABLE 3.2

Chilling Injury in Plants from Warm Climates[a]

Species	Time for first injury to appear	Time for complete killing (days)
Episcia bicolor Hook.	18 hr	5
Sciadocalyx warcewitzii Regel	24 hr	5
Eranthemum tricolor Nichols	48 hr	4–5
Eranthemum couperi Hook.	3–5 days	10
Boehmeria argentea Linden	8 days	20
Iresine acuminata	11 days	19
Uhdea bipinnatifida Kunth	15 days	16
Eranthemum nervosum R.Br.	20 days	30–35

[a]Exposed continuously to 1.4°–3.7°C in diffuse light and covered to prevent transpiration. From 28 species listed by Molisch (1897).

meability at the chilling temperature. In contrast to this explanation for chilling resistant plants, the above evidence appears to favor disruption of an ion uptake process in the case of the chilling sensitive plants. It, therefore, appears possible that the slower, indirect injury is due to a metabolic upset. Several kinds of metabolic disturbances have been described.

b. STARVATION

In the case of plants from temperate climates, a low temperature above freezing produces only an elastic (reversible) strain and the plant is normally uninjured. A similar elastic strain was early suggested as a possible cause of injury in chilling-sensitive plants from tropical or subtropical climates. At chilling temperatures, respiration rate may exceed the rate of photosynthesis, and this may conceivably lead eventually to starvation (Molisch, 1896).

One reason for the sharp drop in photosynthesis is damage to the chloroplast thylakoids at the chilling temperature (4°C). In the case of cucumber cotyledons, inactivation of the thylakoids is greater in the light than in the dark (Garber, 1977), and this occurs both *in vivo* and *in vitro*. The chilling resistant spinach chloroplasts were unaffected in vivo but behaved like the cucumber thylakoids in vitro. Thus, thylakoid membranes of chilling-sensitive species are less able to maintain a light-induced high energy state at chilling temperatures (Melcarek and Brown, 1977). Photooxidation at chilling temperatures may actually bring photosynthesis to a stop (see below). However, all observed cases of chilling injury appear to occur long before the reserves are used up. There is, therefore, no direct experimental confirmation of the starvation hypothesis.

Nevertheless, this concept must not be discarded too hurriedly. Starvation of nonphotosynthesizing plant parts may, for instance, result from the inhibition of translocation by chilling temperatures (Geiger, 1969). In sugar cane, translocation ceases completely at 5°C (Hartt—see Lyons, 1973). The inhibition of translocation may be due to an effect on the sink or on the translocatory path. Recovery of translocation occurs rapidly in sugar beet and in a northern ecotype of the Canada thistle, and very slowly in bean and in a southern ecotype of the thistle. In the case of the tropical grass, *Digitaria decumbens,* the high starch content accumulated during daylight disappears at night at normal temperatures but remains in the leaves when the temperature drops to 10°C (Hilliard and West, 1970). This inhibition of starch translocation out of the chloroplasts by low night temperature appears to account for the decreased photosynthesis and growth even at day temperatures of 30°C. In this case, root starvation is conceivable. Thus, exposure of maize to 2°C caused a sharp and continuing fall of soluble sugar in the root tips, and this was accompanied by reduced respiration and cessation of growth (Crawford and Huxter, 1977). Both respiration and growth were stimulated by a supply of exogenous sugar.

c. RESPIRATORY UPSET

The respiration rate of cucumber fruit increased to a plateau at the chilling temperature (41°F) then decreased (Eaks and Morris, 1956). The increase occurred at the onset of injury, the decrease on death. A change in oxygen concentration (1–10%) had little effect on the respiration rate at 41°F, although it had the usual effect at 59°F. Conversely, CO_2 increased the injury at 41°F but had no effect at 59°F. Similar results were obtained with sweet potato (Lewis and Morris, 1956). It was uninjured at 59°F injured at 50°F, and its respiration accelerated. The same result was obtained with the cotyledons of cacao seed (Ibanez, 1964). Chilled cotyledons showed a higher initial respiration than normal, but after 6 hr it was below normal. Eight minutes at 6°C reduced seed viability; 10 min at 4°C inhibited it completely.

The above results appear at first sight to be contradictory. They can all be explained, however, by an inhibition of the aerobic phase, without any inhibition of the anaerobic phase of the overall process of respiration. Several lines of evidence point to this conclusion.

1. Banga (1936) showed that in green tomatoes and bananas, the ratio of CO_2 evolution in nitrogen (anaerobic respiration) to that in air (aerobic respiration) was 0.7 at 13°C and 1.3–1.4 at 5°–6°C. Therefore, the chilling temperature has a greater inhibiting effect on the aerobic portion of respiration, for it decreases aerobic respiration twice as much as it decreases anaerobic respiration. After some time at the chilling temperature, aerobic respiration might be so greatly inhibited that anaerobic respiration would take over.

2. Murata (1969) observed the accumulation of the products of anaerobic respiration in banana fruit at 4°–6°C.

3. When mitochondria were isolated from healthy but chilling-sensitive plants and then exposed to a chilling temperature, the immediate and direct effect was to suppress respiration before the onset of injury, in opposition to the initial increase in whole tissues (Lyons, 1973). Since the aerobic (but not the anaerobic) phase of respiration takes place in the mitochondria, chilling must inhibit this aerobic phase.

4. The above observation that changing the O_2 concentration from 1–100% had little effect on the respiration rate of chilling-sensitive cucumber fruit at chilling temperatures, is explainable by the absence of aerobic respiration.

5. The respiratory upset due to chilling is in many respects analogous to injury due to low O_2 (see Lyons, 1973). Thus, low O_2 at normal temperatures may actually increase the rate of CO_2 evolution, as much as at normal O_2 concentrations in plants injured by chilling. This is due to the Pasteur effect—an increased rate of breakdown of substrate under anaerobic conditions, due to the inhibition of the aerobic phase of respiration and the consequent acceleration of the anaerobic phase. Since the Pasteur effect varies quantitatively from plant to plant, and even the anaerobic process may differ as to end-products, the effect of the inhibition of the aerobic part of respiration may be either an increase or a decrease in rate of CO_2 evolution. This explains not only the above apparent discrepancies, but also the failure of some investigators to detect a change in the respiratory quotient (Lyons, 1973).

In opposition to these chilling-sensitive plants, chilling resistant potatoes show a different kind of respiratory change at chilling temperatures (1.7°C). They develop a greater involvement of the pentose phosphate pathway (Dwelle and Stallknecht, 1978).

In the case of fruit, there is another possible cause of the respiratory rise associated with chilling injury—an increased production of ethylene, which is known to stimulate respiration. When citrus fruit were chilled by a temperature regime of 20°Day/5°Night, the ethylene in the fruit increased as much as 20–25 times (Cooper *et al.*, 1969). In the case of avocado, two varieties showed an increase and were injured, a third showed no increase and was uninjured.

Effects of chilling on cytoplasmic streaming may possibly be related to these respiratory changes, since streaming depends on utilization of respiratory energy. In all the chilling-sensitive plants tested (tomato, watermelon, honeydew, tobacco, sweet potato), streaming was just perceptible in the trichomes or ceased completely after 1–2 min at 10°C (Lewis, 1956). In the chilling-resistant plants, on the other hand (radish, carrot, filaree—*Erodium cicutarium*), streaming continued even at 0°–2.5°C.

Just what phase of the aerobic process is inhibited by the chilling has not yet been demonstrated, although some attempts have been made to measure individual steps in the process. Lewis and Workman (1964) exposed mature green tomato fruit to 0°C for 4 weeks. Within 12 days, the tomato lost two-thirds of its capacity to esterify phosphate at 20°C, and one-half of its original capacity at 0°C. In contrast, chilling-resistant cabbage leaves showed a steady rise in capacity to esterify phosphate during a 5-week exposure to the same chilling temperature. Stewart and Guinn (1969, 1971) have observed a similar decrease in oxidative phosphorylation in chilled cotton seedlings.

In the case of sweet potatoes, however, the chilling injury affects the oxidative activity as much as the phosphorylating capacity, and the P/O ratio remains constant (Minamikawa *et al.*, 1961). The respiratory activity increased slightly during the first 10 days at 0°C, then declined sharply. They concluded that the declining respiratory rate was limited by the oxidative system and not by the concentration of ADP and ATP. However, even if the P/O ratio is unchanged, this simply means that *both* phosphorylation and oxidative activity are inhibited by the chilling stress. There will, therefore, have to be a decrease in ATP, compared to that in an unstressed plant with normal aerobic respiration.

d. TOXINS

Plank (see Smith, 1954) has explained injury by the accumulation of a cell toxin due to disturbances in the normal balance of biochemical processes. Injury would then depend on whether the rate of accumulation of the toxin exceeds the rate of its dispersal. In plums, the evidence is in agreement with this concept (Smith, 1954). Thus, injury due to 21 days' storage at 31°F was prevented if the fruit were warmed to 65°F for 1 to 2 days at about the seventeenth day. Furthermore, the rate of injury was more rapid at 40° than at 34°F. Similar results have been obtained by others (Pentzer and Heinze, 1954).

The toxin concept is explained by the above-described inhibition of aerobic respiration on chilling, as follows:

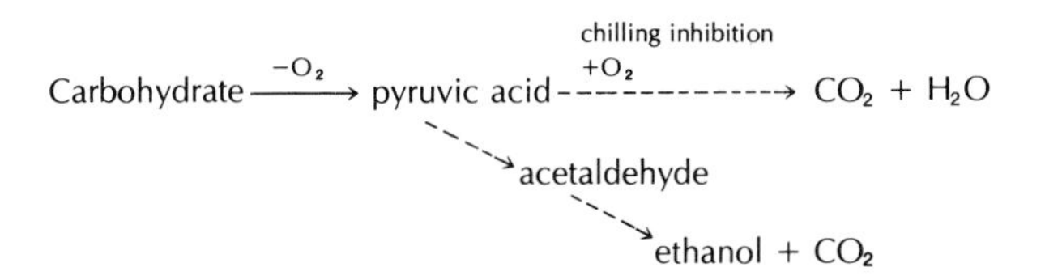

Any inhibition of the aerobic phase of respiration allows the anaerobic respiration to go to completion, producing toxic intermediates such as acetaldehyde, and end products such as ethanol. The initial increase in rate of respiration due to the Pasteur effect, would then be followed by an inhibition due to accumulation of toxins. It has long been known that continued anaerobic respiration kills most higher plants within 24–48 hr at room temperature. At chilling temperatures, it would undoubtedly take longer. Direct support of this concept has been produced by Murata (1969), who investigated chilling injury in banana fruit at 4°–6°C. Acetaldehyde and ethanol contents increased in the chilled fruits after transfer to 20°C. There was an accumulation of α-keto acids in the peel and of browning substances (polyphenols) around the vascular tissue. Similarly, apple scald is induced at chilling temperatures by the accumulated acetaldehyde and is inhibited by the antioxidant ethoxyquin (Lyons, 1973).

Other toxic products may also conceivably arise at the chilling temperatures; for the inhibition of aerobic respiration must leave a higher concentration of O_2 than normal in the tissues. This O_2 would be available to oxidases other than the cytochrome system normally metabolizing O_2 in an aerobically respiring unstressed plant. Peroxides would be among the products formed by some of these oxidases. Inhibition of photosynthesis may also lead to peroxide formation. Cucumber leaves in a moist chamber at 0°C showed injury within 5 hr when exposed to light (Kislyuk, 1964). The injury was proportional to the irradiation, and the leaves did not photosynthesize. Photoperoxides were formed which oxidized the substrate. A superficial scald of apples after prolonged storage at 0–4°C, is caused by conjugated triene hydroperoxides, which are oxidation products of α-farnesene (Lyons, 1973).

e. PROTEIN BREAKDOWN

Protein breakdown at low temperatures without an equally rapid resynthesis has been suggested as a cause of injury, either due to a deficiency of proteins or a toxicity by the products of hydrolysis (amino acids, NH_3). Wilhelm (1935) produced evidence of such hydrolysis in the case of beans and tomato plants exposed to low-temperature stress, but Seible (1939) pointed out that an even greater hydrolysis occurs in control (unchilled) plants kept in the dark (Table 3.3). Minamikawa *et al.* (1961) failed to detect any hydrolysis of proteins in the mitochondria of chilled sweet potatoes (10 days at 0°C) although their respiratory activity declined. They concluded that proteins are not likely to be degraded during chilling injury. Razaev (1965), however, did observe proteolysis in chilling-sensitive plants, but not in others resistant to chilling injury (corn and wheat, respectively).

Although the above experimental evidence is insufficient to prove a

TABLE 3.3

Protein Breakdown in Tomato Plants Exposed to Chilling Temperatures[a]

		Ratio of protein N: soluble N		
Leaf no.	Number days treated	Light controls	Coldroom plants	Dark controls
1	2	15.9	9.7	8.2
2	4	19.8	9.2	6.7
3	7	16.9	8.2	4.6
4	11	—	13.3	2.0

[a]From Seible, 1939.

chilling-induced protein breakdown, it is to be expected on the basis of the respiratory upset. Any decrease in the aerobic phase of respiration must result in a decreased oxidative phosphorylation. This would decrease the supply of ATP and, therefore, the rate of protein synthesis, resulting in a net protein breakdown.

f. BIOCHEMICAL LESIONS

This source of injury has been investigated much more thoroughly in animals than in plants (Peters, 1963). A biochemical lesion is an abnormality in the metabolism of an organism leading to a deficiency of an essential intermediate metabolite. Theoretically, the most important biochemical lesion would be a deficiency of ATP, due to the inhibition of aerobic respiration; for such a deficiency could lead to ion leakage and protein breakdown. Unfortunately, adequate evidence of this deficiency due to chilling is lacking (see above). Other biochemical lesions have, however, been indicated.

Chilling-sensitive bacteria have provided particularly clear cases of biochemical lesions. A cold-sensitive mutant of *E. coli* ceases growth at about 20°C, versus 10°C in the more normal strain (Ingraham, 1969). When the mutant was supplied with histidine, the growth curves of the two strains were identical. It has been suggested (Ketellapper and Bonner, 1961) that chilling injury in higher plants is also due to a biochemical lesion. Attempts were, therefore, made to prevent the injury by supplying the organism with a number of possible intermediates that might be deficient. Some partial success was claimed in preliminary reports, but no clear-cut results have as yet appeared. Other investigators, however, have produced evidence which may be interpreted in this way. Podin (1966) was able to show that a specific reaction (transformation of lutein to violaxanthin) does not occur below 10°C in the chilling-sensitive banana, but continues down to 0°C in the chilling-resistant *Bergenia*. Similarly, there was an accelerated loss of ascorbic acid in chilled sweet potato and pineapple (Miller—see Lyons, 1973). Chilled tobacco plants exhibited symptoms of nitrogen deficiency and there was a four- to-fivefold increase in chlorogenic acid, and lesser increases in other substances (Koeppe *et al.*, 1970). Similar increases in chlorogenic acid were observed by Lieberman (see Lyons, 1973) at chilling temperatures. In tomato fruit stored at temperatures below 10°C, (but not at 15° or 20°C), there was a large increase in activity of an enzyme involved in chlorogenic acid metabolism (Rhodes and Wooltorton, 1977), hydroxycinnamyl transferase. Consequently, the chilling temperature may have shifted the plant's metabolism from the normal pathway to an abnormal one, leading to a deficiency in some intermediates of nitrogen metabolism. A parallel explanation has been proposed for cottonseed killed by exposure to 5°C for 12 hr during hydration (Christiansen, 1968). Even a 30-min exposure may induce subsequent root abnormalities. Sensitivity to chilling persists during the initial 2–4 hr of hydration. If, however, the seeds are imbibed for 4 hr at 31°C and then dried, they are immune to chilling injury. Christiansen suggests that an irreversible event occurs during early hydration at normal germinating temperatures, but that it is blocked or disrupted by chilling. Chilling injury in the case of lima beans is apparently due, at least partly, to leaching of compounds absorbing at 264 mμ (Pollock, 1969), and, therefore, to a deficiency. The injury occurs if they are exposed to 5°–15°C during the initial stages of imbibition. It is avoided if the axes are just allowed to absorb water vapor; but longer exposures to chilling temperatures injure even these high moisture axes.

Unfortunately, such indirect evidence may frequently be interpreted in more than one way. A mutant of *Zea mays* has an extremely high chilling temperature (Millerd and McWilliam, 1968). It is able to produce chlorophyll only at temperatures above 17°C. When photooxidation is minimized, chlorophyll accumulates and seedlings can photosynthesize efficiently at a low temperature. This result points to the interrelationships between the different metabolic upsets and the difficulty in drawing a sharp line between them. Thus the photooxidation may be thought of as producing a toxic product (see above) or as preventing the synthesis of an essential component (chlorophyll) and, therefore, producing a biochemical lesion. It may also damage membranes (see above).

3. Secondary Stress Injury

The development of a secondary water stress at low temperatures was first indicated by Sachs (1864; see Molisch, 1896). He found that plants such as tobacco and cucumber begin to wilt if their roots are cooled to temperatures just above zero. Sugarcane is even more sensitive. It wilts if its root temperature drops to 15°C. This may lead, eventually, to death by desiccation. Molisch (1896) confirmed these results. He also covered some plants with bell jars containing large pieces of wet filter paper before cooling them, and immersed the leaves of others in snow and ice water. In these ways he was able to expose the plants to temperatures just above freezing in the absence of transpiration. This prevented the wilting and, therefore, the water stress injury. It is also possible to prevent the secondary stress injury at chilling temperatures by severing the shoot from the root system and standing it in water, or sometimes simply by spraying the leaves with water. The seat of the injury is, therefore, the roots, which cannot absorb water sufficiently rapidly at chilling temperatures to keep up with the transpirational loss, even though this is also decreased. In some cases, of course, there may be a temperature differential between the leaves and the roots. Although this may explain the daytime effect, when the leaves may be warmer than the soil, it cannot explain the inability of the plant to recover at night when transpiration is minimal. It is even less able to explain the water stress that develops in *Phaseolus vulgaris* when exposed to 5°C for a single night, provided that the roots as well as the shoots are cooled (Crookston *et al.*, 1974), or the fact that cotton roots are unable to absorb water at temperatures below 10°–12°C (St. John and Christiansen, 1976).

Kramer (1942) found that low temperature decreased the water absorption more in chilling-sensitive than in chilling-resistant crops. He concluded that this decreased absorption was due to a chilling-induced decrease in permeability. Evidence in favor of this interpretation has now been produced by Kaufmann (1975). In the cold sensitive citrus plants, cooling markedly lowered the permeability of the roots to water which then became limiting; but no such a lowering was found in the roots of the cold tolerant spruce. The secondary water stress may also sometimes mimic the indirect chilling injury, by producing a metabolic upset, for instance, a decrease in photosynthesis due to stomatal closure (Crookston *et al.*, 1974).

Some evidence indicates that a secondary O_2 deficiency stress may also develop at chilling temperatures. For instance, lima beans are injured when

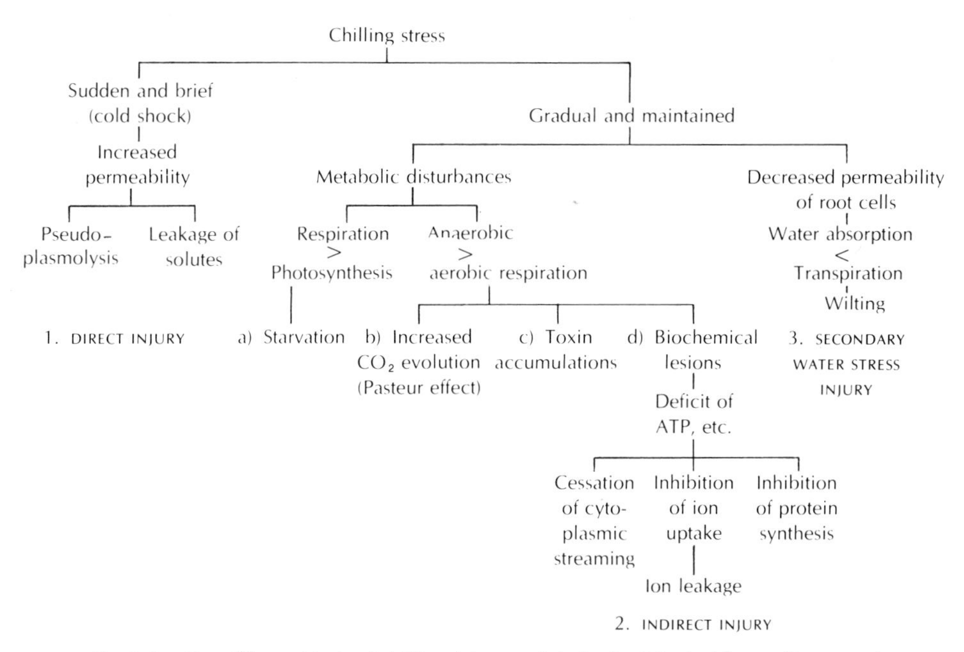

Fig. 3.1. The different kinds of chilling injury and their physiological bases. (> = greater than; < = less than).

allowed to imbibe water at 5°–15°C. This chilling injury can be prevented if the beans are first allowed to absorb water vapor up to a content of 20% (Pollock, 1969). The liquid water would interfere with oxygen uptake, the water vapor would not.

Fig. 3.1 summarizes the different kinds of chilling injury and the physiological changes that appear to produce the injuries.

4. Mechanism of Chilling Injury

The above results appear to indicate three basic changes responsible for the three different kinds of chilling injury: (*a*) metabolic disturbances leading to indirect injury, (*b*) increased permeability leading to direct injury, and (*c*) decreased permeability leading to secondary water stress injury. How can chilling temperatures give rise to these changes?

a. METABOLIC DISTURBANCES.

i. Differences in Energies of Activation. It has long been known that a marked temperature change may alter the end products of certain metabolic paths. Many plants, for instance, accumulate starch at normal growing temperatures but replace the starch by sugars at low temperatures. At least some of these changes can be explained by the different energies of activation of different chemical reactions, and therefore the differences in slopes of the Arrhenius plots (Fig. 3.2). For instance, Dear (1973) showed that lowering the temperature of cabbage leaves induced an immediate (within 15–30 min.) hydrolysis of starch. This rapidity cannot be explained by protein (enzyme) synthesis or even by activation of an already formed enzyme, since the direct effect of low temperature would be to slow down these processes, and any indirect effect would require an induction period. It can, however, be explained by the proposal of Selwyn (1966) that (due to their larger energies of activation) some enzymes are inhibited more than others by the low temperature, resulting in a relative enhancement of the latter, and leading to an increase in a specific metabolic pathway.

In the case of chilling resistant plants, such shifts in metabolism induced by chilling temperatures are noninjurious and reversible (elastic strains). For instance, Pollock and Ap Rees (1975a) obtained evidence of a differential inhibition of enzymes at low temperatures in potato tubers. At 2°C there was a greater inhibition of some of the glycolytic enzymes than of the sucrose synthesizing enzymes, leading to a net sucrose accumulation. This accumulation exceeded the activity of the sucrose synthetase, but was less than that of sucrose phosphate synthetase (Pollock and Ap Rees, 1975b).

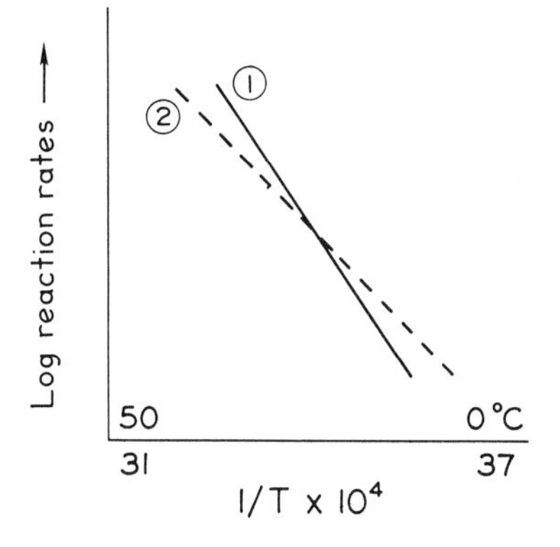

Fig. 3.2. Two metabolic reactions with different energies of activation and, therefore, differences in slopes of Arrhenius plots. Reaction 1 is more rapid than reaction 2 at high temperatures, and less rapid at low temperatures.

Wheat, broadbean, and spinach plants acclimatized to high temperatures (20°–25°C) incorporated much more CO_2 into sucrose and less into glycine and serine in comparison with similar plants grown at 5–7°C (Sawada and Miyachi, 1974). The carboxylation of RuDP was suppressed in plants acclimatized to low temperatures. Instead, the C_1 and C_2 moieties of RuDP were converted to glycine and serine via the glycolate pathway.

Whether or not injurious and irreversible (plastic) changes can be produced in chilling sensitive plants by such differences in activation energies has not been demonstrated.

ii. Membrane-Bound Enzyme Inactivation by Lipid-Phase Transition. It was long ago suggested that chilling injury may be in some way related to a solidification of the membrane lipids (see Lyons,1973). Now, both this solidification and its effect on metabolism are better understood (Raison,1973).

Although Arrhenius plots of ordinary chemical reactions remain straight lines throughout the temperature range tested, this is not true of biological processes involving metabolic reactions (Fig. 3.3). It has long been known that such plots of biological processes show precipitous downward bends at both ends, when the temperatures attained are too high or too low for the living processes to continue in a normal manner. In the case of soluble enzymes, the low temperature bends may be explained by their reversible denaturation (and, therefore, inactivation) due to a low temperature-induced weakening of the hydrophobic bonds, leading to unfolding of the protein molecule (Chapter 10), or to dissociation of large enzymes into subunits, e.g., prolyl-tRNA synthetase (Norris and Fowden, 1974).

In the case of chilling-sensitive plants, however, similar bends occur at

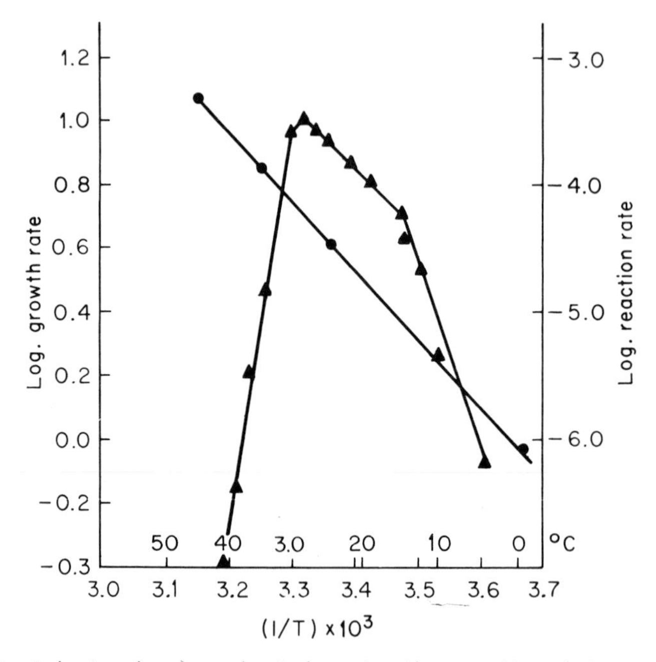

Fig. 3.3. Arrhenius plots for a chemical reaction (decomposition of nitrogen pentoxide ●—●) and for growth (of roots of *Lepidium sativum*, ▲—▲). From Getman and Daniels (1937) and Talma (1918), respectively. From Levitt (1969b)

chilling temperatures that are too high (10°–12°C) to induce such a protein denaturation or dissociation. Furthermore, it is not the soluble enzymes that show bends in their Arrhenius plots at these relatively high chilling temperatures, but the membrane-bound enzymes, whose activities depend on their association with the membrane lipids (Sechi *et al.*, 1973). The lipid phase changes have been shown to occur in the membranes of glyoxysomes, mitochondria, and proplastids (Wade *et al.* 1974). Even the amylolytic activity of amyloplasts is lowered at 10°C night temperature in a tropical but not in a temperate grass, (Carter *et al.*, 1974). Similarly, efficiency of the nuclear DNA polymerase activity was decreased at 2°C in chilling-sensitive germinated cotton seed, and this was associated with the fatty acid content of the nuclear membrane (Clay *et al.*, 1976). The lipid changes at chilling temperatures must, therefore, be understood before attempting to explain the metabolic disturbances.

On cooling, previously warm lipids undergo a phase transition at a temperature well below the true melting point (Lee, 1975). At this cool, phase transition temperature, a change of state occurs from the liquid crystalline to the solid crystalline (or gel) state, associated with decreased conformational freedom. The phase transition temperature is not an exact point as in the case of true freezing or melting points, but a temperature zone with a lower and upper limit.

Like the freezing point, however, the transition is rapid and completely reversible. The rate, however, is dependent on the temperature. In bilayer vesicles, structural defects induced by sonication below the phase transition temperature are annihilated in 10 min at 10° above the phase transition temperature, but require 1 hr at 3° above this temperature (Lawaczek *et al.*, 1976). The phase transition temperature increases with the length of the fatty acid chain, and decreases with increasing unsaturation. On cooling to the transition temperature, the phase change to the solid state is accompanied by a 33% decrease in the surface area occupied by the lipid. What are the effects of such a phase transition on the living cell?

The membrane lipids must be fluid in order for the membrane-bound enzymes such as ATPase to function (Grisham and Barnett, 1973). This does not mean that the greater the fluidity of the membrane lipids, the greater the activity of the associated enzymes. On the contrary, there is an optimum fluidity of the membrane phospholipids. When far from this optimum, conditions for functional activity may be lacking (Sechi *et al.*, 1973). This optimum may be explained as follows. When a chilling-sensitive plant is cooled to a chilling temperature, the membrane lipids undergo a phase transition from the fluid to the solid state, and this results in contraction of the membrane layer. The membrane-bound protein molecules therefore are probably compressed and suffer a conformational change (Lyons, 1973). One immediate effect is a marked increase in the energy of activation of enzymes associated with the lipid membrane. This is observed by the sharp downward bend at the phase transition temperature of the Arrhenius plot (Fig. 3.4). Presumably, there is an optimum protein conformation for normal enzyme activity. Just as solidification of the membrane lipids at chilling temperatures compresses the protein to an enzymatically less active state, so also too fluid a lipid membrane at high temperatures would permit the protein to unfold to too great a degree for normal enzyme activity.

Yamaki and Uritani (1973b) suggest that the membrane protein is bound to the lipid by hydrophobic bonds which are weakened by the chilling, resulting in dissociation and therefore cold denaturation. This concept, however, is unable to explain the optimum fluidity.

Lyons and Raison (1970a,b) showed that the mitochondria of chilling-sensitive plants exhibit a downward bend at 10°–12°C in the Arrhenius plot of their oxidative activity. Chilling resistant plants, on the other hand, exhibit a continuous straight line all the way from 1° to 25°C, and threfore a constant activation energy (Fig. 3.4). The bend may occur at higher temperatures, for instance at 17°C in the case of rice (Nishiyama, 1975). This sensitivity to

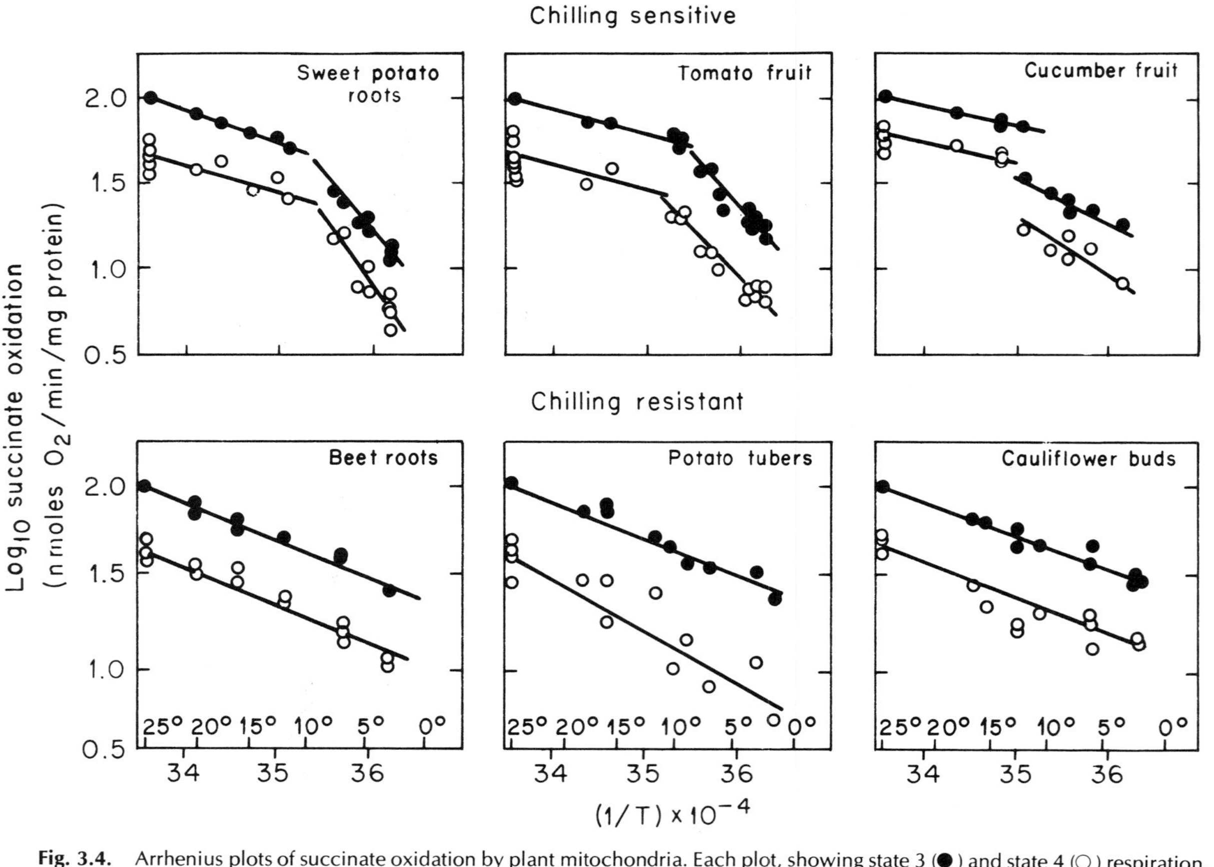

Fig. 3.4. Arrhenius plots of succinate oxidation by plant mitochondria. Each plot, showing state 3 (●) and state 4 (○) respiration, represents three or more mitochondrial preparations. The log values were adjusted by a factor to a common value at 25°C in order to compensate for differences in the rates (nanomoles of 0_2 per min per mg of protein) between the different preparations. From Lyons and Raison (1970a).

chilling holds for membranes other than the mitochondrial, for instance the chloroplast (Shneyour *et al.*, 1973, Murata *et al.*, 1975), and the vacuolar membranes (Yamaki and Uritani, 1973a). That the break in the Arrhenius plot is due to the lipid, is shown by its disappearance when the enzyme is freed from the associated lipid by treatment with non-ionic detergent (Watson *et al.*, 1975).

Not all plants, however, show such a clearcut relation between the bend in the Arrhenius plot and chilling resistance. Sugar beet is a chilling resistant plant, yet a membrane enzyme, ($Na^+ + K^+ + Mg^{2+}$) ATPase, from the roots shows biphasic Arrhenius plots with the transition between 15°–18°C (Lindberg, 1976). Similarly, Pomeroy and Andrew (1975) have obtained a bend in the Arrhenius plot at 10°–12°C for wheat which are fully chilling resistant. In agreement with them, the Ea for the Hill reaction increased as much in chloroplasts of chilling resistant pea below 14°C, as in the case of chilling sensitive maize (below 17°C) and mung bean (below 11°C). Nolan and Smillie (1977) therefore conclude that the change in chloroplast membranes at chilling temperatures is not necessarily correlated with chilling sensitivity. Raison *et al.*, 1977, however, consistently obtain the phase change for chilling resistant wheat mitochondria at a much lower temperature—0°C, and it was somewhat lower in the more freezing resistant than in the less resistant variety. They list some of the factors that yield an *apparent* increase in activation energy at higher temperatures. Erroneous breaks, for instance, may occur if the activity is assayed at a fixed substrate concentration (Silvius *et al.*, 1978).

In support of the concept that the bends in the Arrhenius plots are due to a phase transition of the membrane lipids, ESR studies using spin-labeled compounds have shown that membranes of intact mitochondria from chilling sensitive plants do undergo the physical phase transition at 10°–12°C (Lyons, 1973). More recent measurements have identified not only the lower limit of the phase transition, but also the upper limit. In the presence of an uncoupler, for instance, Nolan and Smillie (1976) have observed temperature-induced changes in Hill reaction activity of barley chloroplasts at 9° and 29°C, which coincided with changes in fluidity of the chloroplast thylakoid membranes, as detected by ESR. The ESR method, however, measures the phase separation in specific regions of the membrane, rather than the phase transition of the bulk membrane lipids, which can be reliably measured by DTA (McElhaney and Souza, 1976). Nevertheless, the DTA method also corroborated the phase transition in sensitive cells at chilling temperatures.

However, although the membrane lipids are entirely or nearly entirely in the liquid crystalline state over most of the growth temperature range, the upper and lower boundaries of the phase transition temperature may not directly determine the maximum and minimum temperatures for growth. The discrepancies may be due to several factors, for instance cluster formation (Lee *et al.*, 1974), charge density (Traeuble and Eibl, 1974), specific ions (Jacobson and Papahadjopoulos, 1975), pH and salt concentration (Traeuble *et al.*, 1976), and even a basic protein (Verklei *et al.*, 1974). A complication arises from the existence of several lipids in the membrane. It has been suggested that between the characteristic transition temperatures of two such lipids in the membrane, the bilayer contains clusters of gel and liquid crystalline lipid which coexist within the phase of the membrane (Pagano *et al.*, 1973). It has, therefore, been proposed that different membrane-associated enzymes are associated in different microenvironments having different lipid compositions and therefore different fluidities (Sechi *et al.*, 1973).

ATPases often show nonlinear relations in Arrhenius plots; yet Kuiper (1972) could not find a distinct transition point for the ATPase from bean roots. Nevertheless, the Q_{10} decreased when the assay temperature increased. Similarly, in the case of wheat and oats, the activation energy in the high temperature range was lower than in the low temperature range (Kähr and Møller, 1976). An actual bend in the Arrhenius plot was much less pronounced in wheat than in oats, for the Mg^{2+} stimulated ATPase. The reverse was true for the Ca^{2+} stimulated ATPase. Since the Mg^{2+} stimulated ATPase dominates in oats and the Ca^{2+} stimulated ATPase dominates in wheat, the bend in both cases was more distinct in the dominating ATPase. They concluded that Mg^{2+} and Ca^{2+} may influence the phase transition temperature directly, as was shown by Träuble and Eibl (1974) for two lipids.

It appears from the above results that at least some of the negative results are due to factors that obscure the phase transition in chilling sensitive plants or produce a transition artefactually in chilling-resistant plants and not because of a true absence of the transition. The phase transition of the membrane lipids fully explains the above-described metabolic disturbances at chilling temperatures (Fig. 3.1):

1. Starvation due to an excess of respiration over photosynthesis at chilling temperatures could occur if the phase transition temperature of the membrane lipids of the chloroplasts occurred at a higher chilling temperature than that of the mitochondria. This possibility has received some support. Chilling can apparently injure the chloroplast-synthesizing apparatus in mesophyll cells of C_4 grasses (*Sorghum* and *Papsalum*) without affecting other organelles or preventing leaf growth. An exposure to low temperature (2 to 4°C) for one night was sufficient to produce chlorotic bands and to decrease the activity of the C_4 enzymes. The bundle-sheath cells, however, were normal (Slack *et al.*, 1974).

2. Respiratory upset. The evidence of an increase in anaerobic and a decrease in aerobic respiration at chilling temperatures would be explained by the location of the aerobic reactions in the mitochondria and their dependence on the normal functioning of the mitochondrial membranes. At chilling temperatures, as shown above, the mitochondrial membranes of chilling-sensitive cells are impaired due to the phase transition of the membrane lipids. Anaerobic reactions, on the other hand, occur free in the cytoplasm, unconnected with membranes, and, therefore, would not show the bends in Arrhenius plots found for the mitochondrial, aerobic processes. The phase transition of the membrane lipids, therefore, explains (*a*) the disturbed respiration, (*b*) the Pasteur effect, (*c*) the accumulation of toxins, and (*d*) the biochemical lesions in chilling sensitive plants at chilling temperatures.

3. Protein hydrolysis. The suppressed mitochondrial activity would reduce the rate of oxidative phosphorylation and, therefore, of ATP production. Stewart and Guinn (1969, 1971) have, in fact, demonstrated this decrease in chilled cotton seedlings. Since ATP is required for protein synthesis but not for its breakdown, this ATP deficiency would shift the balance between the two, producing a net protein hydrolysis.

4. ATP is also needed for ion uptake. Its deficiency would, therefore, lead to a net efflux (or leakage) of ions. The leakage could also result from inactivation of the ion pump enzymes due to a conformational change in the enzyme protein, for instance as a result of compression of the membrane on transition of its lipids to the solid phase.

5. Cytoplasmic streaming would be similarly inhibited by the decrease in ATP.

b. PERMEABILITY CHANGES

It was pointed out above, that the chill-induced phase transition of the membrane lipids from the liquid crystalline to the solid (gel) form involves greater order of the lipid molecules and therefore a contraction of the membrane. These changes are known to restrict the movement of small molecules through the membrane and, therefore, to decrease the permeability of the membrane to water and aqueous solutes. Experiments with artificial liposomes, in fact, have demonstrated a decrease to one-third in their permeability to water (Blok *et al.*, 1976). Similarly, permeability of chloroplasts to glycerol and erythritol showed a break at 11°C in chilling-sensitive plants (*Phaseolus vulgaris* and *Lycopersicon esculentum*) but not in chilling resistant plants (*Pisum sativum* and *Spinacea oleracea*-Nobel, 1974).

If, however, the chilling is sudden (cold shock), the contraction of the membrane and its contents may not be uniform and this could induce mechanical stresses, leading to the formation of fractures in the membrane.

The membrane would, therefore, become much more permeable, permitting the leakage and other injurious phenomena described above for direct injury. Dehydration may conceivably produce similar fractures, which would be resealed at temperatures above the phase transition temperature but not below it. This would explain the reconstitution of yeast cells in water at 38°–42°C after drying to 7% water content, their loss of viability when rehydrated in cold water (Van Steveninck and Ledeboer, 1974).

These two opposite effects of the lipid phase transition on membrane permeability may both be expected to play a role in chilling injury. (*a*) Due to the large soil mass in which the roots are imbedded, chilling must be very gradual, and the root cells have time to adapt by a gradual, uniform contraction of the membrane without the formation of "cracks". Therefore, the effect is a marked decrease in the permeability of the water-absorbing root cells. (*b*) Due to the relatively rapid changes in air temperature, and the even more rapid leaf cooling by radiation at night, the phase transition of the membrane lipids may be rapid enough to produce "cracks" in the membrane due to the incompressibility of the cell water coupled to the slow diffusion of the water through the gelled plasma membrane.

A more severe effect on the membrane has been suggested to occur when chilling was combined with osmotic shock by transfer from 0.5M sucrose at 25°C to H_2O at 2°C (Amar and Reinhold, 1973). This led to the loss of a small amount (3.5%) of the protein of the leaf cells. Amar and Reinhold suggest that this small amount of protein lost may be closely involved in the transport mechanism.

A gradual, prolonged chilling, on the other hand, can be expected to lead only to a uniform solidification and contraction of the membrane, resulting in the above-mentioned decrease in permeability, observed in artificial liposomes. The metabolic disturbances of indirect injury must, therefore, be associated with a decreased cell permeability. Nevertheless, this slow, indirect injury has been reported to include increased cell leakage (see above). This leakage obviously cannot be due to the *decrease* in permeability. It can, however be explained by an inactivation of the enzymes of the ion pumps, rather than a true increase in permeability. Thus, Rb uptake was strongly inhibited below 10°C (with a Q_{10} of 5–8 below 10°C versus 1.3–2.0 above it) in barley and corn roots (Carey and Berry, 1978). In the case of the leaves of *Phaseolus vulgaris,* the increase in rate of electrolyte leakage was observed only when the chilling treatment (5°C) was combined with partial dehydration of the leaves (Wright, 1974). Either treatment alone failed to induce this leakage.

The decreased permeability due to the phase transition is probably the basic cause of the secondary, water stress injury. If only the roots are cooled below the phase transition temperature (for instance on a sunny but chilly day), this would produce the decrease in water absorption without a decrease in transpiration, which is characteristic of secondary water stress injury. Even if both are at the same temperature, since roots are commonly less hardy than tops (see Chapter 7), it is reasonable to postulate that the temperature of the phase transition is slightly higher in the roots than in the shoot. This possibility has not been investigated.

The permeability change due to the phase transition of the membrane lipids may also conceivably lead to a metabolic inhibition. Exposure of chilling-sensitive cotton seedlings to 5°C appeared to lower the permeability of the glyoxysomal membrane to succinate, leading to its accumulation in the organelles, and its inhibition of isocitratase activity (Smith and Fites 1973). This impedes the conversion of stored lipid to carbohydrate and, therefore, decreases seedling growth and viability.

All three kinds of injury can, therefore, be explained by the initial chill-induced but reversible strain—the phase transition of the membrane lipids (Fig. 3.5). A cold shock would induce direct injury by fracturing the solid membrane, permitting leakage of vital cell contents. A gradual chilling would induce the metabolic disturbances characteristic of indirect injury by sharply decreasing the activity of chloroplast and mitochondrial membrane-associated enzymes. The secondary water-stress injury would be due to the marked decrease in permeability of the plasma membrane of the root cells to water, reducing the rate of water absorption below that of water loss from the shoot.

c. CHILLING INJURY NOT ASSOCIATED WITH LIPID PHASE TRANSITION.

Although the phase transition of membrane lipids appears capable of explaining all the known kinds of chilling injury, it is possible that other mechanisms of chilling injury also exist. Two possibilities have been mentioned above: (*a*) a metabolic disturbance due to innate differences between enzymes in energies of activation and (*b*) protein denaturation due to the low temperature- induced weakening of the hydrophobic bonds. The first of these two kinds of chilling injury apparently occurs in tomatoes stored at chilling temperatures (Rhodes and Wooltorton, 1977). Two of seven tested enzymes, involved in chlorogenic acid metabolism (hydroxycinnamyl-transferase and PAL) actually show a large, reversible increase in activity below 10°C. Simon *et al.* (1976) have produced evidence in favor of the second of these kinds of injury. Germination of chilling-sensitive cucumber and mung bean seeds decreased with temperature drop from 20°–14°C with a Q_{10} of 2. Between 14° and 11°C, the Q_{10} rose to 60–1600 for cucumber, 87 for mung bean. Below 11°C germination was slight, at 10°C it was nonexistent. There was little indication of leakage from the seeds, and the Arrhenius plots

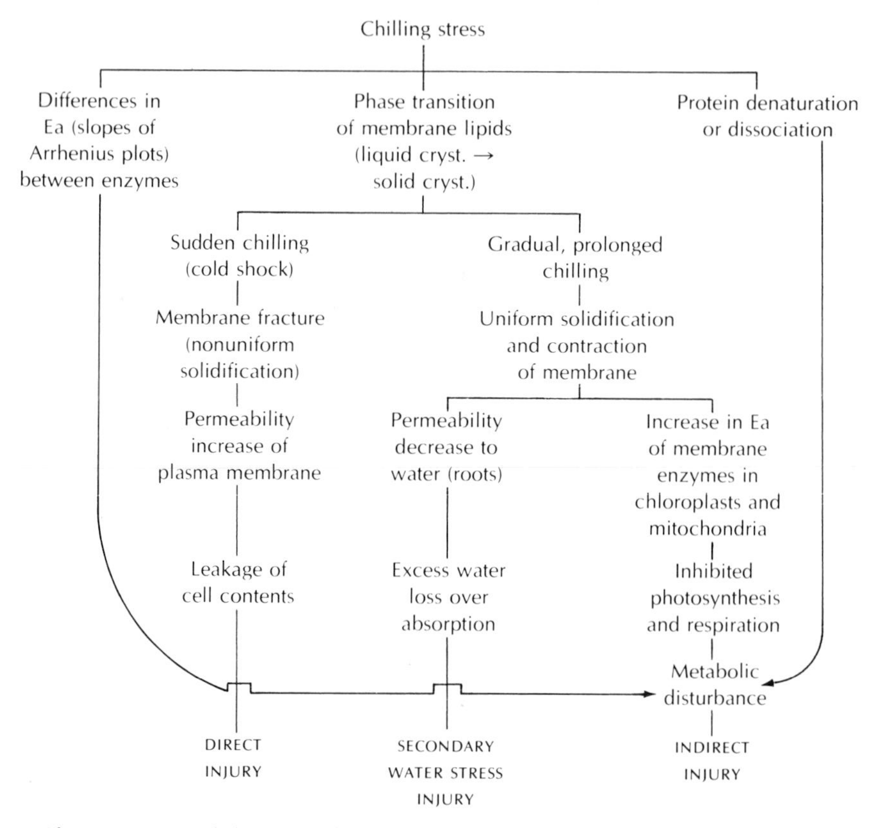

Fig. 3.5. Proposed changes leading to three kinds of chilling injury on exposure of chilling sensitive plants to a chilling stress.

of respiration were linear without bends, and therefore presumably no phase transition of the lipids occurred. Since no known process other than protein denaturation has such high Q_{10}'s, they concluded that this kind of chilling injury is due to protein denaturation.

The possible mechanisms of chilling injury are summarized in Fig. 3.5.

C. CHILLING RESISTANCE

1. Measurement and Development

All plants from temperate climates and all psychrophiles among microorganisms routinely survive exposure to chilling temperatures and, therefore, are fully chilling resistant. Plants from tropical and semitropical climates, on the other hand, show varying degrees of resistance. Attempts have, therefore, been made to develop methods of measuring chilling resistance quantitatively. The usual method is simply to expose the plants to an arbitrary chilling temperature for an arbitrary time and to observe for injury. Another method is to observe for seed germination at chilling temperatures, for instance among soybean cultivars (Littlejohns and Tanner, 1976). Zsoldos (1971) measures the injury by the change in Rb influx.

Since plants are poikilotherms, chilling resistance must be due to tolerance, and no avoidance-related structural change is expected. In agreement with this conclusion, the ultrastructure of five psychrophilic strains of *Cryptococcus albidus* did not differ from that of a mesophilic strain, except for the location of the ER—running parallel to the nuclear membrane (Srivastava and Smith, 1976).

Early evidence indicated that plants which are susceptible to chilling injury may harden in the field, becoming more resistant and therefore capable of surviving chilling temperatures (Sellschop and Salmon, 1928). Similarly, greenhouse plants normally susceptible to chilling injury when grown at 17°C become hardy if grown at 12°C for 2 months. Wheaton and Morris (1967) hardened tomato seedlings by as little as 3 hr at 12.5°C, though maximum protection was achieved in 48 hr. Sweet potato roots, however, failed to harden during this treatment, although effects on the respiratory behavior were observed. Pea seedlings, which are injured by exposure for 3 hr at −3°C (without freezing) may be hardened to survive this treatment by three daily exposures to 5°C for 3 hr (Kuraishi *et al.*, 1968). Kushmrenko and Morozova (1963) succeeded in hardening cucumber seedlings against chilling injury at 3°–12°C, by cooling them to 22°C for 18 hr a day during 4 days. Solov'ev and Nezgovorov (1968) hardened cucumber leaves at 10°C, against chilling injury at 3°–5°C. Cell suspension cultures may also be used to investigate chilling resistance, the results being comparable to those with the seedlings (Breidenbach and Waring, 1977). However, cell lines selected for enhanced chilling resistance did not retain this resistance when allowed to develop into plants (Dix, 1977).

It must be emphasized, however, that physiological age may affect the plant's susceptibility to chilling injury (Lyons, 1973). Therefore the developmental changes may mimic exposures to low temperature and may harden (or deharden) the plant by inducing a change in the phase transition temperature. In the case of the pulp of *Mangifera indica* and *Musa sapientum* discontinuities occurred during the preclimacteric but not during the climacteric stage of the fruit, nor even during the preclimacteric of *M.sapientum* if it was stored at low temperature (13°C) for 5 days (Nagaraja and Patwardham, 1974). Conversely, senescence of bean cotyledons is accompanied by a decreased fluidity of smooth, microsomal membranes (McKersie *et al.*,

1976). In both chloroplast and microsomal membranes of the primary leaves of beans, portions of lipid became crystalline as the tissues senesced (McKersie and Thompson, 1978), the phase transition temperature rising 2–3 weeks after planting, to 38°C for microsomes. The appearance of the gel phase at physiological temperatures coincides with the initiation of decline in total protein and loss of chlorophyll. Similarly, in rose petals, microviscosity of the plasmalemma increased with age (Borochov *et al.*, 1978). In algae, the phase transition temperature rises from a low of 0°C in young cultures to a high of about 70°C for 140-day-old cultures (Thompson *et al.*, 1978), at a time of chlorophyll and protein loss.

The Russian investigators have reported increases in chilling resistance of corn plants as a result of a variety of treatments. Al'tergot and Bukhol'tsev (1967) hardened corn seedlings by alternating a gradual drop in temperature from the optimum to zero, with a gradual rise back to the optimum. Treatment of the seed of corn or cotton with tetramethylthiuram disulfide (TMTD) before planting, considerably increased the survival in cold and wet weather (e.g., 7°–10°C, Radchenko *et al.*, 1964; Nezgovorov and Solov'ev, 1965). This may, of course, be a protection against microorganisms in the soil, rather than against true chilling injury. Other substances used in sprays (e.g., 2,4-D and KCl + NH_4NO_3 + boric acid) have been reported to protect cucumber leaves against chilling injury (Solomonovskii and Pomazova, 1967).

Artificially induced chilling resistance has been reported by application of the chemical substances picolinic acid and Dexon to cotton plants (Amin, 1969). The treated plants recovered better after chilling exposures to 15°C. Since these substances are respiratory inhibitors they are thought to prevent the respiratory disturbances normally induced by the chilling, due to a protection of specific systems by the inhibitors.

Cytokinin prevented chilling injury to peas when sprayed four times (every four days) on the plants before exposure to chilling (−2°C for 3 hr, unfrozen). A day later, the apices of the controls began to turn yellow and lost fresh weight (Kuraishi *et al.*, 1966). Similar results were obtained with five other species. This effect of cytokinin was obtained only during the cold season. ABA applied to cucumber seedlings prior to chilling significantly ameliorated the injury (Rikin and Richmond, 1976; Rikin *et al.*, 1976). This may have been due to improved water balance, and therefore to an effect on the secondary water stress injury.

More recently, rice seedlings injured by a 10hr exposure to 10°C recovered when treated with thiourea and to a lesser extent with potassium thiocyanate and cystine (Ghosh and Chatterjee, 1975).

Chilling resistance of citrus may also be affected by nutrient treatment, a deficiency decreasing resistance (Del Rivero, 1966). Buckwheat shows an increased resistance to both chilling and cold shock (−3°C for 5 min) when potassium supply is increased, but an increase in calcium increases resistance to the former and decreases resistance to the latter (Korovin and Frolov, 1968).

The changes that occur during hardening to chilling temperatures are not too well understood. Attempts have been made to relate resistance to a number of factors—respiration rate and conversion of stored carbohydrates (Mishustina, 1967), the concentration of amino acids and proteins (Petrova, 1967), fat content, and metabolic rate on sprouting (Beletskaya, 1967). Kuraishi *et al.* (1968), found an increase in the ratio of NADPH:NADP. This would seem to indicate that the chilling treatment tends to lower the reduction potential of the living cells and the hardening treatment counteracts this by developing a greater reduction potential. Cotton seedlings chilled at 5°C showed a continuing decrease in ATP concentration (Stewart and Guinn, 1969). If returned to optimum temperature after 1 day, the initial ATP concentration was restored, but not after 2 days of chilling. The decrease in ATP on chilling was prevented by hardening for 2 days at 15°C immediately before chilling. The ATP level of hardened was higher than that of unhardened seedlings, increasing more in the roots than in the leaves. When the hardened plants were exposed to chilling (5°C), the ATP level increased in the leaves and decreased in the roots, the increase leveling off after 2 days of chilling. It has, in fact, been possible to protect cottonseed against chilling injury in the field by application of AMP to the seed (McDaniel and Taylor, 1976).

The above changes during hardening to chilling temperatures have not been widely established. If they do occur, they may be dependent on changes in the membrane lipids. Therefore, the change most commonly associated with the hardening process is an increase in unsaturation of fatty acids. This factor will be discussed below in connection with the mechanism of chilling resistance.

2. Mechanisms

a. INCREASED UNSATURATION OF FATTY ACIDS

From all the above and earlier results, it may be postulated that chilling resistance is due to an ability to maintain the membrane lipids in the liquid crystalline state at chilling temperatures. How can this be brought about? Early workers observed that plants of warm climates contained more saturated fatty acids than plants of cooler climates (see Lyons, 1973).

Lyons *et al.* (1964) found a higher content of polyunsaturated fatty acids in the mitochondria of chilling resistant- than in those of chilling-sensitive plants. Similarly, when fed acetate-2-^{14}C at 10°–40°C for 5 hr, seeds of castor bean, sunflower, and flax showed an increased formation of unsaturated fatty acids at the low temperatures (Harris and James, 1969). Using artificial systems, Lyons and Asmundson (1965) showed that the phase transition points of mixtures of palmitic and linoleic, or palmitic and linolenic acids (the predominant fatty acids in plants) decrease slowly as the unsaturated fatty acid is increased to 60 mole%. Beyond this percentage, the phase transition point is depressed more markedly by each addition. The differences are pronounced at percentages that approximate the composition in plant membrane lipids.

Plants such as cotton and bean, become resistant to chilling temperatures on exposure to temperatures slightly above chilling. This hardening is presumably due to the observed increase in unsaturation of their fatty acids (Wilson and Crawford, 1974a), which lowers the phase transition temperature below the previously injurious chilling temperature. Both the membrane fluidity and its normal high permeability to water are therefore retained at chilling temperatures. Chilling-sensitive plants, on the other hand, when chilled to 5°C, undergo a rapid decrease in the percent of linolenic (or palmitoleic—Kane *et al.*, 1978) acid (Wilson and Crawford, 1974b), and, therefore in the degree of unsaturation. Consequently, they must also lose the fluidity of their membranes and their high permeability to water. Among four alfalfa varieties, the decrease in photosynthesis at chilling temperatures 10°C) was inversely related to the double bond index of the chloroplast membranes (Peoples *et al.*, 1978). Similarly, the leaves of several species show a parallel decrease in unsaturation of lipids and chilling resistance accompanying the rise in the phase transition temperature during physiological ageing at 25°C (see above).

A direct confirmation of the relation between unsaturation of fatty acids and chilling resistance has been produced by St. John and Christiansen (1976). As they lowered the temperature to which cotton seedlings were exposed, the linolenic acid content of the polar lipid fraction increased. Sandoz 9785 decreased this temperature-induced accumulation of linolenic acid and also decreased the seedlings' ability to withstand chilling at 8°C. This was true of the developing root tips but not of the hypocotyl tissue. Apparently, the low temperature alters the fatty acid unsaturation only in newly developing tissue, and there is little or no resynthesis of membrane lipids in mature tissue.

The relation between unsaturation of fatty acids, a lower phase transition temperature, and chilling resistance has been most thoroughly established in the case of microorganisms (Lyons, 1973). Thus, the greater the degree of unsaturation of the membrane lipids, the lower the phase transition temperature of the mitochondrial membranes of yeast cells (Watson *et al.*, 1975). Similarly, the degree of unsaturation in the acyl group of the phospholipids of the fungus *Fusarium oxysporum* was inversely related to the growth temperature at 15°, 25°, and 37°C (Barran *et al.*, 1976; Miller and de la Roche, 1976). The same was true of the blue-green bacterium *Agmenellum quadruplicatum* between 20°–43°C (Olson and Ingram, 1975).

The fatty acid content of the membranes of *E. coli* can be controlled by its nutrition. Membranes with low unsaturated fatty acid content (8–11%) are fragile. Higher contents stabilize the membranes against pressure by maintaining their lipids in the liquid-crystalline state (Akamatsu, 1974). Low temperature hardening of *Candida lipolytica* increased the activity of its fatty acid desaturating enzyme (Pugh and Kates, 1975). Similarly, yeast cells showed phase transition temperatures of 7.7, 10.2 and 21.8°, respectively, when grown on linoleic, oleic, and elaidic acids (Ainsworth *et al.*, 1974).

It may, therefore, be concluded that (*a*) chilling resistance is dependent on a downward shift in the phase transition temperature, and (*b*) this is commonly brought about by an increase in the membrane content of unsaturated fatty acids. One of the major factors, according to Harris and James (1969), was an increase in availability of oxygen, which is the rate limiting factor for desaturation. The chemical mechanisms involved in such changes of unsaturation are not known.

Nevertheless, some analyses have revealed that the correlation between the degree of unsaturation of the fatty acids and chilling resistance is not always precise in higher plants (Lyons, 1973). For instance, the phase transition temperature did not correlate with the susceptibility to chilling of 6 cultivars of apple fruit (McGlasson and Raison, 1973). Similarly, there are chilling resistant plants which show no change in unsaturation of their lipids when exposed to 5°C (Wilson and Crawford 1974b). In the tubers of Jerusalem artichokes, the phase transition temperatures of the mitochondrial membranes drop from 22° and 3°C to 9° and −5°C during initiation of dormancy (Chapman *et al.*, 1979). Yet the fatty acid composition did not change (Hannan and Raison, in press 1979). Passiflora also showed no change in unsaturation of fatty acids with chilling resistance. Leaves of *Phaseolus vulgaris* can be drought-hardened at 25°C and 40% R.H. by withholding water from the roots for 4 days (Wilson, 1976). This drought hardening was just as effective in preventing chilling injury to leaves as 4 days of chill hardening at 12°C and 85% R.H. Yet the drought hardening resulted in no increase in unsaturation of the phospholipids or glycolipids. Wilson therefore suggests that the phase transition of the lipids is not the primary cause of chilling injury.

Another possible reason for the negative results is suggested by Wilson and Crawford (1974a). The acclimatization produced no effect on the composition of the glycolipids or of the total fatty acids. It was the phospholipid fraction alone, representing only 25% of the total leaf fatty acids, that had its degree of unsaturation positively related to the chilling tolerance of the species. Furthermore, when chilled at 5°C, the percent of linolenic acid and the total weight of fatty acid decreased rapidly in the chilling sensitive but not in the chilling resistant species (Wilson and Crawford, 1974b). In agreement with this concept, the ratio of unsaturated to saturated fatty acids was higher in the PC of the chilling resistant imbibed broad beans than in the chilling sensitive lima beans (Dogras *et al.*, 1977). In contrast to these results, all species showed a large decrease in degree of unsaturation on transition from the imbibed seed to the seedling stage. Yet there was no change in sensitivity of the broad bean and pea seedlings, and lima beans actually lost their sensitivity to chilling during seedling formation. This indicates that the degree of unsaturation of the PL is not the only factor governing sensitivity to chilling. This is true also of the phase transition temperature; for it depends on the head group of the PL, as well as on the hydrocarbon chain length and the degree of saturation. Thus, PC with the same hydrocarbon chain length and degree of saturation as PE, changes phase from the liquid-crystalline to the solid-gel state at a lower temperature. Similarly, in tubers of Jerusalem artichoke, PC and PG as well as lyso-PL of the mitochondrial membranes decreased, and PE and PA increased during the decrease in membrane fluidity (Hannan and Raison, in press 1979). Further evidence of the role of PL in chilling injury and resistance is the lowering of the break in the Arrhenius curve of membrane-bound enzymes of yeast, from 27° to 12°C by the addition of adamantane, a quasi-spherical molecule, which hinders axial ordering of phospholipid alkyl chains (Eleter *et al.*, 1974).

b. OTHER LIPID FACTORS

Evidence with animal cells indicates that a downward shift in the phase transition temperature may occur due to the presence of cholesterol in the membrane. X-Ray diffraction patterns of erythrocyte membranes indicated a liquid crystalline to solid-gel transition over a range of 2°–20°C if the cholesterol was removed. With a cholesterol content of 7.3% or more, there was no rigid crystalline pattern above −10°C. None occurred even down to −20°C in the unmodified erythrocyte membrane (Gottlieb and Eanes, 1974). Cholesterol also lowers the phase transition temperature of lecithin (Erdei *et al.*, 1975). A similar effect has been reported for the fungus *Fusarium oxysporum* (Miller and de la Roche, 1976). Growth at a chilling temperature (15°C) increased the sterol content of the membrane relative to the phospholipid content. In higher plants, the solidification of the membrane lipids which accompanies senescence is correlated with a 4 × increase in the ratio of sterol: phospholipid in the chloroplast and microsome membranes of primary bean leaves (McKersie *et al.*, 1976; Thompson *et al.*, 1978) and a similar increase in the plasmalemma of rose petals (Borochov *et al.*, 1978).

The importance of factors other than unsaturation, in lowering the phase transition temperature of the membrane lipids, may be due to the fact that unsaturation is a double-edged weapon. According to Christopherson (1969), the formation of lipid peroxides from the unsaturated fatty acids of membranes is probably highly injurious. He showed that glutathione peroxidase can catalyze the reduction of the hydroperoxides of all the polyunsaturated fatty acids that occur in the subcellular membranes of rat liver. It is, therefore, possible that the membrane lipids are normally protected against such peroxidation by the high reducing power maintained in actively metabolizing cells. When, however, the cells are cooled to chilling temperatures, their metabolism slows down and their reducing power decreases.

That the unsaturated fatty acids of plants are, indeed sensitive to oxidation was shown by van Hasselt (1974). The unsaturated lipids of *Cucumis* leaf discs at 1°C undergo a photooxidative degradation. This damages the reducing capacity of the leaf discs (van Hasselt, 1973). This photooxidation was at least partly prevented by several quinones and other substances (Van Hasselt, 1976), and was promoted by some substances. The mechanism of protection of Cucumis leaf chloroplasts against low-temperature photo-oxidative damage was explained by a protection of the electron transport pathway from the reducing to the oxidizing side of the photosystems. Protection against photooxidation by α-tocopherol has also been suggested (De Kok *et al.*, 1978).

Possible evidence of protection against such peroxidation has been obtained with cotton plants. When hardened for two or more days at 15°C day and 10°C night temperatures, and subsequently chilled at 5°C for 2 days, they showed an increase in protein sulfhydryl content, compared to unhardened and chilled tissue (Cothren and Guinn, 1975). This may indicate an increased reducing capacity due to hardening.

Conversely, at high temperatures it may not be possible for the cell to protect the unsaturated bonds of its fatty acids against peroxidation. This may explain why plants from tropical climates must maintain a low degree of unsaturation of their fatty acids and therefore a high phase transition temperature for their membrane lipids. This has been shown most clearly in the case of thermophilic microorganisms (McElhaney and Souza, 1976).

c. SOLUBLE PROTEINS

Increased unsaturation of the membrane lipids and other mechanisms of lowering their phase transition temperature can explain the hardening mechanism only if chilling injury is due to the phase transition of the membrane lipids from the liquid crystalline to the solid (gel) state. Although this appears to be the predominant injury mechanism, both theory and experiment have indicated that the activity of soluble enzymes (not associated with membranes) may also be involved (see above). In such cases, of course, the unsaturation of the fatty acids of the membrane lipids will be irrelevant to the mechanism of chilling resistance. Possible examples are the adaptation to chilling temperatures by the replacement of labile forms of an enzyme [e.g., glutamate dehydrogenase in corn roots, peroxidase in corn leaves (Brach *et al.*, 1976), by a more stable isozyme (Alekhina and Sokolova, 1974), or more simply the counteraction of the lowered activity of an enzyme (per unit protein) by an increase in concentration of the protein (glutamine synthetase in corn roots-Alekhina *et al.*, 1975]. Peas adapt to chilling temperatures by lowering the Km value for invertase, but maize are unable to adapt in this way (Crawford and Huxter, 1977). Similarly, in soybean, a single mitochond-

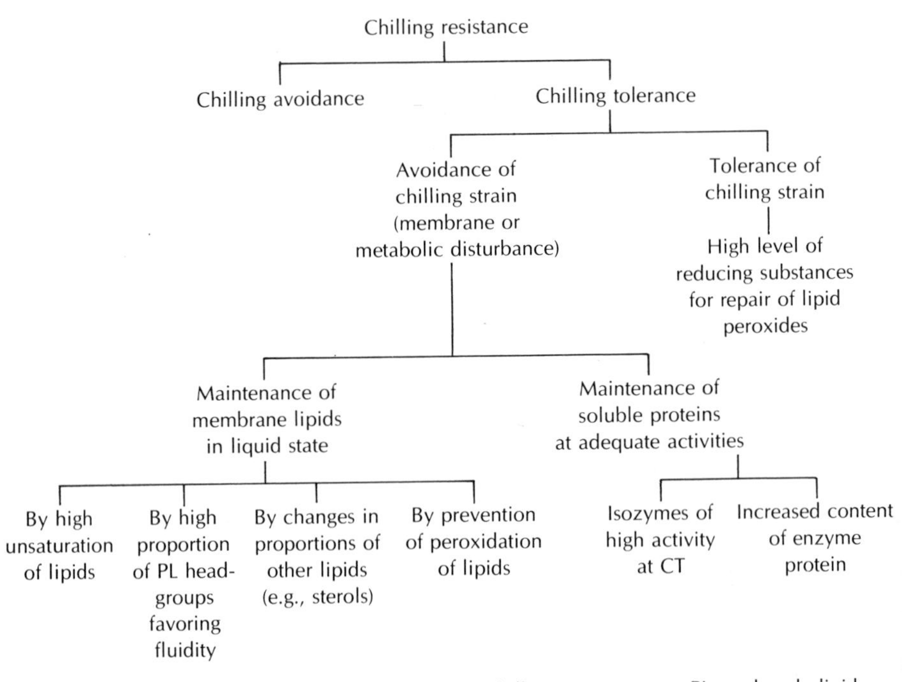

Fig. 3.6. Possible mechanisms of resistance to chilling temperatures. PL = phospholipid; CT = chilling temperature.

rial enzyme (NADP-ICDH) is apparently responsible for the respiratory control at chilling temperatures (Duke *et al.*, 1977). Perhaps the occurrence of this kind of resistance may account for some of the negative results, for instance, when chilling resistance is not related to unsaturation of the fatty acids.

The relations of these mechanisms to chilling resistance are indicated in Fig. 3.6.

Bibliography

(for Chapter 3)

Adir, C. R. (1968). Testing rice seedlings for cold water tolerance. *Crop. Sci.* **8,** 264–265.

Ainsworth, P. J., Janki, R. M. Tustanoff, E. R. and Ball, A. J. S. (1974). The incorporation of cytochrome oxidase into newlyforming yeast mitochondrial membranes. *J. Bioenerg.* **6,** 135–150.

Akamatsu, Y. (1974). Osmotic stabilization of unsaturated fatty acid auxotrophs of *Escherichia coli. J. Biochem,* **76,** 553–561.

Alekhina, N. D., and Sokolova, S. A. (1974). Change of the isozyme composition of glutamate dehydrogenase in connection with the growing temperature of plants. *Dokl. Akad. Nauk. SSSR. Ser. Biol.* **216,** 682–685.

Alekhina, N. D., Shirshova, E. D. and Andreenko, S. S. (1975). Glutamine synthetase activity in corn roots grown at different temperatures in the rhizosphere. *Biol. Nauki.* **18,** 103–106.

Al'tergot, V. F., and Bukhol'tsev, A. N. (1967). Induced resistance to cold in corn shoots. *Izv. sib. otd. Akad. Nauk. SSSR Ser. Biol. Med. Nauk.* **2,** 49–56.

Amar, L. and Reinhold, L. (1973). Loss of membrane transport ability in leaf cells and release of proteins as a result of osmotic shock. *Plant Physiol.* **51,** 620–625.

Amin. J. V. (1969). Some aspects of respiration and respiration inhibitors in low temperature effects of the cotton plant. *Physiol. Plant.* **22,** 1184–1191.

Banga, O. (1936). Physiologische Symptomen van Lage-Temperatuur-Bederf. *Lab. Tuinbouwplant. (Wageningen)* **24,** 3–143.

Barran, L. R., Miller, R. W. and De la Roche, I. (1976). Temperature-induced alterations in phospholipids of *Fusarium oxysporum* f. sp. *lycopersici. Can. J. Microbiol.* **22,** 557–562.

Bartetzko, H. (1909). Untersuchungen über das Erfrieren von Schimmelpilzen. *Jahrb. Wiss. Bot.* **47,** 57–98.

Beletskaya, E. K. (1967). The deposition of fat in the external layer of the endosperm of corn grain differing in their resistance to cold. *Rost. Ustoich. Rast.* **3,** 223–226.

Biebl, R. (1939). Über die Temperaturresistenz von Meeresalgen verschiedener Klimazonen and verschieden tiefer Standorte. Jahrb. Wiss. Bot. 88, 389–420.

Blok, M. C., Van Deenen, L. L. M. and De Gier, J. (1976). Effect of the gel to liquid crystalline phase transition on the osmotic behavior of phosphatidylcholine liposomes. *Biochim. Biophys. Acta* **433,** 1–12.

Borochov, A., Halevy, A. H., Borochov, H., and Shinitzky, M. (1978). Microviscosity of plasmalemmas in rose petals as affected by age and environmental factors. *Plant Physiol.* **61,** 812–815.

Boroughs, H., and Hunter, J. R. (1963). Effect of temperature on the growth of Cacao seeds. *Proc. Am. Soc. Hort. Sci.* **82,** 222–224.

Bramlage, W. J., Leopold, A. C., and Parrish, D. J. (1978). Chilling stress to soybeans during imbibition. *Plant Physiol.* **61,** 525–529.

Breidenbach, R. W. and Waring, A. J. (1977). Response to chilling of tomato seedlings and cells in suspension cultures. *Plant Physiol.* **60,** 190–192.

Carey, R. W. and Berry, J. A. (1978). Effects of low temperature on respiration and uptake of rubidium ions by excised barley and corn roots. *Plant Physiol.* **61,** 858–860.

Carter, J. L., Garrard, L. A. and West, S. H. (1974). Amylolytic activity of orchardgrass and starch and sucrose contents of orchardgrass vs. Pangola digitgrass leaf blades as influenced by night temperature and gibberellic acid. *Crop Sci.* **14,** 384–387.

Casas, I. A., Redshaw, E. S., and Ibanez, M. L. (1965). Respiratory changes in the cacao seed cotyledon coincident with seed (cold) death. *Nature (London)* **208,** 1348–1349.

Chapman, E., Wright, L. C., and Raison, J. K. (1979). Seasonal changes in the structure and function of mitochondrial membranes of artichoke tubers: a requisite for surviving low temperatures during dormancy. *Plant Physiol.* **63,** 363–366.

Christiansen, M. N. (1968). Induction and prevention of chilling injury to radicle tips of imbibing cotton seed. *Plant Physiol.* **43,** 743–746.

Christophersen, B. O. (1969). Reduction of linolenic acid hydroperoxide by a glutathione peroxidase. *Biochim. Biophys. Acta* **176,** 463–470.

Clay, W. F., Bartkowski, E. J. and Katterman, F. R. H. (1976). Nuclear deoxyribonucleic acid metabolism and membrane fatty acid content related to chilling resistance in germinating cotton (*Gossypium barbadense*). *Physiol. Plant.* **8,** 171–175.

Cooper, W. C., Rasmussen, G. K., and Waldon, E. S. (1969). Ethylene evolution stimulated by chilling in citrus and Persea sp. *Plant Physiol.* **44,** 1194–1196.

Cothren, J. T., and Gene Guinn. (1975). Protein-sulfhydryl content and electrophoretic protein patterns in young cotton plants as affected by low temperatures. *Phyton. Rev. Int. Bot. Exp.* **33,** 131–138.

Crawford, R. M. M., and Huxter, T. J. (1977). Root growth and carbohydrate metabolism at low temperatures. *J. Exp. Bot.* **28,** 917–925.

Crookston, R. Kent, O'Toole, J., Lee, R., Ozbun, J. L., and Wallace, D. H. (1974). Photosynthetic depression in beans after exposure to cold for one night. *Crop Sci.* **14,** 457–464.

Dear, J. (1973). A rapid degradation of starch at hardening temperatures. *Cryobiology,* 10, 78–81.

De Kok, L. J., Van Hasselt, P. R., and Kuiper, P. J. C. (1978). Photooxidative degradation of chlorophyll-a, and unsaturated lipids in liposomal dispersions at low-temperature. *Physiol. Plant.* **43,** 7–12.

Del Rivero, J. M. (1966). Importance of the macro- and micro-elements in the citrus tolerance to and recovery from cold weather. *Bol. Patol. Veg. Entomol. Agr.* **29,** 405–411.

Dix, P. J. (1977). Chilling resistance is not transmitted sexually in plants regenerated from *Nicotiana sylvestris* cell lines. *Z. Pflanzenphysiol.* **8,** 223–226.

Dogras, C. C., Dilley, D. R., and Herner, R. C. (1977). Phopholipid biosynthesis and fatty acid germination of seeds. *Plant Physiol.* **60,** 897–902.

Duke, S. H., Schrader, L. E., and Miller, M. G. (1977). Low temperature effects on soybean (*Glycine max.* L.) Merr. cv. Wells) mitochondrial respiration and several dehydrogenases during imbibition and germination. *Plant Physiol.* **60,** 716–722.

Dwelle, R. B. and Stallknecht, G. F. (1978). Pentose phosphate metabolism of potato tuber discs as influenced by prior storage temperature. *Plant Physiol.* **61,** 252–253.

Eaks, I. L., and Morris, L. L., (1956). Respiration of cucumber fruits associated with physiological injury at chilling temperatures. *Plant Physiol.* **31,** 308–314.

Eletr, S., Williams, M. A., Watkins, T., and Keith, A. D. (1974). Perturbations of the dynamics of lipid alkyl chains in membrane systems: Effect on the activity of membrane-bound enzymes. *Biochim. Biophys. Acta.* **339,** 190–201.

Eng-Wilmot, D. L., Hitchock, W. S., and Martin, D. F. 1977. Effect of temperature on the proliferation of *Gymmodinium breve* and *Gomphosphaeria aponina*. *Mar. Biol.* (Berl.) **41,** 71–78.

Erdei, L., Csorba, I. and Thuyen, H. X. (1975). Simple, rapid method for detecting phase transitions of lipids. *Lipids,* **10,** 115–117.

Franco, C. M. (1958). "Influence of Temperature on Growth of Coffee Plant", IBEC Res. Inst., New York.

Garber, M. P. (1977). Effect of light and chilling temperatures on chilling-sensitive and chilling-resistant plants. Pretreatment of cucumber and spinach thylakoids *in vivo* and *in vitro*. *Plant Physiol.* **59,** 981–985.

Geiger, D. R. (1969). Chilling and translocation inhibition. *Ohio J. Sci.* **69,** 356–366.

Getman, F. H., and Daniels, F. (1937). "Outlines of Theoretical Chemistry." Wiley, New York.

Ghosh, B., and Chatterjee, S. K. (1975). Repairing of low-temperature induced injury in rice seedlings by thiourea, cystine and potassium thiocyanate. *Geobios. (Jodhpur)* **2,** 31–32.

Göppert, H. R. (1830). "Über die Wärme-Entwickelung in den Pflanzen, deren Gefrieren und die Schützmittel gegen dasselbe." Max and Comp., Berlin.

Gottlieb, M. H., and Eanes, E. D. (1974). On phase transitions in erythrocyte membranes and extracted membrane lipids. *Biochem. Biophys. Acta.* **373,** 519–522.

Greeley, A. W. (1901). On the analogy between the effect of loss of water and lowering of temperature. *Am. J. Physiol.* **6,** 122–128.

Grisham, C. M. and Barnett, R. E. (1973). The role of lipidphase transitions in the regulation of the (sodium-potassium) adenosine triphosphatase. *Biochemistry,* **12,** 2635–2637.

Hannan, G. N. and Raison, J. K. In Press (1979). Seasonal changes in the structure and function of mitochondrial membrane of artichoke tubers: lipid composition and the effect of growth conditions.

Harris, P., and James, A. T. (1969). Effect of low temperature on fatty acid biosynthesis in seeds. *Biochim. Biophys. Acta* **187,** 13–18.

Hilliard, J. H., and West, S. H. (1970). Starch accumulation associated with growth reduction at low temperatures in a tropical plant. *Science* **168,** 494–496.

Ibanez, M. L. (1964). Role of cotyledon in sensitivity to cold of Cacao seed. *Nature (London)* **201,** 414–415.

Ingraham, J. L. (1969). Factors which preclude growth of bacteria at low temperature. *Cryobiology* **6,** 188–193.

Jacobson, K., and Papahadjopoulos, D. (1975). Phase transitions and phase separations in phospholipid membranes induced by changes in temperature, pH, and concentration of bivalent cations. *Biochemistry,* **14,** 152–161.

Kähr, M., and Møller, I. M. (1976). Temperature response and effect of Ca^{2+} and Mg^{2+} on ATPases from roots of oats and wheat as influenced by growth temperature and nutritional status. *Physiol. Plant.* **38,** 153–158.

Kane, O., Marcellin, P., and Mazliak, P. (1978). Incidence of ripening and chilling injury on the oxidative activities and fatty acid compositions of the mitochondria from mango fruits. *Plant Physiol.* **6,** 634–638.

Katz, S., and Reinhold, L. (1965). Changes in the electrical conductivity of Coleus tissue as a response to chilling temperatures. *Isr. J. Bot.* **13,** 105–114.

Kaufmann, M. R. (1975). Leaf water stress in Engelmann spruce: Influence of the root and shoot environments. *Plant Physiol.* **56,** 841–844.

Ketellapper, H. J. and Bonner, J. (1961). The chemical basis of temperature responses in plants. *Plant Physiol. Suppl.* **36,** XXI.

Kislyuk, I. M. (1964). Influence of light on injury of *Cucumis sativus* leaves during cooling. *Dokl. Akad. Nauk. SSSR* **158,** 1434–1436.

Koeppe, D. E., Rohrbaugh, L. M., Rice, E. L., and Wender, S. H. (1970). The effect of age and chilling temperatures on the concentration of scopolin and caffeoylquinic acids in tobacco. *Physiol. Plant.* **23,** 258–266.

Korovin, A. I., and Frolov, I. N. (1968). The effects of potassium and calcium on the resistance of buckwheat to low soil temperatures and freezing during the initiation of vegatation. *Izv. Sib. Otd. Akad. Nauk. SSSR Ser. Biol. Med. Nauk.* **3,** 48–53.

Kramer, P. (1942). Species differences with respect to water absorption at low soil temperatures. *Am. J. Bot.* **29,** 828–832.

Kuiper, P. J. C. (1972). Temperature response of adenosine triphosphatase of bean roots as related to growth temperature and to lipid requirement of the adenosine triphosphatase. *Physiol. Plant.* **26,** 200–205.

Kuraishi, S., Tezoka, T., Ushijima, T., and Tazaki, T. (1966). Effect of cytokinins on frost hardiness. *Plant Cell Physiol.* **7,** 705–706.

Kuraishi, S., Arai, N., Ushijima, T., and Tazaki, T. (1968). Oxidized and reduced nicotinamide adenine dinucleotide phosphate levels of plants hardened and unhardened against chilling injury. *Plant Physiol.* **43,** 238–242.

Kushmrenko, S. V., and Morozova, R. S. (1963). The effect of positive low temperatures on the structure of plastids in cucumbers hardened to cold. *Bot. Zh. (Leningrad)* **48,** 720–724.

Lawaczeck, R., M. Kainosho, and Chan, S. I. (1976). The formation and annealing of structural defects in lipid bilayer vesicles. *Biochim. Biophys. Acta.* **443,** 313–330.

Lee, A. G. (1975). Interactions with biological membranes. *Endeavour.* **31,** 67–71.

Lee, A. G., Birdsall, N. J. M., Metcalfe, J. C., Toon, P. A., and Warren, G. B. (1974). Clusters in lipid bilayers and the interpretation of thermal effects in biological membranes. *Biochemistry,* **13,** 3699–3705.

Levitt, J. (1969). Growth and survival of plants at extremes of temperature-a unified concept. *Proc. Soc. Exp. Biol.* **23,** 395–448.

Lewis, D. A. (1956). Protoplasmic streaming in plants sensitive and insensitive to chilling temperatures. *Science* **124,** 75–76.

Lewis, D. A., and Morris, L. L. (1956). Effects of chilling storage on respiration and deterioration of several sweet potato varieties. *Proc. Amer. Soc. Hort. Sci.* **68,** 421–428.

Lewis, T. L., and Workman, M. (1964). The effect of low temperature on phosphate esterification and cell membrane permeability in tomato fruit and cabbage leaf tissue. *Aust. J. Biol. Sci.* **17,** 147–152.

Lieberman, M., Craft, C. C., Audia, W. V., and Wilcox, M. S. (1958). Biochemical studies of chilling injury in sweet potatoes. *Plant Physiol.* **33,** 307–311.

Lindberg, S. (1976). Kinetic studies of a ($Na^+ + K^+ + Mg^{2+}$) ATPase in sugar beet roots. II. Activation by Na^+ and K^+. *Physiol. Plant.* **36,** 139–144.

Lindner, J. (1915). Über den Einfluss günstiger Temperaturen auf gefrorene Schimmelpilze. (Zur Kenntnis der Kälteresistenz von *Aspergillus niger*) *Jahrb. Wiss. Bot.* **55,** 1–52.

Littlejohns, D. A., and Tanner, J. W. (1976). Preliminary studies on the cold tolerance of soybean seedlings. *Can. J. Plant. Sci.* **56,** 371–375.

Livingston, B. E. (1903). "The Role of Diffusion and Osmotic Pressure in Plants." Univ. of Chicago Press, Chicago, Illinois.

Lyons, J. M. (1973). Chilling injury in plants. *Annu. Rev. Plant Physiol.* **24,** 445–466.

Lyons, J. M., and Asmundson, C. M. (1965). Solidification of unsaturated-satured fatty acid mixtures and its relationship to chilling sensitivity in plants. *J. Am. Oil Chem. Soc.* **42,** 1056–1058.

Lyons, J. M., and Raison, J. K. (1970a). Oxidative activity of mitochondria isolated from plant tissues sensitive and resistant to chilling injury. *Plant Physiol.* **45,** 386–389.

Lyons, J. M., and Raison, J. K. (1970b). Changes in activation energy of mitochondrial oxidation induced by chilling temperatures in cold sensitive plants and homeothermic animals. *Cryobiology* **6,** 585.

Lyons, J. M., Wheaton, T. A., and Pratt, H. K. (1964). Relationship between the physical nature of mitochondrial membranes and chilling sensitivity in plants. *Plant Physiol.* **39,** 262–268.

McDaniel, R. G., and Taylor, B. B. (1976). AMP treatments improve emergence under cold stress. p.9. In: Cotton, a College of Agriculture report. Ag. Exp. Sta., University of Arizona, Tucson. Series P-37.

MeElhaney, R. N. and Souza. K. A. (1976). The relationship between environmental temperature, cell growth and the fluidity and physical state of the membrane lipids in *Bacillus stearothermophilus. Biochim. Biophys. Acta.* **443,** 348–359.

McGlasson, W. B., and Raison, J. K. (1973). Occurrence of a termperature-induced phase trnasition in mitochondria isolated from apple fruit. *Plant Physiol.* **52,** 390–392.

McKersie, B. D., and Thompson, J. E. (1978). Phase behavior of chloroplast and microsomal membranes during leaf senescence. *Plant Physiol.* **61,** 639–643.

McKersie, B. D., Thompson, J. E., and Brandon, J. K. (1976). X-ray diffraction evidence for decreased lipid fluidity in senescent membranes from cotyledons. *Can. J. Bot.* **54,** 1074–1078.

Melcarek, P. K., and Brown, G. N. (1977). Effects of chill stress on prompt and delayed chlorophyll fluorescence from leaves. *Plant Physiol.* **60,** 822–825.

Miller, R. W. and De La Roche, I. A. (1976). Properties of spin labelled membranes of *Fusarium oxysporum f. sp. lycopersici. Biochim. Biophys. Acta.* **443,** 64–80.

Millerd, A., and McWilliam, J. R. (1968). Studies on a maize mutant sensitive to low-temperature I. Influence of temperature and light on the production of chloroplast pigments. *Plant Physiol.* **43,** 1967–1972.

Millerd, A. D., Goodchild, J., and Spencer, D. (1969). Studies on a maize mutant sensitive to low temperatue. II. Chloroplast structure, development, and physiology. *Plant Physiol.* **44,** 567–583.

Minamikawa, T., Akazawa, T., and Uritani, I. (1961). Mechanism of cold injury in sweet potatoes. II. Biochemical mechanism of cold injury with special reference to mitochondrial activities. *Plant Cell Physiol.* **2,** 301–309.

Minchin, A., and Simon, E. W. (1973). Chilling injury in cucumber leaves in relation to temperature. *J. Exp. Bot.* **24,** 1231–1235.

Mishustina, P. S. (1967). The effect of low temperatures on the respiration rate and carbohydrate metabolism in corn. *Rost. Ustoich. Rast.* **3,** 227–232.

Möbius, M. (1907). Die Erkältung der Pflanzen. *Ber. Deut Bot. Ges.* **25,** 67–70.

Molisch, H. (1896). Das Erfrieren von Pflanzen bei Temperaturen über dem Eispunkt. *Sitzber. Kaiserlichen Akad. Wiss. Wien. Math. Naturwiss. Kl.* **105,** 1–14.

Molisch, H. (1897). "Untersuchungen über das Erfrieren der Pflanzen," pp. 1–73. Fischer, Jena.

Murata, T. (1969). Physiological and biochemical studies of chilling injury in bananas. *Physiol. Plant.* **22,** 401–411.

Murata, N., Troughton, J. H., and Fork, D. C. (1975). Relationships between the transition of the physical phase of membrane lipids and photosynthetic parameters in *Anacystis nidulans* and lettuce and spinach chloroplasts. *Plant Physiol.* **56,** 508–517.

Nagaraja, K. V. and Patwardham, M. V. (1974). Arrhenius plots for mitochondrial succinoxidase and succinate dehydrogenase activities of some plant tissues. *Indian J. Biochem. Biophys.* **11,** 54–56.

Nezgovorov, L. A., and Solov'ev, A. K. (1965). Increase of field cold resistance of maize produced by treating the seeds with large doses of TMTD (tetramethylthiuram disulfide). Fiziol. Rast. **12,** 1093–1103.

Nishiyama, I. (1975). A break on the Arrhenius plot of germination activity in rice seeds. *Plant Cell Physiol.* **16,** 533–536.

Nobel, P. S. (1974). Temperature dependence of the permeability of chloroplasts from chilling-sensitive and chilling-resistant plants. *Planta.* **115,** 369–372.

Nolan, W. G., and Smillie, R. M. (1976). Multi-temperature effects on Hill reaction activity of barley chloroplasts. *Biochim. Biophys. Acta.* **440,** 461–475.

Nolan, W. G., and Smillie, R. M. (1977). Temperature-induced changes in Hill activity of chloroplasts isolated from chilling-sensitive and chilling-resistant plants. *Plant Physiol.* **59,** 1141–1145.

Nordin, A. (1977). Effects of low root temperature on ion uptake and ion translocation in wheat. *Physiol. Plant.* **39,** 305–310.

Norris, R. D., and Fowden, L. (1974). Cold-lability of prolyl-tRNA synthetase from higher plants. *Phytochem.* **13,** 1677–1687.

Olson, G. J. and Ingram, L. O. (1975). Effects of temperature and nutritional changes on the fatty acids of *Agmenellum quadruplicatum. J. Bacteriol.* **124,** 373–379.

Onoda, N. (1937). "Mikroskopische Beobachtungen über das Gefrieren einiger Pflanzenzellen in flüssigem Paraffin." Botan. Inst. der Kaiserlichen Univ. zu Kyoto. *Bot. and Zool.* **5,** 1845–2188.

Pagano, R. E., Cherry, R. J., and Chapman, D. (1973). Phase transitions and heterogeneity in lipid bilayers. *Science.* **181,** 557–559.

Patterson, B. D., Murata, T., and Graham, D. (1976). Electrolyte leakage induced by chilling in *Passiflora* species tolerant to different climates. *Aust. J. Plant Physiol.* **3,** 435–442.

Pentzer, W. T., and Heinze, P. H. (1954). Post-harvest physiology of fruits and vegetables. *Annu. Rev. Plant Physiol.* **5,** 205–224.

Peoples, T. R., Koch, D. W., and Smith, S. C. (1978). Relationship between chloroplast membrane fatty acid composition and photosynthetic response to a chilling temperature in four alfalfa cultivars. *Plant Physiol.* **61,** 472–473.

Peters, R. A. (1963). "Biochemical Lesions and Lethal Synthesis," Macmillan, New York.

Petrova, O. V. (1967). Free amino acids and the form of nitrogen in the leaves during development of corn hybrids differing in resistance to cold. *Rost. Ustoich. Rast.* SB3, 233–240.

Platt-Aloia, K. A., and Thomson, W. W. (1976). An ultrastructural study of two forms of chilling-induced injury to the rind of grapefruit (*Citrus paradisi* Macfed). *Cryobiology* **13,** 95–106.

Podin, V. S. (1966). Comparative study of the xanthophyll transformation reaction of some plants as a factor of temperature in light and darkness. *Izv. Akad. Nauk. Latv. SSSR* **11,** 82–86.

Pollock, B. M. (1969). Imbibition temperature sensitivity of lima bean seeds controlled by initial seed moisture. *Plant Physiol.* **44,** 907–911.

Pollock, C. J. and T. Ap Rees. (1975a). Cold-induced sweetening of tissue cultures of *Solanum tuberosum L. Planta* **122,** 105–107.

Pollock, C. J. and Ap Rees, T. (1975b). Activities of enzymes of sugar metabolism in cold stored tubers of *Solanum tuberosum. Phytochem.* **14,** 613–617.

Pomeroy, K. and Andrew, C. J. (1975). Effect of temperature on respiration of mitochondria and shoot segments from cold-hardened and nonhardened wheat and rye seedlings. *Plant Physiol.* **56,** 703–706.

Pugh, E. L. and M. Kates. 1975. Characterization of a membrane-bound phospholipid desaturase system of *Candida lipolytica. Biochim. Biophys. Acta* **380,** 442–453.

Radchenko, S. I., Konovalov, I. N., and Pozdova, L. M. (1964). Frost resistant corn on the Korelian Isthmus. *Tr. Bot. Inst. Akad. Nauk. SSR. Ser.* **4,** 17:53–72.

Raison, J. K. (1973). The influence of temperature-induced phase changes on the kinetics of respiratory and other membrane-associated enzyme systems. *J. Bioeng.* **4,** 285–309.

Raison, J. K., Chapman, E. A., and White, P. Y. (1977). Wheat mitochondria. Oxidative activity and membrane lipid structure as a function of temperature. *Plant Physiol.* **59,** 623–627.

Razmaev, I. I. (1965). After-effect of low temperatures above 0°C on nitrogen metabolism in wheat and corn. *Izv. Sib. Otol. Akad. Nauk, SSSR Ser. Biol. Med. Nauk.* **1,** 59–63.

Rhodes, M. J. C., and Wooltorton, L. S. O. (1977). Changes in the activity of enzymes of phenylpropanoid metabolism in tomatoes stored at low temperatures. *Phytochem.* **16,** 655–659.

Rikin, A., and Richmond, A. E. (1976). Amelioration of chilling injuries in cucumber seedlings by abscisic acid. *Physiol. Plant.* **38,** 95–97.

Rikin, A., Blumenfeld, A., and Richmond, A. E. (1976). Chilling resistance as affected by stressing environments and abscisic acid. *Bot. Gaz.* **137,** 307–312.

Sachs, J. (1864). Ueber die obere Temperatur-Grenze der Vegetation. *Flora (Jena)* **47,** 5–12, 24–29. 33–39, 65–75.

Sawada, S. I., and Miyachi, S. (1974). Effects of growth temperature on photosynthetic carbon metabolism in green plants. II. Photosynthetic $14CO_2$ incorporation in plants acclimatized to varied temperatures. *Plant Cell Physiol.* **15,** 225–238.

Sechi, A. M., Bertoli, E., Landi, L., Parenti-Castelli, G., Lenaz, G., and Curatola, G. (1973). Temperature dependence of mitochondrial activities and its relation to the physical state of the lipids in the membrane. *Acta Vitaminol. Enzymol.* **27,** 177–190.

Seible, D. (1939). Ein Beitrag zur frage der Kälteschäden an Pflanzen bei Temperaturen über dem Gefrierpunkt. *Beitr. Biol. Pflanz.* **26,** 289–330.

Sellschop, J. P. F., and Salmon, S. C. (1928). The influence of chilling above the freezing point on certain crop plants. *J. Agr. Res.* **37,** 315–338.

Selwyn, M. J. (1966). Temperature and photosynthesis. II. A mechanism for the effects of temperature on carbon dioxide fixation. *Biochim. Biophys. Acta.* **126,** 214–224.

Shneyour, A., Raison, J. K., and Smillie, R. M. (1973). The effect of temperature on the rate of photosynthetic electron transfer in chloroplasts of chilling-sensitive and chilling-resistant plants. *Biochim. Biophys. Acta.* **292,** 152–161.

Silvius, J. R., Read, B. D., and McElhaney, R. N. (1978). Membrane enzymes: artifacts in Arrhenius plots due to temperature dependence of substrate-binding affinity. *Science,* **199,** 902–904.

Simon, E. W., Minchin, A., McMenamin, M. M., and Smith, J. M. (1976). The low temperature limit for seed germination. *New Phytol.* **77,** 301–311.

Slack, C. R., Roughan, P. G., and Bassett, H. C. M. (1974). Selective inhibition of mesophyll chloroplast development in some C_2 pathway species by low night temperature. *Planta* (Berlin) **118,** 57–73.

Smith, W. H. (1954). Non-freezing injury in plant tissues with particular reference to the detached plum fruit. *8th Int. Congr. Bot.* **11,** 280–285.

Smith, E. W. and Fites, R. C. (1973). The influence of chilling temperature alteration of glyoxysomal succinate levels on isocitratase activity from germinating seedlings. *Biochem. Biophys. Res. Commun.* **55,** 647–654.

Solomonovskii, L. Y., and Pomazova, E. N. (1967). The effect of a physiologically active mixture on the water regime and cold resistance of thermophilic plants. *Izv. Sib. Otd. Akad. SSR Ser. Biol. Med. Nauk.* **1,** 72–78.

Solov'ev, A. K., and Nezgovorov, L. A. (1968). Differences in the response of shade plants to

injuring and hardening temperatures as judged by extraction of cellular sap from leaves. *Fiziol. Rast.* **15,** 1045–1054.
Srivastava, K. C., and Smith, D. G. (1976). The ultrastructure of psychrophilic strains of *Cryptococcus albidus*. *Microbios. Lett.* **3,** 175–178.
St. John, J. B. and Christiansen, M. N. (1976). Inhibition of linolenic acid synthesis and modification of chilling resistance in cotton seedlings. *Plant. Physiol.* **57,** 257–259.
Stewart, J. McD., and Guinn, G. (1969). Chilling injury and changes in adenosine triphosphate of cotton seedlings. *Plant Physiol.* **44,** 605–608.
Stewart, J. McD, and Guinn, G. (1971). Chilling injury and nucleotide changes in young cotton plants. *Plant Physiol.* **48,** 166–170.
Talma, E. G. C. (1918). *Recl. Trav. bot. néerl.* **15,** 366 .
Tanczos, O. G. (1977). Influence of chilling on electrolyte permeability, oxygen uptake and 2,4-dinitrophenol stimulated oxygen uptake in leaf discs of the thermophilic Cucumis sativus. *Physiol. Plant.* **41,** 289–292.
Thompson, J. E., Mayfield, C. I., Inniss, W. E., Butler, D. E., and Kruuv, J. (1978). Senescence-related changes in the lipid transition temperature of microsomal membranes from algae. *Physiol. Plant.* **43,** 114–120.
Toda, M. (1962). Studies on the chilling injury in wheat plants-I. Some researchers on the mechanism of occurrence of the sterile phenomenon caused by the low temperature. *Proc. Crop. Sci. Japan* **30,** 241–249.
Traeuble, H. and Eibl, H. (1974). Electrostatic effects on lipid phase transitions: Membrane structure and ionic environment. *Proc. Natl. Acad. Sci.* **71** 214–219.
Traeuble, H., Teubner, M., Woolley, P., and Eibl, H. J. (1976). Electrostatic interactions at charged lipid membranes: I. Effects of pH and univalent cations on membrane structure. *Biophys. Chem.* **4,** 319–342.
Tsunoda, K., Fujimura, K., Nakahari, T., and Oyamado, Z. (1968). Studies on the testing method for cooling tolerance in rice plants. I. An improved method by means of short turn treatment with cool and deep water. *Jap. J. Breed.* **18,** 33–40.
Van Hasselt, P. R. (1973). Photo-oxidative damage to triphenyltetrazoliumchloride (TTC) reducing capacity of Cucumis leaf discs during chilling. *Acta. Bot. Neerl.* **22,** 546–552.
Van Hasselt, P. R. (1974). Photo-oxidation of unsaturated lipids in Cucumis leaf discs during chilling. *Acta. Bot. Neerl.* **23,** 159–169.
Van Hasselt, P. R. (1976). Protection of *Cucumis* leaf pigments against photo-oxidative degradation during chilling. *Acta. Bot. Neerl.* **25,** 41–50.
Van Steveninck, J., and Ledeboer, A. M. (1974). Phase transition in the yeast cell membrane: The influence of temperature on the reconstitution of active dry yeast. *Biochim. Biophys. Acta.* **352,** 64–70.
Verkleij, A. J., deKruyff, B., Ververgaert, P. H. J. Th. Tocanne, J. F., and Van Deenen. L. L. M. (1974). The influence of pH, Ca^{2+} and protein on the thermotropic behavior of the negatively charged phospholipid, phosphatidylglycerol. *Biochim. Biophys, Acta.* **339,** 432–437.
Wade, N. L., Breidenbach, W. R., Lyons, J. M. and Keith, A. D. (1974). Temperature-induced phase changes in the membranes of glyoxysomes, mitochondria, and proplastids from germinating castor bean endosperm. *Plant Physiol.* **54,** 320–323.
Watson, K., Bertoli, E. and Griffiths, D. E. 1975. Phase transitions in yeast mitochondrial activation of the respiratory enzymes of *Saccharomyces cerevisiae*. *Biochem. J.* **146,** 401–407.
Wheaton, T. A., and Morris, L. L. (1967). Modification of chilling sensitivity by temperature condition. *Proc. Amer. Soc. Hort. Sci.* **91,** 529–533.
Wilhelm, A. F. (1935). Untersuchungen über das Verhalten sogennanter nicht eisbeständiger

Kulturpflanzen bei niederen-Temperaturen, unter besonderer Berücksichtigung des Einflusses verschiedener Mineralsalzernährung und des N-Stoffwechsels. *Phytopathol. Z.* **8,** 337–362.

Wilson, J. M. (1976). The mechanism of chill-and drought-hardening of *Phaseolus vulgaris* leaves. *New Phytol.* **76,** 257–260.

Wilson, J. M., and Crawford, R. M. M. (1974a). The acclimatization of plants to chilling temperatures in relation to the fatty-acid composition of leaf polar lipids. *New Phytol.* **73,** 805–820.

Wilson, J. M., and Crawford, R. M. M. (1974b). Leaf fatty-acid content in relation to hardening and chilling injury. *J. Exp. Bot.* **25,** 121–131.

Wright, M. (1974). The effect of chilling on ethylene production, membrane permeability and water loss of leaves of *Phaseolus vulgaris. Planta* **120,** 63–69.

Yamaki, S. and Uritani, I. (1973a). Morphological changes in chilling injured sweet potato root. *Agric. Biol. Chem.* **37,** 183–186.

Yamaki, S. and Uritani, I. (1973b). Mechanism of chilling injury in sweet potato: X. Change in lipid-protein interaction in mitochondria from cold-stored tissue. *Plant Physiol.* **51,** 883–888.

Yelenosky, G. (1978). Freeze survival of citrus trees in Florida. *In* "Plant Cold Hardiness and Freezing Stress." (P. H. Li and A. Sakai eds.), pp. 297–311. Academic Press, New York.

Zsoldos, F. (1971). Isotope technique for investigation of cold resistance in rice and sorghum varieties. *Plant Soil,* **35,** 659–663.

III

FREEZING TEMPERATURES

4. Limits of Low-Temperature Tolerance

The effects of freezing temperatures on living organisms have long interested biologists, both from practical and theoretical points of view. Only in recent years, however, have the practical applications been seriously exploited, with the preservation of food, blood, semen, cultures, and tissues. The resulting explosive increase in research reached a climax in 1958 with the introduction of a new name for the field—*cryobiology,* the biology of freezing temperatures (Parkes, 1964). In 1963, the Society for Cryobiology was organized. A monograph on Cryobiology (Meryman, 1966a) appeared, and later the plant life of snow and ice was treated separately in yet another monograph on Cryovegetation (Kol, 1969). Long before the development of a separate science, biologists attempted to discover whether or not there is a theoretical limit to the low-temperature stress that living cells can survive. This question is now important from a practical as well as a theoretical point of view.

A. DEHYDRATED PROTOPLASM

Low-temperature resistance, as mentioned previously (Chapter 2), is tolerance in nearly all cases. In order to understand low-temperature *resistance* it is, therefore, first necessary to answer the question of whether there is a limit to the low-temperature *tolerance* of plants. This problem was approached long ago by a number of investigators. Since they were searching for maximum tolerance, only the most resistant of plant parts were tested—either lower plants, or higher plants in the dry and dormant states (seeds, spores, pollen grains). These resistant cells and tissues were able to survive the lowest temperatures to which they were exposed, even down to a fraction of a degree above absolute zero (Table 4.1). Since the higher plant tissues were "dry" (i.e., air dry), the small amount of moisture present in them was "bound" and, therefore, was not converted to ice at even the lowest temperatures used. These results, therefore, simply showed that de-

TABLE 4.1

Survival of Very Low Temperatures by Plants[a]

Plant or plant part	Temperature	Exposure time	Reference
Seeds	−100°C	4 days	de Candolle, 1895
Seeds	−190°C	110 hr	Brown and Escombe, 1897
Seeds	−250°C	6 hr	Thiselton-Dyer, 1899
Bacteria and yeast	−190°C	6 months	Macfayden, 1900
Seeds	−190°C	130 hr	Becquerel, 1907
Fungi and algae	−190°C	13 hr	Kärcher, 1931
Seeds	−190°C	60 days	Lipman and Lewis, 1934
Mosses (protonema)	−190°C	50 hr	Lipman, 1936a
Seeds and spores	1–4°K	44 hr	Lipman, 1936b
Fungus mycelium and bacteria	−190°C	48 hr	Lipman, 1937
Spores and pollen grains	−273°C (within a few thousandths of a degree)	2 hr	Becquerel, 1954

[a] All were able to grow after the low temperature exposure. (See also Luyet and Gehenio, 1938.) (−190°C means liquid air was used, −250°C, liquid hydrogen.)

hydrated cells which are unable to freeze at low temperatures can survive the lowest temperatures without suffering any injury.

B. HYDRATED PROTOPLASM

Some lower plant (Table 4.1) and animal (Smith *et al.*, 1951) cells in the normally moist condition can also survive exposure to extremely low temperatures without any special precautions. Yet when the above, fully tolerant dry seeds are allowed to imbibe water, they are killed by very slight freezes (Table 4.2). Similarly, when the water content of pollen grains was high enough (about 36%) to permit X-ray detection of ice crystals at −25°C, the pollen no longer survived the low temperatures (Ching and Slabaugh, 1966). It has long been known, of course, that the normally hydrated protoplasm of most plants (except those that overwinter in cold climates) is normally killed by even slight freezing. What, then, is the limit of low-temperature tolerance in the case of the normally hydrated protoplasm of plants?

Luyet (1937) has shown that even in the normally hydrated state, plant cells that are killed by slight freezes, can nevertheless survive immersion in liquid nitrogen if both the cooling and the rewarming rates are ultrarapid (10,000–100,000°C/sec). He achieved these extremely rapid rates by plung-

TABLE 4.2

Survival of Low Temperatures by Dry and Hydrated Plants

Plant	Low-temperature treatment	Survival: Dry	Survival: Hydrated	Reference
Seeds	−25° to −40°C for 15 hr	All germinated	None germinated	Göppert, 1830
Seeds	−196°C for 130 hr	All germinated	None germinated	Becquerel, 1907
Corn kernels	32°–28°F	25% Moisture all germinated	75–85% Moisture none germinated	Kiesselbach and Ratcliff, 1918[a]
Yeast	−113°C	Unaffected	Vacuolate cells all killed; young (nonvacuolate) uninjured	Schumacher, 1875
Ranunculus tubers	−196°C for 18 days	9% Moisture all survived	30–50% Moisture all killed	Becquerel, 1932
Wheat seed	−196°C for 2 min	10.6% H_2O all germinated	25.1% H_2O none germinated	Lockett and Luyet, 1951
Alfalfa seed	−20°C for 1 day	90% Germinated	None germinated	Tysdal and Pieters, 1934

[a] Similar results were obtained by Jensen, 1925; Steinbauer, 1926; Stuckey and Curtis, 1938; and McRostie, 1939.

ing thin strips of tissue (e.g., one cell thick) directly from room temperature into the liquid nitrogen, followed by direct transfer to a warm (25°–30°C) aqueous solution. In this way, rates of cooling and warming as high as 150,000°C/sec have been obtained (Luyet, 1951). He first considered this a vitrification process, the rapidity of the cooling maintaining the water in the supercooled, noncrystalline or vitrified (i.e., glassy) state. This conclusion was based on the absence of double refraction (which is characteristic of crystals) when viewed under the microscope.

According to the definition suggested by a committee appointed by the National Research Council, "a glass or vitreous substance is a solid giving a typical, amorphous phase X-ray pattern and capable of exhibiting the glass transition" (Angell and Sare, 1970). The glass transition, in turn, is "that

TABLE 4.3

Relation to Low Temperature of Structures Obtained by the Deposition of Water Vapor on a Substrate[a]

Experimental method	Temperature ranges (°C)							Reference
	−180	−160	−140	−120	−100	−80	−60	
X-ray diffraction	Amorphous			Semicrystalline		Hexagonal		Burton and Oliver, 1935
Calorimetric	Amorphous				Crystalline			Staronka, 1939
X-ray diffraction	Small crystals			Intermediate range not investigated		Hexagonal		Vegard and Hillesund, 1942
Electron diffraction	Small crystals			Cubic		Hexagonal		König, 1942
Calorimetric	Vitreous				Crystalline			Pryde and Jones, 1952
Electron diffraction	Crystal growth poor			Cubic		Hexagonal		Honjo *et al.*, 1956
Calorimetric		Amorphous			Crystalline			Ghormley, 1956
Calorimetric		Vitreous			Crystalline			De Nordwall and Staveley, 1956
Electron diffraction		Amorphous or small crystals		Cubic		Hexagonal		Blackman and Lisgarten, 1957

[a] The terms used are those of the respective authors. (From Blackman and Lisgarten, 1958; see Merryman, 1966b.)

phenomenon in which a vitreous phase exhibits with changing temperature a more or less sudden change in the derivative thermodynamic properties, such as heat capacity and expansion coefficient, from crystal- to liquidlike values. The temperature of the transition is called the glass transition temperature." It has also been called the vitrification point. The true vitrification point for water has been calculated to be 162 ± 1°K or −111°C (Miller, 1969). Vitreous (noncrystalline) water has an estimated density of 1.2 g/cm^3 (Venkatesch *et al.*, 1974).

On the basis of this definition, true vitrification was probably not achieved in the above experiments; later investigations, using X-rays, revealed the presence of submicroscopic crystals (Table 4.3), at least in some cases of rapid freezing such as used by Luyet. In contrast to these negative results, enthalpic measurements with a differential scanning calorimeter in the temperature range −110 to 10°C indicated that the cell water of *Saccharomyces cerevisiae* is converted to amorphous water at −38°C and begins to crystallize at −24°C (Gonda and Koga, 1973).

Nevertheless, whether true vitrification is achieved, or a pseudovitrification with crystals too small to be detected microscopically, this method may permit a strip of onion epidermis to survive immersion in liquid N_2 without injury, though the cells are otherwise killed by a very moderate freeze (e.g., −10°C). Algae (twenty-three strains of five genera) have also survived direct immersion in liquid nitrogen (Hwang and Horneland, 1965).

Even if the hydrated cells or tissues survive such ultrarapid freezing and thawing without injury, it does not follow that this absence of injury is independent of the length of time exposed to the low temperature. Meryman (1966a) has demonstrated crystal growth at temperatures as low as −130°C. At -80°C, for instance, they grew from 200–10,000 Å in 8 min, becoming microscopically visible. Another possible source of injury is crystal shape which depends on the temperature and the order of cooling and warming (Table 4.4). Consequently, if the ultrarapid cooling method is successful, maintaining the crystals below microscopic size, the protection would be only temporary unless the cells are stored at temperatures well below −130°C.

Even for short freezes, the method does not work with all living cells. Even when it does succeed in keeping the cells viable, if they contain 70–80% water, the tissue must be less than 0.1 mm thick. Thicker pieces of tissue do not permit the extremely rapid temperature drop (and rise) needed to prevent the growth of crystals large enough to be injurious within the cells. If, however, the water content of the tissue is lower than 70–80%, thicker pieces may be frozen rapidly enough to prevent injury (Luyet, 1951). In the case of wheat grains frozen in liquid nitrogen (–196°C) from seconds to minutes, and then maintained at −120°C, ice can be detected by X-ray analysis if the

TABLE 4.4

Ice Form and Temperature[a]

Temperature	Ice form	
0	Temp. drop ↓	Temp. rise ↑
−80	Hexagonal	Hexagonal ↑
−100	│	Cubic + hexagonal ↑
−140	│	Cubic (diamond) ↑
−192	↓ Hexagonal	Amorphous

[a] Adapted from Shikama, 1963.

moisture content is above 33% (Radzievsky and Shekhtman, 1955). Similar results were obtained by Sun (1958). Small pea seedlings (7–12-mm long), excised from their cotyledons, survive exposure to liquid nitrogen if they are first dried to a moisture content of 27–40%. In the case of larger seedlings, with low water content, only the stem tip survived. This relationship between size and the protective effect of low water content may perhaps explain the failure of other investigators to obtain survival of liquid nitrogen (e.g., Genevès, 1955). Nevertheless, even when the above precautions are taken, some animal cells are killed by too rapid cooling, whereas a slower cooling (e.g., cooled to −79°C in 5 min) permits survival (Smith *et al.*, 1951). The speed of cooling (and warming) is particularly important within a temperature range which may be called the danger zone. Published ranges for this danger zone vary somewhat with both the investigator and the organism used (Table 4.5). In general, it is between the freezing point of the material and about −30° to −40°C. This zone presumably includes the temperature range in which the size of the immediately formed crystals is large enough to damage the cell. Some plants seem able to survive indefinitely even when frozen within this danger zone. Thus, of 291 strains of molds stored for 5 years in a freezer at −17° to −21°C, only 15 were not viable (Carmichael, 1962). This method was successful even in the case of some fungi which do not survive freeze-drying. On the other hand, freeze-drying, which usually involves a slower freezing than used by Luyet, permits the survival of pollen, particularly if stored at −25°C (Snope and Ellison, 1963; Layne, 1963). as well as fungi and mycobacteria (Sarbhoy *et al.*, 1974). Similarly, three species of blue-green algae showed no decline in viability when lyophilized (freeze-dried), even after storage at 25°C for 5

TABLE 4.5

The Critical Low Temperature Zone for Injurious Crystallization of Ice

Living material	Temp. zone (°C)	Reference
General	Freezing point to −30	Luyet, 1940
Red blood cells	−4 to −40	Lovelock, 1953
Pasteurella tularensis	−30 to −45	Mazur *et al.*, 1957
Gill pieces of oyster	−40 to −50	Asahina, 1958
Sucrose solution	Freezing point to −32	Rey, 1961
Denaturation of catalase	−12 to −75	Shikama, 1963
Denaturation of myosin	−20 to −72	Shikama, 1963

years (Holm-Hansen, 1967). In the case of four species of green algae, the same treatment did result in a significant decline in viability.

C. CRYOPROTECTANTS

1. One-Step Freezing

There are two other methods of passing through the danger zone without injury, in addition to the ultrarapid cooling method. If the cells are first treated with protective substances such as glycerol and dimethylsulfoxide (DMSO) in about molar concentrations, they may be cooled slowly (e.g. 1/4–1°C/min) without being killed by liquid nitrogen. Such substances are called *cryoprotectants* if they prevent injury during the freezing. This method works well for animal cells (although not for all; Sherman, 1962) and for microorganisms and cell and tissue cultures of higher plants (Table 4.6). Recent evidence (Robbins and Whitwood, 1973), in fact, suggests that cooling to −196°C may enhance the development of callus tissue and their totipotency. The cryoprotectants normally result in something less than 100% survival. In the case of an alga (*Chlorella*), the cells from the stationary phase survive; those from the exponential phase are damaged (Morris, 1976). In the case of the cell and tissue cultures of higher plants, the small, meristematic cells or cell clumps survive (Nag and Street, 1975a, b; Sugawara and Sakai, 1974; Bajaj, 1976). This explains why it has been possible to obtain 33% survival of −196°C in the case of carnation shoot apices (Seibert 1976). Cold treatment at 4°C for 3 days or more resulted in a doubling of the survival of these shoot apices, and a 6- to 7-fold increase in formation of leaf primordia or shoots (Seibert and Wetherbee, 1977). Cryoprotectants have not proved useful for preserving normal, nonmeristematic plant tissues. Even

TABLE 4.6

Survival of Microorganisms and Tissue Cultures of Higher Plants When Frozen in the Presence of Cryoprotectants. Cooling Rates Usually 1–2°C/min.

Species	Temperature survived (°C)	Time frozen (years)	Survival	Cryoprotectant (%)	Reference
a. Microorganisms and lower plants					
Bacteria 259 strains 32 genera 135 species	−53	1.33	viable counts 10^5/liter in 74–93% aerobes	glycerol 10	Yamasoto *et al.*, 1973
Mycobacteria 179 strains 10 species	lyophilized	8–12 year	100% in six species		Slosarek *et al.*, 1976
Algae green blue-green	−196	0.25	60–85%	glycerol or DMSO 10	Tsuru, 1973
Chlorella			>95%	PVP	Morris, 1976
Chlamydomonas *rheinhardtii* (10 strains)			all	up to 15 DMSO	McGrath and Daggett, 1977
Dinoflagellate (*Crypthecodinium*) *cohnii*	−150	7 days; one strain 6 year	47.7% 68%	glycerol 7.5	Simione and Daggett, 1977
Rust fungi	−196	appx. 1	up to 93%	glycerol 10 or DMSO 5	Cunningham, 1973

Slime molds		appx. 1			Laine *et al.*, 1975
Fungi (molds)	lyophilized		26 of 38 cultures		Sarbhoy *et al.*, 1975
b. Tissue cultures and cell suspensions of higher plants					
Flax, and *Haplopappus gracilis*	−50	1 month		DMSO 10	Quatrano, 1968
Poplar (*P. euramericana*)	−196 (after cold accli- mation)				Sakai and Sugawara, 1973
Sycamore *Acer psuedo- platanus*)	−196		20–25%	DMSO and glucose	Sugawara and Sakai, 1974 Nag and Street, 1975a,b
Carrot (*Daucus carota*)	−196		70–75%	DMSO or glycerol	Dougall and Wetherell, 1974; Nag and Street, 1975a,b
Belladonna *Atropa belladonna*)	−196		30–40%	DMSO or glycerol	Nag and Street, 1975a,b
Datura, tobacco, carrot, soybean	−196		up to 70%	DMSO 5 or 7	Bajaj, 1976

in the case of microorganisms and cell and tissue cultures of higher plants, some species have not survived freezing in the presence of cryoprotectants (Quatrano, 1968; Hwang and Howells, 1968). In the case of cells permeable to glycerol (epidermis of *Campanula* species), 40–50% survive rapid freezing in liquid nitrogen followed by rapid thawing, if they have been allowed to imbibe 15–20% glycerol (Holzl and Bancher, 1968). This method is lethal if cells are stored for some time at −70°C (Richter, 1968a). When the glycerol concentration is increased to over 70%, the solutions solidify in the vitreous state and the cells are uninjured. These cells, however, are unusually indifferent to glycerol, surviving even a transfer to 100% glycerol without apparent injury when returned to mixed solutions (Richter, 1968b). The cells are, however, injured by ethylene glycol and DMSO. Some protective action by glycerol, ethylene glycol, etc., has been reported for sections of collards (Samygin and Matveeva, 1967). Glycerol was also somewhat protective to winter wheat (Trunova, 1968), though high concentrations were injurious.

2. Two-Step Freezing

This method has proved successful for partially resistant plant cells (Sakai, 1958). If a hardened plant twig is frozen slowly down to a temperature which is noninjurious (e.g., −15° to −30°C) and allowed to come to freezing equilibrium at this temperature, it may then be plunged into liquid nitrogen without injury. This method has been confirmed by Tumanov *et al.* (1959) and by Krasavtsev (1961), and the temperature survived was extended to that of liquid hydrogen. The partial dehydration by extracellular freezing to −10°C or below permits these moderately resistant cells to survive the subsequent rapid cooling to the extremely low temperatures of liquid N_2 or H_2. So little unfrozen water remains that the tissues are able to pass rapidly through the temperature zone of growth of intracellular ice crystals without freezing intracellularly (Sakai and Otsuka, 1972). The two-step method has also been used successfully for the alga *Chlorella* (Hatano *et al.*, 1976a), and even for hamster cells (Farrant *et al.*, 1977).

In the case of some extremely hardy plants, such as leaves of *Pinus strobus* in midwinter, survival in liquid nitrogen has been obtained simply by cooling at a rate of 3°C/hr (Parker, 1959b, 1960). Extremely hardy cells from winter twigs of mulberry trees can even survive rapid immersion in liquid nitrogen from room temperature and subsequent rewarming in water at 35°C (Sakai, 1968). Sakai succeeded in using this method, in combination with cryoprotectants even in the case of less tolerant cells, which could not survive the first step freezing below −10°C in the absence of cryoprotectants. These cells were able to survive immersion in liquid nitrogen and subsequent rapid rewarming provided that they were previously treated with an isotonic or slightly hypertonic glucose solution, and were blotted to remove the excess solution before immersion in liquid nitrogen.

From all the above results, it can be concluded that normally hydrated protoplasm of many organisms can survive the lowest freezing temperatures if (1) cooled at rapid enough rates to pass through the danger zone (0 to -40°C) before intracellular ice crystals can grow large enough to injure the cells (2) cooled gradually (1–2°/min.) in the presence of cryoprotectants, or (3) subjected to two-step freezing in the case of moderately resistant cells. In the case of plants that have not responded to such methods, perhaps more sophisticated methods will eventually lead to success.

The mechanism of cryoprotection is still a matter of controversy, and indeed there may very well be more than one mechanism.

1. Intracellular versus extracellular protection. It was assumed at the outset that substances must penetrate the cell in order to produce their cryoprotective effect. Maximov (1912), however, had earlier obtained cryoprotection of plant cells by nonpenetrating substances. In agreement with this result, Leibo *et al.* (1974) concluded that DMSO need not penetrate mouse embryo cells in order to protect them against freezing damage. Similarly, glycerol was a superior protectant against ultrastructural injury when extracellular (Sherman and Liu, 1976). In the case of *E. coli,* however, survival of freezing (at −70 to −196°C) was enhanced by extracellular glycerol and further increased by intracellular glycerol (Nath and Gonda, 1975). Intracellular glycerol alone, on the other hand, failed to increase cell recovery.

2. Solution versus membrane effects. The earlier concepts of cryoprotection by cell penetration were explained by solution effects—colligative effects which prevented injury by preventing an excessive increase in concentration, for instance of the tissue salts. This explanation is still supported by results showing protection of a soluble enzyme (alcohol dehydrogenase) by glycerol (Myers and Jakoby, 1975), apparently by eliciting small conformational changes in the protein. In the case of polymers, cryoprotection appears to result from their ability to alter the physical properties of solutions during the freezing process, rather than from direct effects on the cell (Connor and Ashwood-Smith, 1973a, b). Even in the case of cryoprotectants of low M.W., this explanation may also apply. Thus, both glycerol and DMSO stabilize the amorphous state of water—DMSO to a greater degree, in agreement with its superiority as a cryoprotectant (Boutron and Kaufmann, 1978).

Membrane effects have also been suggested (Pribor 1975). Langmuir trough experiments, however, show that the cryoprotectants do not distribute into the lipids of the membrane (Williams and Harris, 1977) and lead to the conclusion that they exert their effects on the aqueous (internal) side of the membrane.

In all of these examples, the cryoprotectants were present in high concentrations—5 or 10% and therefore 0.5 or 1.0 *M* in the case of DMSO, glycerol, etc. In the case of fruit trees sprayed with cryoprotectants, only small amounts can accumulate in the tissues.

It must be emphasized, however, that the successful survival of low temperatures by protoplasm in the hydrated state does not prove that the hydrated protoplasm possesses plastic resistance to all forms of injury produced by low-temperature stress. The special methods used are successful only if they prevent the occurrence of a specific strain which is normally produced at the freezing temperatures, and which normally kills the hydrated protoplasm of unadapted plants. The adapted plant has developed its own methods of achieving survival of the low-temperature stress (see Chapters 7 and 8) in the absence of the above artificial protective measures.

5. The Freezing Process

A. THE FREEZING STRESS

There are two kinds of low-temperature stress—the chilling low-temperature stress (down to 0°C) and the freezing low-temperature stress (below 0°C). Unfortunately, these terms are cumbersome. Therefore, the short forms *chilling stress* and *freezing stress* have come into general use. *Chilling stress* is an unambiguous term and can only mean the stress due to chilling low temperatures. Strictly speaking, the term *freezing stress* is a misnomer since the low temperature is the stress, and freezing (when it occurs) is the resulting strain. Thus, a plant may remain unfrozen though exposed to a freezing low-temperature stress, but a plant subjected to a freezing strain is frozen. Because of its general occurrence in the literature, the term freezing stress will be accepted as an abbreviation for the longer, more cumbersome term "freezing low-temperature stress". As long as the word freezing is always followed by either stress or strain, the ambiguity will be reduced to a minimum. Perhaps the term *cryostress* will eventually replace freezing stress and remove all ambiguity, since *cryo* means freezing low temperature.

The freezing stress may be defined as the freezing potential of the low-temperature stress. If the plant does freeze, this strain may, of course, induce indirect strains in other components of the plant. In terms of the direct strain (ice formation), a freezing stress can only be measured by the freezing potential of the plant's environment. Since the concentration of a plant's cell sap varies within a wide range, the freezing potential of any one temperature will differ for different plants. For any one plant, however, the lower the temperature, the greater the amount of ice formed at equilibrium until the temperature is reached at which all the freezable water has crystallized. The simplest measure of freezing stress is, therefore, the number of degrees (or kelvins) the environmental temperature is below the freezing point of pure water at atmospheric pressure (i.e., –T°C where T is the temperature in °C).

Once it is initiated, the freezing strain will, therefore, always increase with the freezing stress until a point is reached where no more water can freeze (see below). Instead of the stress being proportional to the strain, as in the case of a physical system, the increase in the freezing strain decreases logarithmically with each equal additional freezing stress (i.e. per degree lowering of freezing temperature).

B. EXTRACELLULAR VERSUS INTRACELLULAR FREEZING

1. Extracellular Freezing

The freezing of plants has long been a controversial subject for scientists from many disciplines. The frost splitting of trees, which occurs with a "crack like that of a gun," led to the belief that the plant tissues expand on freezing and may ultimately rupture because of this expansion (Bobart, 1684; Chomel, 1710; Du Hamel and de Buffon, 1740; Strömer, 1749; Thouin, 1806; Hermbstädt, 1808). Schübler (1827) pointed out that this splitting occurs only in thick trees (1 1/2–2 ft in diameter) and not at all in thin ones (a few inches in diameter), although the temperature drops much lower in the latter. He further states that when a tree is split in this way, no significant injury occurs; on the other hand, it is the youngest twigs that are injured first, although they never suffer such splits. Nevertheless, even the limpness of thawed herbaceous plants was thought to be due to cell rupture (Senebier, 1800; Thouin, 1806; Hermbstädt, 1808). Plants that survive freezing were believed able to prevent such expansion or even to contract due to the presence of oils that shrink on freezing (Bobart, 1684; Du Hamel and de Buffon, 1740; Strömer, 1749; Reum, 1835).

This "rupture theory" has been accepted even relatively recently (Goodale, 1885; Kerner von Marilaun, 1894; West and Edlefsen, 1917; Goetz and Goetz, 1938; Bugaevsky, 1939a), perhaps because it is based on the sound physical fact that water expands on freezing. Since the plant consists mostly of water, the assumption that it, too, expands on freezing was a logical one. Others besides Schübler began to doubt the theory. Du Petit-Thouars (1817) found it difficult to believe that some plants could survive such expansion, yet he knew that many survive freezing. Soon after, Göppert (1830) microscopically examined literally thousands of plants in search of cell rupture. In agreement with previous observations, he found that the juice could be easily squeezed out of freeze-injured leaves, but he was unable to find any torn cell walls, although the cells were somewhat collapsed. Others confirmed his observations (Morren, 1838; Lindley, 1842; Schacht, 1857; Martens, 1872; see Prillieux, 1872; Schumacher, 1875). Nägeli (1861) pointed out, in fact, that cell walls can stretch much more than the small amount that would occur even if all the cell's water froze. He also showed that freeze-killed *Spirogyra* cells will collapse if transferred to glycerin after thawing. This could not happen if there were any tears in the wall. The final death blow to the rupture theory was the discovery that tissues actually contract, instead of expanding, on freezing (Table 5.1). Even freeze splitting of trees can be explained by an asymmetrical contraction (Caspary, 1857).

The explanation of plant contraction during freezing required direct microscopic examination of frozen tissues. This showed that the rupture theory was unsound at the outset, for instead of the ice forming inside the cell, as had been tacitly assumed, it normally occurs outside it—in the intercellular spaces or on the surface of the tissues (Table 5.2). The ice crystals grow to masses larger than the cells (Fig. 5.1), and the cells contract and even collapse (Figs. 5.1 and 5.2). This contraction may be so severe that, in the case of the epidermal cells with colored vacuoles, the opposite sides of the cell can be seen to come in contact with each other (Iljin, 1933b, 1934; Siminovitch and Scarth, 1938). Since the cells are firmly connected with each other, this results in a contraction of the tissues and of the organ as a whole; at the same time, the air from the intercellular spaces is squeezed out (Lindley, 1842; Wiegand, 1906b). The ice crystals may be confined to certain regions, forming such large masses (as much as 1000 times the size of a cell; Müller-Thurgau, 1886) that tissues are pushed apart (Caspary, 1854; Schacht, 1857; Fig. 5.1). Even moss leaves that have no intercellular spaces form ice crystals between the walls which split them apart and grow at the expense of water from the cells (Modlibowska and Rogers, 1955).

On thawing, the tissues are limp and have a water-soaked appearance (Prillieux, 1869); when the ice in the intercellular spaces is converted to

TABLE 5.1

Volume Changes on Freezing of Plant Tissues

Plant part	Percentage change in volume	Reference
Leaves	−25	Hoffman, 1857
Petioles and midribs	−1 to −3½ (length)	Sachs, 1860
Beet root and pumpkin fruit	0	Sachs, 1860
Bark of twigs	−13.5	Wiegand, 1906a
Wood of twigs	−2.5	Wiegand, 1906a
Yeast cells	−10	Molisch, 1897
Spirogyra cells	Diameter reduced to 1/3	Molisch, 1897
Cladophora cells	Diameter reduced by 20%	Molisch, 1897

[a] See also Müller-Thurgau, 1880.

TABLE 5.2

Observations of Extracellular Ice Formation in the Plant

Plant	Reference	Plant	Reference
Daphne, hydrangea, iris, fritillaria	du Petit-Thouars, 1817	Twigs of *Acer negundo*	Dalmer, 1895
Many	Göppert, 1830	Algae, agave, aloe, beet	Molisch, 1897
Exotic plants	Caspary, 1854	Algae, potatoes, beets	Wiegand, 1906a
Potatoes, beets	Schacht, 1857	Tree twigs, buds	Wiegand, 1906b
Pumpkin slices	Sachs, 1860	Cabbage, fungi	Schander and Scaffnit, 1919
Iris germanica, etc.	Prillieux, 1869	Fucaceae	Kylin, 1917
Nitella syncarpa	Kunisch, 1880	*Buxus sempervirens*	Steiner, 1933
Roots and tubers	Müller-Thurgau, 1880	Peach buds	Dorsey, 1934
		Cortical cells of trees	Siminovitch and Scarth, 1938

water, the spaces are essentially free of air, and the cells are flaccid due to the loss of water to the ice masses. This led earlier workers to believe that cell rupture had occurred. If the tissues are uninjured, the intercellular water is soon reabsorbed by the cells. As they regain their turgor, air enters the spaces, and the water-soaked appearance is quickly lost. If thawing is gradual enough, the water may be reabsorbed by the cells as soon as the ice

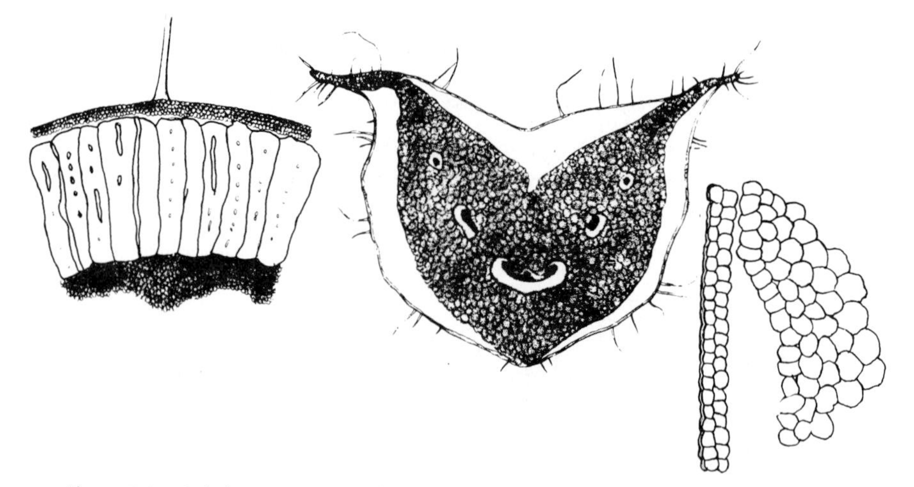

Figure 5.1. *Left:* large ice masses formed between the much smaller cells. *Right:* contraction of tissue caused by ice masses formed beneath the epidermis. From Prillieux, 1869.

melts, and no infiltration with water occurs. Injured cells are unable to reabsorb the water (Wiegand, 1906b).

The freeze-killed cells characteristically show "frost plasmolysis"—a contraction of the dead protoplast, leaving a large space between it and its cell wall (Fig. 5.2). This phenomenon was mentioned by several of the early investigators and studied carefully by Buhlert (1906). The observations of "frost plasmolysis" were always made on dead cells after they were thawed. The term is, therefore, an unfortunate one, since it has led some to believe that this "plasmolysis" is supposed to occur while the cell is frozen (Becquerel, 1949). Actually, it is due to the cell contraction (cell wall as well as protoplast) during extracellular ice formation and the inability of the dead

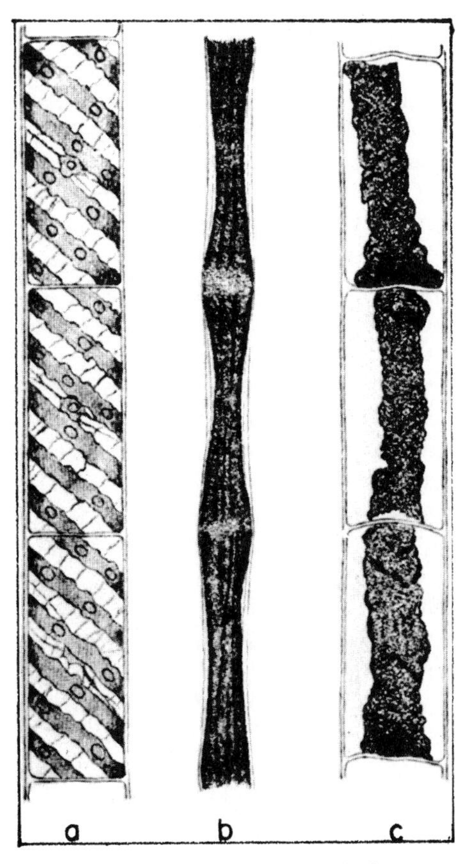

Figure 5.2. *Spirogyra* (X 300): (a) Normal, unfrozen; (b) frozen extracellularly, showing cell collapse without any ice inside the cell; (c) thawed, showing "frost plasmolysis." From Molisch, 1897.

protoplast to reabsorb the water formed in the intercellular spaces on thawing of the extracellular ice. As a result, the cell wall expands back to nearly its original shape, while the dead protoplast remains contracted, giving a false appearance of plasmolysis. This would seem to indicate that the injury had occurred during the freezing process and the cells were nonfunctional by the time the ice began to thaw.

2. Intracellular Freezing

Although ice is normally formed extracellularly, many observations of intracellular ice formation have been made in the laboratory (Table 5.3; Fig. 5.3). The freezing occurs in sudden flashes in one cell at a time, the crystals appearing to be both within and outside the vacuole (e.g., in onion epidermis; Chambers and Hale, 1932). When freezing is relatively slow, the ice may form between the wall and the tonoplast (Asahina, 1956). This may be the same phenomenon as the "thin hull of ice inside the wall" previously described by Schander and Schaffnit (1919). Onoda (1937) explains this exceptional type of ice formation as follows. The protoplasm of the cells is strongly dehydrated by ice growth in adjacent cells. As a result, it separates from the wall. The space formed becomes filled with ice, resulting in pseudoplasmolytic freezing. This description, however, is not in accord with the cell collapse observed by Iljin and others. Ice formation between the wall and the protoplasm in *Nitella* had previously been described by Cohn and David (1871) and in *Conferva fracta* by Göppert (1883). In cortical cells of trees, with thick protoplasm layers, Siminovitch and Scarth (1938) saw the ice form first at one end of the protoplast, and from there spread around each side of the vacuole to the other end. Only when the freezing of the protoplasm was complete did the vacuole freeze. In one case, in fact, the protoplasm froze but the vacuole remained unfrozen. Stuckey and Curtis (1938) also state that the cytoplasm appears to freeze before the vacuole. The flashlike, cell-by-cell freezing that occurs when ice forms intracellularly has been strikingly recorded by the moving picture camera (Luyet, Modlibowska). It has also been demonstrated by videotape micrography (Brown and Reuter, 1974). In cucumber and other fruit tissue, nucleation and growth of ice crystals extracellularly was followed by nucleation and growth of ice crystals in individual, supercooled cells. This was revealed by the occurrence of small spikes in the freezing curves, due to the release of small amounts of heat of crystallization during the freezing of individual cells (Brown *et al.*, 1974).

In nearly all cases (Table 5.3), the intracellular ice formation was brought about by rapid freezing, e.g., by supercooling well below the freezing point

TABLE 5.3

Observations of Intracellular Ice Formation in the Plant

Plant	Temperature at which frozen	Reference
Apple		Morren, 1838
Pumpkin	−12° to −20°C	Sachs, 1860
Nitella syncarpa	−3° to −4°C	Cohn and David, 1871
Roots and tubers	−10°C	Müller-Thurgau, 1880
Conferva fracta		Göppert, 1883
Codium bursa	−11°C	Molisch, 1897
Tradescantia crassula (staminal hairs)	−6.5°C	Molisch, 1897
Agave, aloe, beet	Rapid cooling	Molisch, 1897
Tradescantia discolor and *T. guianensis*	Slow freezing	Molisch, 1897
Tradescantia and *Vallisneria spiralis*	−4 to −5°C	Schaffnit, 1910
Cabbage	−5°C	Schander and Schaffnit, 1919
Fungi	Rapid freezing	Schander and Schaffnit, 1919
Cabbage, tree cortex	−12°C	Siminovitch and Scarth, 1938

of the tissue and then inducing or waiting for ice formation. Not only was freezing more rapid than normally occurs in nature, but, to facilitate observations, sections of tissues were used. That intracellular ice formation is not due to sectioning was shown by Siminovitch and Scarth (1938), who observed it in whole potted plants cooled from 0° to −10°C in one-half hour. This resulted in death of the plant. Such spontaneous intracellular nucleation generally does not occur unless the cells are undercooled to at least −10°C (Mazur, 1977). It may, however occur in large plant cells at higher temperatures (−7°C in cabbage).

The fact that ice can form intracellularly in sections of tissue frozen in the laboratory leads to the suspicion that this kind of ice formation may also occur in nature, if the freezing is sufficiently rapid. It is possible, for instance, that the so-called "sunscald" may be due to this. On a cold winter day, the sun shining on one side of a tree may raise its temperature so high (25°–30°C above the shady side) that thawing occurs. When the sun suddenly disappears behind a cloud (or another object) the temperature of the thawed tissue drops rapidly, and refreezing may perhaps occur so suddenly as to result in intracellular freezing and death. This explanation has never been confirmed experimentally, and there are other equally valid explanations of sunscald (see Godman, 1959). Nevertheless, some observations

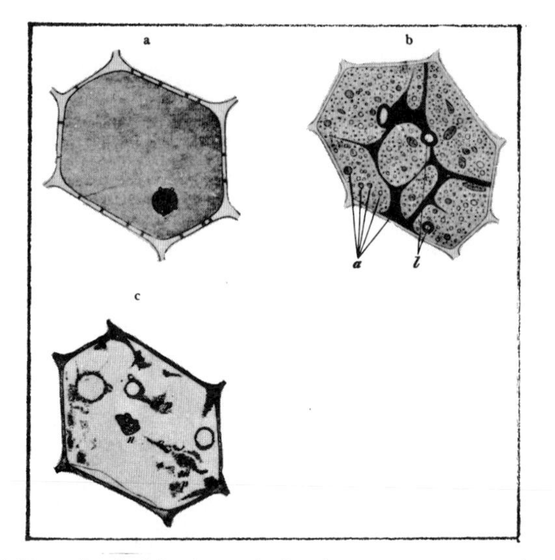

Figure 5.3. Epidermal cell of *Tradescantia discolor* (X 300). (a) Normal cell filled with red cell sap; (b) frozen intracellularly. Most of cell sap converted to ice, leaving concentrated, dark red unfrozen sap at *a* . Gas bubbles at *l* . (c) Thawed, dead cell. Wall and nucleus stained red, coagulated protoplasm at *p*. From Molisch, 1897.

are in agreement with it. Painting tree trunks white on the sunlit side has long been used to prevent "sunscald," and it lowers the trunk temperature as much as 30°C in comparison with unpainted trunks in winter (Martsolf *et al.* 1975).

According to Weiser (1970), evergreen foliage, which is not injured at −87°C during slow freezing in winter, is killed at −10°C when thawed tissues are frozen rapidly (8°–10°C/min). Foliage on the southwest side of these plants were actually found to cool at these rates on sunny winter days in Minnesota, when the sun moved behind an obstruction. He, therefore, concluded that intracellular freezing is the cause of sunscald. An exceptional kind of intracellular freezing with death of the cells may occur in apple flowers in spring (Modlibowska, 1968). During the first phase of freezing, ice forms extracellularly, lifting the skin of the receptacle. This may result in some mechanical damage which heals rapidly on return to favorable conditions. During the second phase of freezing, ice forms intracellularly in the placenta, styles, and ovules, causing death.

A carefully applied electrometric method (Olien, 1961) seems capable of distinguishing between intracellular and extracellular freezing, by enabling calculation of the relative content of liquid extracellular water. According to these results, leaves of hardened barley show only extracellular freezing and are eventually injured by it. Leaves of unhardened barley show the reverse pattern indicating a sudden loss of solutes before extracellular freezing can occur, perhaps due to intracellular freezing. A simpler method is by following the change in permeability of the tissues to gases (Le Saint, 1957), which can only be explained by the replacement of the intercellular gases by ice crystals. The process of ice formation in sections of tissue under laboratory conditions has been thoroughly described and illustrated by Asahina (1956).

Artificial freezing has produced another phenomenon not normally occurring in nature. Many fruits and vegetables, when frozen for purposes of preservation, have their cell walls ruptured (Woodroof, 1938; Mohr and Stein, 1969). The freezing process in this case is, of course, initiated at a much lower external temperature than occurs in nature, and is rapid enough to permit intracellular freezing. Furthermore, the cells in most fruit and some vegetables are very large, thin-walled, and have a very high water content. They are, therefore, more likely to show wall rupture than the smaller, drier, thicker-walled cells of overwintering plants. Nevertheless, if frozen slowly, even artificially frozen tissues may show the kind of extracellular freezing observed in nature. In the case of fruits and vegetables frozen artificially, the size of ice crystals may be increased as much as 500 times by slowing down the rate of freezing (Woodroof, 1938). The largest crystals may be up to 1000 times the size of the cells in asparagus or spinach. Such large crystals result in crushing and distortion of the cells, just as in nature (see Fig. 5.1).

3. The Cause of Extracellular Freezing

It has long been a question as to why ice does form extracellularly under normal conditions. On the basis of the available evidence, the process may be suggested to occur as follows. It has been shown that ice normally crystallizes first in the large vessels both in the case of leaves (Asahina, 1956) and even in defoliated mulberry trees (Kitaura, 1967). Freezing proceeds along the vessels from a few nucleation points and reaches all parts of the shoots at a relatively high velocity (about 34 cm/min/°C), proportional to the supercooling (also called undercooling or subcooling). Freezing in the vessels is to be expected since their large diameter does not favor undercooling, and their dilute sap probably has the highest freezing point of any of the plant's water. Once ice forms in the vessels, it will spread throughout the plant body. Each living cell is surrounded by a highly lipid plasma membrane. It has been shown that ice crystals are barred by this membrane from inoculating the cell contents (Chambers and Hale, 1932). Consequently, the ice from the vessels spreads only throughout the part of the plant external to the living cells, i.e., the intercellular spaces. The crystals will form here at the expense of the water vapor in the air and of the surface film of water on the cell walls.

Freezing is far less likely to extend into the cell walls (except, of course, into the pits in the walls), because their water-containing microcapillaries are less than 0.1 μm in diameter (since they cannot be resolved by the optical microscope). The vapor pressure of water in capillaries can be calculated from the equation (Moor, 1960):

$$p = p_o e^{-c/r}$$

where p = vapor pressure of water in capillaries; p_o = normal vapor pressure of pure water; $c = 10^{-7}$ cm; r = radius of capillary. Such calculations reveal that the vapor pressure of the water in the microcapillaries of the cell wall will be lowered by more than 1%. This will, of course, lower the freezing point slightly. More important, undercooling of water is greatly increased in capillaries.

At the instant before ice formation in the intercellular spaces, the cell contents must be practically at temperature and vapor pressure equilibrium with the immediately adjacent wall and intercellular spaces. When intercellular ice begins to form (as a result of seeding from the vessels), the vapor pressure in the intercellular spaces drops sharply (Fig. 5.4). Due to the capillary size of the cells, undercooling of their contents will be favored, and due to protection by the lipid plasma membrane, seeding with the extracellular ice crystals will be prevented. As soon as the tissue temperature drops below the freezing point of the cell contents to the smallest degree, their vapor pressure will be higher than that of the intercellular ice at the same temperature (Fig. 5.4). Consequently, the cell water at the higher vapor pressure will diffuse through the semipermeable lipid plasma membrane to the intercellular ice at the lower vapor pressure. In this way, these ice crystals on the external surface of the cell wall will grow at their bases, and the cell itself will contract due to the water loss. If the temperature continues to drop at a gradual enough rate, this diffusion of cell water to the external ice loci will continue, and will steadily increase the cell sap concentration. If temperature equilibrium is then attained, vapor pressure equilibrium will also soon occur. However, when the cell contents are at the same temperature and vapor pressure as the intercellular ice, they must be exactly at their freezing point (Fig. 5.4). Consequently, the plant can be kept frozen indefinitely at this constant temperature without any possiblity of ice forming in the remaining cell contents. The major part of the tissue will now consist of intercellular spaces filled with ice. There will be little air in these spaces, since during the temperature drop the solid matter of the tissues contracts at the same time as the freezing water expands into the intercellular spaces. This squeezes and pushes the intercellular air out of the tissues. Such a simultaneous contraction of the structural components of the plant and expansion of the contained water on freezing has long ago been given as the

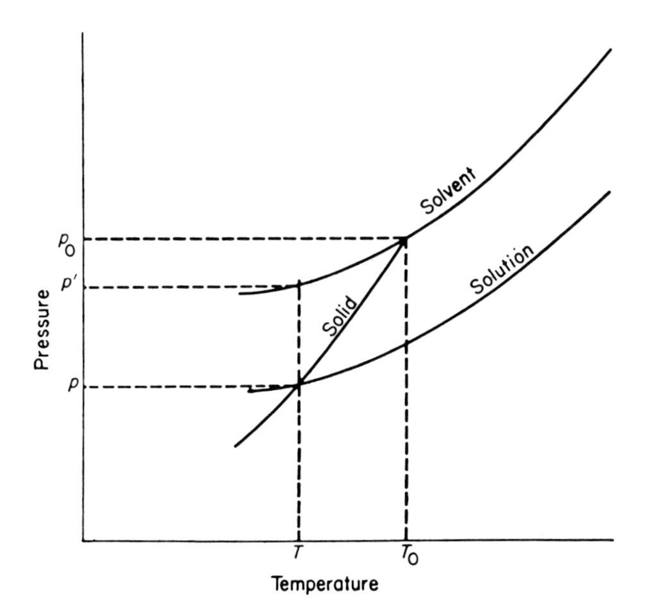

Figure 5.4. Accelerated decrease in vapor pressure of water in the frozen state with drop in temperature. At temperatures below the freezing point, both pure solvent and solution in the supercooled state have higher vapor pressures than that of the solid (ice) at the same temperature. po = vapor pressure of pure water at the freezing point; p = lower vapor pressure of solution at its freezing point T. p' = higher vapor pressure of supercooled water at the same temperature T. (From Getman and Daniels 1937).

explanation for the winter splitting of tree trunks that sometimes occurs with a noise like the crack of a gun (see above).

It must be realized that the rate of this diffusion of water to ice loci outside the cells is limited by the permeability of the lipid plasma membrane surrounding the living cell. Although this membrane is highly permeable to water (relative to solutes), it significantly slows down water movement, so that this is not as rapid as the movement of water through the aqueous cell wall. Therefore, if the temperature drop is rapid enough, this diffusion to the extracellular ice cannot occur with sufficient speed to increase the concentration of the cell contents as rapidly as the temperature drops. The rapidly cooled cell may, therefore, eventually reach a temperature sufficiently below its freezing point to induce spontaneous ice formation intracellularly.

The above description of "slow" extracellular freezing assumes a rapid enough temperature drop to permit a spread of the ice front throughout the plant body, so that the ice crystals contact all or nearly all the cells. This kind of freezing is more likely to occur when relatively rapid, such as in artificial "slow" freezes. Under natural conditions, the freeze may be gradual enough to prevent the ice front from spreading throughout the plant body. It may then be confined to specific regions, growing slowly at the expense of water diffusing to it from relatively distant unfrozen regions (see Fig. 5.1).

If hoarfrost forms on the leaves, as commonly occurs in fall and spring (Kitaura, 1967), this may inoculate the internal tissues rather than spontaneous ice formation in the vessels, particularly in the case of tissues with narrow vessels or tracheids; but the net result would be the same since the plasma membrane would again block seeding of the cell contents. Ice nuclei may, of course, form without inoculation from the outside. Salt and Kaku (1967) suggest that such ice nucleation takes place at sites associated with the cell walls and not on nucleators suspended in the water.

It must be realized that even the slow freezing of sections under the microscope differs, in some respects, from the freezing of a normal whole plant. If the section is immersed in an aqueous medium this will, of course, freeze first, and the subsequent series of events will be very different from that described above (Genevès, 1955).

The ability of the plasma membrane to act as a barrier against inoculation of the cell contents by extracellular ice crystals, is undoubtedly due to its lipid nature. Even nonliving membranes that are freely permeable to liquid water may be completely impermeable to growing ice crystals (Lusena and Cook, 1953). According to Mazur's (1963) calculations, if ice forms at a sufficiently low temperature due to rapid cooling, the ice crystals may then be small enough to penetrate the plasma membrane and therefore to induce intracellular freezing.

C. FREEZING, UNDERCOOLING, AND EUTECTIC POINTS

1. Definitions

The freezing point of a plant is always below that of pure water. When the plant is cooled slowly, and continuously, its temperature normally drops somewhat below its freezing point without ice formation (Table 5.4). This is called undercooling (also supercooling), and the *undercooling* point is the lowest subfreezing temperature attained before ice formation. Even pure water and its solutions undercool. The undercooling point is not, however, a constant value but may vary even for repeated tests on a single solution, although the freezing point remains constant. At some undercooling point, ice begins to crystallize, with the release of the heat of fusion. Since this is large for water (80 cal/g), the previously downward cooling curve shows a marked reversal—a bend upward due to warming by heat of fusion (Fig.

TABLE 5.4

Undercooling and Freezing Points of Some Tissues (°C)

Plant	Undercooling point	Freezing point	Reference
Potatoes	−6.1	−0.98	Müller-Thurgau, 1880
Aspergillus niger			
in 1% dextrose	−6	−0.29	Bartetzko, 1909
in 20% dextrose	−8	−2.89	Bartetzko, 1909
in 50% dextrose	−14	−8.9	Bartetzko, 1909
Red beet root	−6.83	−1.88	Maximov, 1914
Tussilago farfara	−2.70	−1.83	Maximov, 1914
Potato tuber	−1.60 to −2.42	−0.85 to −1.57	Maximov, 1914

5.5). Such upward bends in the cooling curve are called *exotherms* (heat releases). If the undercooling is not too great, the maximum temperature attained by this exotherm is a constant for a specific solution and represents the *freezing point*. Although the freezing point of a pure solution is a constant, the freezing point of a plant may vary slightly depending on the conditions of measurement. It is lower in living than in dead tissues (Table 5.5), and varies somewhat with the rate of temperature drop (Maximov, 1914; Walter and Weismann, 1935; Luyet and Gehenio, 1937). However, if the cooling rate is slow enough, the freezing point of the living tissue is identical with that of dead tissue (Luyet and Galos, 1940). In most cases the true freezing point of the living tissue also coincides with the freezing point of the expressed juice, though in some cases it may be 0.5°C below it (Marshall, 1961). One of the reasons for the discrepancy is variations in cell turgor.

TABLE 5.5

The Freezing Points of Living and Dead Tissues (°C)

Plant	Living	Dead	Reference
Phaseolus leaves	−1.1	−0.5	Müller-Thurgau, 1880
Cypripedium leaves	−2.0	−0.5	Müller-Thurgau, 1880
Red beet roots	−2.15 to −2.55	−1.25	Maximov, 1914
Tussilago farfara	−2.03 to −3.35	−0.97	Maximov, 1914
Helleborus viridus	−4.07 to −5.88	−2.07 to −2.15	Maximov, 1914
Potato tubers	−1.0	−0.5	Müller-Thurgau, 1880
	−1.22	−0.63	Maximov, 1914
	−0.87 to −2.00	−0.69 to 0.80	Walter and Weismann, 1935
	−1.2 to −2.0	−0.5 to −0.75	Luyet and Gehenio, 1937

In nature, the degree of undercooling must be generally relatively small (perhaps 1–3°C, although there are exceptions. The reason for this is the unstable state of undercooled water, and the common occurrence of homogeneous and heterogeneous nucleators. Once initiated, this crystallization will be more rapid the lower the undercooling point (Diller, 1975). In the case of most cells, an undercooling of 10°C, followed by crystallization will result in *intracellular* freezing and death (Mazur, 1977).

If a solution continues to be cooled below the first exotherm, at some point below the freezing point, a second exotherm may occur, usually much smaller than the first. This is called the *eutectic* (or cryohydric) *point* and is the temperature at which all the solute crystallizes and, therefore, the remaining water also solidifies. The importance that such a eutectic point would have is illustrated by Asahina's (1962a) experiments with sea urchin eggs. When frozen in KNO_3 solution with a eutectic point at −2.9°C, the cells were killed at this temperature, although they normally survived much more severe freezing. Similarly, during the freezing of thylakoid membranes in the presence of neutral, nontoxic substances, membrane damage did not occur until the eutectic temperature was reached (Santarius, 1973). Such a eutectic point, however, is clear-cut only in pure solutions. A mixture such as Earle's salt solution containing glucose and glycerine has no real eutectic point but a eutectic zone in which a slow and progressive melting occurs as the temperature of the deep frozen solution rises, becoming marked from −38°C upward (Rey, 1961). Some solutes, such as sugars, tend to prevent the crystallization of others; and since there are many solutes in the cell sap and sugars among them, the plant cannot be expected to have a specific eutectic point.

2. The Double Freezing Point

Actual observation of the freezing process in plant tissues has indeed corroborated this expectation. In many cases, a "double freezing point" has been found and the two points were, at first, interpreted as the freezing point and the eutectic point, respectively. The second freezing point was readily eliminated by either killing the tissues or blotting the surface of the cut tissues and accentuated by soaking in water (Luyet and Gehenio, 1937); it was, in any case, at too high a temperature (−2° to −3°C) for the eutectic points of any of the substances known to be present in large quantities in the plant. Although this double freezing point cannot be due to a eutectic point because both freezing points are too high, it is still in need of an explanation. Luyet and Gehenio suggest that the first freezing point is due to freezing of the extracellular water, while the second is due to freezing of intracellular water.

Other explanations have also been proposed—that the first freezing point is due to freezing of the damaged surface layer of tissue and the second due to freezing of the inner tissues (Aoki, 1950). Levitt (1958) has proposed a turgor theory to explain it. Ice forms first at the cut surface of the piece of tissue. Since it is so small the whole piece of tissue is essentially in vapor pressure equilibrium; and since there is no barrier to the propagation of the ice throughout the intercellular spaces, the ice spreads forming a continuous thin layer on the cell surfaces. If the tissue has been rinsed with water before freezing, it is in the turgid state at the instant of freezing and, therefore, has a higher water potential than that of its expressed sap. The freezing point of the cells is therefore *higher than the freezing point of the cell sap at atmospheric pressure*. This results in a high first freezing point somewhere between 0°C and the freezing point of the cell sap depending on the balance between the temperature gradient and the amount of water able to freeze in the tissues. This initial, relatively rapid, ice formation at the cell surfaces removes

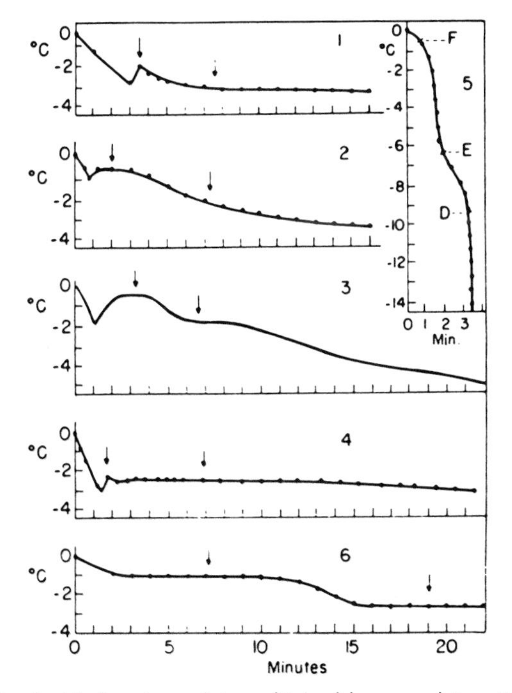

Figure 5.5 The double freezing point as obtained by several investigators. (From Luyet and Gehenio, 1937.)

enough water from the living cells to cause loss of turgor. The result is a *marked* lowering of the freezing point due to the double effect of turgor loss and a slight increase in cell sap concentration. From this point down, much larger quantities of ice can form. As an example of this, a fully turgid piece of tissue with a freezing point lowering (of the cell sap) of −2°C would have about 10% of its water frozen between 0° and −2°C and 45% frozen between −2° and −4°C. This would give the appearance of two freezing points. The double freezing point would, therefore, occur whether freezing is intracellular or extracellular, although showing up more sharply in the former case due to greater undercooling. It should be pointed out that if Aoki's explanation is correct, the double freezing point is an artifact that occurs only in cut tissue. If the turgor theory is correct, it can occur also in normal, turgid, undamaged tissue, though perhaps depending on the rate of temperature drop. Salt and Kaku (1967) have reinvestigated this question. However, since the double freezing point is apparently dependent on artificial conditions of freezing, there may be more than one explanation of the phenomenon, depending on just what these artificial conditions are. This plurality of explanations appears to be borne out in the case of stem sections of *Cornus stolonifera*. The number of freezing points varied with the season (McLeester *et al.*, 1968; Weiser 1970). During summer and winter, there were two distinct points, in early autumn three distinct points, and in spring one prominent first freezing point which tended to mask the second point.

It is now known, however, that there are normally two exotherms in the case of many woody plants, the first occurring not far below the true freezing point of the tissues and signaling the beginning of extracellular freezing in most of the tissues, the second as low as −38° to −47°C, due to intracellular freezing of normally undercooled wood parenchyma cells (Burke *et al.*, 1976)—the only cells in the trees that are capable of this extreme undercooling. In the case of twigs, therefore, the double freezing point consists of the true freezing point in the living bark cells, followed by the undercooling point of the wood parenchyma.

D. FREEZE-DEHYDRATION

Extracellular freezing leads to a secondary water stress, resulting in a dehydration strain, due to the efflux of the cell's water to the ice nuclei in the intercellular spaces. The cell, therefore, desiccates in exactly the same manner as if the water were removed by evaporation. The cell, in fact, does not know that ice is forming outside of it, since the living protoplasm is insulated from the ice by its dead cell wall. It can only know that it is being subjected to a water stress and is losing water. This, in turn, produces two other strains.

1. Concentration of Cell Sap

As the temperature drops and the freeze-dehydration increases, the cell contents become more and more concentrated. In nature, cooling occurs slowly, and the freeze-dehydration steadily lowers the freezing point of the remaining cell sap, maintaining it at approximately the prevailing temperature, in this way avoiding intracellular freezing. The increase in concentration can be calculated quantitatively by an adaptation of the equation of Gusta *et al.* (1975). The degree of concentration is given by the ratio between the original amount of liquid water in the cell (L_o), and the amount remaining after extracellular freezing has occurred at any one freezing temperature (L_T). This ratio can be obtained from Eq (5.1)

$$\frac{L_o}{L_T} = \frac{T}{\Delta T_m} \tag{5.1}$$

where T = the prevailing freezing temperature (°C) and ΔT_m = the freezing point of the original cell sap (°C).

This simple equation holds only if the amount of unfreezable water (K) is negligible. If it is not, then it must be included in the amount of unfrozen water (Gusta *et al.*, 1975) [Eqs. (5.2 and 5.3)]

$$L_T = L_o \frac{\Delta T_m}{T} + K \tag{5.2}$$

and

$$\frac{L_o}{L_T} = \frac{T}{\Delta T_m}\left(1 - \frac{K}{L_T}\right) \tag{5.3}$$

This equation, however, assumes that the unfreezable water (K) cannot act as a solvent—an assumption of doubtful validity. If it is effective as a solvent, it may decrease the concentration of the unfrozen solution to a measurable degree.

From the above equation, as T drops, L_T also decreases by diffusion of the cell water to the extracellular ice loci, and the $(\Delta T_m)_T$ (the freezing point of the concentrated cell sap in equilibrium with the prevailing temperature T) becomes increasingly negative, approaching at all times the prevailing temperature (T).

2. Cell Collapse

Since there is no liquid external to the cell, freeze-dehydration cannot plasmolyze the cell in the true sense of the term—a separation of the dehydrated protoplast from the cell wall. Therefore, the cell wall must adhere to the protoplast and both must collapse together as the freeze–dehydration progresses. This expected cell collapse has, indeed, been observed.

All the described effects of the freezing process must be considered when attempting to explain freezing injury.

E. MEASUREMENT OF ICE FORMATION IN PLANTS

Theoretically, according to Eqs. (5.1 and 5.3), more and more ice continues to form in the plant as the temperature drops farther and farther below the freezing point. In the case of nearly all plants, the equation predicts that nearly all the cell's freezable water will be frozen at about −20 to −30°C. Nevertheless, no end-point for *all* the freezable water can be predicted. If there is a definite eutectic point, *all* the freezable water will freeze at this point; but no true eutectic point can be expected under normal conditions. The question remains, whether or not there is a temperature at which all the freezable water solidifies.

In view of these uncertainties, and as a test of Eqs. (5.1 and 5.3), it is essential to measure the ice formation in freezing plants. The older measurements involved three main methods: (1) the dilatometer method which measures the ice by its expansion on freezing (Fig. 5.6); (2) the calorimeter method which measures the ice formed by the heat of crystallization released (Fig. 5.6), and (3) the ice flotation method which measures the change in specific gravity on freezing (Scholander *et al.*, 1953). The second was the most commonly used since it does not require the use of possibly toxic liquids, and it can be applied at lower temperatures than the first method, i.e., at temperatures at which the expansion is too small to be easily measured. The third method can be used only for air-free tissues. By use of these methods, measurements of ice formation in the plant have repeatedly shown that crystallization continues down to at least −30°C (Scholander *et al.*, 1953), or even down to −72°C in yeast (Wood and Rosenberg, 1957). The freezing curve parallels that of a pure solution (Fig. 5.7) although a smaller fraction of the water freezes due to binding by the tissue colloids. There are, of course, difficulties in the use of these methods, and some assumptions must be made in calculating the amount of ice formed (see Wood and Rosenberg, 1957). Krasavtsev (1968), for instance, has used the calorimetric method to measure ice formation down to temperatures as low as −60°C, but the calorimetric method requires accurate calculations of heat exchange due to processes other than the heat of fusion of ice (Johansson, 1970). Since heat capacity and other values are not accurately known for temperatures as

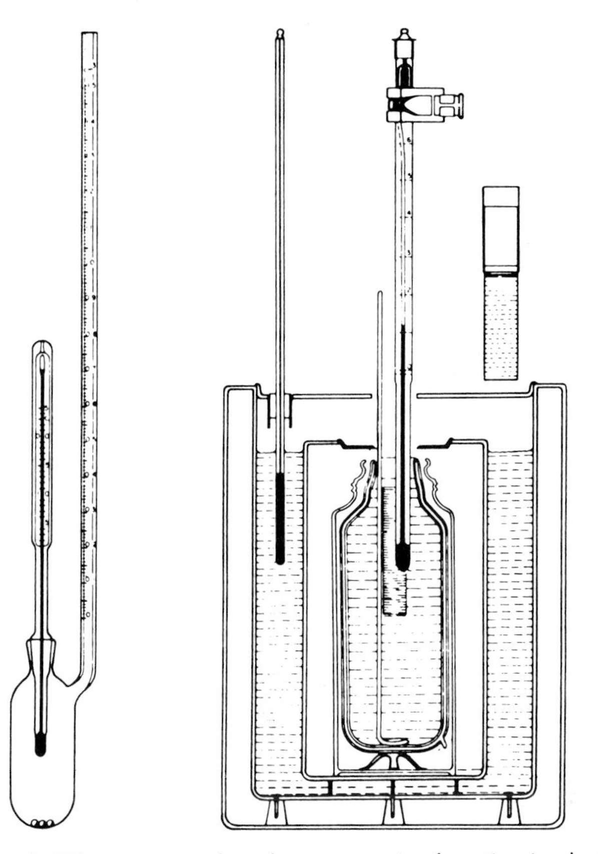

Figure 5.6 *Left:* Dilatometer vessel used to measure ice formation in plant tissues by measuring the expansion due to freezing when submerged in oil. From Lebedincev, 1930. *Right:* Calorimeter for measuring ice formation in tissues by the heat of fusion. (From Levitt, 1939).

low as −60°C, the calculations of the amounts of ice formed at these low temperatures are unreliable. That the results are essentially correct for moderate freezing temperatures is attested to by similar results obtained by use of an indirect dehydration-melting point method. This method of estimating ice formation eliminates the assumptions adopted in the above methods although replacing them by others (Salt, 1955). In the past few years, these methods have been largely replaced by two newer techniques, which are simpler to use and have yielded a vast amount of fundamental information, especially by Burke and his coworkers (Burke *et al.*, 1976). (4) Nuclear magnetic resonance (NMR) spectroscopy employs radio frequency radiation and is able to measure the amounts of ice and liquid water due to the large

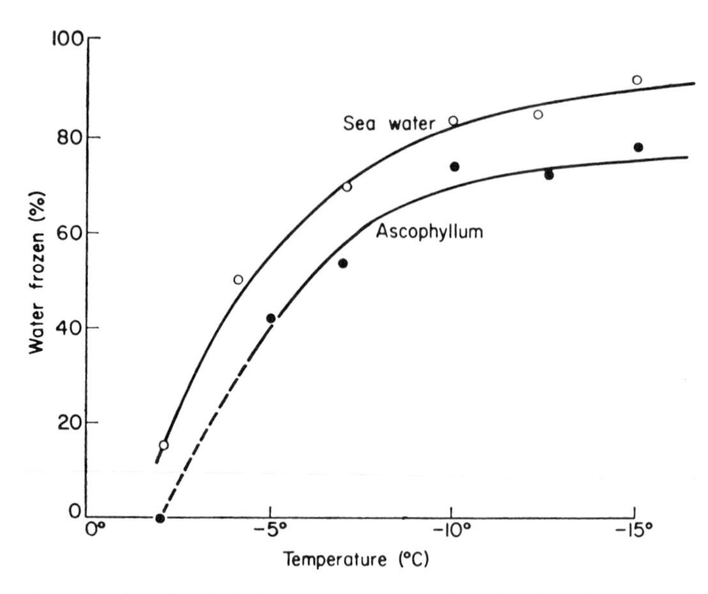

Figure 5.7 The freezing of algal water compared with the fraction of sea water that freezes at the same temperature. (From Kanwisher, 1957).

spectral differences between the two. (5) The above-described calorimetric method has been improved by use of differential thermal analysis (DTA) and differential scanning calorimetry (DSC). By means of two thermocouples, they measure the differences in heat evolution or absorption between a normally moist sample and a reference dry sample during cooling or warming. The amount of water frozen is calculated (as in ordinary calorimetry discussed above) from the heat of fusion of water and the heat capacities of ice and liquid water. A sixth method developed by Olien and Chao (1973) evaluates the liquid water content of cell walls by electrophoresis of indicators. These and other specialized methods are discussed by Burke *et al.*, 1976.

6. Freezing Injury

A. OCCURRENCE

In contrast to chilling injury, which occurs in tropical and subtropical plants, but not in plants from temperate zones, freezing (frost or cryo-) injury may occur in all plants. It is, therefore, far more prevalent than chilling injury and has been more intensively studied. It may simply appear as tissue, organ, or plant death, or it may lead to specific disease symptoms such as frost cankers and burls in conifers and hardwoods (Zalasky, 1975a,b). Many descriptions of it are found in both the old and the new literature (see Levitt, 1956). Those plants that are subject to chilling injury (as well as some that are not) are usually killed by the first touch of frost (Molisch, 1897). On the other hand, some that are native to cold climates may be frozen solid at the lowest temperatures without injury (Scholander *et al.*, 1953). Between these two extremes all gradations occur (Table 6.1). Even for a single plant the range of freeze-killing temperatures may be large, depending on its physiological state (Table 6.1). Does the freezing stress always produce injury in the same way, or is there more than one kind of injury?

B. PRIMARY DIRECT FREEZING INJURY

By definition, direct injury due to the freezing process, can occur only as a result of intracellular (or more correctly intraprotoplasmal) freezing. Extracellular freezing does not permit direct contact between the ice and the protoplasm and therefore cannot produce direct freezing injury. Intraprotoplasmal freezing, on the other hand, always kills the cells, if the ice crystals are large enough to be detected microscopically (see Chapters 4 and 5).

Yet, as mentioned earlier (Chap. 4), under specific, artificial conditions of freezing, very small crystals may form without injuring the cell. The exact critical size of the intracellular ice crystals is unknown. In mammalian cells with crystals less than 0.05 um, as seen under the electron microscope, the organelles seemed to maintain their original structure (Shimada and Asahina, 1975). Recrystallization of such small intracellular crystals occurs

TABLE 6.1

Killing Temperatures for Plants in the Frozen State[a]

Species	Killing temperature (°C) when frozen		Reference
	Unhardened	Hardened	
Potato tuber	−1.53		Maximov, 1914
Red beet root	−2.15		Maximov, 1914
Wheat		−12-15	Tumanov and Borodin, 1930
Cabbage	−2.1	−5.6-20	Levitt, 1939 Kohn and Levitt, 1965
Vaccinium vitis idea	−2.15	−22	Ulmer, 1937
Erica carnea	−3-4	−18-19	Ulmer, 1937
Sempervivum glaucum	−3.0	−25	Kessler, 1935
Rhododendron ferrugineum	−4	−28	Ulmer, 1937
Globularia nudicaulis	−4	−19	Ulmer, 1937
Globularia cordifolia	−4	−18-19	Ulmer, 1937
Saxifraga caesia	−4	−29-30	Ulmer, 1937
Homogyne alpina	−4	−18	Ulmer, 1937
Saxifraga aizoon	−4	−18-19	Ulmer, 1937
Hedera helix	−4.5	−18.5	Kessler, 1935
Rhododendron hirsutum	−5	−28-29	Ulmer, 1937
Saxifraga cordifolia	−5	−19	Kessler, 1935
Carex firma	−5-6	−29-30	Ulmer, 1937
Pinus mugo	−6	−40-41	Ulmer, 1937
Empetrum nigrum	−6	−29	Ulmer, 1937
Juniperus nana	−6-8	−26	Ulmer, 1937
Pinus cembra	−9	−38	Ulmer, 1937
Pinus cembra	−10	−40	Pisek, 1950

[a] From Levitt, 1956.

in yeast cells at as low as −45°C, forming crystals large enough to cause injury (Bank, 1973). It must, therefore, be concluded that at temperatures occurring in nature, intracellular freezing is always fatal, even if the crystals are initially too small to produce injury, since this must be followed by crystal growth. The crystals, presumably, damage the protoplasmic structure, perhaps by lacerating the membranes and destroying their semipermeability. Intracellular freezing injury has rarely been observed in nature, as for instance the injury to apple blossoms (Chapter 5). As mentioned above (Chapter 5), it may possibly explain the sunscald injury on the sunny side of trees. The explanation by Olien *et al.* (1968) for the winterkilling of barley in Michigan and in the northeastern United States, apparently also involves a direct freezing injury. The damage most commonly occurs during cold weather following a midwinter thaw. They propose that it may depend on the type of ice structures which develop in the lower crown tissue. Since ice structure can be important only if the ice penetrates the protoplasm, this would certainly be a direct freezing injury.

C. THE TIME FACTOR IN RELATION TO INJURY

The best evidence of primary direct freezing injury is, therefore, the observation of intracellular freezing. Such observations are difficult and are rarely, if ever, made in whole frozen plants. The next best criterion is the speed of the process, since direct injury due to any stress is rapid, compared to indirect injury by the same stress, and since rapid freezing has been shown to favor intracellular freezing (Chapter 5). Rapid cooling of free protoplasts from winter rye, for instance, was obtained by a cooling rate of 8°/min, leading to supercooling and freezing at −12°C (Siminovitch *et al.*, 1978). The ice formation occurred inside the protoplasts, with lethal disruption of cell and membrane organization. Slow freezing (23 min to cool to −12°C) was extracellular and was followed by complete recovery on thawing. Direct observation of human cells (Hela cells and lymphocytes) has also supported this conclusion (McGrath *et al.*, 1975; Walter *et al.*, 1975). Survival decreased at supraoptimal cooling rates, in proportion to the linear increase in percentage of the cells containing intracellular ice, from no intracellular freezing at a cooling rate of 16°C/min to 100% intracellular freezing at 128°C/min.

1. Rates of Freezing and Thawing

In the above experiments, it is not the rates of freezing and thawing that are measured, but merely the rates of cooling and warming. The actual rate of freezing will depend not only on the cooling rate, but also on the degree of supercooling, the properties of the tissue, etc. In order to eliminate one of these complications (supercooling), rate of freezing has been defined as the ratio of the temperature difference to the time difference between the beginning of the freezing process and the time of the final temperature (Rottenberg, 1968). Even if used as a relative measure, this gives only the average value and may fail to detect an extremely rapid initial rate following supercooling.

The actual rate of freezing has rarely been measured. In the lemon, a subcooling to −5.2°C induced a rate of spread of freezing of 15 cm/min (Lucas, 1954). In a qualitative way, however, rate of cooling may be an indirect measure of the relative rate of freezing in a specific plant part or tissue frozen under identical conditions, and rate of warming an indirect

TABLE 6.2

Approximate Ranges for Different Speeds of Cooling

Rate of cooling	Sections of tissue	Whole plant
Slow	1-2°C/min or less	1-2°C/hr or less
Rapid	5-20°C/min	5-20°C/hr
Ultrarapid for relatively dry tissues	100°C/sec or more	
Ultrarapid for fully moist tissues	10,000°C/sec or more	

measure of the relative rate of thawing. The quantitative ranges for "slow" and "rapid" cooling are several orders below the range for protective ultrarapid cooling (Table 6.2). As a general rule slow cooling of whole plants in a freezing chamber means no more than 1° −2°C/hr. In nature, the rate is usually less, but there are exceptions. Biel *et al.* (1955) inserted thermocouples into the stolons of ladino clover and found maximum cooling rates of 10°F/hr. Even these rates were fast enough to produce rapid cooling injury. Sprague (1955), observed more injury in ladino clover and alfalfa when cooled from 36°–10° or 5°F at a rate of 7° to 10°F/hr than at slower rates. Rates of cooling of 5°C/hr have frequently been found to produce rapid cooling injury (see Levitt, 1958). This rate is incapable of producing rapid freezing injury in the case of sections (Levitt, 1957a), since it is slow enough to permit the small amount of ice formation to occur outside the cells. When freezing cells under the microscope, 1°–2°C/min may be considered slow freezing since, if the tissue is inoculated with ice, it will probably still freeze extracellularly at these rates of cooling. On the other hand, a drop of only 1/3°C/min is rapid enough to induce intracellular freezing in whole plants that are not inoculated (Siminovitch and Scarth, 1938).

In general, then, sections must be cooled about 60 times as rapidly as whole plants in order to induce rapid freezing injury. This can be explained if the following two assumptions are made: (1) rapid freezing injury is due to intracellular freezing and (2) whether or not intracellular freezing occurs depends on whether the rate of diffusion of water from the cell interior can keep up with the rate of temperature drop. Some evidence for both these assumptions has already been presented and more is given below. In the case of sections one living cell thick, the total distance that water must diffuse from the cell interior to the ice is one-half the cell width, since ice crystals occur on both surfaces of the section. In the plant as a whole, ice forms first in and around the xylem vessels. Since some living cells in a leaf may be as much as 1 mm from the nearest vessel, the maximum distance that the water has to diffuse at the moment of freezing would be 50 times that in the sections, if the cells are 40 μm wide, and this would slow down the diffusion of the water from the cell by a factor of 50. This factor is therefore of the right order to account for the fact that the freezing rate required to produce injury in sections is approximately 60 times more rapid than that for whole plants.

If we take into account this difference between sections and whole plants, we must conclude that in both cases, rapid cooling increases the freezing injury and, therefore, raises the killing temperature (Levitt, 1956). In some cases, rate of cooling has no effect on the freezing injury. Warming rates also may or may not affect freezing injury, although increased injury due to rapid warming has been found less frequently than increased injury due to rapid cooling (Levitt, 1956). Many of the negative results are readily understood. A tender plant is killed by the first touch of frost no matter how slowly it is cooled or warmed. Similarly, a plant already killed by a severe freeze cannot be brought back to life by slow warming.

All these results with different rates of freezing and thawing indicate that there is a critical freezing zone for any one hardy plant (Fig. 6.1). Above and below this zone, rates of cooling and warming have no effect. In the upper

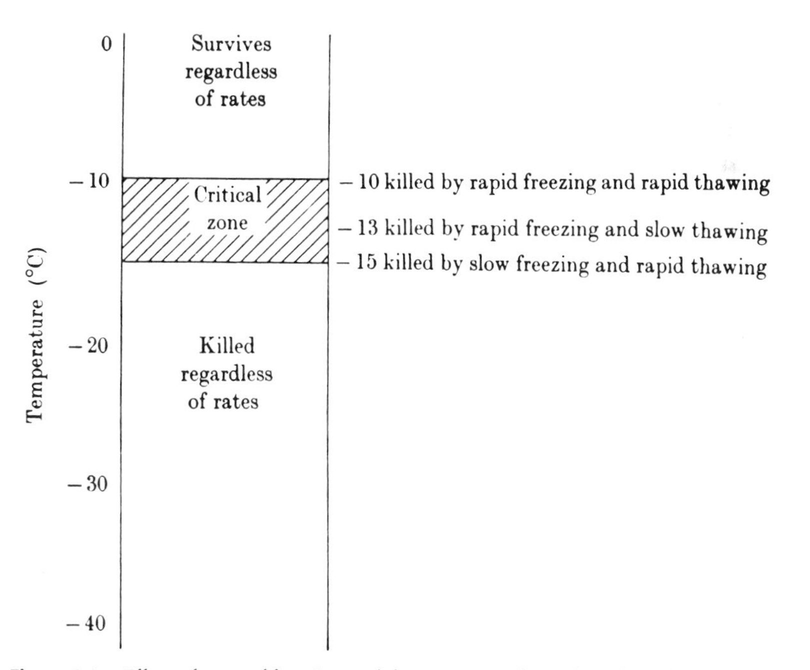

Figure 6.1. Effect of rates of freezing and thawing on a plant of moderate hardiness. (From Levitt, 1966b.)

temperature zone no injury occurs regardless of the rates of cooling and warming. In the lowest zone, the plant is killed no matter how slowly it is cooled or warmed. In the critical zone, however, both rates are important. Death occurs at the top of the zone when both cooling and warming are rapid, at the middle or bottom when one is rapid, the other slow, below it when both are slow. In the case of grape branchlets slowly frozen, rapid warming appears to kill within two temperature zones: −20° to −10°C (in warm water), and −5° to 0°C (Pogosyan and Sakai, 1972b).

In the case of cryopreservation, animal cells frozen rapidly have long been known to respond to rates of thawing in the opposite manner to the above plants. Thus, deepfrozen red blood cells (*frozen rapidly*) show low survival if this is followed by slow warming, increased survival when the warming rate is increased. Now, however, it has been shown that if red blood cells are frozen slowly at moderate freezing temperatures, survival is better following slow than rapid thawing (Miller and Mazur, 1976).

2. Length of Time Frozen

The earlier quantitative evidence on the injurious effect of the length of time the plant is kept frozen is not very conclusive. For relatively short periods of time (2-24 hr) the injury appears to be independent of the time, once equilibrium has been attained (Levitt, 1956). Onion bulbs, in fact, show no greater injury after storage at a constant temperature for 6–12 days than immediately on reaching freeze-equilibrium (Palta et al., 1977a). For longer periods of time (1–30 days) injury appears to increase with the length of time frozen. Unfortunately, however, the freezing temperatures are usually not kept sufficiently constant, and it is conceivable that, at least in some cases, the injury is due to repeated partial thawing and refreezing (see below). Sakai (1956a) has clearly shown the direct relationship between length of time frozen and amount of injury. The time at −10°C survived by mulberry and poplar increased with increase in hardiness, reaching 30 days for mulberry and 90 for poplar. Freezing at −5°C was fully survived 360 days by poplar. Browning due to injury occurred after 60, 30, and 10 days at −10°, −20°, and −30°C, respectively (Table 6.3). According to Sagisaka (1972), poplar twigs show no injury after storage at −6 or −20°C for 6 months, and then transferred to 20°C. After such storage for 1 or 2 years, however, symptoms of metabolic lesions developed, resulting in browning within the living bark. In the case of sea urchin eggs, the freezing injury increases with the time frozen at −10° or −20°C, during freezes of 1–72 hr (Asahina, 1967). Here too, however, there was little change for the first 6 hr, and sometimes for the first 24 hr.

3. Repeated Freezing and Thawing

As in the case of rates of cooling and thawing, repeated freezing and thawing can be expected to increase the injury only within a critical temperature zone that will be specific for the plant and for its physiological state. In one case, for instance, when the freezing and thawing was repeated seven times, the killing temperature was raised from −21°C for a single freeze to −13°C (Winkler, 1913). Some grass varieties seem to be more sensitive to repeated freezing than others (Thomas and Lazenby, 1968). Conversely, repeated freezing of plant material as diverse as *Fucus* eggs (Bird and McLachlan, 1974), shoots of trees (Krasavtsev, 1973), and onion bulbs (Palta *et al.*, 1977a) have not proved harmful.

These effects of the time factor demonstrate that in some cases freezing injury occurs rapidly and is, therefore, presumably direct injury; in other cases, it occurs too slowly for direct injury. These conclusions may be examined more logically on the basis of the moment of injury.

D. THE MOMENT OF FREEZING INJURY

1. During Freezing

When freezing injury occurs during rapid cooling to temperatures that fail to injure when slowly cooled (see above), this is indirect evidence that the injury must occur during the freezing process. More direct evidence has been obtained in a few cases. Some plants undergo a color change or develop a specific odor when killed by any method. Some of these plants show these changes when frozen, and, therefore, must be dead while in the frozen state (Molisch, 1897). Furthermore, these changes occur essentially immediately on freezing. Unfortunately, it was not determined whether this

TABLE 6.3

The Time (Days) Survived by Twigs Frozen at Different Temperatures

	Temperature frozen and stored at (°C)			
Tree	−5	−10	−20	−30
Mulberry	180	30	10	½
Poplar	360	90	50	20

[a] From Sakai, 1956a.

color or odor characteristic of dead cells was accompanied by extracellular or intracellular freezing. Cells of spruce and cherry, however, showed a yellow fluorescence when killed by either flash (1°/min) or slow freezing, during which intracellular and extracellular freezing were observed, respectively (Krasavtsev, 1962). According to Le Saint (1966), when freezing is fatal the evolution of CO_2 ceases, leading to the conclusion that irreversible changes during the freezing cause death.

2. While Frozen

Since injury may increase with time after 24 hr at the same freezing temperature (see above), there must be a second moment of injury—while frozen but after freezing equilibrium has been attained. This, therefore, cannot be a direct freezing injury.

3. During Thawing

The fact that the injury may be increased by increasing the speed of thawing is explainable only if some injury may also occur during the thawing process. Yoshida and Sakai (1968) suggest that the important factor may be the rate of rehydration of the cells at temperatures near the freezing point of the tissues. Leaves frozen at −15°C were killed by thawing in water at 20°C, but survived if first warmed for 2–3 min at −2° to −5°C before transfer to the water at 20°C. Whatever the explanation, if injury occurs during rapid thawing, it cannot be due to direct freezing injury.

4. Post-Thawing Injury and Repair

It has long been known that the death of extracellularly frozen plants may not occur until hours or even days after thawing and that the injury may sometimes be reversible (repairable). It is, therefore, standard procedure to keep the thawed plant cool (0°–5°C) for about 24 hr after thawing. Further evidence of post-thawing injury can be obtained by observing the cells immediately on thawing. They may still be alive and show normal plasmolysis, cytoplasmic streaming, etc., even though the plant may be dead several days later. This was observed in the case of cabbage (Levitt, 1957a) and more recently in onion bulbs (Palta *et al.*, 1977a). Such cells have been injured by the freezing, though alive on thawing, and the injury may either progress or be repaired, depending on the post-thawing treatment. For instance, if sections of cabbage tissue are transferred to $CaC1_2$ solutions after thawing, they show less injury than if transferred to H_2O (Levitt, 1957a).

Similarly, the removal of extracellular K^+ from the thawed onion bulbs may improve post-thawing repair (Palta *et al.,* 1977b).

Reparable damage has also been reported at the organelle level. During overwintering of winter wheat under field conditions, chloroplasts suffer appreciable structural damage (Petrovskaya-Baranova, 1972). Structural integrity is restored during spring. Similar reversible damage to the photosynthetic apparatus has been observed in a freezing resistant potato species but not in the nonresistant species (Nyuppieva *et al.,* 1972). Changes in the rate of the Hill reaction were also reversible.

In opposition to plants, in which post-thawing repair is enhanced by low temperatures (0°–5°C) and decreased by high temperatures (20°C), deep frozen mammalian cells show optimum post-thawing repair at 37°C. The repair decreases with temperature drop until it is virtually nonexistent at 5°C (McGann *et al.,* 1975). This repair is associated with the cell surface and involves carbohydrate and phospholipid (Robinson, 1973). It is also inhibited by $10^{-3} M$ ouabain, an inhibitor of the (Na^+, K^+) ATPase (McGann *et al.,* 1974). This seems to indicate that the ATPase is necessary to avoid additional post-thawing damage to mammalian cells. In the case of microorganisms, post-thawing repair does not seem to require the synthesis of protein, nucleic acid, or cell wall mucopeptide, but does require metabolic energy (Janssen and Busta, 1973). The recovery was, however, greatly increased by addition of cysteine, due to its reducing activity. In contrast, recovery of mammalian cells was not affected by lowering the O_2 concentration (McGann *et al.,* 1975).

One possible kind of purely physical repair is suggested by sonication of a phospholipid-water dispersion (Lawaczeck *et al.,* 1976). When sonicated below the phase transition temperature of the phospholipid, bilayer vesicles were produced with structural defects. These defects were annihilated simply by annealing above the phase transition temperature. The rate of annealing was slow (1 hr) at 3°C above the phase transition temperature, rapid (10 min) at 10°C above it.

E. PRIMARY INDIRECT FREEZING INJURY

Since intracellular freezing, with few exceptions, always causes instant death, it appears unlikely that it can ever lead to indirect injury. However, what of the exceptional cells that have been reported to survive intracellular freezing? If survival is due to "vitrification" or to the formation of intracellular crystals too small to be observed microscopically, the cell apparently remains alive as long as this condition is maintained (see above). If, however, the crystals are allowed to grow large enough, exactly the same kind of injury occurs as when these larger crystals are formed at the instant of freezing—primary direct freezing injury. Observations of tumor cells (Asahina *et al.*, 1970) have revealed that they may remain in the translucent state, with intracellular crystals too small to interfere with ordinary transmitted light even at temperatures above −30°C. It, therefore, appears possible that the exceptional cases of survival of intracellular freezing that have been reported in the literature are all of this type. We must, therefore, conclude that when intracellular freezing injures the cell, it is always due to direct injury, and that there is no evidence for the existence of indirect injury due to the primary freezing process.

F. SECONDARY FREEZING INJURY

Although the uniformly fatal direct intracellular freezing strain is avoided, extracellular freezing introduces other indirect strains associated with secondary stresses. These must be less damaging than the intracellular freezing strain, or the extracellular freezing adaptation would be of no use to the plant. Nevertheless, these indirect strains and secondary stresses are potentially (and often actually) damaging, and the adapted plant must possess avoidance and/or tolerance of them. What are they?

1. Ice Pressure

Whether or not the formation of ice in the intercellular spaces can subject the cells to pressure is still a moot point. The older evidence opposes the concept for many reasons (see Chapter 5), especially since those tissues that have been measured show a contraction during freezing rather than an expansion. It is this very contraction combined with the expansion of the water as it freezes in the intercellular spaces that presumably squeezes the air out of the tissues (see Sec 2). Mechanical damage certainly may occur, causing frost cracks in trees. However, the very fact that this occurs without injuring any of the cells not within the crack, eliminates direct ice pressure as a cause of freezing injury in noncracked tissues. Further evidence in favor of this conclusion is the frequently confirmed direct relation between freezing tolerance and desiccation tolerance. Nevertheless, there may be exceptions and Olien (1973, 1974) is convinced that direct ice-pressure injury occurs in the case of the crowns of grain seedlings.

2. Air Expulsion

The intercellular air is squeezed out of the tissues by the expanding ice. Since the ice is pure water in the solid state, cells that are uninjured by the freezing would be expected to reabsorb the intercellular thaw-water osmotically, as soon as the intercellular ice thaws. If, on the contrary, a thawed leaf remains infiltrated, this may be partly due to the closed stomata, preventing the reentry of air. Prevention of this kind, however, can only be temporary, since the leaf is not hermetically sealed. Futhermore, since pieces of leaf tissue with cut edges also remain infiltrated after thawing (Cox and Levitt, 1972), the reentry of air cannot be a factor; the lack of reabsorption of water must be due to injury during freezing or thawing.

3. Freeze Smothering

In opposition to plant parts frozen in air, ice pressure and air expulsion may conceivably be decisive causes of freezing injury in the case of plants covered or imbedded in a sheet of ice. This kind of injury is called *freeze-smothering*. The term implies a disruption of normal respiration due to an O_2 deficiency stress rather than a primary stress effect of the ice. There is considerable evidence in favor of this interpretation. Thus killing may occur at higher temperatures if the plants are covered by an ice sheet for a considerable period of time, than in the case of control plants frozen in air. In the case of alfalfa, injury was manifest only after 7-10 days and it could not be ascribed to any direct mechanical effect of the ice (Sprague and Graber, 1940). After 20 days under the ice sheet, mortality was high. Sprague and Graber (1940) attributed this to the accumulation of toxic products of aerobic and anaerobic respiration, since ice inhibits the diffusion of CO_2 and other respiratory products.

In favor of this explanation, plants immersed in water at 1°C were injured at approximately the same rate and to the same degree as those frozen in ice at −4°C. When CO_2 was bubbled through the water, survival was more rapidly reduced and injuries were more intense. However, CO_2 was only one of several gases that appeared toxic. Later tests (Sprague and Graber, 1943) showed that even at the highest concentrations of CO_2 used, a longer storage time was needed to cause injury than when the plants were frozen in blocks of ice. This led them to conclude that the external concentration of CO_2 was not directly toxic, but that respiratory compounds accumulated internally until they reached a toxic concentration. Thus, injury similar to that produced by storage in blocks of ice could be induced by removal of both CO_2 and O_2.

Not all plants show the same sensitivity to encasement in ice. When frozen in this way at a temperature that is noninjurious when frozen in air (−3°C) ladino clover died within 12–14 days and white clover survived 4 weeks or more (Smith, 1949). A relationship to CO_2 concentration was again indicated, since this was consistently higher for the ladino clover. In general, Smith (1952) found that the survival of ice encasement among legumes was approximately in the same order as their winter hardiness. He suggests that this may be due to a higher level of metabolic activity in the less hardy legumes. This is in agreement with Bula and Smith (1954), who found that the rate of loss of carbohydrates in legumes during winter dormancy was inversely proportional to hardiness.

All the above results have been confirmed for winter cereals (Andrews and Pomeroy, 1977). Winter wheat and rye were killed during total encasement in ice at −1°C, at rates inversely proportional to their hardiness. Mitochondrial activity was little impaired when intact plants were 50% killed, and the mitochondrial structure was not disrupted until 3 weeks of ice encasement. Ethanol accumulated, but only at nontoxic levels. Lactic acid also accumulated and the rate of O_2 consumption declined in the ice-encased plants (Pomeroy and Andrews, 1978). Artificial applications of these substances proved that the injury was not due to any one, but could be duplicated by applying CO_2, ethanol, and lactic acid in combination (Andrews and Pomeroy, 1978).

4. Freeze-Desiccation

Under natural conditions, winter injury due to desiccation has frequently been reported or inferred from the lower transpiration rates of the more northerly distributed species (Bates, 1923; Iwanoff, 1924; Walter, 1929; Thren, 1934; Michaelis, 1934; Rouschal, 1939). It has even been suggested that the higher concentrations of the cell sap in alpine plants than in individuals of the same species from the plains indicates a better ability to remove water from the soil (Senn, 1922). Direct determinations have failed to produce any evidence of winter injury due to freeze-desiccation in agricultural areas of very cold climates (Hildreth, 1926). Twigs do not dry out appreciably at temperatures below 41°F, and above this temperature the rate of water transfer to the twigs is quite adequate to prevent injury from the transpirational loss (Wilner, 1952). Although desiccation does not seem to be a cause of winter injury in the severe climates of Minnesota (Hildreth) or of Canada (Wilner), it may possibly be a factor in other climates characterized by a loss of water from thawed leaves at a time when translocation of water to them is impossible due to freezing of other parts of the plant. This is apparently the cause of the development of water potentials below −90 bars in the krummholz form of Engelmann spruce in frozen soil and under windy conditions (Nyuppieva, 1973). It also expalins the low-tufted, convex growth of *Diapensia lapponica* in the alpine habitat of the mountains of New Hampshire (Tiffney, 1972). Another possible example is the commonly observed death of evergreens in regions above the snow line when exposed to the wind for a few days without snow cover (Pisek, 1962).

Similarly, the northern boundary for the geographical distribution of species of Viburnum, Populus, and Lonicera seemed to be related to winter transpiration (Bylinska, 1975). Near the timberline, however, the relation may be due to the shortened cooler vegetation periods, during which the plants are less able to reduce their water loss (Tranquillini, 1974), leading to a summer desiccation rather than a winter desiccation. Nevertheless, when the injury occurs during winter, it may depend on the transpiration rate (Larcher, 1957). There is a decrease in the cuticle thickness of spruce and pine needles with increasing altitude and wind exposure at the timberline, correlated with increased transpiration (Baig and Tranquillini, 1976). These factors may lead to desiccation damage and may control the upper timberline.

In eastern Hokkaido (Japan), minimum temperatures reached −30°C and soil temperatures at 10 cm depth remained below zero for 3.5 months even on the southern slopes (Sakai, 1970a). Temperatures of stems and leaves of young conifers in winter rose to about 17° and 9°C at midday on the southern and northern slopes, respectively. They remained unfrozen for 6 and 2 hr, respectively, during daytime. Under these conditions, the conifers on the southern slopes were intensely desiccated toward the end of February. Damage was observed as browning of the stem bark. Artificial desiccation induced a similar browning. Fir, spruce, and arborvitae which were damaged by this kind of desiccation were able to survive freezing at −50°C (fir) or even −120°C (spruce and arborvitae).

In contrast to all these positive relations, measurements of leaf water potential, leaf diffusion resistance, and net photosynthesis of *Eucalyptus pauciflora* suggest that winter desiccation is not a factor in limiting tree distribution at the timberline in the Snowy Mountains of Australia (Slatyer, 1976).

5. Nonevaporative Freeze-Dehydration

The above-described freeze-desiccation is due to a combination of sublimation of ice and evaporation of liquid water from the tissues and is no different in its desiccating effect from the evaporative dehydration that occurs in the absence of freezing at normal, growing temperatures. In both cases, dehydration of the cells involves loss of water from the tissues, and at a certain point characteristic of the cell, injury occurs. An equal nonevaporative freeze-dehydration by diffusion of the cell's water to the extracellular ice centers and retention of this ice within the tissues (in the intercellular spaces) must equally injure the cells and therefore cannot fail to be a factor (or *the* factor) in freezing injury. Freeze-dehydration produces two major indirect strains—increased concentration of the cell sap (see Chapter 5) and cell collapse (see Chapter 5).

a. SOLUTION EFFECTS

It has been suggested from time to time that the increased concentration of the cell sap due to freeze–dehydration, in particular the concentration of the salts, may be the cause of the freezing injury. This would imply that the primary freezing strain gives rise to an indirect dehydration strain, and this in turn to a salt concentration strain. An acid strain is also a possibility, since the pH of certain salt solutions is greatly lowered by freezing (Orii and Iizuka, 1975). On this basis, the freezing stress, due to the resulting water stress, would be injurious via a salt or acid stress. Cyrobiologists call all these and other effects of the increased concentration *solution effects*.

b. MECHANICAL EFFECTS

Since the cell collapse (see Chapter 5) subjects the cell wall and protoplasm to a mechanical stress, any injury due to cell collapse must be mechanical injury.

Olien (1974) believes that dehydration is not the only factor in freezing injury and suggests "adhesion energy between ice and hydrophilic polymer systems as they compete for liquid water in a complex interface" as a source of freezing injury. It is difficult to understand how this can be anything more than one component of freeze–dehydration. Furthermore, the ice could not possibly produce a direct adhesion stress on the protoplasm since the two are separated by the cell wall. Nevertheless, it may be suggested that, due to the much lower temperature, a freeze-dehydration may not injure, even though equal in intensity to an evaporative dehydration that injures at higher temperatures. What, then, is the evidence of injury due to freeze-dehydration?

c. EVIDENCE THAT FREEZING INJURY IS DUE TO FREEZE-DEHYDRATION

i. Freezing and Drought Tolerance. There is ample evidence of a correlation between freezing tolerance and drought tolerance, which will be documented in Chapter 12, Vol. 2.

ii. The Four Moments of Injury. In the absence of ice coverings and of long periods of freeze-desiccation, there are only two kinds of freezing injury. Since intracellular freezing injures instantly, the other three moments of injury can occur only as a result of extracellular freezing. On the other hand, the existence of intracellular freezing injury during the first moment does not eliminate the possibility of extracellular freezing injury during the same first moment. The fact that a single slow cooling (which normally prevents intracellular freezing) can injure the plant no matter how briefly it is held in the equilibrium frozen state and no matter how gradually it is thawed, or how carefully it is protected against postthawing injury, is strong evidence for the existence of extracellular freezing injury during the first moment. Evidence of this kind of injury has been produced for winter wheat (Salcheva and Samygin, 1963). It, therefore, seems safe to conclude that the secondary, freeze-dehydration injury due to extracellular freezing, may injure during any one of the four moments of injury: (1) during freezing, (2) while frozen at equilibrium, (3) during thawing, and (4) after thawing.

iii. Frost Plasmolysis. Intracellular freezing does not remove the water from the cell and, therefore, does not lead to cell collapse as it does in the case of cell desiccation. This difference provides a method of distinguishing between primary direct freezing injury (due to intracellular freezing), and secondary freeze-dehydration injury (due to extracellular freezing). Plants killed by freezing were long ago observed to show "frost plasmolysis" (see Chapter 5). It is difficult to explain this phenomenon except by the death of the extracellularly frozen, dehydrated, and, therefore, contracted cell. On thawing, the dead protoplast is freely permeable and, therefore, unable to reabsorb the thaw water osmotically. The cell wall, on the other hand, is elastic and, therefore, snaps back to nearly its original position as water enters between it and the dead, contracted protoplast. The result is a "frost plasmolysis" of the dead cell.

iv. Extracellular Freezing. Direct observation of plant tissues in nature has repeatedly revealed that the ice formation is extracellular and has failed to detect intracellular freezing. It can, of course, be suggested that as long as ice formation is extracellular no injury results, but as soon as a small amount of intracellular freezing occurs (perhaps too small to be detected) the cell is killed. There is neither evidence of this amount of intracellular freezing, nor evidence that its occurrence would be fatal. On the contrary, all observations of killing due to intracellular freezing have been confined to large amounts of intracellular ice formation. More important, however, are the many records of injury in the presence of extracellular freezing and in the complete absence of intracellular freezing. Müller-Thurgau (1886) long ago succeeded in removing the pieces of ice from frozen potato tubers and showed that they were many times the size of the adjacent cells which were soft and gave no signs of being frozen and yet were dead. In agreement with these results, Terumoto (1960a) observed large ice masses between the cells of table beets frozen at −4° to −5°C. They occurred mostly concentrically in the vascular bundle ring region. The specific conductivity and betacyanin content of this ice increased from the inner to the outer side of the ice masses. This indicated a gradual killing of the living cells external to the ice ring during the growth of the extracellular ice mass. Such a result would have been impossible if ice had formed within the cells, for this would have brought to a stop the growth of the extracellular ice masses; and if the extracellular freezing has not injured the cells, this ice would be uniformly free of electrolyte and betacyanin. Asahina (1954) has also reported injury to potato sprout tissue when frozen extracellularly at −3°C for even a short time. Similarly, a marine alga (*Enteromorpha intestinalis*) is killed only when frozen below −25°C, and the "frost plasmolysis" of the cells indicates that this was due to extracellular freezing (Terumoto, 1961).

Many observers of artificially frozen cells, although agreeing that intracellular freezing is nearly always fatal and that extracellular freezing may be harmless, have nevertheless observed fatal injury to cells clearly frozen only extracellularly (e.g., Molisch, 1897; Iljin, 1933b; Siminovitch and Scarth, 1938; Modlibowska and Rogers, 1955; Asahina, 1956; Terumoto, 1967). Thus when ice forms within cells having colored sap, the colorless crystals are visible between the sap that becomes darker and darker, but when ice forms extracellularly, the cells become colorless at the middle due to collapse. The latter kind of freezing can occur with or without death (Siminovitch and Scarth, 1938). True frost plasmolysis, which involves ice formation between the protoplast and the cell wall, but not within the protoplast, can also occur under artificial conditions with or without injury (Asahina, 1956). Even in the case of insects, extracellular freezing may result in injury. Salt (1962) showed this by the method of freeze-substitution. Certain large cells (e.g., of the fat body) did freeze internally, but the remaining smaller cells of the larvae showed no sign of intracellular freezing, whether they survived the freezing or were killed by it.

v. Intracellular Freezing. The point at which intracellular freezing occurs (even in a single kind of cell of a single plant) is not an exact one, but varies markedly even though the conditions are maintained as nearly constant as possible. Therefore, intracellular freezing is incapable of explaining the relatively exact freeze-killing points (Chapter 7) found by so many investigators for a single variety of plant grown and hardened in a standard way. Due to this, and to the above mentioned correlation between freezing and drought tolerance, it appears obvious that extracellular freezing is the general cause of freeze killing of higher plants in nature.

There may, of course, be exceptions. It has already been mentioned that intracellular freezing may conceivably be the cause of injury in certain specialized cases, e.g., "sunscald" and the rapid freezing of supercooled wood parenchyma and buds. Nevertheless, on the basis of all the above evidence, it must be concluded that in higher plants, the major cause of freezing injury is the secondary water stress, producing freeze-dehydration by extracellular freezing. In microorganisms, such as yeast, which can become air-dried without injury, and therefore cannot be injured by the secondary, freeze-induced water stress, the major cause is apparently the primary freezing stress, producing direct injury by intracellular freezing (Mazur and Schmidt, 1968).

7. Freezing Resistance—Types, Measurement, and Changes

A. POSSIBLE TYPES OF RESISTANCE

Just as the chilling stress may induce a fatal strain in sensitive plants when their tissue temperature is cooled below a threshold *chilling* temperature, so also the freezing stress may induce a fatal strain when the tissues of a plant are frozen below a threshold *freezing* temperature specific for that tissue. In order to survive, the plant must prevent such a fatal strain. Many have, indeed, developed adaptations which protect them to varying degrees. As a result of these adaptations, all gradations of freezing resistance occur in flowering plants (see Table 6.1), from none to an ability to survive the lowest temperatures tested (see Chapter 4). These adaptations differ not only in degree but also in kind. There are, in fact, several options available to the plant for preventing fatal strains when exposed to a freezing stress. As in all stress responses, there are two main classes of freezing resistance—avoidance and tolerance. The avoidance options are simpler, since they do not require an understanding of the injury mechanism, and therefore will be considered first.

1. Avoidance of Low Temperature

The plant's temperature usually closely follows that of its environment, even in the case of bulky, insulated tree trunks, although, of course, lagging behind it during periods of rise or fall (Levitt, 1956). The ineffectiveness of plant structures as insulators against low temperature is readily shown by artificial cooling. Thus, in spite of the presence of surrounding scales, the growing point within the large buds of chestnut (*Aesculus hippocastanum*) cools practically as rapidly as the surrounding atmosphere when the latter's temperature drops from +20° to −20°C in 30 min (Fig. 7.1). Similar results have been obtained by Pisek (1958). In the case of leaves and buds, their temperature may actually drop below that of the air on a cold night: 2.5°C in the case of buds, 3°–5°C in the case of tomato leaves when the sky is clear

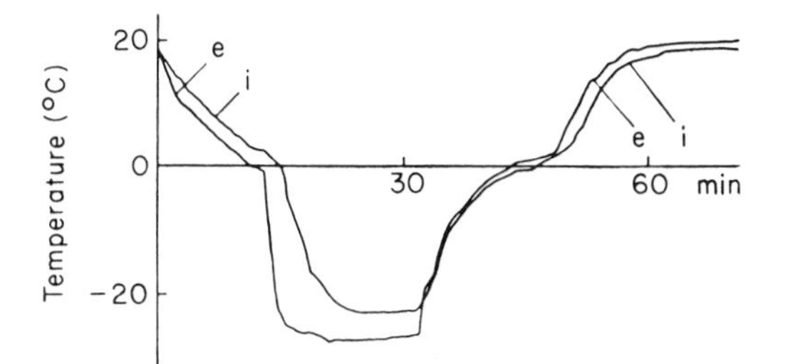

Figure 7.1. Internal temperature of a chestnut bud (i) compared with the external temperature (e), during a 30-min drop in the latter from +20°C to below −20°C. From Genevès, 1957.

(Jenny, 1953; Shaw, 1954). Even pine needles may show this drop (Table 7.1). In full sun, on the other hand, plant temperatures may rise several degrees above that of the air. An extreme example is the marked difference (as much as 30°C) between the cambial temperature of a tree in the sun and that in the shade in midwinter (Harvey, 1923; Eggert, 1944; Sakai, 1966). Both these differences are simply due to radiation from and to the plant, respectively. Unlike warm-blooded animals, plants are poikilotherms, i.e., they are unable to maintain a constant tissue temperature different from that of their environment.

In fact, different parts of a plant may be subject to markedly different freezing stresses, since the microenvironmental temperature adjacent to these parts may differ markedly. Thus, when wheat was exposed to a radia-

TABLE 7.1

Radiation Cooling of *Pinus cembra* Needles Below Air Temperature (°C) at 07.00 Hr[a]

Date (in November)	Air temperature	Degree of cloudiness (increasing from 1-10)	Cooling of needles below air temperature
16	−14.6	1	−6
17	−11.0	9	−2
18	−13.5	10	−1.5
19	−14.4	0	−6.5
20	−10.6	0	−6.5
21	−10.8	5	−4
22	− 9.5	4	−4.5
23	− 8.8	0	−6.5
24	− 9.1	10	−1.5
25	−14.0	10	−1.5
26	−18.5	2	−5.5
27	−13.5	0	−6.5

[a] From Tranquilini, 1958.

tion frost, the temperature near the surface of the crop was coldest—2°C lower than in the middle regions of the canopy (Marcellos and Single, 1975). The temperature just below the soil surface was 4°–6°C warmer than that of the plant or air 5–10 cm above the soil.

Some tissues, e.g., fleshy inflorescences of Araceae, have long been known to raise their temperatures well above that of their environment due to the heat produced by respiration, although this occurs only at temperatures well above freezing. Similarly, some early spring plants have been said to melt the surrounding snow. In neither case is a constant tissue temperature produced. Yet this respiratory production of heat has led to the question of the existence of low-temperature avoidance in plants. Respiratory heat, of course, results from the utilization of reserves. In cold climates, loss of reserves would be more damaging to the overwintering plant than to the animal, since the deciduous plant cannot replenish its reserves during the winter or even during the early spring, and since only those deciduous and evergreen plants that still possess their reserves essentially intact at winter's end can produce the rapid spring growth on which their survival depends. Most important, because of the much larger specific surface of plants than of animals, it would require a much more rapid utilization of reserves in order for the plant to maintain its temperature above that of its environment than in the case of the animal. There is no solution to this problem for the plant as a whole, since the decrease in specific surface needed to reduce the loss of heat (and therefore also the utilization of reserves) would also reduce the rate of accumulation of reserves by decreasing the photosynthetic surface. Adaptation has, in fact, proceeded in the opposite direction; for plants with the best developed low-temperature resistance may possess the lowest respiratory rates (see below), and therefore the lowest heat production at low temperatures.

It is, therefore, useless to look for plants that might have a well developed low-temperature avoidance due to the heat produced by respiration. Other types of low-temperature avoidance may, however, occur to a greater or lesser degree. The cambial cells of tree trunks (and any other living cells of the trunk) are somewhat protected from the extremes of atmospheric temperatures by the insulating effect of the bark; yet this protection may be completely inadequate (see below). Although possessing no true avoidance with respect to their own environments, roots and plants that pass the winter under a snow cover are able to avoid the lower temperatures of the above ground atmosphere. It has, in fact, been shown (Torssell, 1959) that a rare variety with less tolerance may survive better than another with greater tolerance because of its greater ability to remain covered by snow, i.e., its greater low-temperature avoidance. One special case of temporary low-temperature avoidance occurs when freezing commences. The heat released by the crystallization of water may be sufficient to keep plant temperatures above that of the air for a short time, but this must be of very limited value to the plant. That it plays no part in freezing resistance follows from the commonly found inverse relationship between water content and freezing resistance (see Chapter 8).

One case of low-temperature avoidance apparently has survival value during mild, brief freezes in the parmo zone of the Venezuelan Andes. *Espeletia shultzii,* a large rosette species, has strongly nyctinastic leaves. At night, the rosette leaves close around the apical bud, protecting it from freezing temperatures and preventing the rapid warming of young leaves just after sunrise. If nyctinastic leaf movement is prevented, leaf wilting and death occur (Smith, 1974).

Artificial low-temperature avoidance by covering the plants with a liquid foam that rapidly solidifies and liquefies again on rewarming during the day, was introduced by Siminovitch and has been used for plants such as strawberries and citrus (Bartholic, 1972). Both of these methods, the natural as well as the artificial, are successful only against very mild freezing stresses—perhaps two or three degrees below freezing.

2. Avoidance of Freezing

Although the plant is essentially unable to avoid the freezing *stress* (i.e., the freezing temperature of its environment), any one of three adaptations may permit it to avoid the freezing *strain* when cooled below 0°C.

a. BY ANTIFREEZE OR DEHYDRATION

(1) The cells may accumulate enough antifreeze to maintain their freezing point below that of the environment. In some insect larvae, a concentration of glycerol as high as 5 M has been found during hibernation in the fall, and is lost again in spring (Salt, 1958). The freezing point of these larvae is thus depressed to as low as −17.5°C. This extraordinarily high concentration has never been found in normally hydrated parts of higher plants. The highest cell sap concentration ever recorded in normally turgid vegetative plants is equivalent to an osmotic pressure of about 200 bars (and the dependability of these values is doubtful). In nonhalophytes this value seldom exceeds or even equals 50 bars. The latter value would yield a freezing point no lower than about −4°C. Although complete freezing avoidance due to antifreeze is of little importance in plants, partial freezing avoidance due to accumulation of solutes can and does occur, and is definitely protective, at least in combination with another defense factor (see below). (2) The cells may have no

freezable water. This option is available only to plants or plant parts with maximum water stress tolerance, e.g., seeds and pollen in the air-dry state and some buds which dry out during fall hardening.

b. BY SUPERCOOLING (UNDERCOOLING)

Most plants freeze when their tissue temperature drops 1°–2°C below their freezing point because of the presence of (1) homogeneous nucleators either within the plant (e.g. in the vessels) or on its surface (hoarfrost), or (2) heterogeneous nucleators within the plant tissues or on its surface. The necessary qualifications for supercooling are not fully understood but they include (a) small cell size, (b) little or no intercellular space for nucleation, (c) a relatively low moisture content, (d) the absence of internal nucleators, (e) barriers against external nucleators, and (f) the presence of antinucleators—substances which oppose the formation of nucleators (Shearman *et al.*, 1973).

Presumably because of these qualifications many insects survive winter in the supercooled state (Salt, 1950; Asahina *et al.*, 1954). It was, at one time, actually believed that only those plants that avoid ice formation in their tissues survive, and that a frozen plant is invariably a dead plant (see Levitt, 1956). It was even suggested that when an axe bounces off a frozen tree trunk in winter, the trunk actually freezes at the instant the axe touches it! It did not seem to disconcert the proponents of this theory that ice formation throughout a tree trunk takes considerable time, and that this tree (that they could not deny had ice in it), nevertheless remained alive. Continuous recording of the temperature changes during cooling has now clearly demonstrated that the sap of trees is, indeed, frozen in winter (Lybeck, 1959). Actual measurements of the ice formed in the tissues indicated that all the sap was frozen and only the water bound to the wood was unfrozen. Even the trunks of large elm trees (86 cm in diameter) are frozen during winter in northern Japan (Sakai, 1966).

Direct observation in the open under the microscope (Wiegand, 1906b), revealed the presence of ice during winter in the twigs of trees. It could, of course, again be objected (ignoring the appearance and feel of the leaves before sectioning) that they were actually unfrozen until cut by the razor blade. However, the very fact that the ice crystals were always found to occur in the intercellular spaces proved that it had been present before sectioning; for it is now known (see Chapters 5 and 6) that if the ice had formed suddenly at the touch of the razor blade, it would occur intracellularly and not extracellularly (see Chapter 5). In the case of evergreen plants, even without touching them it can be seen that their leaves are frozen at temperatures below freezing, for not only do they become darker in color but they move stiffly in the wind.

Nevertheless, more and more cases of moderate to extreme supercooling in nature have been reported (Table 7.2). It is the *only* resistance mechanism possessed by the leaves of the semi-tropical olive, for they are invariably killed by freezing (Larcher, 1959). Moderate supercooling may also be of some practical importance to other crops. Orange fruit with a freezing point of about −2°C may supercool as much as 3°C below this point before freezing (Hendershott, 1962b). Potato leaves survive −6°C due to supercooling but are frozen and killed at −8°C (Li and Weiser, 1969b). Even the sprouts of potato tubers may remain supercooled for 18 hr at −5.5°C, for 4 hr at −7.5°C (Asahina, 1954).

Such moderate supercooling is not too surprising, for even under artificial conditions, supercooling is commonly observed (Lucas, 1954), and in order to induce freezing under the microscope without excessive supercooling, it is usually necessary to inoculate sections with an ice crystal. Much more unexpected is the pronounced supercooling discovered as a result of the investigations of Weiser, Burke, and their co-workers. Supercooling is now known to be widespread in two kinds of tissues.

1. Measurements by DTA (differential thermal analysis) and NMR (nuclear magnetic resonance) spectroscopy showed that supercooling is the mode of freezing resistance in the meristematic tissues of flower buds of woody plants such as azalea, blueberries, apricots, cherries, and plums (Weiser, 1970; George *et al.*, 1974; and George and Burke, 1977b) and that injury occurs at the moment of freezing. This is, apparently, not true of vegetative buds of woody plants, since Wiegand (1906a) observed numer-

TABLE 7.2

Observations of Pronounced Supercooling of Plants in Nature

Plant	Plant part	Supercooling temperature	Reference
Eight out of 27 species	Buds	−18°C	Wiegand, 1906a
Four out of 27 species	Buds	−26.5°C	Wiegand, 1906a
Plum	Flower buds	−21°F	Dorsey and Strausbaugh, 1923
Pyrola	Leaves	−32°C	Lewis and Tuttle, 1920
Caragana		−21°C	Novikov, 1928
Hedera helix		−20°C	Iljin, 1934
Pine and fir	Needles	−21°F	Clements, 1938
Grain	Roots	−11°C	Zacharowa, 1926
Olive	Leaves	−10°C	Larcher, 1959
Azalea	Flower buds	−43°C	Weiser, 1970

ous crystals at −26.5°C in living buds that showed no ice at −18°C. Their low supercooling points were believed due to their low water contents (20–30%) and their small cell size.

2. Similar methods demonstrated the supercooling of wood ray parenchyma cells of apple (Quamme *et al.*, 1973) and of hardwood trees down to temperatures as low as −38° to −47°C. The wood rays are generally the only part of the stem water that supercools, but shagbark hickory is an exception in which all the water in the stem internodes supercools. The distribution of 49 native species of trees within the United States, was shown to be correlated with the temperature at which the exotherms occurred in fully hardened trees following supercooling (George *et al.*, 1974; George and Burke, 1976). The lower the temperature at which the exotherms occurred, the more northerly was the distribution. There is, however, a limit to this resistance mechanism, since the lowest supercooling point possible for pure water is −38°C and for aqueous solutions of the concentrations found in plants about down to −47°C (Rasmussen *et al.*, 1975; George and Burke, 1976; Baust, 1973).

Yet the supercooled water is extremely stable, even at temperatures only slightly above the spontaneous nucleation temperature (George and Burke, 1977a). Supercooling therefore confines tree species native to the Eastern Deciduous Forest of North America to latitudes where the minimum winter temperature does not drop below −40. It is similarly responsible for the northern limit of pear and apple production in North America (Quamme, 1976). On the other hand, the hardiest tree species of the Boreal Forest of North America (birch, willow, aspen, dogwood) with ranges extending into northern Canada and Alaska (where the temperature may drop below −47°C) had no exotherms and, therefore, have no supercooling mechanism for surviving the extremely low temperatures to which they are exposed. They survive experimental freezing at as low as −196°C. Similarly, even in the case of those plants that do depend on supercooling for survival, their distribution may be dependent on other factors. Thus, *Rosa* survives in climates that cool down below −40°C, even though it undercools only to temperatures somewhat above this (Rajashekar and Burke, 1978). The proposed explanation is survival due to a snow cover, which maintains a higher temperature than that of the atmosphere above it, and above the supercooling point of the plant.

The supercooling mechanism in those plants that use it has not been established. Supercooling of *Eucalyptus urnigera* is possible down to −8° to −10°C if the leaf surface is dry, but only to −2° to −4°C when wet (Thomas and Barber, 1974). This may explain the advantage possessed by the glaucous, water repellent leaves of the plants at high altitudes, compared with the green, wetable leaves at low altitudes. This type of supercooling is, therefore, dependent on the elimination of external, homogeneous nucleators. The importance of external heterogeneous nucleators has been indicated in the case of corn leaves. Dust from other corn leaves, as well as bacteria (*Pseudomonas syringae*) on the leaf surface raised their supercooling temperature, causing damage at −4°C (Arny *et al.*, 1976). Internal nucleators in leaves of *Veronica persica* and *Buxus microphylla* have been implicated for their high nucleating ability (Kaku, 1973). In these and other species, mature leaves appeared to have more effective or higher concentrations of ice nucleators than the immature leaves (Kaku, 1975). In the case of buds of Norway spruce, water freezes in the intercellular spaces of the scales, and the buds dehydrate due to movement of their water to these ice loci (Dereuddre, 1978). Even after rewarming, the dehydration persists, and this is accompanied by a lowering of both the freezing point and the supercooling point.

It is easier to explain the need for the supercooling of the wood ray cells than the mechanism. The xylem rays are the least hardy stem tissue in midwinter. Furthermore, supercooling may be the only resistance mechanism available since there is essentially no intercellular space for extracellular freezing, and none can develop because the cell walls are so rigid (a lignin content of 31% in the rays versus only 26% in the rayfree wood of oak; Furoya, 1974) that cell collapse is all but impossible. Thus if freezing occurs it must be intracellular, and this form of freezing cannot be tolerated by plant cells. Supercooling of these cells is also favored by their small size and the even smaller size of their protoplasts (due to the thick cell walls) and their low water content. Since ice forms elsewhere in the twig (e.g., in the bark of apple twigs; Quamme *et al.*, 1973), there must be some barrier or force preventing the distillation of the ray cell water to these ice centers of lower vapor pressure. Two possibilities exist: (a) an impermeability of the cell walls (except for the plasmodesmata through them) to water or (b) a decrease in vapor pressure (by as much as −500 bars at −40°C) due to development of a high wall tension as soon as a small amount of water distills to the ice centers. This would not, however, be a true supercooling, at least at first. By a process of elimination, the evidence favors the latter explanation since D_2O exchange measurements showed only a weak kinetic barrier to water transport in the xylem rays (George and Burke, 1977a). In agreement with both of these concepts, hickory xylem, which supercools, resists dehydration; dogwood, which does not supercool, does not resist dehydration (George and Burke, 1977a).

Those plants that are capable of supercooling, can actually harden to freezing temperatures by increasing their supercooling ability. This happens normally in the fall, in the case of wood parenchyma of trees and reproductive buds (George and Burke, 1976). It also happens to a lesser degree in the case of citrus trees (Yelenosky, 1978). The wood parenchyma cells, according to the above explanation, presumably undergo cell wall thickening during the hardening period.

In support of this conclusion, it has been shown that the lignin content increases in the wood of grape and apple shoots during hardening (Kantser, 1972). Supercooling is also favored by low water content, and during hardening all cells are known to decrease in water content. Higher concentrations of hydroxyproline (a cell wall amino acid) were correlated with this hardening in the case of citrus (Yelenosky and Gilbert, 1974). Those trees that are native to the most severe climates and therefore, cannot survive in the supercooled state possess wood ray parenchyma cells with thinner and softer cell walls that, therefore, presumably are able to freeze extracellularly and simultaneously collapse.

It must be concluded from the above that supercooling is definitely *the* freezing resistance mechanism for the flowering buds and wood parenchyma of many deciduous woody plants surviving freezing down to as low as −47°C in temperate climates and for the survival of mild freezes by some semitropical plants. Nevertheless, the other living tissues of deciduous woody plants (e.g. the bark tissues) survive severe freezing temperatures only by tolerating extracellular freezing. In all or nearly all other plants, supercooling is confined to a few degrees at most, and its duration is brief. Thus, thousands of plant parts (leaves, roots, stems, flowers, fruits) were all found to freeze readily in a room at −3°C (Whiteman, 1957).

It is easy to understand why supercooling of the whole organism is not the major freezing resistance mechanism among higher plants although common in insects (Riddle and Pugach, 1976). The ability of the water to remain supercooled varies inversely with the diameter of the capillary in which it occurs. The water is, therefore, much more likely to supercool in the case of the smaller insect cells than in the larger cells of higher plants. Even if some of the plant's cells are small enough to favor supercooling, this may not help since ice formation commonly begins in the large, well filled vessels and from here spreads throughout the plant (Asahina, 1956). Furthermore, insects can keep very still and be protected from winds, whereas the leaves and branches of trees are readily moved by any breeze; such movement is one of the best ways to induce ice formation in supercooled water. They may also apparently be inoculated by hoarfrost on leaves (Asahina, 1956). In some cases, in fact, this may be a requirement for freezing, at least during brief freezes. Thus undercooling may be maintained (at −4°C) for some time in etiolated pea shoots even if the intercellular spaces are filled with water, provided that the shoots are paraffin coated and, therefore, protected from external inoculation (Le Saint, 1956). Green shoots, however, are unstable in the undercooled state and freeze spontaneously (Le Saint, 1958). Finally, even in the case of insects, undercooling is not permanent (Salt, 1950), and freezing eventually occurs at irregular intervals over long periods of time.

Freezing, in fact, occurs sooner in moderately cooled than in more severely supercooled systems (until −38° to −47°C below which no supercooling of water is usually possible). It has, therefore, been suggested that suppercooling may be part of the resistance mechanism in cold-acclimated apple trees maintained at −30° to −20°C, but not in the −20° to −10°C range (Bervaes *et al.*, 1977).

Since a sudden freezing after a severe undercooling is far more likely to be fatal than the gradual freezing that occurs when there is no marked undercooling (see below), it is obvious that a marked undercooling of whole plants can do more harm than good and, therefore, is not likely to be selected as a survival factor. It is more likely to be useful during moderate freezes of short duration. This occurs, for instance, in the Venezuelan Andes where there are no thermic seasons (Larcher, 1975). The average temperatures are low but there are large diurnal temperature oscillations. Native plants are adapted to episodic frosts throughout the year by freeze-avoidance mechanisms—insulation, supercooling, and low freezing points of the tissues. Some of these plants supercool to −9° to −11°C, at least for these short periods.

Yet even during the brief night frost of spring, flowers of fruit trees are commonly unable to remain supercooled, and it is necessary to confer artificial avoidance by spraying them with water (Rogers *et al.*, 1954). Such artificial avoidance is itself dangerous, for if the heat released on freezing of the spray water is insufficient to keep the tissue temperature above its freezing point, the cells will freeze even more readily (due to inoculation with the ice crystals) and be killed.

3. Avoidance of Intracellular Freezing

Intracellular freezing is the primary, direct strain that occurs in many living cells when exposed to a freezing stress. This strain is essentially always fatal. There are a few exceptions under artificial conditions, but there is no reason to believe that plant cells ever survive intracellular freezing in nature (see Chapter 6). All plant tissues that survive freezing in nature must, therefore, avoid intracellular freezing by freezing extracellularly.

The first direct evidence of this kind of freezing resistance was obtained by cooling sections under the microscope end inoculating them with ice at different temperatures (Siminovitch and Scarth, 1938). Intracellular freezing occurred at slower rates of freezing (i.e., higher inoculating temperatures) in unhardened than in hardened plants (Table 7.3). In confirmation of these results, the cells of hardened wheat leaves showed intracellular freezing

TABLE 7.3

Avoidance of Intracellular Freezing in Resistant (Hardened) and Nonresistant (Unhardened) Cells[a]

Temp. at inoculation (°C)	Average percentage of cells frozen intracellularly			
	Catalpa		*Cornus*	
	Unhardened	Hardened	Unhardened	Hardened
−2 to −3	75	15	25	0
−3 to −4	90	30	15	0
−4 to −5	100	55	35	0
−5 to −6			90	10

[a] Reproduced by permission of the National Research Council of Canada from Siminovitch and Scarth (1938). *Can. J. Res. C16*, 467–481.

only if cooled at a rate of more than 1°C/3 min, and the meristematic cells of tillering nodes at a rate of more than 1°C/5 min (Salcheva and Samygin, 1963). Intracellular freezing occurred at slower rates of cooling in the unhardened plants, although even in these, the rates required were more rapid than those normally occurring in nature. Aronsson and Eliasson (1970) have successfully used survival of very rapid freezing (by cooling from +4° to −22°C in 10 min or about 3°/min) to measure the seasonal changes in hardiness of pine needles. All these results demonstrate that avoidance of intracellular freezing increases with freezing resistance, but they do not explain how the increases are induced.

Recent measurements (Yelenosky, 1975a) indicate that hardening of orange wood, resulting in more than a 3°C lowering of the freeze-killing temperature, reduces the linear rate of ice spread in the wood. This result suggests that avoidance of intracellular freezing may depend on a mechanism of slowing down the rate of freezing at any one cooling temperature. Olien and his co-workers (see Shearman *et al.*, 1973) ascribe this phenomenon to kinetic freezing inhibitors, which they believe are specific polysaccharides. Whether or not this is a general adaptation is not known, since few attempts have been made to measure the rate of ice spread in plant tissues (see Chapter 6). Whatever the rate of ice spread, however, in order to permit the ice crystals to grow in the intercellular spaces, at the expense of the cell's water, the water must diffuse from the cell rapidly enough to maintain its freezing point below its momentary temperature. The hardy cell accomplishes this by (1) having a large specific surface. This is undoubtedly why hardy cells are commonly smaller than tender cells. (2) It must possess a high permeability to water (Levitt and Scarth, 1936; McKenzie *et al.*,

1974b). These two factors maintain a sufficiently rapid rate of osmosis to the ice centers. (3) A third factor is the lower water content of hardy cells, which, therefore, decreases the amount of water which must leave the cell per unit time. (4) A high solute content decreases the initial (maximum) amount of water that must diffuse to achieve equilibrium.

But there is a very important sixth factor in the avoidance of intracellular freezing: the temperature of the lipid phase transition. In the case of the chilling stress, the injurious effect of a lipid phase transition is clear, due to its twofold effect on the membrane: (1) It lowers the permeability of the root cells to water and therefore produces a secondary water stress injury, and (2) it produces a metabolic imbalance by markedly decreasing the activity of membrane-associated enzymes without having any effect on the soluble enzymes. In the case of the freezing stress, neither of these injurious effects can apply since the xylem is frozen and therefore prevents all transfer of water from roots to top, and the metabolic rates are too slow even in the absence of a phase transition to produce the very rapid freezing injury. By the same token, although increased unsaturation of the lipids during hardening can produce *chilling* resistance by lowering the phase-transition temperature, its role in freezing resistance is unexplained. It is, in fact, reasonable to suggest that the phase transition plays no role in freezing resistance, and this conclusion has been supported by negative results with respect to changes in lipid unsaturation during hardening to freezing temperatures (see Chapter 8). Nevertheless, an increase in percent unsaturation of the fatty acids during hardening has been reported in so many cases that it must be considered as a possible factor in the resistance of *some* plants, even though no relation to varietal hardiness is usually obtained. What does the evidence indicate?

The phase transition of the membrane lipids, from the liquid crystalline to the solid (gel) state, markedly lowers the permeability of the cell to small molecules such as water (Lee, 1975; see Chapter 3), and therefore eliminates the second factor (high permeability to water) in intracellular freezing avoidance. If the transition occurred at, or slightly above, the freezing temperature, it would tend to prevent extracellular freezing by decreasing the rate of exosmosis of water to the intercellular ice loci. Therefore, intracellular freezing would be favored. Consequently, a phase-transition temperature *below* the freezing point of the cells is the first requirement for the development of *any* freezing tolerance. All hardy cells must possess this property, either constitutively as in trees and some hardy herbaceous plants or inductively by hardening at chilling temperatures.

The previously unexplained inductive increase in permeability during hardening (see Chap. 8), may now be related to the above-described lipid change. It is known that an increase in unsaturation of membrane lipids increases the permeability of the membrane to water (Stein, 1967). It is also known, as mentioned above, that an increase in unsaturation lowers the phase-transition temperature and, therefore, prevents the marked *decrease* in permeability which accompanies the phase change. We, therefore, propose that the increase in unsaturation of the membrane lipids, when it occurs during hardening, and the consequent lowering of the phase-transition temperature plays a role in the avoidance of intracellular freezing by maintaining a high enough permeability of the cell to water, in this way permitting all the ice to form extracellularly. Those plants not in danger of intracellular freezing (such as trees that already possess a low phase transition temperature) would not have to increase the unsaturation of their lipids during hardening. As in the case of chilling, the phase transition temperature may also be lowered by changes other than an increase in unsaturation.

In summary, five avoidance options are theoretically available to the plant for preventing freezing injury (Diag. 7.1). The first of these is essentially ineffective except for the mildest of freezing stresses and the second is effective only in the quantitative sense—when accompanied by freezing tolerance. Similarly, the last three avoidance mechanisms are generally ineffective by themselves, since they must be accompanied by freezing tolerance in other tissues or organs in order for the plant as a whole to survive the freezing stress. Therefore, the only resistance option that must be developed by all vegetative plants, in order to survive the freezing stress of temperate climates is *extracellular freezing tolerance,* without which the five avoidance mechanisms would be useless.

4. Freezing Tolerance

Of the two conceivable tolerance options, the first—tolerance of intracellular freezing—is not available to the plant, presumably due to piercing of

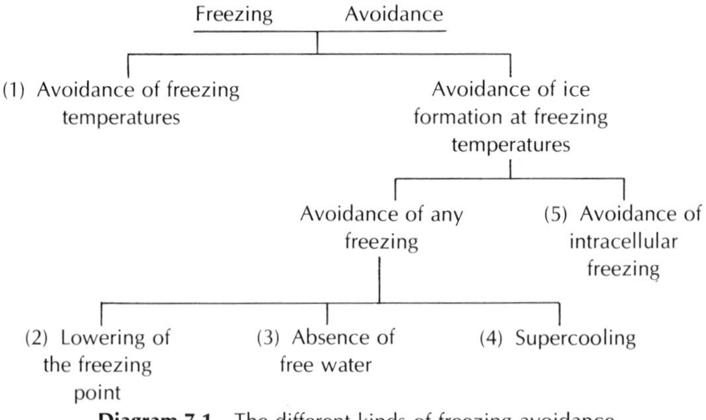

Diagram 7.1 The different kinds of freezing avoidance.

the membranes by the internal ice crystals, with the destruction of their semi-permeability, which is essential to the life of the cell. The only tolerance option developed by the plant is, therefore, tolerance of extracellular freezing. Since freezing tolerance is the major mechanism of freezing resistance in plants, it must be measured and its nature determined. Several methods have been developed for measuring it.

B. MEASUREMENT OF FREEZING TOLERANCE

1. Field Survival

Although sophisticated equipment is required for some kinds of avoidance measurements, in principle only the temperature of the plant and its environment during cooling need be measured. The difference between the two measures temperature avoidance, and the exotherms measure the limits of the other four avoidances. In each case, the exotherm signals death of the cell, so survival need not be measured. In the case of freezing tolerance, measurement is more complicated, partly because survival must be measured.

The ability of plants to survive the severities of winter has long been known as *winter hardiness* and has, therefore, been measured by field survival. The surviving plants were classified as hardy, while those that did not survive, as tender. Gradations between these two extremes, such as "moderately hardy" or "semihardy" were later introduced and developed into a semiquantitative rating (cf. Table 7.4). More recently, a fully quantitative method, adaptable to computer programming has been developed (Bittenbender and Howell, 1974). In all cases, however, hardiness must be measured by field survival of a "test winter"—a winter severe enough to kill the most tender and to damage those of intermediate hardiness in graded degrees. Test winters occur on an average of once every 10 years and, therefore, the need for a quicker and more accurate method of measurement soon became apparent. Furthermore, although freezing tolerance is usually the major component of winter hardiness, it is not always the limiting factor. In some woody plants (See Sec. A2b), winter hardiness is limited by avoidance—the ability to keep their wood parenchyma and buds supercooled.

2. Artificial Freezing

Freezing chambers in which the hardiness of plants could be tested quickly and quantitatively were first introduced by Harvey (1918). By freezing a series of apple varieties at a previously determined temperature, it was possible to obtain a graded series of injuries, depending on the resistance or hardiness of the variety (Table 7.4). This method was developed to a high degree by the Swedish investigators (Åkerman, 1927), who gave a numerical rating for hardiness from 1 to 10 (cf. Table 8.3). The usefulness of this procedure was soon recognized and it was applied particularly as an aid to the breeding and selection of hardy wheat varieties of high quality. Similar numerical ratings are still used for the selection of hardy plants of many kinds (Manis and Knight, 1967). For this purpose, the varieties must be compared when in their hardened state (see below), since some wheat varieties may change their hardiness rank relative to the others as they develop (Roberts and Grant, 1968).

In the several decades since its introduction by Harvey, artificial freezing tests have been used for many different kinds of plants (Emmert and Howlett, 1953; Coffman, 1955; Rachie and Schmid, 1955; Johansson *et al.*, 1955; Amirshahi and Patterson, 1956). This method has even been used in the field, in the case of 2-year-old grapefruit trees (Cooper *et al.*, 1954), as well as

TABLE 7.4

Comparison of Hardiness Rating from Artificial Freezing Tests (in Order of Decreasing Hardiness) and from Field Experience[a]

Order of apple varieties from artificial freezing test		Field experience (Horticultural Society rating, 1 = maximum hardiness)
1	Charlamoff	2
2	Hibernal	1
3	Duchess (Oldenberg)	1
4	Patten	1
5	Wealthy	2
6	Windsor Chief	3
7	McIntosh	4
8	Fameuse	4
9	University	4
10	Northwestern	4
11	Wolf River	5
12	Jonathan	4
13	Delicious	4
14	King David	4
15	Black Ben[b]	6
16	Paragon[b]	6
17	Lansingburg[b]	6

[a] Adapted from Hildreth, 1926.
[b] Not recommended by the Horticultural Society

for winter turnips, rape, and rye (Johansson and Torssell, 1956). In the latter case, temperatures as low as −28°C were obtained over an area of a square meter. Greatest sensitivity to freezing was found at the flowering stage, e.g., rye was injured at 0° to −2°C. Liquid nitrogen has been used as the refrigerant for controlled freezing of dormant twigs (Weaver and Jackson, 1969). Where small leaf samples are used, space and time can be saved by using a temperature gradient bar (Rowley *et al.*, 1975). For a preliminary screening of genotypes, it is also possible to use a single minimum temperature instead of a series of temperatures (Fowler *et al.*, 1973).

Even the highly quantitative numerical rating of the Swedish investigators is only relative, and does not permit a direct comparison with the varieties of another species that may require a different freezing temperature for such a rating. As in all scientific measurements, the ultimate goal must be an absolute rather than a relative measure of the quantity. This is especially necessary for the solution of theoretical problems, such as the quantitative relationships between freezing and drought resistance (see chapter 12, Vol. 2).

The need for an absolute measurement is most simply met by determining the frost (or freeze) killing point, the freezing temperature required to kill 50% of the plant (Fig. 7.2). The temperature determined has sometimes been either the "ultimate frost-killing point" resulting in 100% killing or the "incipient frost-killing point" that just begins to cause injury (Pisek, 1958). The curve in Fig. 7.2 illustrates that the most readily measured point is the 50% killing point (the T_{k50} or LD_{50}) which has now become standard.

The frost-killing point must be determined under standard conditions. Although no single uniform standardization has been adopted by all inves-

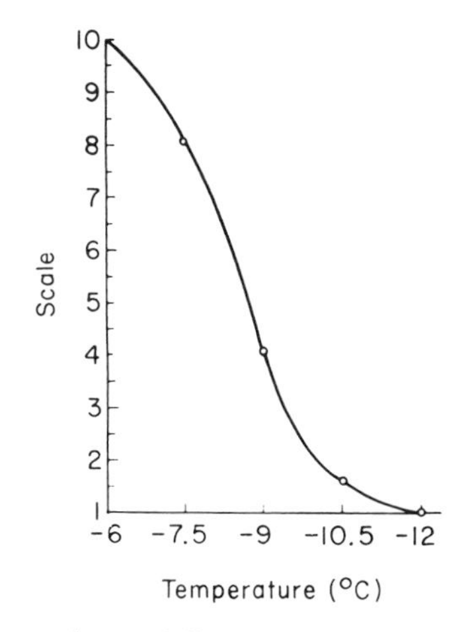

Figure 7.2. Determination of 50% killing point (in this case, −9°C). Adapted from Johansson *et al.*, 1955.

tigators, the following five steps have been proposed as basic requirements (Levitt, 1956):

1. The plants must be inoculated to ensure freezing
2. Cooling must be at a standard rate.
3. A single freeze must be used for a standard length of time.
4. Thawing must be at a standard rate of warming.
5. Post-thawing conditions must be standardized.

In order to achieve these basic requirements, the following procedure has been used. A series of shoots or leaves are taken from a number of plants of a genetically pure strain which have been grown under identical conditions. These are suspended in a freezing chamber at 0°C and after reaching temperature equilibrium the cut surfaces are sprinkled with snow to prevent appreciable supercooling. The temperature is then dropped at the rate of 2°C/hr. At each of a series of temperatures ranging above and below the suspected killing temperature, several shoots or leaves are quickly transferred to a room at a temperature just above freezing, where they are allowed to thaw and recover for 24 hr. They are then transferred to room temperature where their cut bases are stood in water. The percentage injury is estimated as soon as it becomes clear-cut (2–7 days depending on the species). The selected temperature is that which produces 50% killing—the frost-killing point.

Artificial freezing tests have usually been found to give excellent agreement with winter survival in the field (Table 7.4). Yet some differences between the two may be expected since some of the field conditions are not duplicated by the freezing test. Torssell (1959) has shown that a less frost-hardy variety may have a greater "field hardiness," using the latter term for both low-temperature avoidance and freezing tolerance. This is to be expected since the freezing test measures only freezing tolerance, and ignores the ability of the plant to avoid low-temperature injury by, for instance, remaining under the snow. Another complication may arise in those cases where freezing injury occurs mainly in the spring (Till, 1956). The order of hardiness may then conceivably be different from that in midwinter or after artificial hardening. In apples, only the fall measurements of hardiness may agree with field experience in some climates (Emmert and Howlett, 1953). In some cases, the length of the rest period or the ability to reharden after temporary losses of hardiness during midwinter warm spells may be the deciding factors (Brierley and Landon, 1946).

In other cases (for instance, winter wheat), insufficient hardening may occur under field conditions during a mild winter. Additional hardening is, therefore, recommended at the end of autumn in the laboratory, accompanied by top dressing with a solution of sucrose. This permits easier establishment of differences in freezing tolerance (Tumanov *et al.*, 1976a).

In summary, if the artificial freezing tests are intended to give a complete picture of the plant's freezing tolerance, measurements must be made at frequent intervals throughout the year. According to Scheumann (1968), there are at least four important aspects of this annual freezing curve. (1) The plant's readiness for hardening is important for tolerance of early frosts. (2) The extent of its hardening potential is important for winter hardiness. (3) The stability of its hardiness is important during periods of widely fluctuating temperatures. (4) The time of bud bursting and flowering is important for tolerance of late frosts. As pointed out by Larcher (1968) a complete picture of the plant's hardiness further requires measurements on different parts of the plant, since injury to these parts may range from 0 to 100%.

Even the method of using the freezing test may conceivably introduce differences between artificial and field testing. If estimates of injury are based on percentage killing of the plant's foliage (e.g., in winter annuals), this may show little relationship to subsequent yield, since a plant with all or nearly all its foliage killed may still yield well. Yield is, in these cases, more closely related to the percentage of plants completely killed. Therefore, in the case of wheat, instead of foliage survival, crown freezing survival of varieties is well correlated with field observations and with results obtained in different years (Landi, 1974). It must also be remembered that measurements of freezing resistance by the artificial freezing method determine the resistance of the plant under standard conditions of slow freezing and slow thawing with optimum conditions for extracellular freezing and post-thawing recovery. It has never really been adequately determined whether or not the relative ratings of varieties would be altered by changing these conditions.

A departure from this standardized procedure has been introduced by Aronsson and Eliasson (1970). Shoots of 1 to 2-year-old seedlings of Scots pine (*Pinus sylvestris*) were first stored at 4°C, then plunged into a deep-freeze at −12°, −22°, −32°, or −44°C. The temperature of the needles dropped to the deep-freeze temperature in about 10 min, a cooling rate of about 1.5-5°C/min. Although the lowest temperatures were too severe, at −22°C they obtained the same kind of seasonal curve as is normally obtained by the slow-freezing method. Yet this method presumably measures avoidance of primary direct freezing injury, i.e., avoidance of intracellular freezing.

Artificial freezing tests have generally given such good agreement with field experience (and have even revealed errors in the latter: Hildreth, 1926) that it must be concluded that the freezing tolerance, which is measured in these tests, is by far the major factor in the overwintering of plants. For most purposes, the frost-killing point is all that is needed. For some theoretical considerations, it has been proposed to define hardiness (and therefore tolerance) as the numerical difference between the frost-killing point and the freezing point (Levitt, 1956):

$$H = (T_{k50} - T_{\Delta})$$

where H = frost hardiness; T_{k50} = frost-killing point for 50% killing; T_{Δ}= freezing point. The advantage of this quantity over the frost-killing point is that it yields a value of zero for plants killed at the first touch of frost, and increasingly positive values for those with lower (negative) frost-killing points.

3. Methods of Estimating Freezing Injury

One weakness in the above methods is that the estimate of injury is subjective. This can be overcome by the electrical conductivity method originated by Dexter (1932) and used regularly by many investigators (e.g., Wilner, 1955, 1960). Since cell injury results in greater loss of electrolytes from the cell sap, the greater the injury the greater the conductivity of the extract. The conductivity value of the tissues or an extract of them is, therefore, a direct measure of the injury produced by freezing. Since it is not affected by dormancy, conductivity may prove more reliable even than estimates of recovery, for instance for strawberry cultivars (Harris, 1973).

Other substances besides inorganic electrolytes will, of course, also diffuse out of the injured cells. The release of amino acids and other ninhydrin reacting substances was used by Siminovitch *et al.* (1962) as a sensitive measure of freezing injury.

An interesting variant of the conductivity method has been described by Greenham and Daday (1957, 1960). The ratio of conductance of a high frequency to that of a low frequency current drops from 10 in the living plant to 1 when a tissue is killed. With this simple method, a fine electrode can be inserted directly into the leaf and injury can be determined repeatedly in different parts of the leaf at different times after thawing.

Many investigators have adopted this method of applying alternating instead of direct currents to plant parts and measuring their impedance. Unlike the conductivity method, however, impedance measurements have been used to *predict* injury before exposure to freezing (see below), as well as to *measure* injury after freezing. Impedance has been correlated with hardiness of forty-three peach-bearing trees (Weaver *et al.*, 1968) and with increases in apparent freezing resistance of alfalfa (Hayden *et al.*, 1969). However, scion diameter was also correlated with hardiness in the same forty-three peach trees, and impedance was not correlated with the hardiness of non-bearing trees (Weaver *et al.*, 1968). Similarly, impedance readings did not separate the winter hardy from the tender cultivars of red raspberry (Craig *et al.*, 1970), although they did indicate when the rest period began in the fall and the resumption of growth in the spring.

More recently, the impedance has been found to decrease with increasing freezing injury in a variety of plants: conifers (Glerum, 1973; Van den Driessche, 1973), grapes (Golodriga and Osipov, 1972), blueberry buds (Doughty and Hemerick, 1975), *Forsythia* (Blazich *et al.*, 1974), wheat (Voinikov *et al.*, 1974), and even brown algae (Macdonald *et al.*, 1974). In the case of alfalfa (Walton, 1973a) and grasses (Walton, 1973b), however, the method proved of little value for measuring the differences between species and varieties. Some of the difficulties with the method may be due to fluctuations in impedance values in the absence of injury, for instance, due to diurnal rhythms (Brach *et al.*, 1976). Later experiments, however, demonstrated a close correlation between the hardiness and impedance of six alfalfa cultivars in the presence of added sucrose (Walton, 1974; 1975). The differences between the hardening capacities of roots and shoots of several woody species were quantitatively related to the ratios of impedance (Wilner and Brach, 1974). In the case of young apple trees, impedance proved reliable for indicating winter damage but not for predicting it (Fejer, 1976). Impedance of stem tissue of apricots was correlated with the LD_{50} of the flower buds (Hewett, 1976).

In most of the earlier investigations, the conductivity method was used only to measure the relative injuries suffered by a series of varieties frozen under identical conditions. The results are far more useful when the freezing temperature is determined that produces the conductivity corresponding to 50% killing (e.g., Aronsson and Eliasson, 1970). One difference between the subjective method of estimating injury and the conductivity method is that the former determines the injury several days after freezing and the latter immediately or a few hours afterward. The two values may perhaps differ somewhat due to postthawing injury or recovery. Lapins (1962), for instance, was able to estimate recovery more reliably after a 3-week forcing period in a growth chamber than by the electrical conductivity method.

Estimations of injury may also be made immediately after freezing by vital staining of sections with neutral red and observation of the percentage of living (stained) cells. Simple plasmolysis may also be used either alone or in combination with vital staining. The reduction of colorless triphenyltetrazolium chloride (TTC) to the red form on the cut surface of tissues has also been used successfully (Parker, 1953, 1958; Torssell and Hellstrom, 1955; Larcher and Eggarter, 1960; Larcher, 1969). A 25% decrease in TTC absorbance following the freeze treatment correlated closely with visual survival checks in the case of bermudagrass, and served as a good index of viability (Ahring and Irving, 1969).

The TTC method also gave a quantitative measure of freezing injury for plant tissue cultures (Bannier and Steponkus, 1976) and for citrus (Yelenosky, 1975b). Although satisfactory for grape, it was not as reliable as the conductivity test for cherry and raspberry (Stergios and Howell, 1973).

Xylem pressure potential and chlorophyll fluorescence have been used as indicators of freezing survival in black locust and western hemlock seedlings (Brown *et al.*, 1977). In the case of tissues or organs that owe their resistance to supercooling, hardiness is most readily and exactly measured by determining the exotherm temperature (Quamme *et al.*, 1975). This, however, is a measure of freezing avoidance due to supercooling, not of freezing tolerance.

For many years, investigators have searched for an indirect "measuring stick" to evaluate hardiness without having to freeze the plant, e.g., sugar content, bound water (see below). To this day, the search continues, and sometimes the methods work beautifully for one series of plants but not for another, e.g., sugar content is directly proportional to hardiness in a series of twelve wheat varieties, but not in the case of other varieties or other grains (see Chapter 8). Frequently, methods that seem promising at first, fail in later tests. Thus, attempts to evaluate the hardiness of alfalfa varieties by the extent of germination in solutions of graded osmotic pressures gave variable results and, in contrast to earlier reports, proved unreliable (Heinrichs, 1959). An indirect method that has worked satisfactorily so far, but has not been tried by many investigators, is the deplasmolysis method (Scarth and Levitt, 1937; Siminovitch and Levitt, 1941; Siminovitch and Briggs, 1953a; Sakai, 1955b, 1956a). It must be cautioned, however, that this method works only if plasmolysis is maintained for a sufficiently long period (e.g., some hours). It is not precise enough for small varietal differences.

In the case of 87 wheat varieties and 15 barley varieties a correlation coefficient of 0.83 to 0.85 has been obtained between SH content of the homogenate and winter hardiness (Schmuetz *et al.*, 1961; Schmuetz, 1962, 1969). The SH measurements, however, had to be made under rigidly standardized conditions (see Chapter 8). An equally good correlation has been obtained between ascorbic acid content and winter hardiness in the case of several wheat varieties (Schmuetz, 1969; see Chapter 8).

In spite of the promising results described above, past experience indicates that no one indirect "measuring stick" can be trusted as a measure of relative freezing resistance in all plants. Sakharova and Yakupov (1969) have, in fact, recommended measuring a long list of variables. In the final analysis, however, direct freezing tests are essential for fully reliable measurements of freezing resistance, whether it is due to tolerance or avoidance.

C. CHANGES IN FREEZING TOLERANCE

1. Seasonal Changes

Some plants are killed by the first touch of frost. For instance, a leaf temperature of −1.5°C for 10 to 15 min produces enough freezing to kill banana leaves (Shmueli, 1960). Others survive the lowest winter temperatures without injury, and all gradations exist between these extremes. Yet the frost-killing point, although determined in a standard way, is not a constant even for a genetically pure strain, but varies markedly with the stage of development and with several environmental factors that can alter resistance. In the case of even the most resistant species, the actual tolerance varies from a minimum of practically none in the new, spring growth to the maximum value of midwinter (Fig. 7.3). The hardiest trees (for instance, white birch, poplar, and willow) retain this maximum hardiness throughout the winter (Sakai, 1974), whereas in less hardy plants such as winter wheat, the hardening capacity decreases as winter progresses (Tumanov *et al.*, 1976a).

This increase in freezing tolerance is called frost (or freeze) hardening, cold hardening, cold acclimation, etc. Due to this seasonal change, even conifers that are among the most hardy of plants may be severely injured by light summer frost (Pomerleau and Ray, 1957). Very tender species (e.g., those from tropical climates), on the other hand, never develop any frost resistance no matter what the stage of growth or time of the year. Under artificial conditions, tropical species of willows are exceptions to this rule (Sakai, 1970b). When grown in Sapporo and hardened in the same manner as the northern species, they developed as high a degree of tolerance, surviving −50°C or even liquid nitrogen. Nevertheless, in the vast majority of cases, when hardiness is at its maximum, it also varies markedly from species to species (Table 6.1) and from variety to variety (Tables 7.4 and 8.14). It does not follow that a species or variety with the maximum tolerance at its full development necessarily is also the most tolerant of a series of species or varieties at all times of the year. The rates of hardening may vary independently of the maximum attained. A hardy apple variety, for instance, shows far greater hardiness in the fall than the less hardy variety, but the difference between the two is slight in midwinter (Fig. 7.3).

Some plants, in fact, are "resistance stable" and their hardiness fails to change much with external conditions (Larcher, 1954). These may, therefore, be the most freezing tolerant of their community in summer (e.g., a killing point of −12° to −16°C in *Citrus trifoliata*) and the least tolerant in winter. High-mountain plants may retain extreme resistance even in summer, the rhizome of *Saxifraga,* for instance, surviving −80°C (Kainmüller, 1975). The most extreme examples of resistance-stable plants are to be found among lower plants. *Porphyridium cruentum* shows normal optimum growth above 27°C, but no damage results from rhythmic alternation of periods of freezing at −22°C in the dark with periods of 27°C in the light (Rieth, 1966). Whole taxonomic groups may be resistance stable. Thus, the freezing tolerance of arctic mosses and liverworts was not found to differ from that of tropical mosses (Biebl, 1967b). Many algae, however, are resistance labile (Schölm, 1968). In the case of higher plants of temperate climates, the autumn rise in freezing tolerance is a universal phenomenon and has been reported in innumerable species and varieties tested by investigators throughout the world. The same is true of the spring drop in freezing tolerance. The more developed the bud, the sooner it loses its resistance in spring (Mair, 1968). Yet it must not be concluded that development and freezing tolerance in higher plants are indissolubly related. The genes for reaction to cold in the coleoptile stage of wheat are not linked to the genes of development (Goujon *et al.,* 1968).

The pronounced seasonal change in freezing tolerance is even found in overwintering insects. Although the insect is usually resistant only in the pupal state, the adult carabid beetle has been found to tolerate −35°C

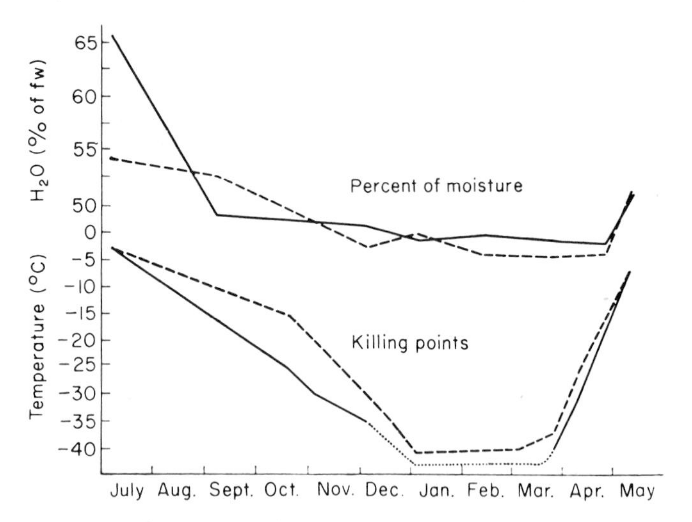

Figure 7.3. Seasonal variations in frost-killing points and in moisture contents of Duchess (solid line) and Jonathan (broken line) apple twigs. (From Hildreth, 1926.)

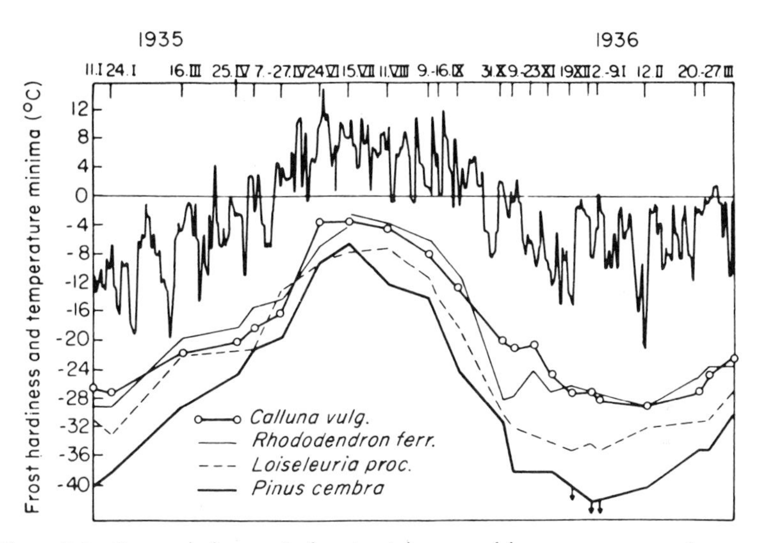

Figure 7.4. Seasonal changes in freezing tolerance of four evergreen species compared with the daily temperature minima. (From Ulmer, 1937.)

during winter (Miller, 1969). In summer, the beetle is killed if frozen at −6.6°C. Even in winter, however, relatively slow cooling is required to avoid damage.

Many marine algae are highly tolerant of freezing, but show no seasonal change in tolerance (Terumoto, 1967). Due to their marine environment, they are exposed to relatively small seasonal temperature changes. Land plants, on the other hand, are exposed to marked seasonal temperature changes and these parallel the seasonal changes in freezing tolerance (Fig. 7.4).

2. Environmental Factors

a. TEMPERATURE

Since the environmental factors also change seasonally, one of the first questions asked was the relation of these factors to the seasonal changes in freezing tolerance. The first to be investigated was temperature. The importance of this factor was quickly established. A decrease in freezing tolerance occurs during winter when the plants are exposed to warm weather for 2 weeks (Göppert, 1830) or even a few hours (mosses, Irmscher, 1912; evergreen trees, Pisek, 1950) and an increase in freezing tolerance, when they are exposed to low temperatures (Haberlandt, 1875; Schaffnit, 1910; Irmscher, 1912; Chandler, 1913; Gassner and Grimme, 1913). It is now standard procedure to "harden off" plants by exposing them for a week or two to temperatures a few degrees above the freezing point. The "threshold temperature" above which hardening does not occur is usually 5°–10°C (Harvey, 1922). The precise threshold temperature for hardening or dehardening is difficult to determine, and varies with the species and probably the variety. Some hardening of cabbage may occur at 12°C, but none occurs at 18°C (Le Saint, 1966). Similar results have been obtained for *Mentha viridis* (Codaccioni, 1968) in the case of artificial tissue cultures. These survived −7°C for 10 hr after growth at 12°C, but not after growth at 20°C; hardening occurred only if the cultures were supplied with 2% glucose. In the case of some young trees, any temperature above 13°C led to a loss in freezing tolerance (Sakai, 1967). Nevertheless, poplars held at 15°C increased in tolerance from a killing temperature of −2° to −30°C over a 2-month period (Sakai and Yoshida, 1968a). The more hardy of eighteen citrus types were able to harden at higher temperatures (70°F D/50°F N or 60°F D/40°F N) than the less hardy types (Young, 1969b).

A temperature of 0° to 5°C will induce greater hardening than 5°–10°C. Once the maximum hardening possible at this temperature has been attained, a second stage increase may still occur at a temperature just below 0°C: −4°C in the case of barley at all stages and in sprouting wheat (Dantuma and Andrews, 1960), −2.5° to −5°C in the case of mulberry (Sakai, 1956b). Tumanov and his co-workers in the USSR have long emphasized the "second stage" hardening that may occur at temperatures well below the freezing point. He lists three periods in the preparation of plants for hibernation (Tumanov, 1969): (1) the onset of dormancy, (2) the first stage of hardening at about 0°C, and (3) the second stage of hardening during a gradual lowering of the temperature below 0°C. The freezing tolerance of cherry and apple twigs was markedly increased by prolonged (5–20 day) exposure to −5° and subsequently −10°C (Krasavtsev, 1969). Further hardening occurred during 1 day at −20° and −30°C. Birch and poplar twigs did not require this preliminary gradual freezing treatment. They survived −60°C if cooled gradually and −30° (in the case of birch) and −50°C (in the case of poplar) if cooled rapidly.

These "periods" or "stages" are not necessarily qualitatively different. At least in the case of cabbage, the differences are purely quantitative (H. Kohn, unpublished.) When hardened for successive 2-week-periods, hardening at 5°C (day and night) led to a drop in the killing temperature which then remained constant. Lowering the temperature to 5°C D/0°N lowered the killing point further, again reaching a plateau. A third hardening at 0°D/0°N further lowered the killing point to a third plateau. Finally, after exposure to −3°C, a fourth lowering occurred, to a killing temperature of −20°C. This demonstrates that one can obtain as many "stages" of hardening as desired, simply by means of a graded series of hardening treatments. It is, in fact, just as reasonable to consider low but nonhardening temperatures (e.g. 15°C day/0°C night) as the first stage of hardening, since a preliminary growth at such temperatures, though nonhardening in itself, leads to subsequent more rapid hardening at lower temperatures (Pomeroy *et al.*, 1975; Bervaes and Kylin, 1972). Nevertheless, the Soviet investigators prefer to consider the hardening of cabbage and other Crucifers as occurring in two stages (Shpota and Bochkareva, 1974).

Some results have indicated that alternating warm and cold temperatures are at least as effective as constant low temperatures (Harvey, 1918, 1930; Tumanov, 1931; Tysdal, 1933; Angelo *et al.*, 1939). Others have failed to obtain as hardy plants with alternating temperatures (Peltier and Kiesselbach, 1934; Suneson and Peltier, 1934; Day and Peace, 1937). Suneson and Peltier (1938) seem to have resolved these differences, for they obtained maximum resistance by exposure to alternating temperatures during November and December, followed by sustained low temperature for 3 weeks.

Even callus tissue can be hardened by exposure to low temperature—6 weeks at 4.5°C lowered the killing temperature from −6.6°C (unhardened) to −16.1° in the case of chrysanthemum (Bannier and Steponkus, 1976). A similar hardening lowered it from −10°C to −25°C in the case of cherry and apple (Ogolevets, 1976a). This ability to harden decreased with callus age.

The prevailing temperature under natural conditions also markedly affects the freezing tolerance. The above ground parts of the plant may have 11.5°C greater freezing tolerance than the below ground parts (Till, 1956). Even among the above ground parts, the most exposed and therefore colder parts of the plant are more freezing tolerant than parts covered by snow (Brierley and Landon, 1954). Among the roots, the deeper ones are killed by freezing temperatures that fail to injure the shallower ones (Smirnova, 1959). Roots may, however, achieve a high degree of hardiness if exposed to hardening temperatures—for instance in the case of container grown ornamentals, whose killing temperatures range from −5° to −23.3°C (Havis, 1976). Similarly, a colder winter has frequently been observed to result in greater freezing tolerance (Kohn, 1959), and midwinter thaws to result in a partial loss of freezing tolerance. Finally, the cortical cells on the south (warmer) side of trees in northern Japan are less tolerant of freezing than those on the north (cooler) side (Sakai, 1966). In general, freezing tolerance has been found to fluctuate throughout the winter, increasing as the temperature drops, decreasing as it rises (e.g., Proebsting, 1959).

b. LIGHT

Low temperature by itself is incapable of inducing hardening, at least in the case of winter annuals, biennials, and seedlings of perennials. The hardening of plants when alternating low and high temperatures are used has been shown to occur only if light is supplied during the high-temperature period (Dexter, 1933; Tysdal, 1933). Even continuous low temperatures in the absence of light are incapable of inducing hardiness in winter annuals (Tumanov, 1931; Dexter, 1933; Pfeiffer, 1933; Constantinescu, 1933; Andersson, 1944). Many have also found a reduction in hardiness as a result of darkening the plant (Lidforss, 1907; Weimer, 1929; Angelo *et al.,* 1939). No hardening of cabbage seedlings occurred in the dark at +4°C or in the light at +18°C (Le Saint, 1966), but normal hardening occurred when they were exposed to both low temperature and light. A threshold illumination of about 1000 fc was required for grains (see above) and also for young conifers (McGuire and Flint, 1962; Scheumann and Börtitz, 1965). Douglas fir seedlings failed to harden in the dark at 2.5°C even after several weeks (van den Driessche, 1969b), but they did harden at low light intensities (40 or 100 fc). An exceptional slight hardening in the dark occurred after growth in high-intensity light with a 16-hr photoperiod. Light enhances the rate of hardening of *Hedera helix,* but it is not essential for the hardening process (Steponkus and Lanphear, 1968a). The second stage of hardening, on exposure to temperatures below 0°C, may occur in the dark (Tumanov and Trunova, 1963; Kohn and Levitt, 1965). Even at this stage, however, light is necessary in the case of conifers (Scheumann and Börtitz, 1965). In the case of cabbage, 2000 lux was sufficient for hardening to a tolerance of −7°C, and some hardening was obtained even at 1250 lux (Le Saint, 1966). Much greater hardening of cabbage (survival of −20°C) was obtained when illuminated with 1000 fc, and cooled to a series of successively lower hardening temperatures (Kohn and Levitt, 1965).

The major need for light is apparently due to a need for photosynthesis, since if the leaves are chlorotic, the plants are unable to harden even when exposed to light, although when allowed to become green by spraying with ferrous sulfate, they harden normally (Rosa, 1921). Similarly, if exposed to CO_2-free air, hardening does not occur even in the light (Dexter, 1933). Finally, the freezing tolerance of *Lolium perenne* depended on the total light energy received (Lawrence *et al.,* 1973).

Plants with abundant organic reserves are the exceptions, since they harden markedly at 0°C even in the dark (Dexter, 1933). It is undoubtedly for this reason that seedlings of winter wheat grown on moist filter paper are able to harden at low temperatures in the dark (Andrews *et al.,* 1974b). This is not true of cabbage seedlings kept in the dark at the hardening temperature (Le Saint, 1966). Nevertheless, once hardened in the light, they maintained their hardiness in the dark for at least 2 weeks. Furthermore, if the part of the shoot that is capable of hardening is kept in the dark, and the remaining leaves are illuminated, the darkened part will harden due to translocation from the illuminated part. Since only the younger leaves are capable of hardening, this method works if the upper leaves are darkened and the lower ones illuminated, but if this is reversed, the darkened (older) leaves do not harden. Steponkus and Lanphear (1967a), in agreement with Le Saint, found that light results in the production of a promoter of hardiness in *Hedera helix,* which could be translocated to a darkened receptor. Labeling with ^{14}C indicated that the translocatable promoter was sucrose.

In spite of all this evidence that the light effect is due to photosynthesis, it has been concluded that light is also needed for the maintenance of ultrastructure in wheat plants (Tumanov *et al.,* 1976c). Light also affects plant development, which in its turn may be controlled by photoperiod (see Sec. 3C). It is not surprising, therefore, that a light effect via phytochrome may be important, at least in some plants. Dark-period interruptions with red radiation suppressed the short-day enhancement of cold acclimation in twigs of *Cornus* and *Weigela* (Williams *et al.,* 1972). When red light was followed by far red light, the suppression was relieved. Acclimation of *Pyrocantha,* however, was not controlled by radiation. The increased hardiness of 3-day-old winter rape seedlings on exposure to light was reversed by far-red light (Kacperska-Palacz *et al.,* 1975). Similarly, a short term illumination (of winter wheat) for 15 min. or even 5 min., or the illumination of only part of the leaves for a longer time increased the freezing tolerance of winter wheat plants kept in the dark (Tumanov *et al.,* 1976c).

Even in the case of roots, which are commonly much less resistant than the shoots, light may be important. Thus the roots of apple trees in winter may be killed by −15° to −18°C, although the shoots survive until frozen at −40°C (Tumanov and Khvalin, 1967). Improvement of the hardening conditions increased root hardiness very little. If, however, they were kept in the air and light for 3–5 months before hardening, they hardened to the same degree as the shoots.

Temperature and light are thus the two main environmental factors controlling the development of freezing tolerance in plants. Artificial hardening by control of these two factors is, in fact, capable of producing a degree of hardiness equal to that developed under natural conditions, e.g., in the cases of cabbage (Kohn and Levitt, 1965) and *Hedera helix* (Steponkus and Lanphear, 1967b). This is true even of extremely hardy plants such as dogwood, which is capable of surviving −100°F or lower after either natural or artificial hardening (Van Huystee *et al.,* 1967), but to achieve this extreme degree of hardening under artificial conditions, it was necessary also to control the photoperiod (see below).

c. WATER

The role of water in the hardening process is often overlooked. Yet it has long been known that the hardening of trees in the fall depends on "maturing" of the twigs and buds, involving a slow drying out. Conversely, dehardening of buds occurs due to increased moisture content (Bittenbender and Howell, 1975). This kind of hardening and dehardening may, however, affect only the ability of the buds to supercool, since this ability has been shown to increase during hardening (Timmis and Worrall, 1975; and Burke *et al.*, 1976). The role of moisture in such cases may, therefore, be unrelated to freezing tolerance. Nevertheless, it does play a role also in plants and plant parts incapable of appreciable supercooling. Apple roots were hardier under conditions of greatly lowered rainfall, leading to a lower level of root hydration (Wildung *et al.*, 1973). Under artificial conditions it has long been known that a deficiency of water may induce some freezing tolerance in herbaceous plants (Levitt, 1956). A water stress for 7 days lowered the plant water potential and increased the freezing tolerance (from −3°C in the control to −11°C in red osier dogwood (*Cornus stolonifera*) plants grown for 21 days under controlled conditions (Chen *et al.*, 1975, 1977). A further water stress treatment had little effect. Winter wheat plants have been shown to acquire maximum tolerance at a soil moisture of 30% although maximum productivity required 60% (Bondarenko *et al.*, 1973, 1975). The strict requirement of a water deficit for the hardening of cabbage seedlings has recently been strikingly demonstrated. When kept at full turgor, the seedlings were unable to harden at all, even under optimum conditions of temperature and light (Cox and Levitt, 1976). Lowering the xylem water potential of Douglas-fir seedlings from −6.5 to −12 bars by restricting water during the growth period increased their subsequent ability to cold-harden (Timmis and Tanaka, 1976).

d. MINERAL NUTRIENTS

Other factors must, of course, be optimum in order for the plant to harden maximally on exposure to hardening levels of temperature and light. Many reports (see Levitt, 1956) indicate that full hardening is not obtained in the presence of excess nitrogen even in the case of plants as hardy as *Picea abies* (Puempel *et al.*, 1975) or of insufficient potassium, phosphorus, or even calcium. Other attempts to discover an effect of mineral nutrients on freezing resistance have met with little or no success, except for the effect of N (Christersson, 1973, 1975a,b; Cook and Duff, 1976; Hart and Van der Molen, 1972; Beattie and Flint, 1973; Mathias *et al.*, 1973; Pellett, 1973).

Contradictory results have been obtained in the case of the other deficiencies (Levitt, 1956). Furthermore, cabbage seedlings that have grown excessively due to an excess of nitrogen may attain essentially the same degree of hardiness as normal plants, provided that they are exposed to optimum light and temperature regimes for hardening (see Kohn and Levitt, 1966). This will be discussed in connection with drought resistance.

3. Relationship to Growth and Development

a. DEVELOPMENT

It has frequently been reported that hardiness is directly related to the degree of "ripening" of the twigs (e.g., as judged by cork formation (Sakai, 1955a). On the other hand, at some stages of development, low temperature and adequate light intensity are unable to induce the hardening of potentially hardy plants. Newly formed buds of evergreens, for instance, fail to harden at low temperatures and with normal light, even though they may survive −30°C during the subsequent winter (Winkler, 1913). The buds of fruit trees also lose their ability to harden when they begin to develop into shoots during spring (West and Edlefsen, 1917, 1921; Roberts, 1922; Knowlton and Dorsey, 1927; Field, 1939; Geslin, 1939). Similarly, the changes in tolerance may not follow the temperature changes at certain times of the year (Ulmer, 1937). This is clearly shown by the effects of exposure to low and high temperatures for 1 day on the hardiness of the plant at different times of the year (Pisek, 1953).

Similarly, the optimal hardening temperatures for grape plants in the spring differ from those in the winter (Pogosyan and Sakai, 1972b). Even the green alga, *Chlorella,* when grown synchronously, showed highest survival of freezing after hardening, when at the L_2 stage (ripening phase) in their life cycle, during which nuclear division was insignificant (Hatano *et al.,* 1976b).

Even if kept constantly at the hardening temperature, the plant does not retain its maximum hardiness indefinitely. Sprouting winter wheat, for instance, reaches its maximum hardening at 1.5°C in the dark after about 5 weeks, and hardiness decreases rapidly between the seventh and eleventh weeks (Andrews *et al;* 1960). Although this drop in hardiness may sometimes be due to loss of reserves (Jung and Smith, 1960), it may also occur without an appreciable loss, e.g., the earlier spring development in the less tolerant species or varieties may be indicated by an earlier appearance of starch (Sergeev *et al;* 1959). The metabolic changes that occur in preparation for spring growth apparently lead to loss of freezing tolerance even though the plants are exposed to optimum hardening temperature and light. The mere cessation of growth and development in the fall may itself confer some hardiness without the aid of hardening temperatures (e.g., at 20°–30°C in mediterranean evergreens (Larcher, 1954) and in mulberry (Sakai, 1955a), or it may permit earlier hardening, as in the case of the hardy raspberry cultivar "Carnival" (Jennings *et al.,* 1972).

In many cases the plant actually enters a nongrowing "rest period," and freezing tolerance has been frequently related to the depth or length of this period (Levitt, 1956). In European black currant and white birch, this is due to a greater hardening in response to low temperature when in the state of deep dormancy than when growing (Tumanov *et al.*, 1973). Similarly, the reduction in freezing tolerance of plants brought indoors during winter occurs only if they are no longer in their rest period (Lidforss, 1907; Meyer, 1932; Kessler, 1935). This has been confirmed in the case of *Acer negundo* and *Viburnum plicatum*. When in the naturally hardened state, their dormant condition retarded the loss of tolerance on exposure to 70°F (Irving and Lanphear, 1967b). Similarly, among ten provenances of Douglas fir, for each additional week by which bud set preceded frost, the proportion of freeze-damaged seedlings decreased by 25% (Campbell and Sorensen, 1973).

This correlation between rest period and freezing tolerance does not occur in all plants (Pojarkova, 1924). Some plants may be dormant although possessing no freezing tolerance (Clements, 1938), and others may survive the winter without a rest period (Walter, 1949). In the case of prevernal, remoral ephemerals, the dormant stages were the least resistant to cold (Goryshina and Kovaleva, 1967). Even some hardy woody plants (*Acer negundo, Viburnum plicatum tomentosum*) develop tolerance independently of bud dormancy (Irving and Lanphear, 1967b). In the case of very hardy plants, such as dogwood, the high degree of freezing tolerance may be maintained in winter long after its rest period is over (van Huystee *et al.*, 1967). In many cases, the importance of the rest period is believed due to prevention of growth and the accompanying loss of freezing tolerance during winter warm spells (Brierley and Landon, 1946). In opposition to this concept, raspberry canes may deharden at high temperatures more rapidly when in the resting than when in the nonresting state (Weiser, 1970).

In the case of *Forsythia,* however, once dormancy had been broken, the temperature required for significant dehardening decreased (Hamilton, 1973). Furthermore, since freezing tolerance depends on more than one factor, it is not surprising that it may not always be related to rest period. Early leaf loss of cherry (*Prunus cerasus*), for instance, delayed acclimation in the fall (Howell and Stockhouse, 1973), though presumably indicating an early rest period. This may have been due to loss of reserves during leaf fall, which would have been translocated to the shoots or buds, if the leaves had remained on the trees for the normal period. Similarly, sprinkling of peach trees lowered the wood temperature by as much as 6.5°C and delayed bloom by 15 days without significantly altering the hardiness (Bauer *et al.*, 1976). Since the sprinkling was performed at the end of the rest period, no true dormancy change was induced.

b. GROWTH

These and other observations have led many investigators to point to growth per se, rather than stage of development as the factor that prevents hardening. In the case of winter annuals, hardiness is inversely related to rate of growth in the fall (Buhlert, 1906; Schaffnit, 1910; Hedlund, 1917; Klages, 1926; Worzella, 1932; Mark, 1936; Kolomycev, 1936; Vassiliev, 1939). In the case of British and North African varieties of tall fescue, there was a clear, inverse relationship between the ability of the varieties to survive low temperature and their ability to grow rapidly during winter (Robson and Jewiss, 1968). It is generally found, in fact, that if plants are growing rapidly, they cannot be frost hardened (Rivera and Corneli, 1931; Dexter *et al.*, 1932), whereas treatments that retard growth increase hardening (Chandler, 1913; Harvey, 1918; Rosa, 1921; Collison and Harlan, 1934; Shmelev, 1935; Kessler and Ruhland, 1938), although exeptions may occur (Kuksa, 1939). An increase in tolerance can frequently be obtained by withholding water from the plant to a sufficient degree to induce some wilting and stunting of growth. Conversely, as mentioned above, a marked decrease in tolerance often results from heavy nitrogen fertilization leading to a rapid, succulent growth, and, as in the case of the normal spring growth, the plants lose their ability to harden normally on exposure to hardening temperatures in the light. Similarly, growth-promoting substances sometimes decrease freezing tolerance, whereas growth-inhibiting substances sometimes increase freezing tolerance (see Chapter 8); but there are exceptions (Tumanov and Trunova, 1958). For instance, the different growth rates of *Picea abies*, resulting from different fertilization, had no effect on the development of frost hardiness (Christersson, 1975b). It has been pointed out, however, that a relation can be expected only if true ("intrinsic") growth is measured, and not merely changes in dry weight which may include metabolic changes independent of growth (MacDowall, 1974). It has also been suggested that the real relation may sometimes be to translocation, which in its turn may control growth (Hoshino *et al.*, 1972).

c. PHOTOPERIOD

The relationship of freezing tolerance to growth and development is clearly demonstrated by controlling the photoperiod. Hardening is improved by short days, both in the case of woody plants (Moschkov, 1935; Bogdanov, 1935) and in herbaceous plants (Dexter, 1933; Timofejeva, 1935; Saprygina, 1935; Saltykovskij and Saprygina, 1935; Sestakov, 1936; Rudorf, 1938; Suneson and Peltier, 1938; Tysdal, 1933; Frischenschlager, 1937; Smith, 1942; Ahring and Irving, 1969). In fact, the annual curve for freezing tolerance is as clearly correlated with the change in photoperiod as with the temperature (Fig. 7.5), and the hardening of peach and nectarine trees (to −12°C) may occur from November to January even if the temperature remains continuously above 15°C (Buchanan *et al.*, 1974). Conversely, photoperiod had no effect on the buds of peach trees at hardening tempera-

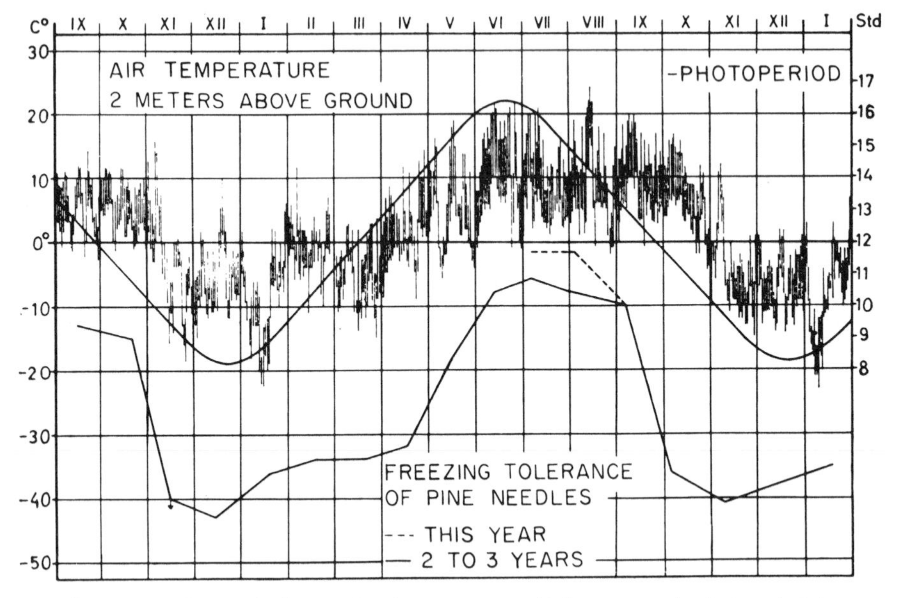

Figure 7.5. Seasonal changes in air temperature (daily extremes), photoperiod (upper curve), and freezing tolerance (lower curve) of pine needles from September 1965 to January 1967. (From Schwarz, 1968.)

tures (Ormrod and Layne, 1974). The effects of red and far-red light on the initiation of cold acclimation in *Cornus stolonifera* indicate that the photoperiodic effect is phytochrome-mediated (McKenzie *et al.*, 1974).

In at least some cases, the normal autumn hardening can be prevented if the plants have previously been induced to continue their growth by maintaining them in a long photoperiod. This can be seen in some modern cities where the shoots of introduced trees and shrubs close to street lights may be winter-killed, whereas the other parts of the plants or others like them are uninjured due to their safe distance from weak lights (Kramer, 1937). Conversely, after too long a period of growth, leading to a deficiency of reserve substances, a prolonged short day decreased the freezing tolerance of 1-year-old black currant and birch seedlings (Tumanov *et al.*, 1972).

In the case of arctic plants, Biebl (1967c) concluded that the fall increase in freezing tolerance is dependent primarily on the shortening of the day length and only secondarily on the decrease in temperature. When different species or varieties are compared, a direct correlation frequently exists between length of the critical photoperiod and freezing tolerance (e.g., in wheat varieties; Rimpau, 1958), and an inverse correlation between the growth effect of a long photoperiod and hardiness (Schmalz, 1957). Adequate hardening may sometimes occur, however, in spite of long photoperiods (Kneen and Blish, 1941). Cabbage seedlings harden as well when grown before and during hardening at any photoperiod from 8 to 24 hr, provided that optimum temperature and light regimes are used for hardening (Kohn and Levitt, 1965). This is also true of potato species that are capable of hardening (Chen and Li, 1976). Photoperiod alone may not affect tolerance of other plants, for instance centipede grass (Johnston and Dickens, 1976), though a short photoperiod in combination with low temperature gives the maximum tolerance.

In the case of Douglas fir seedlings, day length, temperature, and light were all important for the development of freezing tolerance, but only temperature affected the loss of tolerance (van den Driessche 1969b). Long days retarded the development of tolerance in autumn, but not later. Both photoperiod and low temperature appear to affect the hardiness of pine and spruce in the same way. Short days (6–12 hr) enhanced their hardiness though still shorter days were less effective (Aronsson, 1975). Low temperature failed to induce hardening under long days, but increased it under short days. Nevertheless, both the short days and the low temperature appeared to increase hardiness in the same manner via an increase in carbohydrate (particularly sucrose) content (Aronsson *et al.*, 1976). In *Cornus stolonifera,* however, NMR absorption spectra showed that the short-day-induced cold acclimation involves a marked decrease in tissue hydration (McKenzie *et al.*, 1974a). In agreement with this observation, water stress and short days appeared to induce the same physiological change and freezing tolerance in this red osier dogwood (Chen and Li, 1977).

In summary, photoperiod definitely controls the first stage of hardening in many woody plants in which it controls growth and development, but has no effect on the hardening of herbaceous biennials and some other herbaceous plants, whose growth and development are unaffected by photoperiod. When it is effective, photoperiod appears to control hardening by controlling the growth, reserves, and hydration of the tissues. It has also been suggested, however, that phytochrome (and therefore photoperiod) regulates the synthesis of linolenic acid (Tremolières *et al.*, 1973). Since the synthesis of unsaturated fatty acids is sometimes related to hardening (see Chap. 8) this may indicate another role of photoperiod in freezing tolerance.

d. VERNALIZATION

Vernalization of winter annuals (by storage of the imbibed seeds at 0°–5°C for 30–60 days) also stimulates growth and decreases hardiness (Schmalz, 1958), though exceptions occur. Since the low temperature used to vernalize plants does induce some hardening, the vernalized may be more hardy than the nonvernalized plants if these have not been subjected to hardening temperatures (Saltykovskij and Saprygina, 1935; Timofejeva,

1935; Vetuhova, 1936). If both are exposed to hardening temperatures, the nonvernalized always become more tolerant of freezing than the vernalized plants (Vetuhova, 1936, 1938, 1939). In many cases, those varieties that require the longest cold treatment for vernalization are the most hardy; though there are many exceptions (Hayes and Aamodt, 1927; Martin, 1932; Quisenberry and Bayles, 1939; Saltykovskij and Saprygina, 1935; Straib, 1946). Furthermore, the longer the vernalization period, the greater the effect of an increased photoperiod in lowering the freezing tolerance (Rimpau, 1958).

e. INTERRELATIONS BETWEEN GROWTH, DEVELOPMENT, AND FREEZING TOLERANCE

The following evidence, therefore, indicates that freezing tolerance is inversely related to growth and development.

1. The rapid spring growth is essentially unable to harden.
2. Preparation for spring growth is accompanied by a loss of freezing tolerance, even at hardening temperatures.
3. The fall cessation of growth is accompanied by an increase in freezing tolerance.
4. The relative growth rate of winter annuals in the fall is inversely related to their relative hardiness.
5. Artificial stimulation of growth by excess nitrogen fertilization, by long days, by vernalization, or by growth regulators (see below) is accompanied by a loss of tolerance or of ability to harden. Artificial retardation of growth by wilting or by growth inhibitors is accompanied by an increase in freezing tolerance.

This evidence has led to the conclusion that freezing tolerance (T_F) is inversely related to growth:

$$T_F \propto 1/\text{growth}$$

A more careful examination of the results, however, reveals that even in the above cases, there are many exceptions. Furthermore, a distinction must be made between growth *per se* and developmental stage. At the same time as growth occurs and tolerance decreases due to photoperiodic inductions, the length of the embryonic spike increases (Sestakov and Smirnova, 1936; Sestakov and Sergeev, 1937), and the plant passes through the "second stage of development." In nearly all the above cases, a developmental change accompanied the change in freezing tolerance, e.g., the rapid spring growth includes reproductive development, the varieties of winter annuals that grow more rapidly in the fall also are less dependent on vernalization for their development. Conversely, the smallest effects on freezing tolerance are produced by those treatments that affect growth but have little effect on development (wilting, nitrogen fertilization, certain growth regulators, etc.)

It, therefore, appears likely that the true inverse relation is between freezing tolerance and developmental stage, rather than growth, so that

$$T_F \propto 1/\text{development}$$

Evidence in favor of this interpretation has been produced (Cox and Levitt, 1969). When the growth rate of individual cabbage leaves was measured, the freezing tolerance achieved, when exposed to hardening temperature and light, was *directly* proportional to growth rate, whether the latter was measured at the moment of transfer to hardening conditions or during the hardening period (Fig. 7.6). Similarly, although tubers of Jerusalem artichoke are able to survive freezing down to −5°C, when in the dormant but not when in the active state, actively growing callus from these tubers are even hardier, some surviving even −15 to −20°C (Sugawara and Sakai, 1976, 1978). Cabbage is a biennial and does not undergo reproductive development until transferred back to warm temperatures after exposure to a low (vernalizing) temperature.

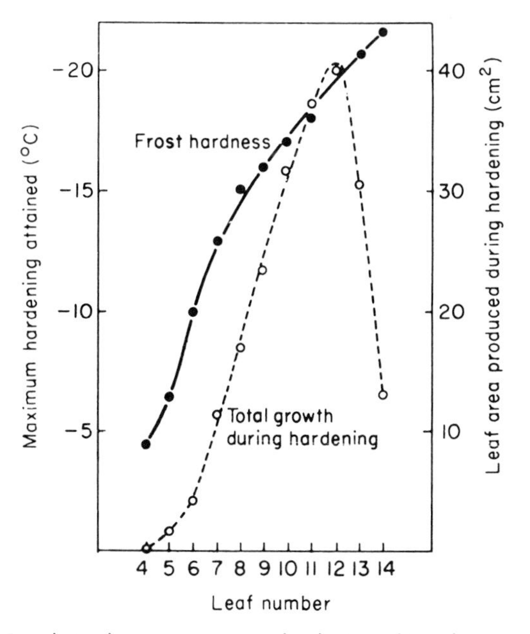

Figure 7.6. Correlation between maximum hardening achieved (●——●) and total growth (---) produced during hardening period in leaves 4–12 of cabbage plants. (From Cox and Levitt, 1969.)

f. AGE

Many attempts have been made to relate the age of a plant to its freezing tolerance (see Levitt, 1956). In two pine species, for instance, 36-day-old seedlings were injured more severely by a freezing temperature than were 22-day-old seedlings (Cochran and Berntsen, 1973). A similar result was obtained with rootlets of swede turnips (Jonassen, 1973). However, black locust seedlings were not able to attain maximum hardiness during the first two months of growth (Brown and Bixby, 1976). It is difficult to separate age from developmental stage, and the age factor may simply be considered as another method of determining the developmental state. In the case of *Quercus ilex,* for instance, the greatest increase in hardening occurs within the first 5 years (Larcher, 1969), but the full capacity for freezing tolerance is not attained until the tree enters its reproductive phase. The roots, on the other hand, do not change their tolerance pattern with age.

As mentioned above, however, there is evidence that development and freezing tolerance are not indissolubly related. Siminovitch *et al.* (1967b) have pointed to "seasonal rhythms" in the tree as being more important than

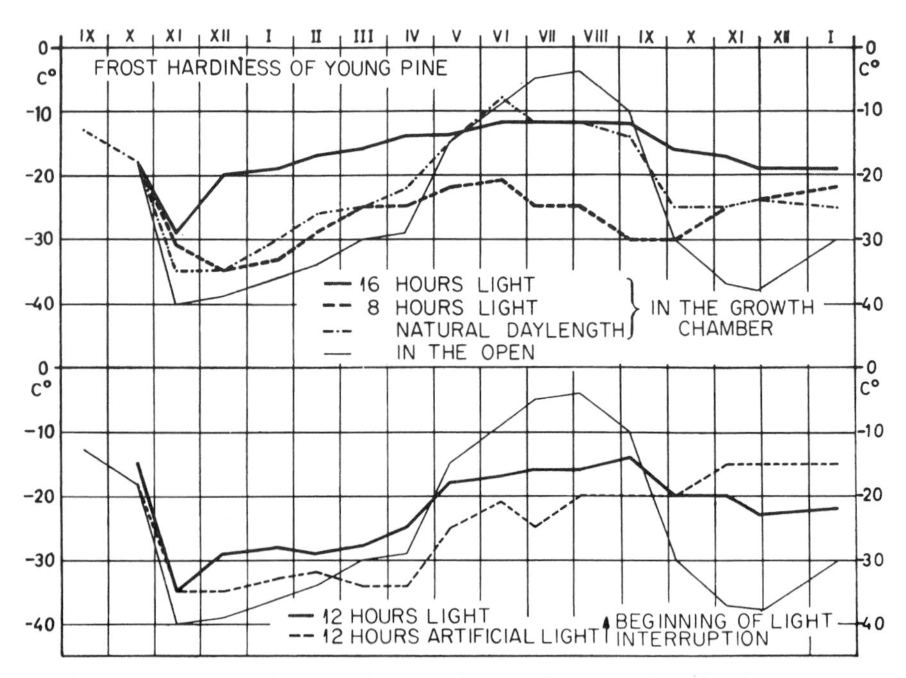

Figure 7.7. Seasonal changes in freezing tolerance of *Pinus cembra* when kept at a constant temperature (15°C) in a growth chamber and when left under natural conditions (thin, continuous line). (From Schwarz, 1968.)

temperature *per se*. Schwarz (1968) believes that it is these rhythms, rather than the developmental stage, that control freezing tolerance. Thus, when he maintained *Pinus cembra* at a constant temperature (15°C) in a growth chamber throughout the year, tolerance varied in exactly the same way as in others kept in the open (Fig. 7.7). In the second year, however, the annual amplitude of freezing tolerance decreased at the constant temperature. Similarly, when the photoperiod was kept constant, the annual fluctuations in freezing tolerance were much smaller (Fig. 7.7). Schwarz, therefore, concluded that the annual change in freezing tolerance depends on three factors: temperature, photoperiod, and the internal rhythm of the plant.

It must be realized, however, that growth and development, as well as the above factors that may control growth and development, are all indirect factors in freezing tolerance. They can affect it only by way of their effects on the cell and molecular properties of the plant. Therefore, we cannot expect an invariable relation of freezing tolerance to either growth or development. Nor can we expect to explain the relations when they do occur until the associated cell and molecular changes are fully understood.

D. THE NATURE OF FREEZING TOLERANCE

As indicated above, the only freezing tolerance developed by the plant is tolerance of the secondary water stress induced by extracellular freezing. From the basic concepts of stress tolerance (Chapter 2) this can conceivably be of two kinds:

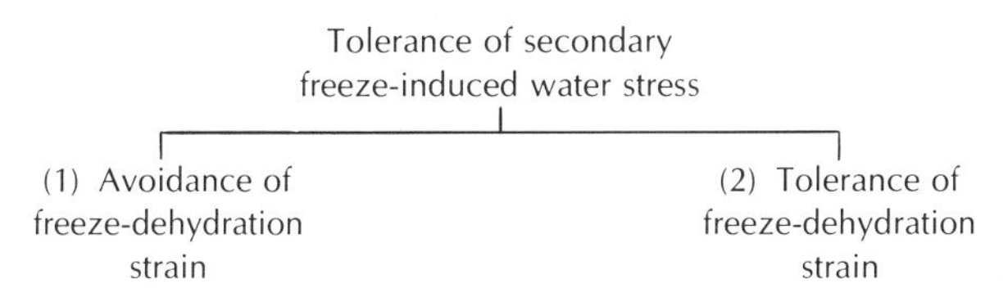

It is of fundamental importance to examine the evidence for these two types of freezing tolerance.

1. Avoidance of Dehydration Strain

As shown above, the plant soon comes to thermodynamic equilibrium with the secondary freeze-induced water stress, and its water potential drops to that of its environment. It can, however, decrease the amount of ice formed in its tissues, and therefore, the dehydration strain. Stress tolerance of this kind is proportional to the fraction of water kept in the unfrozen state at any one freezing temperature, and plants possessing such stress tolerance would require a lower freezing temperature to produce an injurious dehydration strain. Assuming that the plant sap acts as an ideal solution, this kind of tolerance would be proportional to the freezing point lowering of the plant sap. Many plants do, indeed, show a direct relationship between freezing point lowering and freezing tolerance (Chapter 8), indicating that at least part of their tolerance is due to an increased avoidance of dehydration strain.

Since the cell sap may not act as an ideal solution, more direct evidence is necessary. The amount of water unfrozen at freezing temperatures has been measured calorimetrically by several investigators. In the case of two molluscs differing in freezing tolerance, the difference was completely accounted for by avoidance of dehydration strain, both being killed when 64% of their water was frozen (Table 7.5). Human red blood cells, with a smaller freezing tolerance, were also killed at the same percentage dehydration, though muscle cells required a greater degree of freezing dehydration (Table 7.5). In the case of two wheat varieties (Johansson and Krull, 1970), each increase in freezing tolerance during hardening was accompanied by a calorimetrically measurable decrease in percentage of freeze dehydration at −9.0°C. These calorimetrically measured values all agreed perfectly with the cryoscopically calculated values (from the freezing point lowerings) throughout the hardening period (Table 7.6). The hardening achieved was, unfortunately, too small to reveal the difference in hardiness between the two varieties. The hardier variety, under field conditions, is capable of developing two to three times the maximum freezing tolerance obtained artificially by Johansson and Krull. It is interesting to note that the observed degrees of dehydration were identical in the two varieties (82.7%), when they possessed freezing tolerance essentially equal to the temperature at

TABLE 7.5

Percentage Dehydration in a Number of Animal Cells at Their Freeze-Killing Temperatures

Species or tissue	Freeze-killing temperature (°C)	Percentage water frozen at killing temperature	Reference
Muscle	−2	78	Moran, 1929
Molluscs			(see Salt, 1955)
Venus mercenaria	−6	64	Williams and Meryman, 1970
Mytilus edulis	−10	64	Williams and Meryman, 1970
Human red cells	−3	64	Meryman, 1970

TABLE 7.6

Percentage Water Frozen in a Hardy (Sammetsvete) and a Tender (Capelle Desprez) Wheat Variety at Different Degrees of Hardening[a]

			Percentage of water frozen at −9.0°C			
			Observed calorimetrically			Calculated cryoscopically
Days hardened	Freezing point depression (°C)	Frost-killing temperature (°C)	Intact plants	Expressed sap	Residue	Intact plants
Sammetsvete						
0	0.97	° 6.5	92.0	93.6	85.5	91.8
10	1.38	− 9.0	82.7	84.4	79.1	83.0
20	1.64	−11.5	79.2	81.2	75.6	78.2
Capelle Desprez						
0	0.78	above − 6.0	93.4	96.2	84.3	93.2
10	1.15	− 7.5	87.4	89.7	81.9	87.4
20	1.42	−10.0	82.7	84.0	80.4	82.7

[a] From Johansson and Krull, 1970.

which the measurements of ice formation were made (−9.0°C and −10.0°C respectively).

Following up these results, Johansson (1970) investigated eleven varieties of three species calorimetrically: wheat (7), rye (2), and turnip rape (2). In all cases, increased freezing tolerance was accompanied by an ability to decrease the freeze-dehydration at any one temperature; the percentage of water frozen at the lower killing temperature of the more tolerant (hardened) plants was about the same as the percentage at the higher killing temperature of the less tolerant (unhardened) plants (Table 7.7). The 11 different grain varieties, therefore, all possessed the same tolerance of freeze-dehydration, and the varietal differences in freezing tolerance (from −5.5° to −13°C) were all due to differences in avoidance of freeze-dehydration.

It should be possible, from these measurements, to determine whether or not there is a specific freeze-dehydration at which all cells are killed. The above values for eleven varieties of three plant species all fall between 84–89%. These values are considerably higher than for the four kinds of animal cells given above (Table 7.5). This difference is undoubtedly at least partly due to the large vacuoles in plant cells. The percentage dehydration of the protoplasm must be much less, since it consists largely of hydrophilic colloids as opposed to the essentially pure solution of the vacuole. Johansson and Krull (1970) expressed the sap (mainly vacuolar) and measured the water frozen in the sap and residue separately. Since the residue consists essentially of vacuole-free material, it should give an average value for cell wall and protoplasm. The percentage dehydration of the residue at the frost-killing temperature was 79 and 80.5% for the two wheat varieties, respectively. This value is close to the one for muscle, but considerably larger than for the other animal cells.

2. Tolerance of Dehydration Strain

In the above investigations by Johansson, the increase in stress tolerance appears to be fully accounted for by an increased avoidance of freeze-dehydration. In some cases, in fact, there seems to be an overshoot, and the more stress-tolerant plants cannot survive as high a freeze-dehydration at their lower killing temperatures as survived by the less stress-tolerant plants at their higher killing temperatures (Table 7.7). This relationship, however, may not necessarily hold true for all plants. Johansson and Krull (1970) showed that calculations for cryoscopic values of the amount of water frozen at the frost-killing point give values identical with the calorimetrically measured ones (Table 7.6). Since many cryoscopic values are available in the literature, it is, therefore possible to find out how widespread this type of tolerance is.

Table 7.7

Relationship of Freezing Tolerance to Freezing Dehydration[a]

Plant	Variety	Days hardened	Frost-killing temperature	Percentage of water frozen at frost-killing temperature
Wheat	Dalavårvete	0	− 5.5	87.0
		13	− 9.0	87.2
	Norre	0	− 6.5	86.7
		12	−10.5	87.3
		25	−12.5	86.6
	Odin	0	− 7.5	85.5
		11	−10.5	87.2
		18	−12.0	87.6
		36	−13.5	84.4
	Skandia IIIB	0	− 7.5	85.9
		11	− 9.0	87.9
		21	−10.0	87.3
	Sammetsvete	0	− 8.5	85.6
		11	−12.0	87.0
		20	−12.0	87.1
		37	−13.5	84.0
	Starke	0	− 9.0	85.5
		12	−11.0	87.1
		19	−12.0	88.0
	Virtus	14	−11.5	86.3
Rye	Kungs II	8	− 7.5	86.3
		12	−12.5	86.9
		14	−13.0	87.3
Turnip rape	Storrybs	0	− 5.5	(88.8)
		13	− 9.5	87.4
	Rapids II	0	− 6.5	87.4
		13	− 8.0	(89.0)
		32	−10.5	86.7
		38	−11.0	84.5
Wheat	Capelle Desprez	0	− 7.5	87.8
		23	−11.0	86.3
	Eka Nowa	0	− 8.0	87.9
		14	−13.0	88.0
		35	−13.5	85.9
	Sammetsvete	0	− 8.5	87.6
		13	−11.5	87.8
	Starke	0	− 9.0	86.5
		20	−12.0	86.2
	Mironowskaja 808	0	−10.0	86.7
		14	−15.0	87.0
		34	−16.0	86.3
Rye	Kungs II	0	−10.5	85.8
		20	−12.5	86.9
	Ensi	0	− 9.5	88.0
		22	−16.0	87.5

[a] From Johansson, 1970.

Calculations from values obtained by Tumanov and Borodin (1930; see Levitt, 1956) for eight hardy wheat varieties not used by Johansson, yield considerably higher freeze-dehydrations (90–92%) at the frost-killing temperatures, than those obtained by Johansson (84–89%). In fact, when a large number of wheat varieties from all over the world are compared, there is relatively little relation between freezing point lowering and varietal hardiness (W. Schmuetz, personal communication). In the case of some of the hardiest plants, this lack of a correlation is even more pronounced. Needles of *Pinus cembra* do show a steady lowering of their freezing point during hardening, but the change is so small that the calculated percentage dehydration at the frost-killing point rises steadily from a low of 84.5%, in the least tolerant state, to a high of 95.7%, in the most tolerant state (Table 7.8). An objection may be made that these calculated values for pine are not comparable to the measured values in wheat. A comparison of the two, however, reveals that the calculated percentage freeze-dehydration at the frost-killing temperatures of pine that fell within the range of Johansson's wheats (−10° to −15°C) are 84.5 and 89%, in perfect agreement with Johansson's measured values for wheat (84–89%). In the case of wheat itself, the calorimetrically observed and the cryoscopically calculated values agreed perfectly throughout the range tested, including as high as 93.4% (Table 7.6). There is no escaping the conclusion therefore, that in the case of pine needles, the increase in freezing tolerance during hardening is mainly due to an increase in tolerance of the freeze-induced dehydration strain.

That hardening may involve both factors, even in the case of herbaceous plants, was shown by direct calorimetric measurements of ice formation in cabbage plants (Levitt, 1939). Although the hardening process involved a

TABLE 7.8

Seasonal Changes in Current Needles of *Pinus cembra*[a]

Date	Percentage water	Freezing point lowering	Frost-killing temperature	Calculated percentage water frozen at frost-killing temperature	Calculated gram water unfrozen/gram dry matter
9/23/42	59	1.55	−10	84.5	0.22
10/8/42	56	1.64	−15	89.0	0.14
11/27/42	57	1.78	−35	94.9	0.06
1/11/43	57	1.73	−40	95.7	0.06
3/28/43	55	1.69	−30	94.4	0.07
5/3/43	50	1.96	−13	85.0	0.15

[a] Adapted from Pisek, 1950.

TABLE 7.9

Nature of Increase in Freezing Tolerance of Cabbage during Hardening[a]

Hardening	Freezing point lowering (°C)	Frost-killing temperature	Percentage water frozen at frost killing temperature: Measured (calorimetrically)	Percentage water frozen at frost killing temperature: Calculated (cryoscopically)
Unhardened	0.81	−2.1	60	61.5
Hardened	1.257	−5.6	75	77.5

[a] From Levitt, 1939.

large increase in avoidance of dehydration strain (a 50% increase in freezing point lowering) there was also a very definite increase in tolerance of the freeze-induced dehydration strain, from 60% in the unhardened to 75% in the hardened (Table 7.9). The cryoscopically calculated values agreed almost perfectly with the calorimetrically measured values (Table 7.9). These measurements also agree almost perfectly with Johansson's; he found 62% of the water frozen in unhardened wheat leaves at −2.65°C, compared to the above 60% in the cabbage at −2.1°C. Results from a third laboratory are also in agreement with these two values. Williams and Meryman (1970) concluded that a removal of 65% of the osmotically active water is the limit tolerated by grana isolated from unhardened spinach chloroplasts. It is, of course, doubtful whether chloroplast grana can be expected to show the same tolerance of dehydration as in the whole cell, or even the whole protoplasm. Furthermore, their tolerance when isolated may differ from their normal tolerance when protected within the protoplasm. Nevertheless, the agreement encourages the acceptance of the above results as valid.

More recent measurements of ice formation in frozen plant tissue have used NMR spectroscopy instead of calorimetry. By means of this method, the crowns of acclimated winter cereals (wheat and rye) were shown to tolerate the freezing of a larger fraction of their water than those of nonacclimated plants or of acclimated, less tolerant spring wheat (Gusta *et al.*, 1975). This disagreement with Johansson's results may be due to the difference in varieties used, or to the use of crowns rather than leaves. Similar measurments on dogwood (*Cornus stolonifera*) stems also proved that their ability to survive low temperatures depends on an ability to tolerate diminished quantities of liquid water (Chen, *et al.*, 1977). The amount of liquid water in the leaf tissues of the susceptible *Solanum tuberosum* at its killing temperature was 41–47%, whereas it was 43% in Alaska Frostless, and only 22–26% in the more hardy species (Chen *et al.*, 1976). The excised leaf tissues of all species froze as ideal solutions. There was no correlation between the freeze-killing temperature and tissue water content or the freezing point depression of the cell sap. The major difference between the tender and hardy tissues was the ability of the hardy tissue to tolerate more frozen water at its freeze-killing temperature.

3. Interactions between Avoidance and Tolerance of Dehydration Strain

Since all these results from different laboratories agree with each other quantitatively whenever comparisons can be made, it must be concluded that all are equally trustworthy. An increase in tolerance of the freezing stress is, therefore, due to an increase in (1) avoidance of the dehydration strain, or (2) tolerance of the dehydration strain, or both. It is of fundamental importance to understand the quantitative interrelationships between these two components of freezing tolerance.

On the basis of the above conclusions, if the tolerance of the dehydration

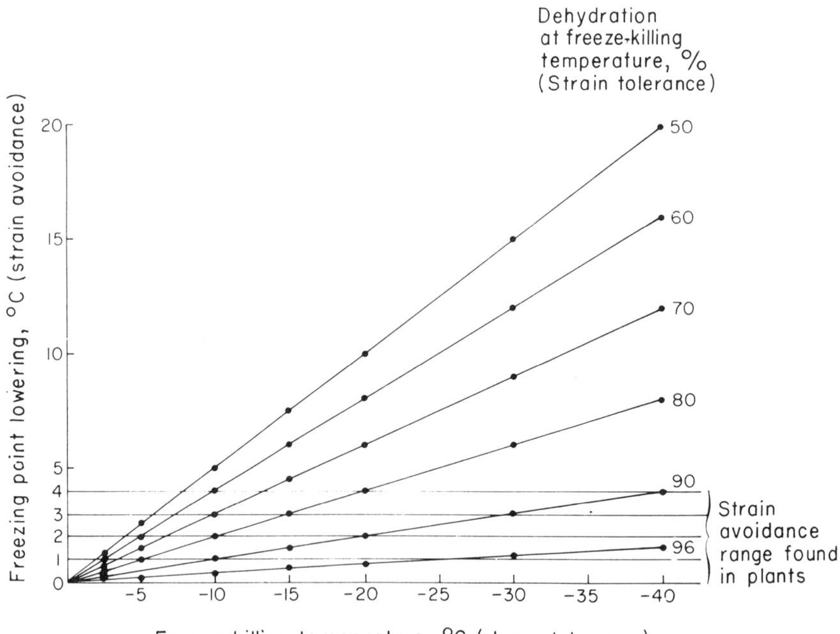

Figure 7.8. Dependence of freezing stress tolerance (abscissa) on dehydration strain avoidance (freezing-point lowering of plant sap—ordinate) and dehydration strain tolerance (percentage dehydration at frost-killing temperature).

strain is known, the stress tolerance can be calculated from the freezing point lowering. The values for a series of dehydration strains are given in Fig. 7.8 for a range of stress tolerance from 0° to −40°C. Since the lowest freezing points recorded in the literature for the hardiest plants are about −4.0°C, the possibilities open to the plant range between this value and 0°C. There are several obvious conclusions (Table 7.10).

1. Plants with freezing point lowerings of 0.5°C (osmotic potential of −6 bars) will be killed by the "first touch of frost" (i.e., not lower than −1.6°C) unless their tolerance of the freeze-induced dehydration strain is higher than 70%. This explains the extreme tenderness of all freshwater aquatic plants, succulents, and many other plants.

2. Plants with freezing point lowerings of 2°C will not be killed by any temperature above −5.5°C unless their tolerance of freeze-induced dehydration is less than 70%. This explains the ability of even the completely unhardened trees to survive slight freezes in summer.

3. Plants with a tolerance of freeze-induced dehydration in the neighborhood of 80%, can increase their stress tolerance from as little as −2.5°C to as much as −20°C, by simply increasing their freezing point lowering from 0.5° to 4.0°C. This simple method of hardening has apparently been adopted by several varieties of grains and other winter annuals.

4. Plants with a tolerance of freeze-induced dehydration that does not exceed 50%, cannot harden markedly by increasing their freezing point lowering. This is why plants such as potato tubers, which undergo a marked increase in freezing point lowering on exposure to hardening temperatures, nevertheless, fail to harden appreciably.

5. In order to tolerate the severe freezing stresses of northern climates plants must develop at least a 90% tolerance of freeze-induced dehydration.

TABLE 7.10

Calculated Stress Tolerance of Plants with Specific Freezing-Point Lowerings and Specific Tolerances of Freeze-Induced Dehydration Strain

Freezing-point lowering (°C)	Freeze-killing temperatures (°C) at dehydration tolerances of 50–96%					
	50	60	70	80	90	96
4	−8	−10	−13.5	−20	−40	
3	−6	− 7.5	−10	−15	−30	−70 (apprx.)
2	−3.5	− 4	− 5.5	− 8.5	−16.5	−42.5
1	−2	− 2.5	− 3.5	− 5	−10	−25
0.5	−1	− 1.3	− 1.6	− 2.5	− 5	−13

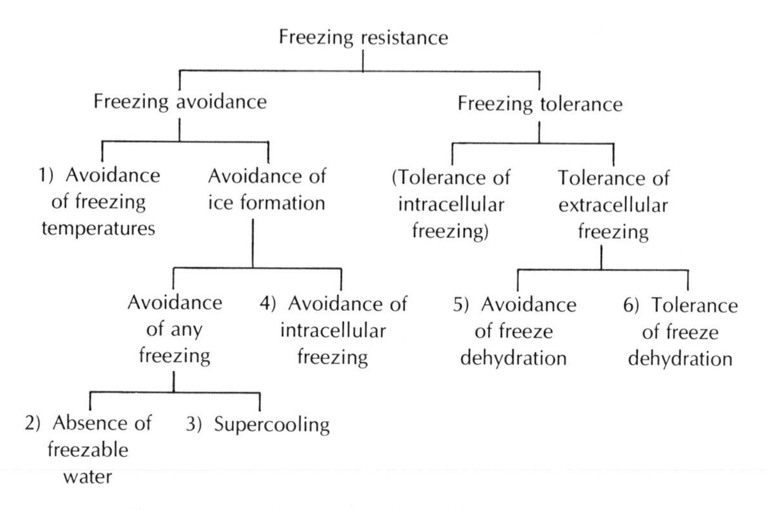

Diagram 7.2 The six mechanisms of freezing resistance.

In all the above cases, tolerance of freeze-induced dehydration is the primary factor in freezing tolerance. When the tolerance of freeze-induced dehydration is low, no amount of avoidance of freeze-induced dehydration can help the plant. When the tolerance of freeze-induced dehydration is high, the plant can harden markedly by a relatively small increase in avoidance of freeze-induced dehydration, which is, therefore, a secondary factor in freezing tolerance. Nevertheless, both factors are required for marked freezing tolerance.

It is now apparent that of the five kinds of freezing avoidance presented in Diag. 7.1, the second—lowering of the freezing point—is more important as a mechanism of partial than of complete avoidance of freeze dehydration, and therefore should be considered a mechanism of extracellular freezing tolerance. If we eliminate it from the avoidance mechanisms, six different kinds of freezing resistance remain (Diag. 7.2).

8. Factors Related to Freezing Tolerance

A. MORPHOLOGY, ANATOMY, ORGANELLES

In view of the above inverse relationships between freezing tolerance and development, the morphological characteristics associated with dormancy or with retarded development might reasonably be sought in hardy plants. Conversely, the characteristics of advanced development might be expected in tender plants. These relationships have, indeed, often been reported. Thus, when varieties of a species differing in hardiness are compared, it is frequently found that hardiness varies inversely with the size characteristics of the plant, e.g., height, leaf length, internode length, and especially cell size (Levitt, 1956). Hardy varieties of flax, for instance, form a rosette during the winter, in contrast to the upright growth of the sensitive lines (Omran *et al.*, 1968) and those varieties of *Lolium perenne* with the smaller leaves and more moderate autumn growth were the more cold resistant (Mytsyk, 1972). This relationship can be expected to hold true only if the measured characters are the result of development during the hardening period, for instance, in the case of winter annuals. It is not surprising, therefore, that species of native herbaceous plants that complete their growth before the hardening period may show the opposite relationship, e.g., the taller the plant the greater the freezing tolerance (Till, 1956). This is presumably due to the selective effect of the lower temperatures to which the taller plants are exposed.

Different organs of the plant may differ in hardiness. It has long been known that roots and other protected parts are normally less hardy than the exposed shoot. For instance, among 23 species of plants with bulbs or other underground structures, the basal part of the plant and the roots had the least hardy tissues (Lundquist and Pellett, 1976).

These morphological relationships do not by any means always hold true; nor can they be expected to. When, for instance, a plant at the right stage of development is exposed to low temperature and adequate light intensity, it undergoes a rapid increase in hardiness yet it does not shrink in size, nor do its cells become smaller. In fact, no gross morphological or anatomical changes occur during this hardening period, nor do they occur during the dehardening period that takes place before growth commences.

On the other hand, changes in form and arrangement of the protoplasmic constituents may conceivably occur during hardening. Among the organelles, the chloroplasts show the most striking and easily detected changes, perhaps because of their relatively large size and distinct color. It has long been known, for instance, that they aggregate during the fall hardening period, in both the cells of conifer needles and of the stem cortex of deciduous trees, instead of being uniformly distributed throughout the cell surface as in the growing, nonhardy state. This had been observed in every one of the many overwintering species examined by the author in Canada, the aggregated chloroplasts forming a green belt around the cell. In spruce (Senser *et al.*, 1975) and fir (Chabot and Chabot, 1975), they aggregate together with the other organelles around the nucleus. This is accompanied by some loss of chloroplast structure (they swell and their membrane system becomes disorganized and reduced) and the absence of starch grains, which are replaced in the fir (*Abies balsamea*) by lipids throughout the cell. In the spring, the spruce chloroplasts recover and now function as amyloplasts, providing reserve material for the new shoot (Senser *et al.*, 1975). In summer, they are again converted into typical chloroplasts. These changes in overwintering trees may conceivably be related to dormancy rather than to hardiness, and a marked increase in hardiness may occur in herbaceous plants without any change in chloroplast distribution. In opposition to this interpretation, however, similar changes occur to some extent in the chloroplasts of a hardy winter wheat (Kharkov), which has no winter dormancy. They show a quick and complete disappearance of starch grains on exposure to hardening conditions, accompanied by an undulation of the chloroplast membranes and an increase in osmiophilic granules (Rochat and Therrien, 1975a). In the chloroplasts of the less hardy variety (Selkirk), starch grains persisted even after a month of hardening, the osmiophilic granules diminished in number and intensity, and vesicles formed between the double membrane surrounding the chloroplast. According to Heber (1959a), chloroplasts of wheat increase in size during hardening due to an accumulation of low molecular weight sugars and water-soluble proteins (see Sec. b). In spinach, the chloroplasts change from the lamellar type structure to the vesicle type (Garber and Steponkus, 1973a,b). In grass species, electron micrographs indicated that they are the most sensitive organelle to hardening temperatures (Kimball and Salisbury, 1973). Freeze-fracture electron microscopy has revealed a decreased particle concentration on the inner fracture face of acclimated thylakoids from *Spinacea oleracea* (Garber and

Steponkus, 1976a,b). They were of one size group, intermediate between the two size groups in the nonacclimated thylakoids.

Other organelles may also reveal visible, seasonal changes. When winter arrived, the nuclear pores of a hardy winter wheat gradually disappeared, whereas those of a spring wheat remained open (Chien *et al.*, 1973). The authors suggest that this change may be related to the cessation of growth or to the development of cold resistance. The mitochondria of the winter wheat increased in size and changed in form, and the number of mitochondrial cristae increased. No such mitochondrial changes occurred in the spring wheat. An increased content of mitochondria was also observed in the tillering nodes of winter wheat before winter began (Khokhlova *et al.*, 1974). This was suggested as a compensation for their decreased size and activity. Mitochondria from unhardened wheat possess more clearly defined cristae, less matrix, and are more electron dense than those from hardened (2°C) seedlings (Pomeroy, 1977). On the other hand, the swelling and contraction properties of mitochondria isolated from four varieties of winter wheat and one of rye changed markedly, due to growth at low temperature, but the change was similar regardless of the hardiness of the variety. Mitochondria are most abundant in the fusiform cambial cells of pine when cambial activity is at a minimum, for instance during winter (Barnett, 1975). Freeze-etching revealed that the size of particles within the plasmalemma of cambial zone cells of *Salix fragilis* remained constant between winter and spring (when hardiness decreases), but the number increased considerably (Parish, 1974). The tonoplast showed the opposite changes—the size of the particles increased but the number of particles decreased. Changes in other organelles related to freezing tolerance are difficult to detect, perhaps because the cells become dormant in winter, and many of the organelles are not so easily discernible (Mia, 1972). Ribosomal structure, however, appears to be altered during the induction of hardiness in black locust, as judged by their thermal profiles (Bixby and Brown, 1975). In the cortical cells of mulberry, there was a striking change in the ER (both smooth and granular), vesicles, free ribosomes, Golgi bodies, and vacuoles, apparently associated with changes in freezing tolerance (Otsuka, 1972).

It must be concluded from all these observations that when a relationship between hardiness and morphological or anatomical characteristics occurs this is indirect, due to the accompanying physiological factors. The aggregation of chloroplasts, for instance, is reminiscent of the aggregation of red blood cells by phytoagglutinins or lectins. This phenomenon may, therefore, be due to an accumulation of lectins during hardening, and these lectins may conceivably be more directly related to freezing tolerance than the chloroplast aggregation per se. This is true even of development itself, for the growth of tender plants can be brought to a stop and they can even have a well-defined rest period without developing any hardiness (see above). These facts have long been known and led to many investigations of the individual physiological or physiochemical characters of the plant in relation to hardiness.

B. PHYSIOLOGICAL FACTORS

Unfortunately, the total number of factors involved in freezing tolerance is unknown. Genetic investigations have so far succeeded only in pointing to a multifactor relationship. In the case of *Brassica oleracea,* for instance, the genetic evidence suggested that two dominant, epistatic genes conditioned freezing tolerance (Bouwkamp and Homna, 1969). An essentially unlimited number of factors has been investigated. To some degree, at least, those factors related to adaptation produce additive effects (Rehfeldt, 1977).

The factor most commonly investigated, is the quantity of a specific substance (i.e., its accumulation) in relation to freezing tolerance.

1. Accumulation of Substances

a. TOTAL SOLUTES

A vast number of measurements have revealed that, in general, cell sap concentration increases with freezing tolerance. This parallel has been found (1) during hardening (Fig. 8.1), (2) when species or varieties differing in hardiness are compared (see Levitt, 1956), and (3) when solutes are fed artificially to potentially tolerant plants (Fig. 8.2). Even simple refractometer measurements of total soluble dry matter showed a close correlation with frost resistance in kale (Thompson and Taylor, 1968). This method had earlier been used successfully by other investigators (see Levitt, 1941). In a few cases of supposed varietal differences (Magistad and Truog, 1925; Civinskij, 1934), freezing tolerance is not involved, since the plants are killed by very slight freezes and the so-called "hardier" variety is simply better able to avoid freezing by virtue of a lower freezing point. Such tender plants are usually incapable of undergoing any increase in solutes on exposure to hardening temperatures (Rein, 1908; Pantanelli, 1918; Rosa, 1921; Meindl, 1934; Schlösser, 1936) unlike frost-hardy plants (Fig. 8.1). Potato tubers are exceptions, for although unable to harden, they show a marked increase in solutes at hardening temperatures (Müller-Thurgau, 1882; Apelt, 1907). Negative results have also been obtained, primarily when comparing varieties, both the more hardy and the less hardy varieties showing the same

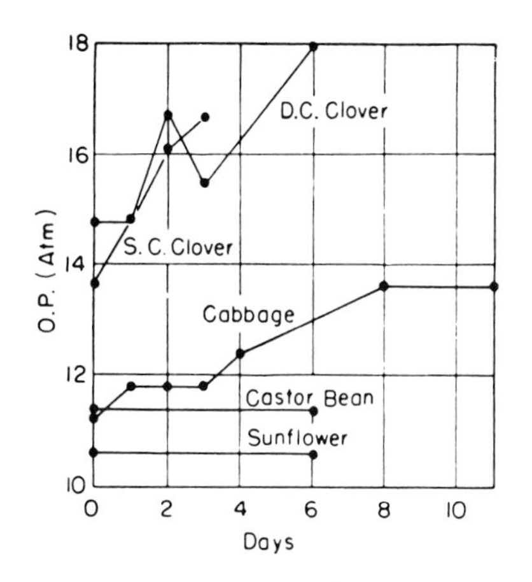

Figure 8.1. Increase in osmotic concentration of hardy (cabbage and clover) leaves during hardening at 5°C, and absence of any increase in tender (castor bean, sunflower) leaves. From Levitt and Scarth (1936). (Reproduced by permission of the National Research Council of Canada from the *Can. J. Res.* C**14**.)

increase in solute concentration with hardening (see Levitt, 1956). Some of the negative results may be due to a passive loss of water (Grahle, 1933; Pisek *et al.*, 1935). This error is possible only when freezing-point determinations are used. The loss of water can, however, be compensated for by measuring the water content and using the product of this times the freezing-point lowering (Tranquillini, 1958). Yet even this corrected value, as well as values obtained by the plasmolytic method, may fail to show a

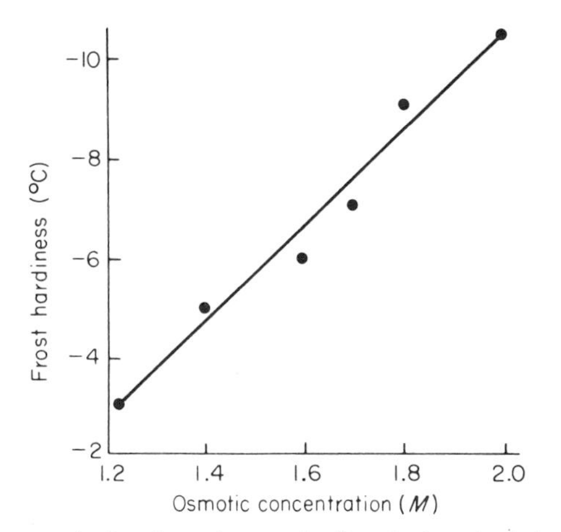

Figure 8.2. Increase in freezing tolerance (ordinate) of gardenia leaves with increase in osmotic concentration due to sugar (glucose) feeding. (From Sakai, 1962.)

correlation with varietal tolerance (see Levitt, 1956). This proves that there are some real exceptions. Furthermore, the very investigators who have succeeded in demonstrating the most striking correlations when some varieties are compared, have completely failed in other cases. Wheat varieties, for instance, sometimes give excellent results; barley varieties usually show no correlation between freezing tolerance and solute concentration; and some of the hardiest evergreens (e.g., *Picea;* see Levitt, 1956) show relatively little change in cell sap concentration although their tolerance changes seasonally from one extreme to the other.

Some attempts have been made to find out whether the relationship between cell sap concentration and freezing tolerance is a direct one. It is easy to show that increasing the concentration of solutes in the nutrient medium markedly increases tolerance (Table 8.1; Maximov, 1908; Chandler, 1913). This certainly produces other changes besides the increase in cell sap concentration (e.g., in cell size and total growth; Todd and Levitt, 1951). When rapidly penetrating substances (e.g., glycerine, urea, ethyl alcohol) are used, such secondary changes are presumably avoided and a small, but definite lowering, of the frost-killing point occurs (Table 8.2; Åkerman, 1927; Iljin, 1935a). In the case of succulents, however, there was little or no increase in freezing tolerance (Kessler, 1935). Even when considerable quantities of sugar are absorbed from a solution, there is only a slight increase in the freezing tolerance of cabbage leaves (Dexter, 1935). On the other hand, it is possible to lower freezing tolerance very markedly by chloroform treatment without affecting the cell sap concentration (Kessler, 1935; Ulmer, 1937).

It seems obvious from these results that the increase in cell sap concentration may sometimes be a factor, but is unable by itself to account for all the increase in freezing tolerance that normally accompanies it. Furthermore, plants with osmotic concentrations as high as 30 atm may be killed by light

TABLE 8.1

Frost Killing of *Aspergillus niger* Grown in Dextrose Solutions of Different Concentrations[a]

Conc. of dextrose (%)	Frost-killing point (°C)	Isotonic $NaNO_3$ (%)	Freezing point of $NaNO_3$ (°C)
1	−2	9	−3.3
10	−4	16	−5.6
20	−9	24	−8.2
30	−14	30.5	−10.2
40	−22	36.5	−11.9
50	living at −26	38.5	−12.5

[a] From Bartetzko, 1909.

TABLE 8.2

Effect of Solute Uptake on Frost Hardiness[a]

Plant	Treatment	Δ of solution used (°C)	Δ of sap after solute uptake (°C)	% Killing
Cabbage	None		0.780	80 at −4°C
	KCl	0.775	1.145	40
	Glycerine	2.82	1.780	20
	NH_4Cl	0.360	0.950	100
Cowpeas	Sucrose	1.570	1.230	0.0 at −3°C
	Glucose	1.740	1.250	49.9
	Glycerine	1.575	1.160	0.0
	KCl	0.730	1.130	66.6
	NH_4Cl	0.725	1.140	41.6
	Water	0.00	0.870	66.6

[a] From Chandler, 1913.

frosts (Walter, 1931). Consequently, the specific solutes involved have been examined in the hope of clearing up these discrepancies.

b. SUGARS, AND RELATED SUBSTANCES

Although the above plasmolytic and freezing-point measurements merely determine total osmotic concentration, there is ample evidence that the major changes are due to changes in concentration of sugars. Thus, when both measurements are made, the seasonal curves run parallel (Fig. 8.3). These and many other measurements made over the years, all clearly show that sugars normally increase in the fall as plants harden, and decrease in the spring as they deharden (Levitt, 1956). Such changes in sugar content occur in both woody and herbaceous perennials as well as in winter annuals in both cultivated and native plants (Markova, 1973), in both pasture grasses (Noble and Lowe, 1974; Hillard, 1975) and weeds (Gleir and Caruso, 1973), and in roots as well as tops, due to a translocation downward (Jennings and Carmichael, 1975).

Even when different tissues of the same woody plant are compared during late fall and winter, the sugar content of the tissue is commonly proportional to its freezing tolerance (Levitt, 1956). In the case of tea plants, for instance, both decrease from the outer cortex to the inner cortex and xylem, and from here to the least tolerant pith (Sugiyama and Simura, 1968a). Sugar content also increases during artificial hardening, for instance in cabbage, the younger leaves showing the greater increase in both sugar content and freezing tolerance (Le Saint and Frotte, 1972). It has even been used as a

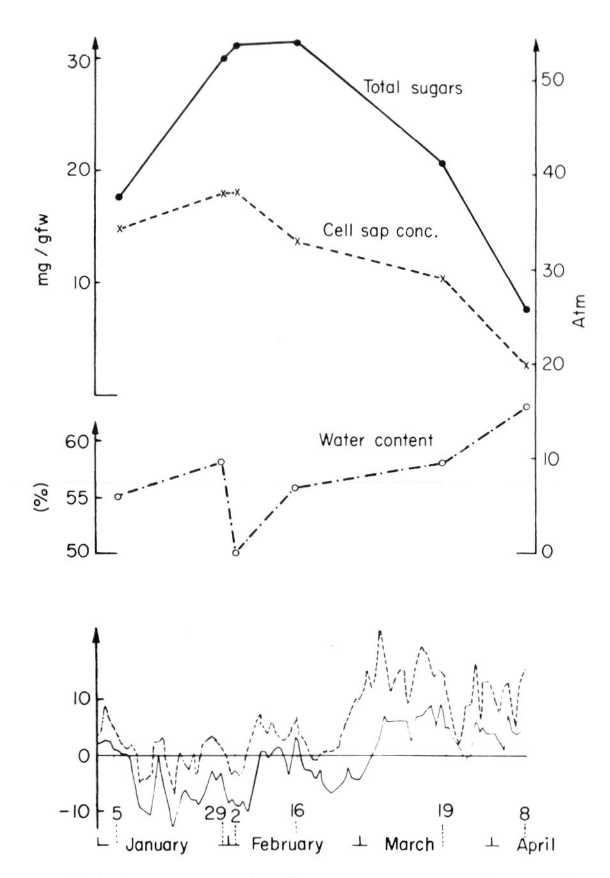

Figure 8.3. Parallel changes in total cell sap concentration (atm.) and in sugar content (mg/gfw) of sycamore during seasonal changes in freezing tolerance. (From Le Saint and Catesson, 1966).

measure of relative varietal hardiness, e.g., in wheat (Table 8.3). Thermophilic cereals, on the other hand, fail to accumulate oligosaccharides at hardening temperatures (Babenko and Gevorkyan, 1967). Many other wheat varieties also fail to show this parallel; when different species of cereals are compared, the relationship of sugar content to freezing tolerance may or may not become evident (Levitt, 1956). It is even possible to find an inverse relation between hardiness and sugar content when a small number of varieties (four) are compared (Green and Ratzlaff, 1975). Similarly, even hardy species of halophytes with high salt contents accumulate considerable quantities of sugars in winter, at the expense of a part of their salt content (Kappen, 1969a). Yet the most resistant of the three species showed the smallest sugar increase. In place of the sugars, salts of organic acids (citrate and to a lesser extent malate) accumulated (Kappen and Maier, 1973). Unlike the hardy glycophytes, the osmotic concentration of the sugars in these halophytes was, in all cases, lower than that of the salts.

The relationship between sugar content and freezing tolerance may become more pronounced from late fall to late winter, judging by the inverse relationship between tolerance and respiration rate in grains (Newton *et al.*, 1931), apples (Fig. 8.4), legumes (Bula and Smith, 1954), etc. In the spring, tolerance markedly decreases even at temperatures that permit the development of maximum tolerance in the late fall (e.g., +2° to −10°C), and the sugars are simultaneously converted to starch (Levitt, 1956). In midsummer, exposure to these same hardening temperatures fails to induce a rise in freezing tolerance or an increase in sugar content (e.g., in *Tilia;* see Levitt, 1956). The sugar increase in the fall is frequently accompanied by hydrolysis of accumulated starch, although it may also accumulate directly from photosynthesis (see Sec. 2c). In the case of the slightly hardy Valencia orange, reducing sugars accumulated most rapidly between 15° and 5°C, but there was little starch hydrolysis, indicating that this restricts the ability to harden (Yelenosky and Guy, 1977).

In some plants (e.g., mulberry; Sakai, 1961) any treatment that increases the sugar content increases hardiness and any treatment that decreases the sugar content lowers hardiness. A similar intimate relationship between sugar content and freezing tolerance was observed in black locust at all times of the year (Yoshida and Sakai, 1967). No other factor showed this close relationship. This was not true, however, of poplar. When held at 15°C

TABLE 8.3

Relationship between Sugar Content and Frost Hardiness in Wheat[a]

Variety	Relative sugar content	Relative frost hardiness (I = highest)
Sammet	100	I
Svea II	87	II
Thule II	67	IV
Standard	66	IV
Sol II	65	IV
Pansar II	48	V
Extra Squarehead II	44	VI
Danish small wheat	41	VII
Wilhelmina	39	VIII
Perl summer wheat	29	IX
Halland summer wheat	22	X

[a] From Åkerman, 1927.

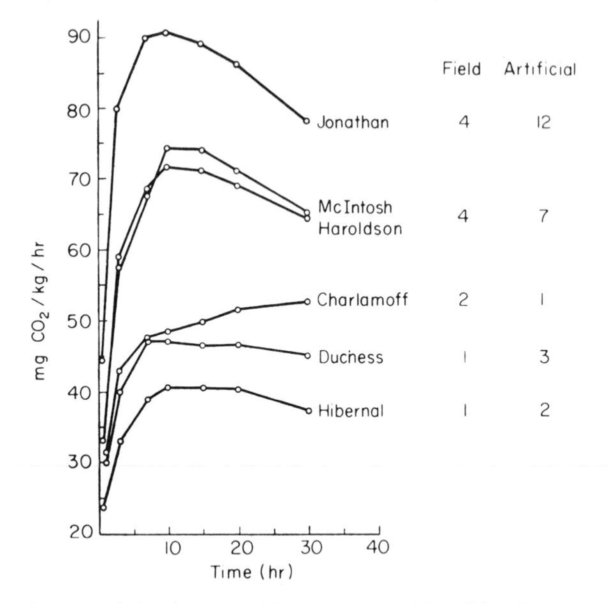

Figure 8.4. Inverse relation between CO_2 output at +6°C and hardiness (see Table 7.4) of apple twigs. (From De Long *et al.*, 1930.)

for 2 months, its freezing tolerance increased from a killing temperature of −2° to −30°C, without an appreciable increase in sugar (Sakai and Yoshida, 1968a). In other experiments, poplar *did* show an accumulation of the trisaccharide, raffinose, during the winter months in a greenhouse at 25°C (Kull, 1973). Therefore, an additional hardening low temperature could have only a quantitative effect on the storage of this sugar.

Conversely, cells that do not increase in freezing tolerance may increase markedly in sugar content (e.g., potatoes kept at low temperatures), and some plants with the highest sugar contents possess little or no hardiness (e.g., sugarcane).

Other investigators have found similar exceptions. Thus, although sugars appear to play a role in the hardening of *Hedera helix,* there was a lack of parallelism between the two (Steponkus and Lanphear, 1968b). In the case of chestnut trees, a positive correlation between hardiness and sugar content occurred during winter, but not from April to October (Sawano, 1965). In the case of wheat, more sucrose accumulated in the roots of the more resistant than in those of the less resistant variety, but there was no difference between the stems (Musich, 1968). In fact, the total sugar content was less in the stems of the more resistant variety. Similarly, some of the most hardy species and varieties have lower contents than less hardy ones; mediterranean evergreens, for instance, show little or no increase during the hardening period (Pisek, 1950; Larcher, 1954). This is also true of another evergreen—*Juniperus chinensis* (Pellett and White, 1969). Finally, some marine algae are highly tolerant of freezing in spite of very low sugar contents, and their freezing tolerance is actually lowered by sugars (Terumoto, 1962).

That sugars can truly increase hardiness was shown by early sugar feeding experiments (Chandler, 1913). Those plants fed sugars were able to survive freezes that killed the controls (Table 8.2); but the differences were small in this as well as in later experiments by other investigators. Some failed to obtain any differences, (e.g., Fuchigami *et al.*, 1973), and in only one case were large differences reported (Fig. 8.2).

These discrepancies can all be explained on the basis of the concepts of freezing tolerance described previously (Chapter 7). If the effect of the sugar increase is purely osmotic, leading to an increase in avoidance of freeze-dehydration, a hardened plant with a higher tolerance of freeze-dehydration will show a greater increase in freezing tolerance per unit sugar added than will an unhardened plant with a lower tolerance of freeze-dehydration. Thus, for tolerance values of 80 and 50%, respectively, the hardened plant will have its killing temperature lowered a full 5°C (from −5° to −10°C) by the same increase in sugar content as causes only a 2°C increase (from −1° to −3°C) in the case of the unhardened plant (Table 8.4). Furthermore, this difference is an underestimate, since as the sugar concentration increases, the effect of an equal addition on the freezing point lowering increases. Thus, the less hardy the plant, the smaller (and, therefore, the more difficult to detect) the increase in hardiness that results from a specific increase in sugar content. In the case of plants already in the hardened condition, it has, in fact, been possible to increase the hardiness significantly by infiltrating with sugars and

TABLE 8.4

Relative Effect on Freezing Tolerance of an Equal Addition of Sugar to the Cell Sap of a Hardened and an Unhardened Plant

	Unhardened		Hardened	
	Original	After addition of 9 gm glucose per 100 ml	Original	After addition of 9 gm glucose per 100 ml
Tolerance of freeze-dehydration (%)	50	50	80	80
Freezing-point lowering (°C)	0.5	1.5	1.0	2.0
Freeze-killing	−1	−3	−5	−10

other solutes, and the actual increase observed was always the expected amount on the basis of the above calculations (Table 8.5). Similarly, Perkins and Andrews (1960) obtained greater protection in the hardier wheat variety than in the less hardy one when both absorbed sugar. Further evidence that the protective effect by the infiltrated sugar was simply osmotic in nature, follows from the results of Sakai (1960b). A variety of sugars, as well as polyhydric alcohols, acetamide, and urea were capable of increasing hardiness when taken up in the transpiration stream, although inorganic salts were ineffective. The order of effectiveness of 0.7 *M* solutions for woody plants is shown in the following scheme:

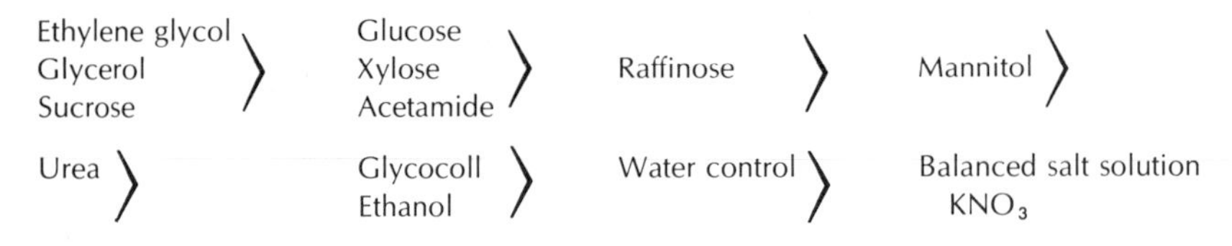

Amino acids were ineffective and so was the sugar galactose (Perkins and Andrews, 1960). *Hedera helix,* on the other hand, increased significantly in freezing tolerance when fed sucrose solutions, but not with equimolar solutions of glucose, galactose, or mannitol (Steponkus and Lanphear, 1968b). Concentrations ranging from 5 to 500 m*M* were equally effective, inducing maximum tolerance in excised leaves. These results do not agree with those of Sakai. In none of the above cases was the amount of sugar that entered the cells determined. Differences in both permeability and active absorption

TABLE 8.5

Effect of Infiltration with Organic Substances on the Hardiness of Plants

Species	Substance	Treatment	Osmotic potential (atm)	Frost-killing temp. (°C) Determined	Frost-killing temp. (°C) Calculated	Reference
Cabbage	Glycerol	Control	−20.25	− 7		
		Infiltrated twice	−27.0	−11	−10	Levitt, 1957b
		Control	−25.2	−12.5		
		Infiltrated once	−32.3	−15.5	−15.5	Levitt, 1957b
Saxifrage		Control	−21.9	−16		
		Infiltrated	−28.7	−20	−20.5	Levitt, 1957b
Cabbage	Dextrose	Control	−16.1	− 7.5		
		Infiltrated	−20.0	− 9.5	− 9	Levitt, 1959a
	Fructose	Control	−15.0	− 7.0		
		Infiltrated	−18.6	− 9.5	− 9	Levitt, 1959a

may account for some of the differences in effectiveness of different substances, as well as the discrepancies found between investigators.

In a later investigation (Sakai, 1962), the osmotic increases were determined and penetration was found to be an important factor. Unlike the infiltration experiments (Levitt, 1957b, 1959a) these results indicate greater increases in hardiness of gardenia leaves by sugar uptake than can be accounted for by the increase in cell sap concentration; for instance, controls with a concentration of 1.2 *M* were killed by −3°C, but after sugar uptake the cell sap concentration was 2.0 *M* and the killing point, −11°C. In these experiments, not only is the amount that enters into the cells unknown and nonuniform (more is probably taken up by the cells adjacent to the bundles), but the amount that does enter may then be metabolized. Infiltration of leaves avoids these difficulties as the exact amount entering the cells is known, and since the whole process can be conducted within 12–24 hr at close to 0°C, metabolism and temperature-induced changes in hardiness are reduced to a minimum.

The importance of metabolism of the sugars in long-term absorption experiments is clearly demonstrated by Tumanov and Trunova (1957, 1963). They increased the freezing tolerance of wheat seedlings by keeping their roots in 12% sucrose solutions for 14 days in the dark. The Soviet investigators, in fact, assert that the maximum hardiness potential of winter cereals, as well as of callus tissue from woody plants, cannot be attained unless they are fed with sucrose, in addition to the low-temperature hardening (Ogolevets, 1976b ; Trunova, 1975). The seedlings were originally somewhat hardy (killing temperature, −10°C), but they became much hardier (killing temperature, −28.5°C) during this period. It was necessary, however, to maintain them at a hardening temperature (+2°C) during the feeding period, in order to obtain this large increase in tolerance. Freezing was gradual (3 days at −4°, 1 day at −7°, 1 day at −10°, and 1 day at −13°C) before the final temperature lowering. The same degree of hardening was obtained by 6 hr of feeding per day as by 24 hr. The optimal sugar concentration for this treatment was 20%. Galactose was ineffective; lactose and maltose were less effective than sucrose; glucose and rhamnose were nearly equally effective; raffinose and fructose were as effective as sucrose. According to Trunova (1963), lactose penetrated the cell but remained unchanged. The total sugars did not increase, but decreased due to utilization in respiration and growth. In the earlier results, however, the sugar content of the tillering nodes increased from 24.7 to 55.1% of the dry weight. Pentose gave no protection.

If the sugar feeding leads to an increase in sugar content and cell sap concentration, the sugars must accumulate in the vacuole. If, however, the sugars are metabolized, they must enter the protoplasm. Sugiyama and Simura (1968b) attempted to decide this question by immersing tea shoots into 0.1 and 0.3 *M* ^{14}C-labeled sucrose solutions at 14°–18°C for 2 days. This led to increases in freezing tolerance, in osmotic concentration, and in content of total sugars. Radioautographs indicated that the ^{14}C was found only in the thin cytoplasmic layer adjacent to the cell wall. Nevertheless, the osmotic concentration of the plant sap increased 25–30% and the total sugar content by about 30–35%. Since the thin cytoplasmic layers appeared to account for only about one-tenth the cell volume, if the original sugar concentration in the protoplasm and vacuole were about the same, this would require an increase in the cytoplasmic sugar content of about three times that in the vacuole, and a corresponding increase in its volume. Since the radioautographs do not indicate any increase in thickness of the cytoplasmic layer, it is difficult to accept their conclusions.

A specific effect of ribose, not obtainable with glucose, has been reported (Jeremias, 1956). The methods used were not valid (Levitt, 1958) and more recent work from the same laboratory has disproved the conclusions of this work (see below).

Just which sugars increase normally during hardening was not clear from the early investigations. In the case of certain trees and vines it is sucrose (Siminovitch and Briggs, 1954; Sakai, 1957; Steponkus and Lanphear, 1967a, 1968b). Some results with grains indicate hexoses (Heber, 1958a), while others indicate sucrose (Hylmö, 1943; Johansson *et al.*, 1955). In the later stages of hardening, wheat accumulates raffinose and perhaps another high molecular weight sugar (Gunar and Sileva, 1954; Johansson *et al.*, 1955). With the use of chromatographic methods, Parker (1959a) found that raffinose and stachyose increased markedly in the bark and leaves of six conifers during fall. Sucrose and sometimes glucose and fructose also increased. Broad-leaved, deciduous trees did not show an increase in raffinose and stachyose although broad-leaved evergreens did. In *Pinus strobus* (Parker, 1959b), the raffinose content (from 0 to 1.5% of fresh weight) followed hardiness changes both in fall and spring, although sucrose, glucose, and fructose failed to show any such close correlation with hardiness. On the other hand (Sakai, 1960a), the proportions of the different sugars were found to vary in eighteen species of woody plants, but no specific sugar was consistently correlated with hardiness. In six species, sucrose, glucose, and fructose showed marked increases during hardening, but raffinose and stachyose did not. In one species (mulberry), the latter two sugars also increased, but less so than the above sugars. Results with some other plants are summarized in Table 8.6.

The kind of carbohydrate accumulated is, of course, dependent on the normal carbohydrate metabolism of the particular plant. Cold storage of

chicory, dandelion roots, and artichoke tubers, caused a breakdown of inulin and of high molecular weight oligosaccharides (polymers of more than ten subunits) to oligosaccharides of a lower degree of polymerization (Rutherford and Weston, 1968; Kakhana and Arasimovich, 1973). In some grass leaves, the major accumulation at low temperatures is due to fructosans (Smith, 1968a). Even alcohols related to carbohydrates may be accumulated. In some insects, for instance, glycerol increases during hardening and this acts as an antifreeze or at least favors undercooling (see above). Salt (1961) has, in fact, suggested that perhpas all freezing-tolerant insects and possibly other invertebrates contain glycerol or some equally protective substances, although exceptions containing glycerol may not be resistant (Salt, 1957). Both kinds of exceptions occur. Thus, in spite of its relatively large glycerol content (2–3%), a carpenter ant was found to be killed by freezing at −10°C, whereas the larvae of two species of butterfly survived freezing at −15°C without any glycerol or other polyhydric alcohol (Takehara and Asahina, 1960b). Even when glycerol was effective in a caterpillar, considerable freezing tolerance developed before glycerol formation, and maximum tolerance was attained with 2% glycerol, although it continued to accumulate beyond this point (Takehara and Asahina, 1961). Furthermore, injection of glycerol to the amount of 3% of the body weight failed to enhance their freezing tolerance. Analyses of plants (Sakai, 1961) failed to reveal any glycerol or other polyhydric alcohol in appreciable quantity in most of the woody plants tested. In a few species (gardenia, apple, mountain ash, pomegranate) polyhydric alcohols (mannitol, sorbitol, glycerol) amount to about 40% of the total sugar content and may therefore play some role in hardiness (Sakai, 1960b; Raese *et al.*, 1977). The accumulation of glycerol during the hardening of insects and its disappearance during dehardening is apparently due to glycogen-glycerol conversions (Takehara and Asahina, 1960a). The lack of glycogen in higher plants may therefore perhaps explain the absence of this metabolic adaptation to low temperature.

What can be concluded from this vast mass of data relating or not relating sugars and related substances to freezing tolerance? The many records of a correlation with freezing tolerance strongly indicate that, at least in some plants, sugars must play some role in the mechanism of tolerance. Although the correlations under natural conditions may be due to some other relationship common to the two, feeding experiments are more conclusive. They indicate that sugars may increase freezing tolerance in two ways. (1) The osmotic effect—by accumulating in the vacuole, sugars can decrease the amount of ice formed, and therefore increase the avoidance of freeze-induced dehydration. On this basis, sugar is a secondary factor, and cannot induce tolerance of freeze-induced dehydration. This would explain the existence of tender plants with high sugar contents (e.g., sugarcane). Only those plants possessing a marked tolerance of freeze-induced dehydration would be able to show a detectable increase in freezing tolerance due to an increase in sugar content. (2) The metabolic effect—sugars, as such, have little or no effect on freezing tolerance, but by being metabolized in the protoplasm at low, hardening temperatures, they produce unknown protective changes. Some of these changes may conceivably increase the tolerance of freeze-induced dehydration. There are several lines of evidence in favor of this explanation. Very hardy plants can survive much lower temperatures than their sugar content can explain, on the basis of avoidance of freeze-induced dehydration. The fact that the hardiest plants commonly accumulate their sugars in the fall before the main increase in freezing tolerance can be explained by either effect.

c. WATER (TOTAL AND BOUND)

Water content is frequently inversely related to hardiness (Fig. 7.3, also see Levitt, 1956). The practical man, in fact, has long related frost hardiness to "maturity" or "ripening" of the tissues, which he usually considers inversely related to water content. For instance, the percentage moisture in both the root and top of *Juniperus chinensis* decreased during the fall, the rate paralleling the increase in freezing tolerance (Pellett and White, 1969). This may be partly due to displacement of the water in the cell during the accumulation of sugars. A more active method is involved in *Cornus stolonifera*. These plants lower their water content during early fall by increasing their water loss and decreasing their absorption of water. This is apparently accomplished by suberization of the root surface and therefore a lowering of the permeability to water and by a decrease in stomatal resistance of the leaves (McKenzie *et al.,* 1974c; Parsons, 1978). There are, again, many exceptions. Some succulents (e.g., species of *Saxifraga* and *Sempervivum*) with water contents as high as 90–95% nevertheless develop a high degree of hardiness; among three species of halophytes, water content was lower in winter than in summer in two of the species, but there was no difference in the third, most tolerant species (Kappen, 1969a). Similarly, the water content of sycamore twigs fluctuates during winter in a manner that does not parallel freezing tolerance (Le Saint and Catesson, 1966). One possible cause of the exceptions is the occurrence of freezing injury followed by drying out. Thus, there is a high positive correlation between freezing survival and water content in the crowns of winter wheat and rye (Gusta and Fowler, 1976).

All of these apparently contradictory results are to be expected, on the basis of the different kinds of freezing resistance discussed above. Those plant parts that survive winter in the supercooled state (buds and wood rays of trees) must lose their excess free water during the fall hardening period, in order to remain in the supercooled state and to avoid fatal *intra*cellular freezing during the subsequent winter. On the other hand, those that tolerate moderate, *extra*cellular freezing (winter annuals, biennials, etc.) must not lose more water by desiccation than the amount which can be frozen without injury.

It was suspected early that only one component of the total water content, the so-called "bound water" is important. This water was presumed to be held so tightly that it could not freeze. Excellent correlations were at first, obtained between this quantity and frost hardiness in grains. These are the same plants that had previously shown an equally good correlation between sugars and hardiness. It was soon realized that sugars themselves bind water, and therefore determinations of bound water on the whole plant may reveal no more than sugar analyses or, more simply, determinations of cell sap concentrations. The reason for this is that the methods used in these early investigations were not valid. Determinations were made at much too high temperatures, and therefore measured the water that was prevented from freezing due to the high concentrations of sugars but that was easily frozen at somewhat lower temperatures. Later attempts avoided the major part of this error by using much lower temperatures. The bound water values are, therefore, also much lower—33% (0.5 gm/gm dry matter) in wheat grain at −120°C (Radzievsky and Shekhtman, 1955) and 9% of the total water (0.2 gm/gdm) in yeast at −72°C (Wood and Rosenberg, 1957). Recent NMR measurements have also yielded values of 0.3 to 0.5 gm of H_2O/gram protein at −35°C (Kuntz *et al.*, 1969) and of 30% in the thylakoid membranes of chloroplasts of *Antirrhinum majus* (Hirtz and Menke, 1973). It has been suggested that this unfreezable water (0.47 gm/gm) may be divided into two more or less distinct species, 0.25 gm/gm being the minimum water content above which the system can gain enough mobility to give rise to a detectable glass transition (Simatos *et al.*, 1975). Nucleic acids, however, proved to be three to five times as hydrated as proteins (Kuntz *et al.*, 1969).

Similar NMR measurements on winter cereals showed that the amount of unfreezable (bound) water/gdm was not strongly dependent on the degree of cold acclimation (Gusta *et al.*, 1975). Even the Soviet investigators who are still reporting an increase in bound water during hardening are unable to establish a direct correlation between the quantity of bound water and freezing resistance (Sharashidze, 1972). Bound water as a percentage of total water is, of course, meaningless, since it varies markedly in an individual plant whether or not hardiness is altered. Only the amount bound per unit of binding substance can have any significance in freezing resistance. In the winter stems of woody plants, the unfreezable (bound) water may amount to 30% of the total tissue water. Nevertheless, Burke and his co-workers have found no relation between bound water and hardiness in either winter cereals or woody plants, with freeze killing temperatures ranging from $-3°$ to $-196°C$.

As Kuntz *et al.* point out, the very fact that the water signals can be observed by high-resolution NMR suggests that the bound water is not "icelike" in any literal sense, although it is clearly less mobile than liquid water at the same temperature. The quantity bound must, of course, depend on the vapor pressure of the surrounding water. In the case of egg albumin, at a relative humidity of 92% at 25°C, the value is 0.3 gm of H_2O/gram of protein, but as the relative humidity approaches 100%, the bound water exceeds 0.5 gm H_2O (Bull and Breese, 1968). The protein hydration is proportional to the sum of the polar residues minus the amides, which apparently inhibit binding of water. About 6 moles H_2O are complexed per mole of polar residue.

Conclusions of a relationship between freezing tolerance and bound water have frequently been based on the most indirect evidence that actually has nothing to do with bound water. For instance, a tacit assumption has been made by some workers that bound water is a mysterious fraction of the plant's water that is unable to freeze in the living plant, but that can freeze as soon as the plant dies (Hatakeyama, 1957, 1960, 1961). This conclusion is based on the difference between the freezing points of living and dead tissues. It has already been shown (Chapter 5) that this difference can be increased or decreased by simply altering the method of freezing. All these results have been explained by the simple principles of the freezing point method and therefore have nothing to do with bound water (Levitt, 1966b). Thus, using pear tissue held at fairly low temperatures, Marshall (1961) was able to obtain a freezing point 2°F below the freezing point of the juice. In the exceptional cases, no more than 1°F difference was obtained. In most cases, however, the true freezing point of living pear tissue, when determined correctly, was shown to coincide with the freezing point of the expressed juice.

The earlier concept of the role of bound water was as a means of protecting by reducing the amount of water frozen. This explanation was also proved incorrect, at least in some cases. Measurements of the amount of ice formed in cabbage showed that the hardened plant survived the freezing of a *larger* fraction (75%) of its water than did the same plant in the unhardened state (60%; Table 7.9). These results have since been confirmed for wheat crowns (Gusta *et al.*, 1975) and potato leaves (Chen *et al.*, 1977). On the other hand, the role of the bound water might be specifically in the protoplasm rather than in the nonliving vacuole and this might be overlooked since the latter occupies most of the cell. In an attempt to measure the water bound by the protoplasm itself, determinations were later made on the isolated chloroplasts, and again showed a relationship between bound water and hardiness. One of the problems in this kind of measurement is the increase in proteins that frequently parallels hardening. If, for instance, the protein content of the chloroplasts is higher in the hardened state, as found by Heber (1959a), an increase in bound water might simply be due to an increase in the proportion of these water-binding proteins without any increase per unit of protein. It is, therefore, essential to measure the water bound by the proteins themselves. This has been done indirectly by measuring a series of properties of the soluble proteins (Brown *et al.*, 1970). The proteins were extracted from the root tissue of a hardy and a nonhardy alfalfa variety sampled in the field in mid-October, mid-November, and mid-January, when large differences in freezing tolerance occur. Measurements were made of the partial specific volume of the protein and water of the extracts, the specific heat capacities, the temperature of spontaneous nucleation, and the expansion on freezing. Absorption isotherms of the lyophilized protein powders were also determined. Essentially no differences were found between the two varieties, indicating that the hydration properties of the soluble proteins were similar.

d. AMINO ACIDS

Many earlier investigators measured the amino acid content of plants in an attempt to find a correlation with hardiness. The results were not convincing (Levitt, 1956). In some cases (Wilding *et al.*, 1960a, b), amino acid content may increase with hardiness (e.g., red clover), in others, it may fail to show any relationship (e.g., alfalfa varieties); and where the relationship exists, it seems to reflect a general increase in storage of organic nitrogen during the fall, either due to the nitrogen-storing amide asparagine or the accompanying even greater increase in nonamino acid nitrogen. Similar results have been obtained by Romanova (1967), Ostaplyuk (1967), Protsenko and Rubanyuk (1967), and Paulsen (1968). It has been suggested (Heber, 1958b), however, that an increase in amino acids and peptides may occur during fall in the field, due to a partial, reversible injury as a result of slight freezing. In the case of two varieties of winter wheat, the amino acid composition was the same and did not change with hardening (Toman and Mitchell, 1968). In contrast to these results, the concentration of most of the amino acids was higher in wheat plants grown at 5°C than at 10°C (Srivastava and Fowden, 1972). In the phloem and cambium of 2- to 3-year-old branches of sycamore, the total amino acids reach a maximum at the end of January when hardiness is presumably maximal, but a second, larger maximum occurs in April when growth is revived and hardiness is largely lost (Le Saint and Catesson, 1966). The changes, therefore, were explained by dependence, during winter, on the temperature and during spring, on the physiological activity of the plant.

Although total amino acid content has not been consistently related to freezing tolerance, the specific amino acid proline has been reported by several investigators to accumulate at hardening temperatures (Heber, 1958b; Le Saint, 1958, 1960, 1966; Markowski *et al.*, 1962; Ostaplyuk, 1967; Protsenko and Rubanyuk, 1967), and to decrease on loss of hardiness or to increase more in more hardy plants. A winter-hardy rose variety had a higher proline content than a nonhardy variety (Ziganigirov, 1968). Proline content appeared to be related to dormancy in peaches and apricots but not in apples (Vajsablova and Benko, 1971). There was no relation, however, between proline and the freezing resistance of any of these trees. Conversely, citrus cultivars with higher concentrations of hydroxyproline before hardening, were more tolerant of freezing after cold hardening (Yelenosky and Gilbert, 1974).

In the case of forage plants, arginine and alanine were chief among the several amino acids that increased more in the hardy than in the nonhardy variety (Smith, 1968b). During hardening of *Lolium perenne,* there was a progressive increase in free amino acids mainly due to an accumulation of glutamine and proline (Draper, 1975). In poplar, arginine was the major amino acid during winter. Glutamine and glutamate became dominant at the time of budding (Sagisaka, 1974). Among the free amino acids that accumulated during hardening of winter rape, proline showed by far the greatest increase, although asparagine also increased somewhat and cysteic acid decreased (Kacperska-Palacz and Wcislinska, 1972).

In cabbage, the content of free proline parallels the freezing tolerance of the different organs in hardened and unhardened plants (Le Saint, 1969a), with a maximum in the hardened terminal shoot. It is the only amino acid that accumulates on hardening, increasing from 2–4% of the total amino acid content of the whole unhardened cabbage plant to 60% in the hardened plant. The correlation between proline and hardiness was, in fact, better than between sugars and hardiness; the basal leaves which have little freezing tolerance may be rich in sugars but have little or no proline. Conversely, the terminal part of the shoot which is richest in proline, is not particularly rich in sugar. When hardiness is transmitted from illuminated (older) leaves to darkened (younger) leaves during exposure to hardening temperature (4°C), proline accumulates in the darkened leaves [Le Saint (-Quervel), 1969b]. Conversely, when hardiness is not transmitted from illuminated (younger) leaves to darkened (older) leaves during exposure to hardening temperature (4°C), proline fails to accumulate in the darkened leaves. Le Saint, therefore, concludes that the chemical effector responsible for the transmission of hardening is either proline itself or a substance closely related to its metabolism. It was even possible to induce some hardening by allowing cabbage shoots to absorb proline from a solution (5 gm/liter) at a nonhardening temperature (18°C; Le Saint, 1966). This was accompanied by the same increase in sugars as occurs on exposure to hardening temperatures.

The relationship between proline and freezing tolerance was not as clear-cut in the case of the phloem and cambium of 2- to 3-year-old branches of sycamore (Le Saint and Catesson, 1966). Proline showed the same two maxima as the total amino acid content, the spring maximum being again about double the winter maximum. The authors suggest that the first maximum is related to the maximum winter cold which preceded it and the spring maximum is due to a regeneration of growth. Somewhat similar results have been obtained for buds of white spruce (Durzan, 1968a, b). Free arginine declined in the fall, as proline and the amides contributed more to the seasonal levels of soluble nitrogen. The high levels of proline extended into the early spring, so that the expanding buds contained a high proportion of proline. At this time, proline levels showed a diurnal periodicity, being maximal at sunset and sunrise (Durzan, 1969). Since these daily changes were extremely large (from a minimum of 50 μg N/gfw to a maximum of 700 μg N/gram fresh weight), proline can hardly contribute to freezing tolerance by its mere presence, for these changes occurred at the end of May when hardening does not occur. Labeled carbon (14) from arginine and citrulline and to a lesser extent from γ-guanidinobutyric acid fed to the plant was recovered in the protein fraction mainly as arginine, glutamic acid, and proline.

In the case of wheat, if seeds are germinated for 2 days at 24°C and then incubated at 2° for 3 weeks, there is a manifold increase in free amino acids,

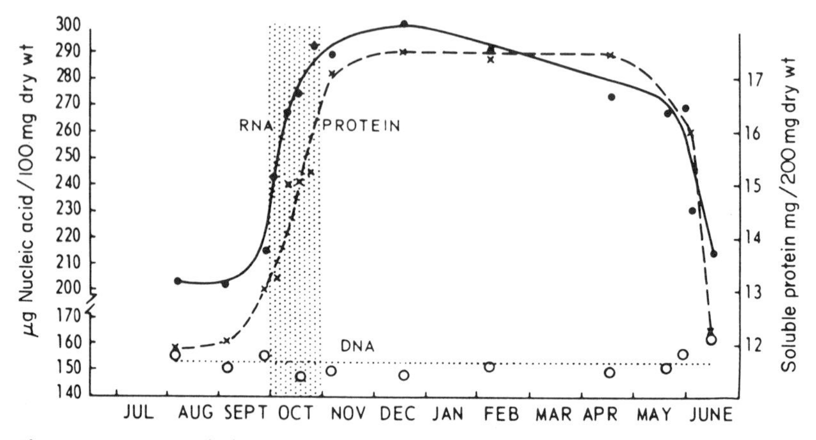

Figure 8.5. Seasonal changes in water-soluble protein, RNA, and DNA in bark of locust trees. Shaded area is the period of maximum rise in freezing tolerance. (From Siminovitch *et al.*, 1967a.)

with outstandingly high concentration of proline (Scheffer and Lorenz, 1968). In germinating spring wheat, proline does not increase as much as in winter wheat varieties. Young leaves of winter wheat varieties grown at 24°C contain relatively high concentrations of amino acids, except for proline. In similar plants of the spring wheat variety the proline concentration is rather high. Even in the cotyledons of the peanut (a nonhardy plant) during germination a marked synthesis of proline occurs (Mazelis and Fowden, 1969). They extracted a soluble enzyme system which converts ornithine into proline.

e. PROTEINS

A striking parallel between soluble protein content and freezing tolerance (Fig. 8.5) has been clearly demonstrated in the cortical cells of trees both during hardening and dehardening (Siminovitch and Briggs, 1949, 1953a, b, 1954; Siminovitch *et al.*, 1967a; Li and Weiser, 1967). This increase in soluble proteins may actually precede the increase in hardiness. After its development in the fall, however, a greater effect is obtained from artificial hardening (Sakai, 1958). A possible factor in the rise of soluble proteins is the increase during hardening in the ability to incorporate glycine into water-soluble proteins (Fig. 8.6). The increase in soluble proteins may sometimes fail to parallel either the main autumn increase in hardiness or the spring decrease or even an artificial increase in mulberry, according to Sakai (1957, 1962). Similarly, the spring swelling of grape buds, which is accompanied by a loss of freezing tolerance, involves an intense accumulation of protein (Cheban, 1968). Finally, under artificial hardening conditions, the increase may not occur; Siminovitch *et al.* (1968) succeeded in hardening bark tissue of locust to a considerable degree (killing temperature −45°C) under conditions preventing an increase in soluble proteins (Table 8.7). The same results have been obtained with poplars kept at 15°C during October and November (Sakai and Yoshida, 1968a). The trees increased in freezing tolerance (the killing point dropped from −2° to −30°C) without any concomitant increase in soluble proteins. On the other hand, when hardened at a more normal temperature of 0°C, during the same two months, the hardening was more extreme (killing point −70°C) and soluble proteins did increase with hardening. Similarly, the normally hardened bark tissue of the locust tree did show a doubling of soluble protein, and survived the temperature of liquid nitrogen (Siminovitch *et al.*, 1968). Brown and Bixby (1973b) confirmed these results with black locust but observed an increase in total protein during the early stages of hardening when soluble protein content remained constant. They, therefore, concluded that insoluble membrane proteins must have accounted for the initial increase in total protein. In grape shoots there was actually a higher content of soluble proteins in the less hardy variety

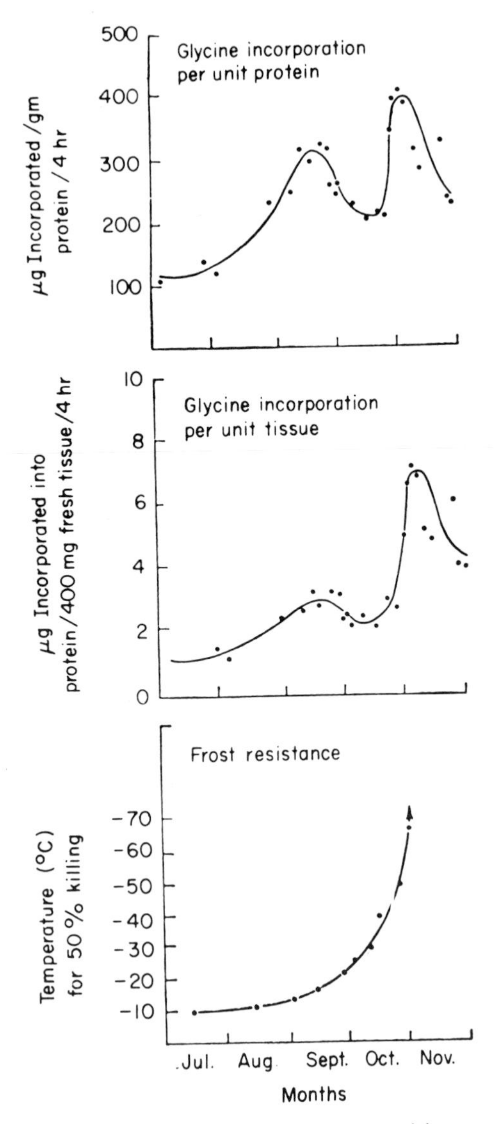

Figure 8.6. Increase in rate of protein synthesis (measured by rate of incorporation of [^{14}C]glycine) during the period of rapid increase in freezing tolerance (frost resistance). (From Siminovitch, 1963.)

cytoplasmic and basic nuclear proteins (Chian and Wu, 1965). Heber (1959a) found that the protein content of chloroplasts increased during hardening of wheat. In agreement with these results, the ratio of chloroplast to total soluble protein was higher in the more resistant than in the less resistant tea varieties (Sugiyama and Simura, 1967a). Pauli and Mitchell (1960) at first failed to detect any increase in soluble proteins (as percentage of total nitrogen) on hardening of wheat plants, although free amino acids and amides did increase. Later results (Zech and Pauli, 1960) revealed a relationship between soluble protein and hardiness in all three wheat varieties tested, but none with other nitrogen fractions. More recent measurements (Harper and Paulsen, 1967) fail to support this relationship. The crown tissue of field grown "Pawnee" wheat showed a decrease in water-soluble protein with hardening and an increase with dehardening. Similarly, there was no relationship between the hardening of two varieties of winter wheat and the amount of protein eluted from the crown tissue with 0.05, 0.1, or 0.2 *M* NaCl (Toman and Mitchell, 1968). In the case of forage legumes, there was an increase during the fall in total nitrogen, water-soluble nonprotein nitrogen, water-insoluble nitrogen, and water-soluble protein nitrogen/gfw. The water-soluble protein nitrogen was the only fraction consistently higher in both the roots and crowns of the more resistant alfalfa than in the less resistant red clover (Jung and Smith, 1962). The proteins in the microsomes were more closely associated with hardiness than those in the nucleus, the mitochondria, or the cytoplasm (Shih *et al.*, 1967). Furthermore, a significant correlation was found between soluble protein content and hardiness of ten alfalfa varieties, although later investigations failed to confirm this relationship (Smith, 1968b). Nevertheless, after hardening for 40 days, the cold tolerant alfalfa variety (Vernal) had a higher content of total soluble protein as well as of 10 individual proteins separated by electrophoresis, than the cold-sensitive variety, Arizona Common (Faw and Jung, 1972). In contrast to the many positive results with legumes, in the case of cabbage, hardening induced a decrease in total nitrogen, protein nitrogen, as well as soluble nitrogen per unit dry matter (Le Saint, 1966). The ratio of protein nitrogen: total nitrogen did not change and no difference was found among the soluble proteins. In our own laboratory, hardening induced a marked increase in soluble proteins if the whole shoot was analyzed after long-continued hardening (Morton, 1969). When individual, rapidly hardened leaves were compared, no increase could be detected (J. Dear, unpublished).

It is now clear that these discrepancies are partly due to the method used for extracting the proteins which may have led, in some cases, to an artificial difference. If lyophilized powders are used, a greater aggregation of the soluble proteins may conceivably occur during the freeze-drying of the less tolerant tissues. If distilled water is the extractant, a higher pH of the hardened tissue (which has been observed in some cases; see Levitt, 1956) may permit extraction of a larger fraction in the soluble state. Thus, in contrast to the above negative results of Le Saint and of Dear, according to Kacperska-Palacz *et al.* (1969), the soluble protein content of hardened cabbage was about 3.2 times as high as in the nonhardened plants, even though only the fourth pair of leaf blades were used in all cases. The leaves were homogenized in distilled water instead of a buffer solution (15 gm/55 ml H_2O), and homogenization was continued for 10 min, instead of the 30 sec, more commonly employed by other investigators. Kacperska-Palacz *et al.* suggest that the difference may be an artifact due to pH changes occurring during hardening. Another possible cause is the greater sensitivity of proteins of nonhardened plants to homogenization (see Schmuetz below), which would result in greater precipitation, during the lengthy (10 min) homogenization period used by them. Similar results have now been obtained in the case of legumes. Faw *et al.*, (1976) found that, in general, more protein was extracted from a given sample as the pH of the extracting solution increased. Varietal differences were not significant when three of nine buffer solutions were used for protein extraction. The total increase, when it was observed, was the result of general increases in many of the fourteen electrophoretically distinguishable regions. The amount determined densitometrically in ten of these regions was highly correlated with hardiness. It must be cautioned, however, that correlations of this kind can be obtained in the absence of hardening. Proteins accumulate in potato tubers stored at low temperatures, even though no freezing tolerance is developed (Levitt, 1954). Both soluble proteins and 4 S RNA may accumulate in potato foliage at hardening temperatures although no true tolerance is developed (Li and Weiser, 1969b). In *Mimulus,* soluble protein increased progressively from a growing temperature of 30°C to growing temperatures of 20° and 10°C (Björkman *et al.,* 1970). These temperatures are not low enough to induce hardening even in hardy plants.

From all these results, the most logical conclusion seems to be that the first stage of hardening (down to a killing temperature of about −30°C) may occur without a detectable increase in soluble protein, but that for freezing tolerance in excess of this, an increase in soluble proteins may be necessary. In some plants, of course, such as alfalfa, it is conceivable that the order may be the reverse of this. It is obvious, however, that the relationship will not be categorically established without a quantitative investigation of the individual proteins. Some attempts have been made to obtain this more specific information.

Indirect evidence of a change in a specific protein was obtained by Soviet investigators and confirmed by Roberts (1967). The energy of activation of invertase from leaves of winter wheat (Kharkov) when grown at 6°C (resulting in a killing temperature of −20°C) was lower than that of the same enzyme obtained from leaves grown at 20°C (killing temperature −10°C). The enzyme from the less hardy (lowest killing temperature −10°C) spring wheat (Rescue) had the identical energy of activation under the above two growing conditions. One explanation suggested by Roberts was that the invertase consists of a number of isozymes with different temperature coefficients. A change in the proportions of these isozymes could then account for the above results.

Coleman *et al.* (1966) investigated the electrophoretic and immunological properties of the soluble proteins of alfalfa roots. A zone of highly charged and/or low molecular weight protein components was more prevalent in the hardened than in the unhardened material. Disc electrophoresis failed to distinguish differences in pattern between varieties differing in hardiness (Gerloff *et al.*, 1967). Certain proteins did increase during hardening, but this occurred in both hardy and nonhardy varieties. Morton (1969) separated some thirty proteins from cabbage leaves by disc electrophoresis but found them all present in both the hardened and unhardened plants. Other investigators have been more successful in detecting specific protein changes.

Of three clones of dianthus, the two winter hardy ones showed a gradual synthesis of two to four new peroxidase isoenzymes from August through November, the tender one showed only a relatively weak initiation of one isoenzyme. The formation of the new isoenzymes preceded the hardening period by several weeks (McCown *et al.*, 1969a). Apple and arborvitae also showed changes in specific protein bands which were separated by disc electrophoresis (Craker *et al.*, 1969). When apple dehardened (from a killing point of −50° to −10°C), the total protein nitrogen remained at about the same value (0.67–0.85% of the dry matter). Of the twenty-seven bands that could be distinguished, only six were observed at all sampling dates. There were four bands present only when the plants were hardy. Most of the other changes were related to bud swell. Arborvitae was hardened during a 60-day period from a killing point of −10° to −50°C. During the first 36 days, the twenty-two bands were very constant, although hardening occurred to a killing temperature of −24°C. During the final 21 days of hardening (from a killing point of −24° to −50°C), four previously present bands disappeared, two appreared and later disappeared, and three appeared and remained. Bermuda grass cultivars, which increased in freezing tolerance by a maximum of 6.0°C showed a loss of density in two protein bands during a 15-day hardening treatment (Davis and Gilbert, 1970). During overwintering, four bands appeared near the origin. Indoleacetic acid oxidase increased ten times in acitivity in wheat seedlings held at 2°C for 4 days (Bolduc *et al.*, 1970). This change was inhibited by puromycin or 6-methylpurine.

Incorporation of [^{14}C] leucine disclosed the synthesis of two new proteins during the hardening of the hardy wheat Kharkov (Rochat and Therrien, 1975b). Some attempts have been made to locate the protein changes in specific cell organelles. Specific, cryoprotective proteins were obtained from the chloroplasts of hardy spinach and Brassicas (Volger and Heber, 1975). The hardening of winter wheat was accompanied by both quantitative and qualitative changes in mitochondrial proteins (Khokhlova *et al.*, 1975). No changes in specific chloroplast and membrane proteins were found during the early stages of hardening of 2 wheat varieties (Rochat and Therrien, 1975b), nor in chloroplast membrane polypeptides during hardening of wheat and rye (Huner and MacDowall, 1976), but subtle changes were evident in the soluble chloroplast protein fraction. At least 17 ribosomal proteins from hardened black locust seedlings were different from those of unhardened seedlings (Bixby and Brown, 1975). Accumulation of a specific protein with R_F value 0.49 was observed during the hardening of winter rape (Kacperska-Palacz and Wcislinska, 1972). Later work identified two protein fractions that increased with R_F values of 0.47 and 0.51. They were of low mol wt. and localized in the cytosol. They seemed to be linked with the hydrophobic compounds of the cytoplasm. They are basic proteins, low in proline and methionine residues. Cysteine and leucine also were somewhat lower. It was suggested that these proteins would not aggregate as readily on freeze-dehydration of the cells.

It is apparent, from the above results, that specific protein changes do occur during the hardening period. This does not, of course, prove a cause and effect relationship.

f. NUCLEIC ACIDS AND SIMPLER NUCLEOTIDES

Increases in RNA accompany the increase in proteins during the fall hardening of locust trees (Fig. 8.5). Similar changes were found in dogwood (Li and Weiser, 1967). Cytoplasmic tests on wheat plants revealed an increase in RNA during late fall and early winter (Chian and Wu, 1965). Chilling temperatures nearly doubled the intensity of rRNA synthesis in winter wheat though not in nonresistant spring wheat (Devay and Paldi, 1977). In all three of the above cases, DNA failed to increase in quantity or in the percent hybridized to RNA. In partial contrast to these results, both nuclear RNA and DNA increased in alfalfa roots (Jung *et al.*, 1967a) and in meristems of winter wheat seedlings (Kallinin and Stasevskaya, 1972) on exposure to hardening temperatures. The lack of an increase in DNA in the first three cases is presumably due to the absence of meristems.

In the case of alfalfa the differences between two varieties differing in hardiness, appeared first in the DNA content during fall hardening, then in RNA content, and finally in water-soluble, TCA precipitable protein content and freezing tolerance (Jung *et al.*, 1967a). These NA changes explain the increased capacity for protein synthesis associated with cold acclimation (Sarhan and D'Aoust, 1975). Similar results and conclusions have been reported for barley (Shiomi and Hori, 1972; Bergman, 1973).

The increase in RNA in winter wheat during hardening has been explained as probably due in part to a decrease in RNase activity (Rochat and Therrien, 1976a). In potato, however, hardening increased RNA synthesis, rather than a decrease in RNase activity (Chen *et al.*, 1977).

In 1-year-old apple twigs, RNA began to increase 1 week prior to the rapid increase in freezing tolerance (Li and Weiser, 1969a). The sRNA increased 38% in 1 week and the light and heavy rRNA increased 41% in 2 weeks, just prior to and during the stage of rapid hardening. DNA showed a possible slight increase during hardening, but during slow dehardening, RNA decreased while DNA increased. The most dramatic decrease during the 3-week dehardening was in the heavy rRNa (4.0%). Both rRNA and sRNA increased markedly, less so in spring than in winter wheat (Sarhan and D'Aoust, 1975). Furthermore, the base composition of sRNA changed in the winter wheat, but not in the spring wheat. Among 11 varieties of wheat with LT_{50} from −2.8 to −14.4°C, the intensity of rRNA synthesis around the freezing point was closely correlated with the LT_{50} (Paldi and Devay, 1977). Rochat and Therrien (1976b), on the other hand, were unable to relate the increase in resistance of winter wheat to a particular class of RNA.

Cytochemical methods have corroborated these nucleic acid changes (Chuvashina, 1962). When radial sections of the bark next to the cambium of 1-year-old apple shoots were stained, hardy varieties showed methylophilic nuclei as early as August, with the maximum between December and February. In the nonhardy varieties, the nuclei were pyroninophilic even in winter. Similar results were obtained with cherry and apple by Sergeeva (1968): histochemical tests showed a correlation between RNA content in winter and the winter hardiness of the variety.

There is some evidence of accumulation of the simpler nucleotides during hardening. According to Trunova (1968), inorganic phosphate is lower and acid-soluble organic phosphate (mainly sugar-phosphate esters) is higher in frost-hardened winter wheat than in the unhardened state. The hardened plants contain more high-energy phosphorus, indicating a high degree of coupling between oxidation and phosphorylation at low positive temperatures. A similar correlation was obtained by Sergeeva (1968) and by Borzakivska and Motruk (1969). In confirmation of such a relationship, infiltration with DNP during the first hardening stage decreased the nucleotide content 25–28% and lowered the freezing tolerance (Trunova, 1969). On the other hand, the nucleotide content greatly decreased during the second hardening phase at −5°C, due presumably to utilization in the hardening process.

Similarly, the application of phosphate fertilizers increased the content of high-energy phosphorus in the tillering nodes of winter wheat and winter rye and raised their freezing tolerance (Kolosha and Reshetnikova, 1967). Lowering the temperature from 0° to −20°C lowered the high-energy phosphorus to a greater degree in the less resistant species. Soaking the plants for 12 days in a 10% sucrose solution caused a rise in high-energy phosphorus, and an increase in reducing sugars as well as in freezing tolerance. In opposition to these results, the content of organic acid-soluble phosphorus decreases markedly in grape buds during the fall transition to the rest period, and increases after the rest period (Cheban, 1968).

As in the case of the proteins, it is again a question whether or not the inverse correlation between nucleic acid content and temperature indicates a role in the hardening process. In the case of four perennial grasses, as the day temperature was increased from 18.3° to 43.8°C, the RNA concentration decreased in a nearly linear manner (Baker and Jung, 1970b). This is the same kind of a correlation as found during hardening, yet it can have no relationship to freezing tolerance since no hardening occurs within this temperature range.

g. LIPIDS

A positive correlation between lipid content and hardiness was often noted in the early literature (see Levitt, 1941). Some of these early results have long been known to be due to erroneous methods of identifying lipids, and these errors were still being made in relatively recent times (see Pieniazek and Wisniewska, 1954). Modern quantitative methods, however, have succeeded in establishing an increase in lipid content during hardening of locust trees (Siminovitch *et al.*, 1968; Siminovitch *et al.*, 1967a, b), as well as in apple shoots (Okanenko, 1974; Ketchie and Burts, 1973) and the tubers of hardy but not of freeze-sensitive species of nutsedge (Stoller and Weber, 1975). There was nearly a doubling of the fatty acid content of alfalfa roots (Gerloff *et al.*, 1966) during hardening. The increased synthesis was more active in a hardier than in a less hardy variety of alfalfa (Grenier *et al.*, 1973, 1975; Grenier and Willemot, 1974). Similar results have been obtained with a less hardy species (*Citrus sp.*), although only in the hardier varieties (Kuiper, 1969). Even the green alga, *Chlorella ellipsoidea* accumulated lipid particles (observable with the electron microscope) during hardening (Hatamo, 1978).

In the case of locust trees, Siminovitch's group found that the lipid increase during the first stage of hardening was confined to the phospholipids.

The fundamental nature of this increase was revealed by isolating bark segments in August (Siminovitch *et al.*, 1968). These "starved" segments showed none of the capacity to synthesize soluble proteins or RNA or to increase in cytoplasmic substance. Yet they tolerated freezing down to −45°C in November or December. They did show a slight increase in [^{14}C] leucine incorporation, but the largest increase was in phospholipid (Table 8.7), which was repeatedly shown to double during hardening (Siminovitch *et al.*, 1975; Singh *et al.*, 1975). It also increased in wheat seedlings, especially in the hardier variety (de la Roche *et al.*, 1972; Willemot, 1975).

A similar result was obtained in the case of citrus (Kuiper, 1969). The two hardy varieties (Dancy tangerine and Satsuma tangerine) had a smaller amount of neutral lipids and a larger amount of phospholipids than the two tender varieties (Marsh grapefruit and Eureka lemon). The glycolipids showed no significant difference.

Yoshida (1969b) also corroborated the increase in phospholipids (PL), during hardening of black locust. Phosphatidylethanolamine (PE), phosphatidylcholine (PC), and an unidentified component increased markedly from autumn to winter, then decreased toward spring. Phosphatidylglycerol (PG), accumulated in late autumn and seemed to be transformed into the other three substances under conditions of both artificial and natural hardening. An active synthesis of lipids also occurred in the cortex of poplar twigs during freeze-hardening, as demonstrated by the active incorporation of labeled glycerol and acetate (Yoshida, 1971). The phospholipids increased, apparently at the expense of the triglycerides (Yoshida and Sakai, 1973). Similarly, a hardier alfalfa variety showed a greater increase in phospholipids and a lesser increase in triglycerides than a less hardy variety (Grenier and Willemot, 1974). The hardening of alfalfa at 1°C actually decreased the incorporation of labeled *P* into the lipids, but less so in the hardier variety (Grenier and Willemot, 1975), and the proportion incorporated into phosphatidylcholine increased. In wheat, however, the phospholipid composition was unchanged, indicating no preferential synthesis of individual phospholipids (de la Roche *et al.*, 1973). It has, in fact, been suggested that increased phospholipid synthesis may not be a prerequisite to hardening of winter wheat, but may be required to maintain resistance (Willemot, 1975). In contrast to these results, a decrease in temperature caused a considerable increase in total content of PL in winter wheat, mainly due to PC and PE, less to PI (phosphatidylinositol) and PG (Pankratova and Khokhlova, 1977). There was a greater content of these four fractions in the hardy variety.

Freezing injury is now known to be accompanied by the degradation of phospholipids to phosphatidic acid in the cortex of poplar twigs (Yoshida and Sakai, 1974), in soybean cotyledons (Wilson and Rinne, 1976), in yeast cells (Souzu, 1973), and in potato leaves (Rodionov *et al.*, 1973). This would seem to reinforce the need for the buildup of a reserve of phospholipids to replace those degraded during freezing.

Other lipids have also been found to increase during the hardening of some plants—DGDG at the expense of MGDG in the case of poplar twigs (Yoshida, 1969a, 1973a) and pine needles (Bervaes *et al.*, 1973), as well as a slight increase in the sterols of poplar twigs. In wheat seedlings, the sterols actually decrease during the initial stages of hardening, but increase later in the root, to a final value equal to or greater than the initial value (Davis and Finkner, 1973). Dehardening reversed all these lipid changes (Yoshida, 1973b). It has long been known (e.g., Malhotra, see Levitt, 1941), that low temperature increases the degree of unsaturation of the fatty acids (as measured by the iodine number). This has been corroborated with more modern methods in the case of alfalfa (Gerloff *et al.*, 1966), which accumulated polyunsaturated fatty acids (linoleic and linolenic) during hardening. This initial stimulation of unsaturation was converted to a repression in a later stage of hardening of alfalfa. An increase in unsaturation of the fatty acids (a rise of linolenic at the expense of linoleic acid) occurs during the hardening of the seedlings of 10 cereal varieties (9 wheat and 1 rye) at 2°C and spin labeling showed that the phase transition of the lipids (see Chapter 4) occurs at a lower temperature in the cold grown seedlings (Miller *et al.*, 1974). The more resistant the variety, the greater the change (Farkas *et al.*, 1975). Similarly, the ratio of unsaturated to saturated fatty acids in three nutsedges followed the same order as their freezing tolerance (Stoller and Weber, 1975). The greater resistance of quackgrass rhizomes (survival of −17°C) than rhizomes of Johnson grass (killed below −9°C) was attributed at least partially to tne higher content of lipids and higher proportion of unsaturated fatty acids in the quackgrass (Stoller, 1977). This change in unsaturation, however, occurred to the same degree in two less hardy as in two hardier varieties of wheat (de la Roche *et al.*, 1975), and was not pronounced in the roots (Willemot *et al.*, 1977). On the other hand, a much larger increase in content of linoleic and linolenic acid occurred at 2°C in a mutant winter wheat genotype than in the parent spring genotype. De la Roche and his co-workers therefore suggest that the increased unsaturation of the lipids at hardening temperatures (2°C) may be associated with the vernalization process rather than with hardening to freezing (De Silva *et al.*, 1975). In opposition to this interpretation, both spring and winter rye seedlings showed this increased unsaturation in the first 4 weeks of vernalization (Thomson and Zalik, 1973).

In support of a role for unsaturation of fatty acids in freezing tolerance, treatment of 12-day old winter wheat plants with BASF 13-338, 36 hr before frost hardening, simultaneously and completely inhibited both the hardening and the accumulation of linolenic acid in the roots during the hardening period (Willemot, 1977). There was, however, a concurrent increase in linoleic acid. The complexity of the relation is further indicated by the increase in linolenic acid in both the leaves and roots of winter rape plants at 5°C, although only the leaves increased in freezing tolerance (Smolenska and Kuiper, 1977). Even in the leaves, however, the freezing tolerance increased only 2–3.5°C.

Similar results have been obtained with microorganisms. Eleven species of *Aspergillus* produced an increased unsaturation of the fatty acids at a lower temperature (24° versus 30°C). Three other species, however, failed to show any change and a fourth changed in the reverse direction (Dart and Stretton, 1976). A mutant of *Neurospora crassa* had double the saturation of the fatty acids found in the wild type, and a higher phase transition temperature for its phospholipid extracts (−11° instead of −31°C in the wild type; Friedman, 1977). These results may be explained by the increased activity of acyl-CoA desaturase in the microsome fraction from cold-grown cells of *Candida lipolytica* than from cells grown at 25°C (Pugh and Kates, 1975).

When a fatty acid auxotroph of *E. coli* is grown with a low content of unsaturated fatty acid (8–11%) in the membrane, the membrane becomes fragile, (Akamatsu, 1974) due presumably to insufficient fluidity. It can, however, be grown in 1.40 OsM medium (due to added glycerol, sucrose, KCl or NaCl), though not in normal basal medium which is only 0.36 OsM.

Unfortunately, in the case of microorganisms it is seldom if ever determined whether intracellular or extracellular freezing is involved. It is probably intracellular and therefore not comparable to the freezing of higher plants. Thus, the enrichment with linolenic acid (m.p. −10°C) of a fatty acid–desaturase auxotroph of yeast resulted in more fluid membranes than with stearoleic acid (m.p.+45°C, Kruuv *et al.*, 1978). Yeast cells grown anaerobically survived freezing in liquid N_2 (−196°C) better when the fatty acid composition was mainly stearoleic than when linolenic acid. When grown aerobically, mitochondria were functional (unlike the anaerobic cultures) and freeze-thaw survival was reversed—cells with linolenic acid in their membranes survived better. Since the cells were frozen rapidly, freezing must have been intracellular, and survival may have depended on ability to repair the membrane on thawing.

h. GROWTH REGULATORS

While some investigators believe that growth regulators are important factors in freezing tolerance (Sulakadze and Rapava, 1973), others conclude that they are, at best, of secondary importance (Reid *et al.*, 1974), or less effective than short days (Tumanov *et al.*, 1974).

The presence of a "translocatable factor" from hardened to unhardened parts of the same plant has been offered as evidence of a hormone effect. At least in some cases, however, the "translocatable factor" is simply sugar (Le Saint and Frotte, 1972). Nevertheless, the results obtained by several investigators with a number of different species indicate that hardening is accompanied by an increase in content of inhibitors (such as ABA) and a decrease in content of auxins and GA-like substances (Table 8.8). Several of the plants investigated, however, develop only very moderate freezing tolerance. In the case of the more resistant trees, it has been suggested that these changes act primarily by increasing the duration of deep dormancy (Sarkisova and Chailakhyan, 1974). In agreement with this conclusion, after deep dormancy was attained in black currant and birch seedlings, the content of inhibitors sharply decreased and the stimulants increased although the hardening capacity remained maximal (Tumanov *et al.,* 1973b). A warm branch of Douglas fir transmitted a factor (which may have been a growth regulator) to a chilled branch which prevented full hardening (Timmis and Worrall, 1974).

i. MISCELLANEOUS SUBSTANCES

Many substances have at one time or another been reported to show a correlation with hardiness, e.g., pentosans, anthocyanins, tannins (see Levitt, 1956). In *Hedera helix,* for instance, anthocyanin (as well as total sugar) content of the leaves paralleled hardiness from August to May (Parker, 1962). In each case, however, negative results are at least as common (see Levitt, 1956).

Ascorbic acid content has been reported to increase in many plants during September when prolonged frost sets in (Shmatok, 1958). The content of wheat seedlings was much higher when germination (in the dark) occurred at a hardening temperature (1.5°C) than when at nonhardening temperatures (Andrews and Roberts, 1961; Waisel *et al.,* 1962). The hardier winter varieties developed higher quantities than the less hardy varieties, but only at hardening temperatures. After 6 weeks both hardiness and ascorbic acid decreased. Winter hardy oat varieties (Bronco and Mustang) contained more ascorbic acid than did winter tender varieties (Alamo and Frazier); similarly lines selected by ascorbic acid content proved more hardy in field tests (Futrell *et al.,* 1962). During the winter months, the ascorbic acid content of Brussels sprouts increased at 2°-4°C, decreased at higher temperatures, but also decreased at temperatures below 0°C if kept at this temperature for a considerable time (Arndt, 1974). According to Polishchak *et al.* (1968), bound ascorbic acid is converted to free ascorbic acid in the bark of the frost-resistant black walnut on transfer from the cold room to the laboratory. The less resistant Persian walnut showed the opposite change—a rise in bound ascorbic acid. Schmuetz (1969) has established a close relationship between content of reduced ascorbic acid and varietal hardiness of wheat and barley.

A small rise in pH has been observed during hardening by several investigators (Levitt, 1956). This may conceivably be due to a decrease in organic acids, which has been reported to occur in wheat at low, although freezing, temperatures (Babenko and Ruzhitskaya, 1973).

2. Metabolic Rates

a. RELATION TO ACCUMULATION OF SUBSTANCES

From the above results, we must conclude that hardening is normally accompanied by an accumulation of one or more substances synthesized by the plant—sugars, amino acids, proteins, nucleic acids, lipids, and perhaps several others. This conclusion is supported by the visible increase in the amount of protoplasm per cell (Fig. 8.7). Two questions follow logically: 1) What metabolic processes lead to the accumulation? 2) What is the role of this accumulation of substances?

The mere fact that substances accumulate in the fall is no proof, of course, that they play a role in freezing tolerance. They may simply serve as reserve substances to be used in the spring burst of growth. In favor of this explanation, is the pronounced (although not full) hardening described above, that can occur without the accumulation of proteins, nucleic acids, and lipids, and the absence of sugar accumulation in many cases of full hardening. It is still conceivable, however, that the accumulation is necessary for full hardening.

How can we account for this accumulation of substances during harden-

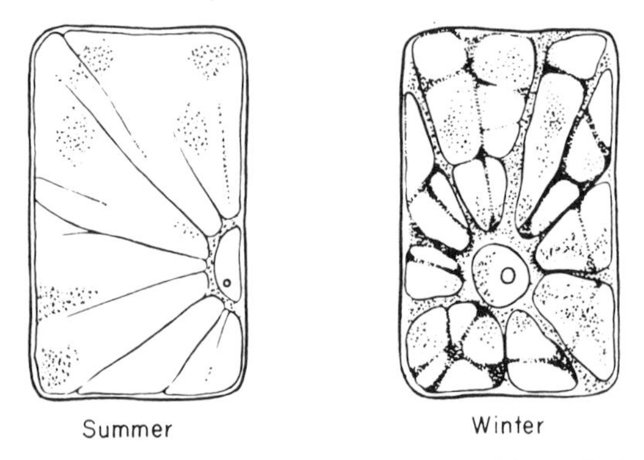

Figure 8.7. Diagrammatic representation of living cells of locust bark, showing higher content of protoplasm during winter. (From Siminovitch *et al.*, 1967b.)

TABLE 8.8

Relation of Growth Regulators to Freezing Tolerance

Species	Relation	Reference
a. Auxin		
Citrus and peach trees	Fall and winter decrease; greater in more tolerant plants	Sulakadze, 1961
Cinnamomum spp.	No activity in the hardier species during Jan.; low activity in the less hardy species	Bregvadze *et al.*, 1975
b. Gibberellin		
Acer negundo	Greatest activity under nonhardening conditions	Irving, 1969a
Alfalfa	Disappears during hardening only of hardy variety. Addition	Waldman *et al.*, 1975
	nullified hardening	Rikin *et al.*, 1975
Cherry	Spraying Aug. 22–Sept. 12 increases injury	Proebsting and Mills, 1974
Wheat	Content of seedlings at 2°C 28–320 × less than at 20°C	Reid, 1974
c. Abscisic acid (ABA)		
Acer negundo	Hardening conditions induced the highest levels	Irving, 1969a
Alfalfa	Application simulates SD and low T (diminishes GA and increases hardiness	Waldman *et al.*, 1975 Rikin *et al.*, 1975
Maple	ABA accumulates in SD and cold night in both northern and southern race	Perry and Helmers, 1973
Tobacco	4 stresses all increased ABA content and increased freezing tolerance; application increased freezing tolerance	Boussiba *et al.*, 1975
Siberian larch	ABA accumulates during frosts	Fedorova *et al.*, 1976
Cinnamomum japonicum *Cinnamomum camphora*	Activity of inhibitors is low during growth period, high during winter rest, and higher in the hardier species	Bregvadze *et al.*, 1975

ing? The accumulation of any substance synthesized by the plant is the net result of an excess of synthesis over breakdown:

$$A = S - B \tag{8.1}$$

where A = amount accumulated; S = amount synthesized; B = amount broken down.

Therefore, the increased content of any substance under hardening conditions over the content under nonhardening conditions must be due to an increased net synthesis:

$$(S - B)_H > (S - B)_{NH} \tag{8.2}$$

where H = hardening conditions; NH = nonhardening conditions. Since hardening temperatures are lower than nonhardening temperatures, both the rates of synthesis and of breakdown may be expected to decrease. The greater accumulation of substances in a hardy plant at hardening temperatures than in a tender plant, must then be due to either (1) a smaller decrease in synthesis, or (2) a larger decrease in breakdown:

$$(S_{NH} - S_H)_{HP} < (S_{NH} - S_H)_{TP} \tag{8.3}$$

or

$$(B_{NH} - B_H)_{HP} < (B_{NH} - B_H)_{TP} \tag{8.4}$$

where HP = hardy plant; TP = tender plant.

Reaction (8.4) consists of two components: (a) breakdown reactions in the fundamental metabolism of the plant (the metabolism required to keep a cell alive), and (b) breakdown reactions in the additional metabolism that supports the growth and development of the plant (cell division, enlargement, and differentiation). Therefore, (8.4)—the greater decrease in the breakdown reactions of hardy than of tender plants at hardening temperatures—may be due to a greater decrease either in its fundamental metabolism (reaction 8.4a) or in the additional metabolism of growth and development (reaction 8.4b):

$$(B_{FNH} - B_{FH})_{HP} > (B_{FNH} - B_{FH})_{TP} \tag{8.4a}$$

$$(B_{GNH} - B_{GH})_{HP} > (B_{GNH} - B_{GH})_{TP} \tag{8.4b}$$

where F = fundamental metabolism; G = additional metabolism for growth and development.

Which of these three theoretical possibilities—Eqs. (8.3), (8.4a) or (8.4b)—is (or are) actually responsible for the accumulation of substances by hardy plants during the hardening process?

Hardening normally begins in the short days of fall (Chapter 7), which according to the available evidence (see above) induce a decrease in growth stimulators (gibberellins and auxins) and an increase in growth inhibitors (abscisic acid). As a result, growth comes to a complete stop in the case of woody plants which enter a period of dormancy. The growth process is apparently uncoupled from metabolism. The fundamental metabolism continues, and the net synthetic product, which would have been used for growth, accumulates. In the case of winter annuals, the hardiest of which do not attain the extreme hardiness of woody plants, no true dormancy occurs, and the seedlings continue to grow slowly. In many cases, the less hardy varieties and species do not have their growth sufficiently curtailed by the short days (due, presumably, to an insufficient change in growth regulators) and due to this excess growth in the fall cannot accumulate the synthesized substances to as high a degree as the more hardy varieties and species.

In both woody perennials and winter annuals we must, therefore, conclude that the accumulation of substances during the fall is at least partly due to (8.4b)—a cessation (woody plants) or a decrease (winter annuals) in the breakdown reactions associated with growth. This effect is so large, that the accumulation will occur even if the breakdown reactions of the fundamental metabolism of the plant (reaction 8.4a) have not been greatly slowed down, and even if the synthesizing reactions (reaction 8.3) are markedly slowed down at hardening temperatures. The accumulation of substances therefore depends on the effect of hardening conditions on both the respiratory and photosynthetic rates.

b. RESPIRATORY RATE

The earlier measurements and several more recent ones [Table 8.9(a)] indicate an inverse relation between freezing tolerance and respiration rate, as would be expected if the accumulation of substances (specifically carbohydrates) is a component of freezing tolerance. Some of the more recent results, however, favor a direct relation [Table 8.9(b)]. One possible explanation for this apparent discrepancy is provided by Barashkova and Udovenko (1973). During fall hardening, the carbohydrate, N, and P metabolism were all more intense in the more freezing tolerant wheat variety. During winter, the relation was reversed. Results with barley were similar. Respiration rate increased from October until near the end of November, but decreased after this date (Svec and Hodges, 1973).

In the case of wheat varieties, there is a close relationship between absorption of micronutrients (^{65}Zn and ^{60}Co) at low temperatures and resistance to freezing (Giosan *et al.*, 1962). The resistant varieties, in fact, were able to maintain a level of absorption at 5° to 7°C close to that at higher temperatures. It is difficult to explain these results except by a greater respiratory rate in the more resistant varieties leading to increased active absorption.

It is, therefore, conceivable that both a decrease and an increase in respiratory rate may occur during different stages of hardening, the former leading to an increased accumulation of organic solutes, the latter to an increased ion uptake.

c. PHOTOSYNTHETIC RATE

As in the case of respiration, the relation between photosynthesis and freezing tolerance is not simple. It has long been known that the low-temperature hardening of plants requires light and CO_2 (see Sec. 2b, chap. 7) and therefore does not occur in the absence of photosynthesis. Photosynthesis continues in excess of respiration at hardening temperatures (Andersson, 1944), and the slow growth rate fails to use up the excess photosynthetic products. This is undoubtedly because the hardening temperature may permit a surprisingly high rate of photosynthesis, e.g., in mimulus at 0°C the rate of light-saturated photosynthesis ranged from 11–20% of the varieties (Andersson, 1944). Similarly, the rate of ^{14}C incorporation into maximum rate at higher temperatures (Milner and Hiesey, 1964). Furthermore, this accumulation of excess photosynthetic products during the hardening period is greater in hardier varieties of wheat than in less hardy

TABLE 8.9

Relation between Respiratory Metabolism and Freezing Tolerance

Species	Reference
a. Inverse relation (respiration $\alpha \frac{1}{\text{tolerance}}$)	
Apples (measured at 6°C)	DeLong *et al.*, 1930
Grains (measured at warm temperature)	Newton *et al.*, 1931
Legumes	Bula and Smith, 1954
Pears	Zotochnika, 1962
Wheat	Andrews *et al.*, 1974a; Bakumenko,, 1974
Lauraceae	Khisamutdinova *et al.*, 1972 Bregvadze, 1974
b. Direct relation (respiration α tolerance)	
Alfalfa	Swanson and Adams, 1959
Fruit and nut trees	Yasmykova and Tolmachov, 1967
Douglas fir	Weise and Polster, 1962 Sorensen and Ferrell, 1973
Wheat (mitochondria)	Kenefick and Swanson, 1963 Karmanenko, 1972
Grapevine (higher ATP concentration)	Dogramadzhyan and Marutyan, 1972
Barley	Svec and Hodges, 1973

sucrose was 50% higher in snow plants when in the hardened state than when in the unhardened state (Zhuravlev and Popova, 1968). There was, in fact, little change in rate of photosynthesis from 0° to 10°C. Similarly, the alga *Chlorella pyrenoidosa* photosynthesises at essentially the same rate at temperatures from 7°–20°C (Steemann-Nielsen and Jorgenson, 1968). In fact, the absolute rate in winter annuals shifted from an optimum at 25°–30°C during summer, to one at 15°C during winter, approaching the rate at the summer maximum (Regehr and Bazzaz, 1977). But this dependence on photosynthesis occurs during the early stages of hardening (late summer or early fall) when the plant is growing and/or accumulating the reserves needed for successful overwintering. What about photosynthesis during late fall and winter? Here the results appear to be contradictory, some indicating that freezing tolerance is correlated with increased, others with decreased photosynthesis. It is now apparent that the specific correlation depends on the severity of the climate to which the plant must adapt (Table 8.10).

1. Plants adapted to a moderate climate depend on their ability to maintain a positive metabolic balance (an excess of photosynthesis over respiration). This group includes algae, overwintering herbaceous plants, evergreen, broadleaved trees and shrubs, and conifers of relatively mild climates. All showed a positive net photosynthesis at 0°C or slightly above, and the algae actually had as high a rate at the low temperature as at 20°C. Even some of these plants, however, may show a decreased rate of photosynthesis during the later stages of hardening (wheat; Anderssen, 1944) or in late winter (holly; Martincic *et al.*, 1975).

2. Extremely hardy plants enter a rest period in the fall during which their net photosynthesis is zero or close to it, even if they are warmed up to growing temperatures [Table 8.10(b)]. This group includes both the hardiest evergreen conifers and the deciduous woody angiosperms whose twigs contain ample chlorophyll that colors the cytoplasm layer of the cortical cells an intense green in fall and winter. Furthermore, the chlorophyll content of the shoots has been found to increase from October to March, followed by a decrease (Smol'skaya, 1964). Similarly, the bark of hardier trees has a higher chlorophyll content during winter than that of less hardy trees (Borzakivs'ka, 1965). In *Populus tremuloides,* the chlorophyll content of the bark is lowest in winter (Barr and Potter, 1974). Nevertheless, on a surface area basis, the chlorophyll concentration is similar to that of the leaves (Foote and Schaedle, 1976a). As a result, the bark photosynthesis reduced the respiratory loss of the twigs, though not achieving a net gain (Foote and Schaedle, 1976b). In the case of apple, it has been suggested that the winter maximum of the sugars may be connected to this photosynthetic activity of the branches and trunk (Savitskii, 1976). Photosynthesis may remain active in silver fir if the

winter is mild enough (Pisek and Kemnitzer, 1967). This may explain the results with *Picea sitchensis* which place it in both groups.

Both respiratory and assimilation rates decreased during hardening in the fall in wheat, barley, spinach, lamb's lettuce, and *Picea excelsa* (Zeller, 1951). Rising temperatures in winter increased both, but to a lower level than the same temperatures in spring, and more so in winter annuals (which had no winter rest period) than in *Picea excelsa* (which did). Assimilation was detected at −2° to −3°C and respiration at −6° to −7°C, or even as low as −30°C (Scholander *et al.*, 1953) although the Q_{10} (temperature coefficient) rose precipitously to 20–50 at temperatures below 0°C (Scholander *et al.*, 1953). *Dryas integrifolia*, an arctic rosaceous plant growing in the North West Territory of Canada appears to be intermediate between these two groups. In a controlled environment, its leaves showed a positive net CO_2 assimilation at −5°C (Hartgerink and Mayo, 1976). With the onset of dormancy, however, respiration and net CO_2 assimilation both decreased, and the optimum leaf temperature for net CO_2 assimilation was between 9° and 14°C.

There is, of course, the possibility of freezing injury to the photosynthetic apparatus. Thus, chloroplasts of winter wheat showed structural damage after overwintering (Petrovskaya-Baranova, 1972). Hardy plants, however, do not show this injury. For instance, chlorophyll content decreases in a sensitive species of potato, but remains constant or increases in a hardy one (Nyuppieva, 1973). The above winter wheat was able to repair the damage. Similarly, winter vetch did show a decrease but had the capacity to synthesize the pigments (Karimov *et al.*, 1974), and the extremely hardy conifers prevented pigment decomposition by their firm binding to proteins under severe winter conditions (Tsaregorodtseva and Novitskaya, 1973).

The net photosynthesis of fir and maple leaves, however, was inhibited by freezing temperatures far above those causing tissue injury (Bauer *et al.*, 1969). In fact, in the case of fir (*Abies alba*), as soon as the needles were shown to form ice (by thermocouple measurements), photosynthesis (measured at 20°C) was inactivated. The depression of photosynthesis following thawing was proportional to the previous cooling. After 2 days of freezing at −8°C, CO_2 was evolved in the light for 2–3 days following thawing. Yet at least 80% of the assimilation tissue was intact, and the leaves finally regained their full photosynthetic capacity. In opposition to photosynthesis, the cooling leads to a reversible rise in respiration (measured at 20°C). The extent of the rise is less in hardened than in unhardened plants—in the case of firs, in the open, it was only half of that in greenhouse plants cooled to the same degree (250 and 500%, respectively, of the original rate). The increase in respiration reaches a peak some 5 hr after thawing, if frozen at −6°C, some 9–12 hr after thawing when the freezing was more severe.

TABLE 8.10

Relation between Photosynthesis and Freezing Tolerance

Plants	Observation	Reference
a. Plants adapted to moderately severe low temperatures		
1. Algae		
Chlorella pyranoidosa	Photosynthesis equal at 7°C and 20°C	Steemann-Nielsen and Jorgenson, 1968
Dunaliella tertiolecta	Photosynthesis greater at 12°C than at 20°C	Morris and Farrell, 1971
2. Herbs		
Alpine snow tussocks (*Chionochloa* spp.)	Maximum net assimilation during brief periods above freezing	Mark 1975
3. Broadleaved evergreens		
Holly (*Ilex aquifolium*)	Net photosynthesis positive above 0°C	Martincic *et al.*, 1975
Quercus ilex and	Net photosynthesis positive above 0°C	Weinmann and Kreeb, 1975
Olea europea	(No latent period even after −8° night)	
4. Conifers		
Picea sitchensis	Photosynthesis unreduced after subzero temperatures	Neilson *et al.*, 1972
b. Plants adapted to extreme freezing temperatures		
1. Deciduous, woody angiosperms after leaf drop		
Linden and lilac buds	During bud dormancy, no photochemical activity of chloroplasts	Yasnikova, 1975
Populus tremula and *Fraxinus americana* (also *Larix decidua*)	Even after warming up for 16 hr, respiration of bark exceeded photosynthesis	Keller, 1973
Populus tremuloides	No net photosynthesis during winter in either winter or summer stems	Foote and Schaedle, 1976
2. Conifers		
Pines	End of October, half the rate in May, beginning Nov., 1/10 original value	Tranquillini, 1957
Pines (white, stone, bristlecone)	Decreases rapidly in fall, to zero in late fall or early winter	Shiroya *et al.*, 1966 Bamberg *et al.*, 1967 Schulz *et al.*, 1967
Silver fir	As above	Pisek and Kemnitzer, 1967
Pine, spruce, juniper	As above	Ungerson and Scherdin, 1965
Picea sitkensis	Marked decrease in rate when soil temperature below 1°C	Turner and Jarvis, 1975
3. Evergreen angiosperms		
Broad-leaved evergreens (at 0° to −20°C)	Steady decrease from October to zero in Dec. or Jan.	Steinhubel and Halas, 1969

It should be pointed out that it is *apparent* photosynthesis which is zero in winter. *Actual* photosynthesis is, on the average, positive in winter (Ungerson and Scherdin, 1965). The plant may even adapt so as to continue photosynthesis in the dehydrated (frozen) state. Possible indirect evidence of such adaptation has been obtained in the case of *Pinus contorta* (Morris and Tranquillini, 1969). In winter and spring (December–April) photosynthesis decreased less with decreasing osmotic potential of the root medium than it did in summer.

Other plants, besides the conifers, are capable of photosynthesizing in the frozen and, therefore, dehydrated state. Unlike the conifers, some of them may show net as well as actual photosynthesis. Lichens, that survive −75°C for several days, detectably photosynthesize at −24°C (Lange, 1962, 1965). The temperature optimum for photosynthesis was 10°C, and it ceased above 20°C. At −5°C, in spite of ice formation, the rate was still half the optimum. Most of the species ceased CO_2 uptake at −7° to −13°C. Tropical lichens showed little CO_2 fixation below 0°C and their optimum temperature for photosynthesis was 20°C. However, the temperature minimum did not always parallel the habitat of the lichen. By means of ^{14}C labeling, two lichens were shown to assimilate carbon at −11°C at rates 1/5 to 1/9 the rates at +15°C (Lange and Metzner, 1965). Even among the higher plants, *Ranunculus glacialis* and *Geum reptens* (which grow at altitudes as high as 3100 m in the Alps) had compensation points below 0°C. A net CO_2 uptake occurred at −3° to −5°C (Moser, 1969).

The winter deficit of conifers may be serious in the subarctic. In *Pinus sylvestris,* although positive photosynthesis occurs only above −4°C, respiration rate is measurable down to −18°C (Ungerson and Scherdin, 1967). As a result, the daily respiration loss amounted to 186 mg CO_2/100 gm dry needles. The first positive apparent photosynthesis in several conifers did not occur until April 12–17 (Ungerson and Scherdin, 1965). In bristlecone pine, a winter loss of 140 mg CO_2/gm dry weight required 117 hr of photosynthesis during summertime at peak rates to compensate for the loss (Schulz *et al.,* 1967).

In summary, the folowing metabolic changes appear to occur during the hardening of plants in the fall. (1) The respiration rate may increase at one stage of hardening, decrease at another. (2) Photosynthetic carbon assimilation may continue at a high rate during the early stages of hardening, but it decreases during later stages, approaching zero during early winter in the hardiest plants. (3) The accumulation of organic substances (sugars, nucleic acids, proteins, phospholipids, etc.) is due to a cessation or a marked retardation of growth during the early fall at a time when photosynthesis still exceeds respiration. (4) During late fall and winter, both respiration and photosynthesis may continue at reduced rates even in the frozen plant.

3. Specific Reaction Rates and Enzyme Activity

The quantitative relations between growth, photosynthesis, and respiration can explain the net accumulation of groups of substances (e.g. carbohydrates). They cannot, however, explain the accumulation of specific substances belonging to these groups, which depends on the rates of specific reactions and the activities of the enzymes controlling them.

a. REACTIONS INVOLVED IN SUGAR ACCUMULATION

Of all the substances that accumulate during hardening, the carbohydrates have been most intensively investigated (see above). The accumulation of sugars during hardening is brought about in two ways.

(1) In the case of winter annuals, the excess of photosynthesis over respiration and growth during fall is accumulated directly as soluble sugars. (2) Perennials and biennials, on the contrary, accumulate most of their carbohydrates at normal growing temperatures during summer and therefore deposit it in the form of starch. Their sugar increase during hardening is, therefore, due to a starch $\rightarrow$ sugar conversion. The cause of this hydrolysis at low, hardening temperatures has been sought for many years. Müller-Thurgau (1882) ascribed the sugar accumulation in potato tubers to a marked reduction in the respiration rate at the low temperature, without as marked a drop in activity of the hydrolytic enzymes. In opposition to this concept, the sugar accumulation is greater at 0°C than at 3°C even though the respiration rate is minimal at 3°C and rises at 0°C (Hopkins, 1924; Schander *et al.*, 1931—see Snell, 1932; Wright, 1932). Another complication is the stimulating effect of the sugar increase on respiration rate (Snell, 1932). In any case, the respiratory utilization of sugars could not possibly account for the decreased sugar content at high temperatures, for at this rate, all the reserves of the tuber would be used up in a very short time and, as shown above, the sugar loss is actually due to starch formation.

Among some of the earlier investigators, the most widely held concept is a shift in the starch $\rightleftharpoons$ sugar equilibrium caused by the temperature drop (Overton, 1899; Czapek, 1901; Rosa, 1921; Fuchs, 1935; Algera, 1936). According to the laws of thermodynamics, since hydrolysis of starch is an exothermic process the concentration of sugars must increase at low temperatures. Direct evidence of this is the fact that when leaves are floated on 2–5% sugar solutions, rich starch accumulation occurs at 16° to 18°C, but little or none occurs at 0° to 2°C (Czapek, 1901). The sugar must be increased to 7% before some starch forms at 0°C. Unfortunately, the sugar formation directly due to the temperature drop would be relatively slight (Algera, 1936). Furthermore, if this were the complete explanation, all the plants (whether frost-tender or frost-hardy), would show the same sugar accumulation at low temperatures, and it would happen at all times of the year. Neither of these predictions agrees with the facts.

Several other suggestions have been made without any evidence to back them: (1) a weakening of the (starch) synthetic mechanism (Michel-Durand, 1919; Doyle and Clinch, 1927); (2) a dependence of starch synthesis on the temperature, and of starch solution on the time of the year (Weber, 1909); (3) an increased permeability of the starch "membrane" to the enzyme (Coville, 1920); (4) a pH control (Mitra, 1921; but this could not be confirmed by Hopkins, 1924); and (5) a hormone control (Lewis and Tuttle, 1923).

The most logical explanation of the starch → sugar conversion is an activation of the starch hydrolases. Unfortunately, most attempts to prove this have ended in failure (see Levitt, 1972b). The sugar synthetases were not known at the time of this early work and were, therefore, not investigated. Two recent investigations, however, point to a different explanation. Dear (1973), showed that when cabbage leaves are cooled to a hardening temperature, starch hydrolysis begins as soon as measurements can be made—within a half hour and perhaps instantly. Since a mere lowering of the temperature can only reduce reaction rates, no increased enzyme activity can explain this essentially instantaneous effect. Selwyn and Walker (1967) proposed that some enzymes are inhibited more than others by the low temperature, resulting in a relative enhancement of the latter, and leading to an increase in a specific metabolic pathway. This may occur due to a difference in slope of the Arrhenius plot. In favor of this concept, wheat, bean, and spinach plants adapted to high temperatures (20°–25°C) incorporated much more $^{14}CO_2$ into sucrose and less into glycine and serine in comparison with plants grown in the cold (5°–7°C; Sawada and Miyachi, 1974b). Similarly, Pollock and Ap Rees (see Chap. 3) obtained evidence of a differential inhibition of enzymes at low temperatures in potato tubers. At 2°C there was a greater inhibition of some of the glycolytic enzymes than of the sucrose synthesizing enzymes, leading to sugar accumulation. This accumulation exceeded the activity of the sucrose synthetase, but was less than that of sucrose phosphate synthetase (Pollock and Ap Rees; see Chap. 3). The conversion of starch to sugar at low temperatures may, therefore, be explained by an unequal inhibition of the hydrolyzing and synthesizing enzymes.

b. CHANGES IN ENZYME ACTIVITY DURING HARDENING

Many early investigations of enzyme activities failed to reveal any clearcut relation to freezing tolerance (Levitt, 1972b).

A number of investigations have obtained an inverse relationship between temperature and peroxidase activity. Peroxidase was more active at 2° than

at 25° and 37°C in the bulbils and tubers of *Ficaria verna* (Augusten, 1
Two types of peroxidase were found in wheat embryos. One predomi
in embryos germinating at 20°–24°C and the other in embryos germinat
0°–4°C (Antoniani and Lanzani, 1963). Aging pea stem sections for
results in a quantitative increase in existing peroxidases (Highkin, 1967
increase was greater at the lower (22°C) than at the higher (34°C) tem
ture. The isozyme pattern also depended on the aging temperature.
tobacco plant cells grown in suspension secrete a peroxidase isozy
13°C which is absent at 25° or 35°C (De Jong *et al.*, 1968). In all these
the low temperature stimulation of peroxidase activity seems unrelat
freezing resistance. Gerloff *et al.* (1967) detected increases in
peroxidase and catalase activity (per unit protein or dry weight) durin
hardening of alfalfa root. Although the quantities did not clearly differe
between varieties, peroxidase increased sooner in the hardier varietie

The Wisconsin group has investigated peroxidase in many plants in relation to freezing tolerance. They obtained evidence of a relationship between peroxidase activity and hardiness in overwintering carnation plants (Hall *et al.*, 1969). When separated by disc electrophoresis, the peroxidase bands from field-hardened plants showed an increase in number and density in three species (*Sedum, Mitchella,* and *Salix* spp.) as compared with those from unhardened greenhouse plants (McCown *et al.*, 1969b,c). In *Cornus,* however, there was a decrease. In *Salix fragilis,* they also failed to observe a relationship, for the peroxidase activity which appeared during hardening, remained after dehardening (Hall *et al.*, 1969). Roberts (1969a) also found a greater intensity of staining of the fastest moving peroxidase isozyme in the leaves of wheat plants grown at 6°C than in similar leaves from plants grown at 20°C. This increase in peroxidase was apparently not associated with freezing tolerance, since there was no difference between hardy and sensitive varieties. According to Polishchak *et al.* (1968), the reverse change occurs in the bark of walnut trees. On transfer from −1.8° to −15°C, the oxidoreductases (ascorbic acid oxidase, polyphenol oxidase, and peroxidase) all decreased. On return to 20°C, peroxidase activity rose again. The other enzymes failed to show this change.

These results all lead to the conclusion that some enzymes increase in activity during the fall hardening, but that this may have little or nothing to do with freezing tolerance. This is not surprising, particularly in the case of peroxidase, since so many factors may lead to changes in its activity: it may be either inhibited or promoted by indoleacetic acid (IAA) and inhibited by kinetin (Lavee and Galston, 1968), and the IAA may induce one of eight isoperoxidases and repress another (Stuber and Levings, 1969). Nevertheless, there does seem to be a general increase in peroxidase activity at low temperatures—even at low temperatures above the freeze hardening range.

TABLE 8.11

Recent Investigations of Enzyme Activity in Relation to Freezing Tolerance

Enzyme	Plant	Observation	Reference
a. Enzymes whose activity *increased* at hardening temperatures			
1. Dehydrogenases	Alfalfa	In both hardier and less hardy varieties, and decreased during dehardening	Krasnuk *et al.*, 1975, 1976
2. G-6-PDH,6-P-Gluconate-DH	Poplar bark	Activity rose in Sept.	Sagisaka, 1974
3. RuDPC	*Dunaliella tertiolecta*	Higher at 12°C than at 20°C	Morris and Farrell, 1971
4. Ferredoxin-NADP reductase	Wheat	Increased during hardening, more so in the hardier variety	Riov and Brown, 1976
5. Glycerol kinase	*Neurospora crassa*	Increased activity from 26° to 12°, 9°, 6°, 4°, 0°C	North, 1973
6. Catalase	Grapevine		Mininberg and Shumik, 1972
7. Amylase	*Eragrostis curvula*	Increase in total amount of enzyme	Hillard, 1975
b. Enzymes whose activity *decreased* at hardening temperatures			
8. Isocitrate DH	Poplar bark	Activity decreased in Sept.	Sagisaka, 1974
9. RuDPC	Wheat	Chloroplasts from 5-7°C plants less than ½ activity of those from 20-25°C plants	Sawada *et al.*, 1974
10. RuDPC	Spring wheat	Inactivated at higher temp. in more resistant var.	Weidner and Salisbury, 1974
11. RNase	*Albizzia julibrissin*	Decreased during induction of hardiness	Brown and Bixby, 1973a
c. Enzymes whose activity is higher in more tolerant varieties			
12. Catalase	Mulberry trees		Prilutskii *et al.*, 1974
13. Invertase	Wheat	Change in ratio of isozymes	Roberts, 1973, 1975
14. β-fructo-furanosidase	Wheat		
15. Proteases			Kolosha and Kostenko, 1973
16. Catalase			
d. Enzymes whose activity is unaffected by hardening temps.			
17. Glycolate oxidase	Wheat		Sawada *et al.*, 1974

Possibly, this is a hardening to chilling rather than freezing. The reducing system that normally protects the membrane lipids against injurious peroxidation (see Chapter 5) may be unable to do so at chilling temperatures. An increase in peroxidase activity would destroy the peroxides rapidly enough to prevent this kind of chilling injury.

Roberts (1969b) suggests "the substitution at hardening temperatures of a modified form of a protein for the form of the functionally identical protein present at higher temperatures," i.e., the formation of isozymes during hardening.

More recent results have established that enzyme changes do occur during hardening (Table 8.11). In alfalfa, in fact, all the enzymes tested increased in winter, as freezing tolerance and content of soluble proteins increased (Krasnuk *et al.*, 1975). This total enzyme activation is most simply explained by the general increase in protein content, and would not be expected to produce a qualitative metabolic change, such as the hydrolysis of starch to sugar. Several specific enzymes increase in activity in a variety of species at hardening temperatures—for instance some dehydrogenases and catalase. Some are more active in varieties of higher freezing tolerance. Still others decrease in activity or are unaffected. A correlation between catalase activity and hardening has been found in a variety of different plants (grape, wheat, mulberry), agreeing with the earlier results with wheat and pine (Newton and Brown, 1931; Langlet, 1934). The one exception is Citrus (Ivanov, 1939), but this genus possesses only the very minimum of freezing tolerance.

4. Protoplasmic Properties

The anatomical characteristics of protoplasm associated with freezing tolerance have been described above. Several physiological characteristics have also been investigated.

a. PERMEABILITY

On exposure to hardening temperatures, hardy plants that increase in freezing tolerance also increase in permeability to polar substances; tender plants do not (Fig. 8.8). Later investigations (Granhall, 1943) of a series of wheat varieties differing in hardiness failed to show any difference in this factor. This is to be expected since the differences in hardiness had earlier been accounted for by sugar content (see above). The wheat varieties may, therefore, differ in freezing tolerance due to avoidance of freeze-

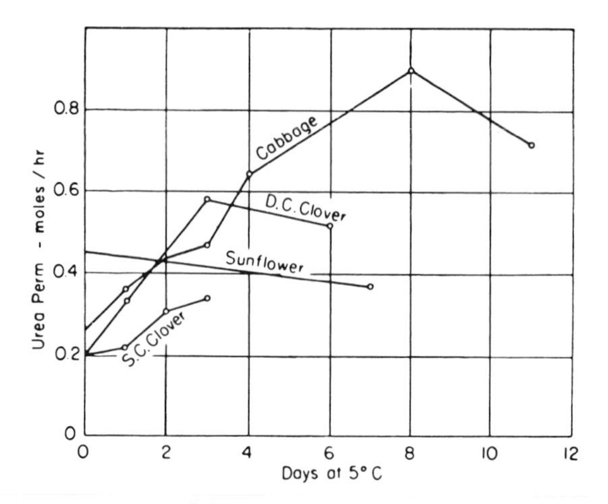

Figure 8.8. Increase in permeability to urea of hardy (cabbage and clover leaf) cells during hardening at 5°C and absence of any increase in tender (sunflower leaf) cells. From Levitt and Scarth, 1936. (Reproduced by permission of the National Research Council of Canada from the *Can. J. Res.* C**14**,297.)

dehydration, and not due to tolerance of freeze-dehydration, or to avoidance of intracellular freezing. In the mulberry tree (Sakai, 1955b), permeability to water rises with hardiness from October to January, but the maximum is then retained even up to May, although freezing tolerance drops markedly. That the difference in permeability exists during freezing and thawing was shown by Siminovitch and Scarth (1938), for the hardy cells both lost water and gained it more rapidly than the nonhardy cells. Similar results have been recently reported by Asahina (1956). He also showed (1962b) that 1 *M* urea increases the permeability to water of unfertilized arbacia eggs, enabling them to survive rapid freezing by preventing intracellular ice formation. Similarly, anaerobically grown cells of *E. coli* have a lower permeability to water than the aerobically grown cells, and are more susceptible to intracellular freezing (Nei *et al.*, 1967). Akabane (see Kuiper, 1969) observed that permeability was directly related to the freezing tolerance of the pistil cells of apple flowers.

In the case of the stems of red osier dogwood, the permeability to water as determined by the diffusion of tritiated water from phloem and cortical parenchyma cells increased during the initial (photoperiodically induced) phase of cold acclimation (McKenzie *et al.*, 1974b). This accompanied an increase in hardiness from −3° to −12°C. Little, if any, further increase occurred during subsequent acclimation to −65°C. This is in agreement with the explanation of the role of cell permeability to water in freezing stress resistance (see also Fig. 8.8). A high permeability to water can be useful only at moderate freezing temperatures when large quantities of water must exosmose quickly to the intercellular ice loci. Any further increase in cell permeability to water would be useless, since a much slower rate of exosmosis is adequate to prevent intracellular freezing at lower freezing temperatures after most of the cell's water has already moved out to the intercellular ice loci. Preliminary NMR tests on the bark cells of *Hedera helix* failed to detect any difference in permeability between acclimated and nonacclimated cells (Stout *et al.*, 1973), presumably because the permeability of the unacclimated cells was already adequate to prevent intracellular freezing at the normal cooling rates. It must be realized that both of the above species possess relatively small cells and therefore a high enough specific surface to maintain rapid rates of exosmosis even with moderate rates of cell permeability.

Hardening of winter rape was accompanied by a temporary increase in electrolyte leakage, which was assumed to indicate an increase in cell permeability (Kacperska-Palacz *et al.*, 1977a). Continued hardening, however, led to a marked decrease in leakage. These results are more logically explained by changes in the active uptake mechanism, the hardening temperature at first decreasing the uptake rate more than the efflux rate since the former has a higher activation energy. This was presumably followed by metabolic adaptations leading to increased influx rate at the hardening temperature, due perhaps to increased respiration rate or ATP concentration (see Chapter 10).

Unlike the other factors discussed in this chapter, protoplasmic permeability is an avoidance factor—avoidance of intracellular freezing. There is no evidence that it is a factor in freezing tolerance.

b. PROTOPLASMIC VISCOSITY

Many measurements have been made of the effect of hardening on protoplasmic viscosity. This is, unfortunately, a complex property, since protoplasm is not a simple liquid. Protoplasmic "viscosity" is, therefore, actually structural or non-Newtonian viscosity. It has also been given the more general name "consistency." Two diametrically opposite results have been obtained. Some investigators (Kessler and Ruhland, 1938, 1942; Granhall, 1943; Johansson *et al.*, 1955) reported a direct relationship between this protoplasmic consistency and hardiness, others (Levitt and Scarth, 1936; Scarth and Levitt, 1937; Siminovitch and Levitt, 1941; Levitt and Siminovitch, 1940) an inverse relationship. Parker (1958) reports a gel to sol conversion in the vacuole of *Pinus* cells during hardening in the fall but the protoplasm may, of course, behave differently. The differences are at least partly due to errors inherent in the methods used. A common method of measuring protoplasmic consistency is to determine the rate of displacement

TABLE 8.12

Evidence of Stiffening due to Dehydration of the Protoplasm or Its Outermost Layer in Unhardened but not in Hardened Cells[a]

Cell property	Unhardened state	Hardened state
1. Plasmolysis shape	Concave	Convex
2. Cytoplasmic strands	Stiffen and rupture on plasmolysis	Remain fluid on plasmolysis
3. Shape of oil drop injected into cytoplasm of plasmolyzed cell	Flattened on deplasmolysis	Convex on deplasmolysis
4. Microdissection tests	Stiff	Fluid
5. Deplasmolysis after prolonged plasmolysis	Rupture of ectoplasm	Normal deplasmolysis
6. Pseudoplasmolysis on thawing	Common	Rare

[a] Numbers 1-5, see Levitt, 1956; number 6, see Asahina.

of the included chloroplasts when subjected to centrifugation. The plastids of nonhardy plants frequently contain large starch grains and are, therefore, heavier and more easily displaced than those of the hardy plant, since the hardening process leads to hydrolysis of starch to sugar and the accumulation of the latter in the vacuole. The difference is enhanced by the larger size of the chloroplasts in the hardy state (Heber, 1959a). These differences would indicate a spurious increase in viscosity on hardening. The more direct methods of microdissection, though not quantitative, failed to reveal any significant difference between the two. However, a small difference could exist without being detected by microdissection. Furthermore, due to the higher vacuole concentration in the hardy cell, its protoplasm would be slightly more dehydrated than that of the unhardy cell when both are unplasmolyzed. Consequently, it is reasonable to expect a slight increase in protoplasmic viscosity on hardening simply due to this reduced water con-

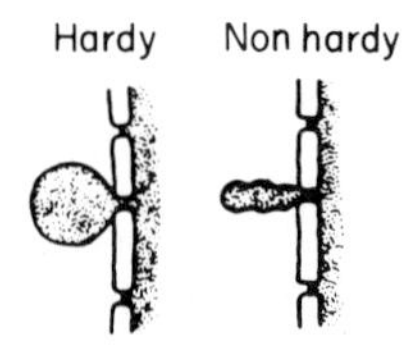

Figure 8.9. Shape of cytoplasm when forced through a punctured pit in a hardened (left) and an unhardened (right) cell of catalpa in a balanced solution of NaCl + $CaCl_2$ (9:1) with an osmotic potential of −50 atm. (From Scarth, 1941.)

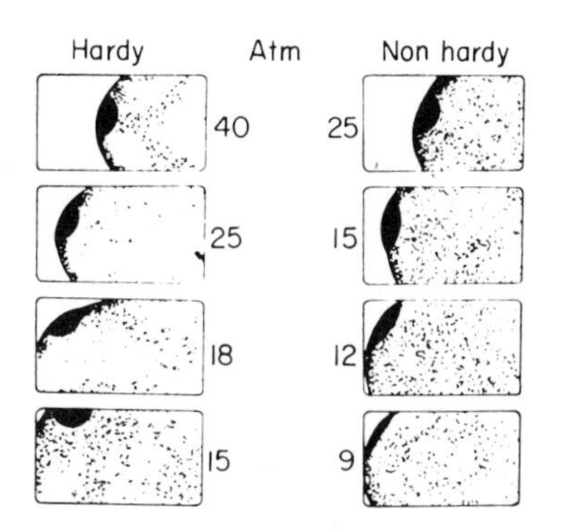

Figure 8.10. Shape assumed by oil drops injected into cytoplasm of hardened (left) and unhardened (right) cortical cells of cornus during progressive deplasmolysis. Osmotic values (atm) of solutions in equilibrium with cells given by the numbers. (From Scarth, 1941.)

tent. The pronounced difference between the two is seen when the cells are dehydrated due to plasmolysis or freezing, and this difference is in the opposite direction. The dehydration leads to a marked stiffening of the protoplasm (or at least its outermost layer) in the unhardened cell, but not in the hardened one. This is clearly shown in many ways (Table 8.12 and Figs. 8.9–8.11).

This protoplasmic stiffening is so closely related to hardiness that it can be used quantitatively as a measure of hardiness by determining relative resistance to plasmolysis and deplasmolysis injury (Scarth and Levitt, 1937; Siminovitch and Briggs, 1953a). Sakai (1957) corroborated this fact but obtained relatively small differences that could be at least partly accounted for

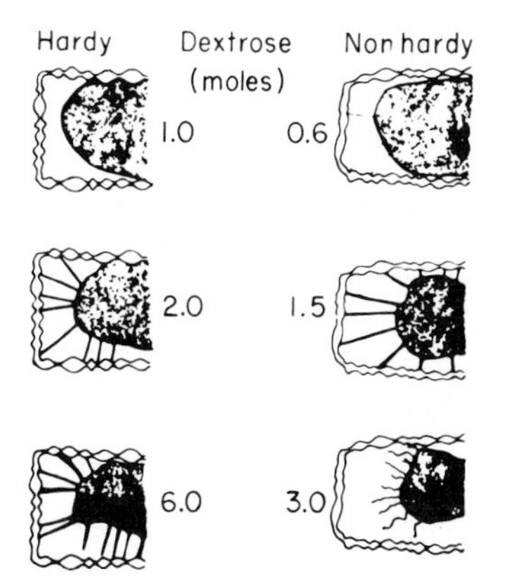

Figure 8.11. Rupture of cytoplasmic strands during progressive plasmolysis of unhardened cortical cells of hydrangea (right), absence of rupture of hardened cells (left). Solutions in equilibrium with cells from top to bottom: Right (unhardened): 0.6, 1.5, 3.0 *M* dextrose; left (hardened): 1.0, 2.0, 6.0 *M* dextrose. (From Scarth, 1941.)

by the higher cell sap concentration of the hardy cells. Such results are obtained only if the plasmolysis time is brief. After prolonged plasmolysis, the hardy cells can be deplasmolyzed successfully from the same degree of plasmolysis as is sufficient to kill nonhardy cells (Table 8.13), and therefore this factor cannot be accounted for by any differences in cell sap concentration. Terumoto (1967) found that marine algae having a high resistance to plasmolysis were also highly tolerant of freezing. Suspension tissue culture cells of *Haplopappus gracilis* are extremely sensitive to freezing, most surviving −1°C, but nearly all being killed at −4°C. This is close to the equivalent concentration of sucrose which causes plasmolysis injury (Towill *et al.*, 1973; Towill and Mazur, 1976). Similar results were obtained even with bovine erythrocytes, using glycerol as the solute (Leibo, 1976).

These protoplasmal differences may be interpreted as evidence of a greater hydrophily of protoplasm in the hardened than in the unhardened state. Siminovitch (1963), however, has concluded that all these protoplasmal changes during hardening, together with the increase in soluble proteins which he showed was due to synthesis during hardening, are part of a general augmentation of protoplasm during the hardening period. He and his co-workers later (1967b) gave visual evidence that the augmentation, indeed, does occur. Furthermore, as already mentioned, the greater mobility (and, therefore, presumably hydration) of the protoplasm in the hardened state is not demonstrable in the normally hydrated protoplasm, but arises as a result of the dehydration process. In other words, dehydration by any method so changes the protoplasmic proteins as to reduce their hydrophily to a greater degree the less hardy the plant. This may explain the many contradictory results in the literature when attempts have been made to relate "bound water" to hardiness.

TABLE 8.13

Comparison of Deplasmolysis in Hardened and Unhardened Cabbage Cells after Plasmolysis in Twice Isotonic $CaCl_2$; Deplasmolyzed in Distilled Water[a]

Condition of plant	Osmotic value (M $CaCl_2$)	Percentage of surviving cells: In epidermis and chlorenchyma	Percentage of surviving cells: In pith
Nonhardened	0.16	Trace	0
5-Day hardened	0.23	Most	0
Nonhardened	0.17	Few	0
10-Day hardened	0.25	All	Many

[a] From Scarth and Levitt, 1937.

Scarth and Levitt's (1937) discovery that many hardy cells show convex plasmolysis and nonhardy cells, concave plasmolysis, has been confirmed by the Russian workers (e.g., Genkel and Oknina, 1954). They have interpreted this as due to the absorption of plasmodesmata into the protoplasm during the winter rest, the protoplasm losing contact with the cell wall. Although their evidence of plasmodesmal differences was based on observations of fixed and stained cells and is, therefore, unreliable as an indication of the condition in the living cell, it has been partially confirmed by observations of living cells with the aid of phase-contrast microscopy (Pieniazek and Wisniewska, 1954). They call this phenomenon "protoplasm isolation" (Genkel and Zhivukhnia, 1959). They have published numerous papers on the subject, culminating (Akad. Nauk., 1968) in a symposium on the "depth of the resting state as judged by the separation of the protoplasm" in the case of winter wheat, native overwintering plants, strawberries, raspberries, pears, citrus fruit, apples, and maple.

It is impossible to accept the above interpretation of Genkel and Oknina; for the photomicrographs of Siminovitch and Levitt (1941) clearly demonstrate the existence of as many protoplasmal strands connecting the plasmolyzed protoplasts of hardened cells to their walls as in the case of the unhardened cells. Some of the strands are clearly attached to opposite sides of pits in the walls and, therefore, represent plasmodesmal connections. The observations of Genkel and his co-workers are, therefore, undoubtedly due to an artifact, and do not apply to the normal, living cells. This interpretation has been supported by a thorough investigation in another Soviet laboratory (Alexandrov and Shukhtina, 1964). A microscopic observation of cells (mounted in silicone oil) of numerous herbaceous and woody overwintering plants was carried out in December, January, and February under natural conditions outdoors as well as after thawing in the laboratory. In all the cases studied, the protoplasm was closely attached to the cell walls. There was no evidence of Genkel's "protoplasm isolation." They, therefore, conclude that the theory of the separation of protoplasm from the cell wall as a means of cell protection against the injurious effects of cold is erroneous.

5. Sulfhydryl (SH) Groups

The sulfhydryl group occurs in plant protoplasm mainly as a component of proteins, an essential amino acid (cysteine or CSH), and a peptide (glutathione or GSH). From time to time, this chemical group has been suggested as a factor in freezing tolerance. Probably the first suggestion was by Ivanov (1939), who found an inverse relationship between glutathione content and winter hardiness of citrus varieties. He explained this by the commonly found inverse relationship between growth and hardiness, and the equally common direct relationship between growth and GSH content. Ewart *et al.* (1953) point to qualitative observations by biochemists, which suggest that amylasė activity is dependent on the integrity of the SH groups within the enzyme molecule. They, therefore, investigated the amylase system of the living bark of the black locust tree. It proved to be inhibited by some SH reagents but not by others. The inhibition, when it occurred, was readily reversed by cysteine. Similar results were obtained with phosphorylase. They, therefore, suggested that differential sensitivity of enzymes to natural SH reagents may be involved in the regulation of metabolic processes of plants and, particularly, in controlling the seasonal changes in composition of the protoplasm.

The first two investigations, though pointing to a possible relation between SH groups and freezing tolerance, appear to contradict each other. The first indicates an inverse relationship between the two, the second a direct one. This apparent contradiction arose also in later investigations. Measurements made on homogenates or on their supernatant solutions obtained from the leaves of four species of hardy plants revealed an increase in SH content after a hardening period of a few days that resulted in increased freezing tolerance (Fig. 8.12). In the case of some of these plants, this relationship was reversed during later stages of hardening (also see below). A close correlation between SH content and freezing tolerance was obtained when

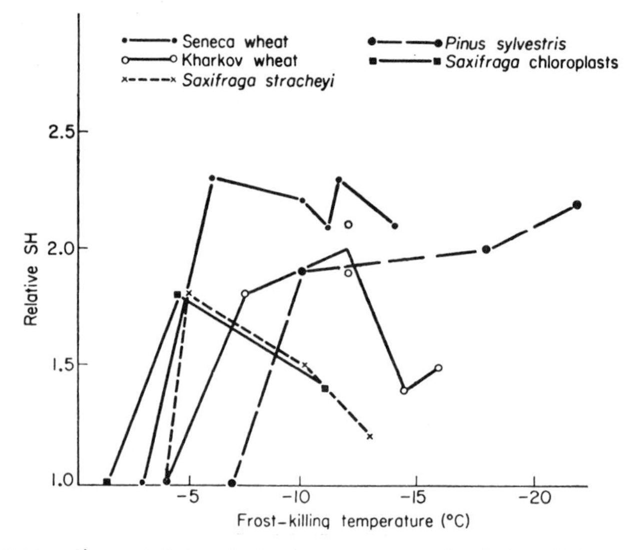

Figure 8.12. Change in SH content of homogenates on hardening. (Adapted from Levitt *et al.*, 1961.)

TABLE 8.14

Frost Hardiness and SH Content in Wheat Varieties[a]

Variety	Hardiness from field experience	Frost-killing temp. (°C)	SH content of supernatant (μmole/gfw leaves)
Anna Migliori	Very hardy	−15	0.84
Carsten VIII		−14	0.93
Eroica II		−12.5	0.84
Criewener 192	Hardy	−12.5	0.81
Derenburger Silber		−12.5	0.77
Austro Bankut		−12.5	0.73
General v. Stocken		−12.5	0.65
Pfeuffers Schernauer	Moderately hardy	−12.5	0.52
Heine VII		−11	0.58
Panter		−11	0.49
Etoile de Choisy	Slightly hardy	−12.5	0.46

[a] From Schmuetz, *et al.*, 1961.

fifteen wheat varieties differing in hardiness were compared (Table 8.14). This relationship held whether SH was determined by the argentometric, amperometric method of analysis or by the completely different colorimetric nitroprusside method (Schmuetz, 1962). In the case of cabbage, the SH changes were detected even in the free chloroplasts, and therefore were undoubtedly due to the proteins (Fig. 8.12). Further evidence was the inability to account for the increase by an increase in GSH, since this was oxidized too rapidly under the experimental conditions used (Levitt *et al.*, 1962). The lower SH content of the unhardened plants was, at least in some cases, associated with a higher SS (disulfide) content and, therefore, due to oxidation (Levitt, 1962).

In the case of vernalizing wheat in the hardened state, the SH content of the proteins was found to be higher relative to SS content (Kohn *et al.*, 1963). In at least one of the wheats, the higher proportion of protein SH to protein SS in the hardened state was an artifact due to a more rapid oxidation of SH to SS in the unhardened material during preparation of the samples for analysis. Similarly, in the case of *Saxifraga,* homogenization in air lowered the SH content, and the degree of this lowering was inversely related to freezing tolerance, due presumably to a more rapid oxidation to SS in the unhardened material (Levitt, 1962).

A thorough investigation by Schmuetz (1969) has now firmly established both the correlation between SH measurements and freezing tolerance, and the artifactual nature of this correlation. He analyzed a total of eighty-seven wheat varieties, including all degrees of winter hardiness to be found in the varieties from western Europe, as well as extremely resistant ones from the U.S.S.R., Finland, and the United States. The SH measurements obtained from three replicates gave a correlation coefficient of 0.85 with the winter-hardiness of the varieties. The wheat seed were germinated in the dark at 20°C in order to eliminate the development of color (which might interfere with the colorimetric determinations) and 6-day-old seedlings were used. Even though the seedlings were not subjected to a hardening treatment, they possessed considerable freezing tolerance (killing temperature −12°C to −15°C) and their relative freezing tolerance agreed with the known winter hardiness of each variety.

In contrast to these results, the seedlings of several barley varieties differing in hardiness all had the same SH contents. After 20 days' hardening at 2°C, however, they showed the same close correlation between SH content and winter hardiness as in the case of the wheat varieties—a coefficient of 0.83 for the eighteen varieties tested. These results agree with practical experience; for winter barley are not as hardy as wheat and are more dependent on a hardening treatment for the attainment of their freezing tolerance.

In order to obtain these high correlations between SH content and freezing tolerance, Schmuetz showed that a rigid standardization of the procedure was necessary. The seedlings had to be homogenized and centrifuged rapidly enough to determine the SH content colorimetrically 7 min after the homogenization. In spite of this speed, an oxidation of the SH groups occurred, which was more or less intense, according to the variety. An exact time plan, therefore, had to be followed. The original SH contents in hardy and nonhardy varieties were the same, and the differences at the time of measurement were due to different degrees of oxidation of the SH groups during preparation of the sample. The oxidation was slow in hardy varieties and more rapid in the less hardy varieties.

The (reduced) ascorbic acid content was also correlated with hardiness in wheat and barley. Furthermore, this correlation was obtained whether the ascorbic acid was measured immediately after preparing the plant sap or some time later. Again, the correlation depended on the different rates of oxidation of ascorbic acid to dehydroascorbic acid in the time between homogenization and titration. This was proved by adding a known amount of ascorbic acid to each homogenate (Table 8.15). It is, thus, not the total ascorbic acid (oxidized + reduced) content that is related to winter hardiness in the cereals tested, but that portion, which at the moment of testing, remains in the reduced form. In the case of vernalizing wheat, on the con-

TABLE 8.15

Relation of (Reduced) Ascorbic Acid Content to Hardiness of Wheat Varieties (Arranged in Order of Decreasing Hardiness)[a]

	Ascorbic acid content		
Wheat variety	Original	After adding the same amount of ascorbic acid to each	Increase
Minhardi	49	91	42
Carsten VIII	44	83	39
Strubes General v. Stocken	39	72	33
Graf Toerring II	33	56	23
Heines Peko	29	51	22

[a] From Schmuetz, 1969.

trary, the total ascorbic (+ dehydroascorbic) acid content does increase (Waisel *et al.*, 1962). Nevertheless, even in this case, there is a relationship between oxidation state and temperature. The temperature optimum of the ascorbic acid oxidizing system shifts during vernalization to lower temperatures, and is 5°C at the 50% vernalization level (Dévay, 1965). Since ascorbic acid is a cofactor for the enzyme GSH reductase, Dévay concludes that new enzymes are synthesized at the low temperature, which lead to an increase in the tissue content of ascorbic acid. This cofactor in its turn regulates the SH content. Dévay's interpretation agrees with earlier determinations of an increase in GSH-oxidizing activity during hardening (Fig. 8.13), although this depended on the hardening stage and the species.

In the case of cabbage seedlings hardened over long periods of time, the artifactual rise in SH was observed only during the early stages of hardening, and was followed by a very marked and steady drop in SH content which paralleled the steady decrease in the killing temperature (Fig. 8.14). In some cases, even the early artifactual rise was not observed, due undoubtedly to homogenization and titration under nitrogen (Kohn and Levitt, 1966). It was at first concluded that this decrease in SH content may be a significant factor in the hardening process, since it paralleled freezing tolerance. Thus, the early artifactual increase in SH with hardening could indicate a protective prevention of oxidation of the SH group, while the later steady decrease in total SH could indicate a decrease in the number of SH groups available for oxidation. Later measurements, however (J. Dear, unpublished), revealed a similar drop in SH content in control plants kept at nonhardening temperatures. The controls showed no appreciable increase in freezing tolerance.

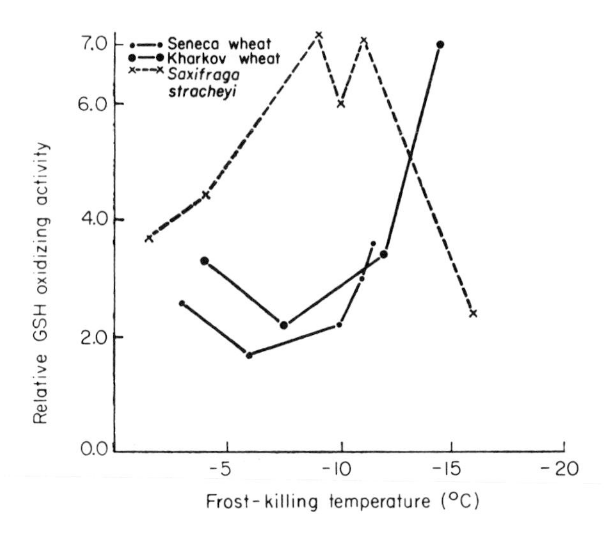

Figure 8.13. Change in GSH oxidizing activity of homogenates on hardening. (Adapted from Levitt *et al.*, 1962.)

The decrease in SH content is, therefore, at least partly, and perhaps solely, an ageing effect.

It can, therefore, be concluded that freezing tolerance involves an increase in resistance toward oxidation of both the SH groups of the proteins and of ascorbic acid. There is no valid evidence that the decrease in SH observed during later stages of hardening is anything other than an ageing effect.

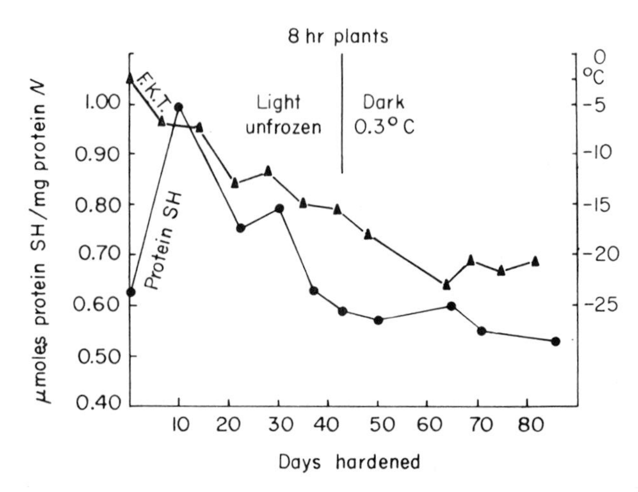

Figure 8.14. Changes in protein SH content per unit nitrogen and in frost killing temperature (FKT) of short day (8 hr) cabbage plants during hardening. (From Kohn and Levitt, 1966.)

C. RESISTANCE INDUCED BY APPLIED SUBSTANCES

1. Sugars, Sugar Alcohols, Salts

Infiltration with sugars and other neutral substances has been shown (see above) to confer a small osmotically accountable increase in freezing tolerance in the case of plants that already possess some hardiness. For instance, both sugars and polyols (glycerol, mannitol, dulcitol) protected the rhizomes of Jerusalem artichokes, when frozen at −5 to −8°C (Leddet, 1974).

A much greater effect of these same substances was first reported by Maximov (1912). Cells were able to survive as low as −30°C if frozen in solutions of nonpenetrating sugars or other nontoxic solutes above their eutectic points (Table 8.16). These results were confirmed by Åkerman (1927) who further showed that the survival depended on plasmolysis of the cells. Iljin (1933a, 1934) was able to extend these experiments and reported survival at extreme low temperatures simply by transferring the frozen, nonhardy cells to the sugar solutions before thawing. This protective effect was also obtained with salts; and in the case of the anion the effectiveness varied with the cation as follows: Ca > Mg > Sr > Ba, Li, Na, K(Terumoto, 1959). The difference between the anions was minor, and oxalate and citrate actually decreased hardiness; Ca^{2+} seemed to prevent intracellular freezing. Sakai (1961) has even reported survival in liquid nitrogen with the use of sugar solutions, but salt solutions were much less effective.

The weakness of all these results is in the use of plasmolysis as the crite-

TABLE 8.16

Frost Killing of Red Cabbage When Frozen in Glucose Solutions of Different Concentrations[a]

Freezing temperature (°C)	Percentage alive in glucose solutions of given molarities						
	0.0 (H_2O)	0.06	0.13	0.25	0.50	1.0	2.0
− 5.2	50	100	100	100	100	100	100
− 7.8	0	Few	25	100	100	100	100
−11.1	0	0	Few	50	100	100	100
−17.3	0	0	0	Few	25	100	100
−22.0	0	0	0	0	Few	100	100
−32.0	0	0	0	0	0	Few	50

[a] From Maximov, 1912.

rion of survival. It has long been known that injured cells may frequently show "tonoplast plasmolysis" even though all the protoplasm except the tonoplast is dead. Iljin admitted that the plasmolyzed cells could not be deplasmolyzed successfully, and this fact was corroborated by others (see Levitt, 1956). Furthermore, as was mentioned earlier, cells that prove to be dead several days after freezing and thawing, may plasmolyze normally immediately after thawing. It is, therefore, difficult to say just how effective these nonpenetrating solutions are.

In the case of some animal cells, it is possible to prevent injury by freezing in solutions of glycerol or DMSO (see Chapter 4). Red blood cells and spermatozoa, for instance, can survive freezing below −40°C in glycerol (Lovelock, 1953). The same substance, as seen above (Tables 8.2 and 8.5), increases the freezing tolerance of higher plant cells by only a few degrees when absorbed. Recent results (Terumoto, 1960b) with a green alga (*Aegagropilla santeri*) resemble those obtained with animal cells. Substances that penetrated the cells (ethylene glycol and propylene glycol) permitted freezing at −35°C for 2 hr without injury. Nonpenetrating substances (including the most effective one for animals—glycerol) failed to protect against freezing injury. Ethylene glycol protects sections of hardened mulberry twigs from slow warming injury after cooling in liquid nitrogen (Sakai, 1961).

Heber (1967) showed that sugar could protect spinach chloroplasts against injury by a rapid freeze at −20°C. These results were confirmed for tea chloroplasts (Sugiyama and Simura, 1967b) and for wheat (Trunova and Zvereva, 1974). Those suspended in 0.1 to 0.3 *M* sugar solutions were not injured by freezing at −20°C, as judged by their Hill reaction activity. The protection occurred whether glucose, fructose, sucrose, xylose, or raffinose were used. Sucrose showed about 20% better protection. They also showed that when the plant was frozen, damage to the chloroplasts was proportional to freezing injury. Nevertheless, when the leaves were infiltrated with sugar solutions, this failed to protect either the leaves or the chloroplasts from injury on freezing at −20°C. Chloroplast function can be preserved in the frozen state for at least 6 weeks at −20°C in the presence of 1% bovine serum albumin and 10% glycerol (Wasserman and Fleischer, 1968).

2. Growth Regulators, Cofactors, and Related Substances

Attempts have been made to increase freezing tolerance by applying growth regulators to whole plants, e.g., by spraying. These attempts were not at first successful (Oknina and Markovich, 1951; Moretti, 1953). Kessler and Ruhland (1942) were able to decrease tolerance by inducing growth with heteroauxin and ethylene. Tumanov and Trunova (1958) found that auxin (both free and bound) decreases on hardening. On the other hand, the ability of coleoptiles to harden was appreciably reduced if a high (growth inhibiting) concentration (200 mg/liter) of indoleacetic acid was added to a 12% solution of sucrose. Another exception was the suppression of growth by maleic hydrazide, which failed to induce hardening. Consequently, they concluded that growth rate cannot always be regarded as a reliable inverse index of hardiness. Lona *et al.* (1956; Lona, 1962), however, have reported an increased survival of −3° to −4°C when the applied substances inhibited growth and a decreased survival when they increased growth. In contrast to his results, a fall spray of GA (gibberellic acid 200 ppm), applied in 2 consecutive years to peach trees, allowed the foliage to remain green for 15–20 days longer than the controls, and it delayed the full bloom of the trees, but did not increase the freezing resistance of the flowers (Marlangeon, 1969). In spite of some negative results (see above), maleic hydrazide seems to give some protection. Even when it does, however, the growth following recovery may be abnormal (Hendershott, 1962a) and the increased tolerance is small (Le Saint, 1966). Young winter wheat seedlings have shown an increase in freezing tolerance following root feeding with ascorbic acid (Andrews and Roberts, 1961). Since such high concentrations were usually required (0.25–1.0 *M*), the possibility of a purely osmotic effect cannot be eliminated. Freezing tolerance of alfalfa has been reported to increase following foliar applications of uracil, thiouracil, or guanine (Jung, 1962). At the same time, protein and nucleic acid content as well as tissue pH were higher than in the controls (Jung *et al.*, 1967b).

Le Saint (1966) did not succeed in hardening cabbage seedlings by supplying them with either sugar or proline in the dark. When the seedlings were stood in solutions of proline (5 gm/liter) in the light at 18°C (a nonhardening temperature) for 2 weeks (with five changes), some increase in tolerance resulted, although it was not as good as by hardening in the light at 4°C. The plants that were fed the proline accumulated the same sugars and in the same amount as the hardened plants. Similar results were obtained with Jerusalem artichoke (Leddet and Schaeverbeke, 1975). Applied proline in high concentrations (0.05 to 0.5*M*) increased freezing tolerance from a killing temperature of −3°C in the controls to −5° to −7°C in those treated with 15–38 mmoles proline/100 gfw (grams fresh weight). They concluded that the lower concentrations protected osmotically, the higher ones by a specific process. Some of the many reports in the literature are listed in Table 8.17.

Kuiper (1967) has reported increases in freezing resistance due to application of a completely different group of substances (2-alkenyl succinic acids) which also act as antitranspirants, supposedly by leading to stomatal clo-

TABLE 8.17

Effects of Growth Regulators and Other Applied Substances on Freezing Resistance

Substance	Effect	References[a]
Abscisic acid	Increase in apple seedlings, *Acer negundo*	9,10,12
Alar	No effect on grapevine, peach; increase in alfalfa	25,26,27
2-Alkenyl succinic acids	Slight increase in strawberry beans; no effect in peach, grains	22,25,6
Amo	Increase in *Acer negundo*; others, no retardation of loss	11,12,14,10
B-9	Increase in *Acer negundo,* cabbage, wheat	10,11,15,32
CCC	No effect on trees, vine; increase in *Brassica* sp., tomato; possible increase in wheat, but probably not true tolerance	10,25,26 1,15,17 32,34,21,33
CK	Increase in wheat	33
Dalapan	Increase in Litchi trees, sugar beets; ineffective in clover, alfalfa, beans	5,4 4,
DMSO	No effect on grapevine; increase in apple, pear trees	25,18
GA	Ineffective or decrease in trees; protected by delaying bloom, decrease in rape	9,10,11,12,26 30 16
IAA	No effect on rape	16
MH	Increase in Litchi, citrus, mulberry	5,8,29,31
Phosphon (D,2000)	Increase in alfalfa; others, no effect on grapevine	27,24,25
Proline	Increase in cabbage	23
Purines and pyrimidines	Increase in alfalfa	13,14
PVP, glycerol, ethylene glycol	Increase in fruit trees	18
TCA	Ineffective in clover, alfalfa, beans	4
T-D692	Ineffective in peach trees	26
Trichloropropionic acid	Ineffective in clover, alfalfa	4

[a] Key to references: (1) Chrominski *et al.*, 1969; (2) Cooper *et al.*, 1955; (3) Cooper, 1959; (4) Corns and Schwerdtfeger, 1954; (5) Gaskin, 1959; (6) Green *et al.*, 1970; (7) Griggs *et al.*, 1956; (8)Hendershott, 1962a; (9) Holubowicz and Boe, 1969; (10) Irving, 1969a; (11) Irving, 1969b; (12) Irving and Lanphear, 1968; (13) Jung, 1962; (14) Jung *et al.*, 1967b; (15) Kacperska-Palacz *et al.*, 1969; (16) Kacperska-Palacz *et al.*, 1975; (17) Kentzer, 1967; (18) Ketchie *et al.*, 1973; (19) Khisemutdinova *et al.*, 1975; (20) Konovalov, 1955; (21) Kretschner and Beyer, 1970; (22) Kuiper, 1967; (23) Le Saint, 1966; (24) Lona, 1962; (25) Marlangeon, 1968; (26) Marlangeon, 1969; (27) Paquin *et al.*, 1976; (28) Rogers, 1954; (29)Sakai, 1957; (30) Stembridge and Larue, 1969; (31) Stewart and Leonard, 1960; (32) Toman and Mitchell, 1968; (33) Tumanov *et al.*, 1976b; (34)Varenitsa *et al.*, 1975.

sure. He used completely tender plants which normally are killed by the first touch of frost (e.g., tomato, beans, strawberry flowers). Since the temperatures used were close to the freezing point (usually −3°C or higher) and freezing periods were brief (1 hr in his early experiments) it may have involved freezing avoidance rather than tolerance. An increase in permeability due to the applied substance led him to suggest that this was the resistance factor. If so, this would be a case of avoidance of intracellular freezing. However, the concentration of decenylsuccinic acid (10^{-3} *M*) used to produce the protection was shown to increase the permeability of bean roots by killing them (Newman and Kramer, 1966). Kuiper (1969) applied this concentration of spray to strawberry flowers and 2 hours later exposed them to −6°C for 2 hr. The untreated flowers were all killed, while most of the treated ones survived and set fruit. Hilborn (1966; see Kuiper) confirmed these results but the protection did not last longer than 8 hr. According to Kuiper, the protection seems to last longer for flowers of plum and cherry. Some of the differences, however, are very slight, e.g., 8% survival in the controls, 10–40% in the treated. Other investigators have failed to obtain an increased tolerance (e.g., Marlangeon 1969). On the contrary, Green *et al.* (1970) not only failed to obtain an increase in freezing tolerance of winter wheat treated with decenylsuccinic acid (DSA) in a nonhardening environment, they actually obtained a negative effect. Treated plants were killed by −5.5° (±0.5°), untreated plants by −6.5°C (±0.5°). They also confirmed Newman and Kramer—the increased permeability following DSA treatment was due to injury.

9. Theories of Freezing Injury and Resistance

Theories of freezing injury and resistance can be classified according to the proposed kinds of injury they attempt to explain: (1) primary direct injury due to intracellular freezing, (2) secondary, freeze-induced dehydration injury due to extracellular freezing, and (3) injury due to other secondary or tertiary freeze-induced stresses. Since the second and third kinds are interrelated, they will be combined.

A. PRIMARY DIRECT (INTRACELLULAR) FREEZING INJURY

1. The Rupture Theory

This was the first theory to attempt an explanation of freezing injury. It ascribed the injury to cell rupture due to expansion on freezing. This theory was disposed of by showing that, in nature, cell rupture due to freezing never occurs (see Chapter 6), and that the tissues normally contract rather than expand. Since air is squeezed out of leaves during freezing, a small pressure does develop; this pressure originates due to a contraction external to the living cells, and is symmetrical. It therefore cannot cause cell rupture. Nevertheless, under artificial conditions intracellular freezing can be induced, and if the ice crystals are large enough to be detected by the optical microscope, is always fatal (Chapter 6). It has, therefore, been frequently proposed that this intracellular freezing does occur in nature and is the cause of freezing injury. Cell rupture does not, however, occur except under very special artificial conditions (see Chapter 6).

2. The Intracellular Freezing Theory

Although cell rupture does not occur due to freezing in nature, intracellular freezing does kill the cells and has been suggested as *the* cause of freezing injury. In the case of certain microorganisms that are uninjured by dehydration even to the point of air drying, intracellular freezing would undoubtedly be the only possible source of injury. Mazur *et al.* (1957) and Mazur (1961) have concluded that this is true for yeast and other microorganisms which are unaffected by freezing above −10°C but are killed in a critical temperature range of −10° to −30°C. In the case of some (e.g., *Pasteurella*) the range is −30° to −45°C. Thus, rapid cooling to −30°C results in 99.99% killing of yeast cells; slow cooling to the same temperature results in 50% killing (Mazur, 1960). Freeze substitution of the cells by cold ethanol in order to retain the cell shape that occurred in the frozen state, revealed that the rapidly frozen cells retained their original shape and that the slowly frozen were considerably smaller and more flattened. Even in the case of higher plants and animals, at least under experimental conditions, intracellular freezing is the main cause of killing due to rapid freezing (Asahina, 1961). Such evidence has frequently led to the suggestion that all freezing resistance is due to the avoidance of intracellular freezing (e.g., Stuckey and Curtis, 1938). Scarth (1936) was the first to suggest that the higher permeability of hardy cells to water would favor this avoidance, and later to produce evidence that hardy cells do actually show such an avoidance of intracellular freezing (Siminovitch and Scarth, 1938). Both of these observations have since been amply confirmed by other investigators, especially at the Japanese Low Temperature Station at Hokkaido (e.g., Asahina, 1956). Even in the case of animal cells, it has been found (Asahina, 1961) that sea urchin eggs are much more resistant to rapid freezing injury when fertilized than when unfertilized and this is correlated with a four times higher water permeability. This rapid injury resulted from intracellular ice formation. Both were equally resistant to slow freezing injury.

It might be objected that the permeability increase during hardening would actually favor seeding of the cell interior due to penetration of the larger aqueous pores in the lipid plasma membrane by the external ice crystals. Chambers and Hale (1932) first showed that this membrane is an essentially impermeable barrier to such seeding. This has since been clearly demonstrated by the true frost plasmolysis that occurs on ice formation between the protoplast and the cell wall under certain artificial conditions (Asahina, 1956; Modlibowska and Rogers, 1955). The ice mass grows inward from the cell wall by withdrawing water from the unfrozen protoplast which contracts and retreats before the advancing ice. In spite of this constant contact of the protoplast surface with ice, no seeding of the protoplast interior occurs. Mazur (1961) has concluded from theoretical considerations that only at very low temperatures (e.g., −30° to −40°C) is the radius of curvature of the ice crystal small enough to suggest its penetration of pores the size of those in the plasma membrane.

It was even possible (Asahina, 1962b) to increase the permeability of unfertilized egg cells by means of urea, and this enabled them to avoid intracellular freezing injury. The Japanese workers (Asahina, 1956; Terumoto, 1957a, 1959; Sakai, 1958), therefore, suggest that not only natural hardening, but even the prevention of freezing injury by protective solutions may be due to the prevention of intracellular freezing. Unlike urea, however, most protective solutes do not increase cell permeability and, therefore, their effects cannot be explained in this way.

Olien (1965) has proposed another concept on the basis of electrical measurements made during the freezing of leaves. He concluded that high molecular weight polysaccharides may act as natural inhibitors of ice formation in the crowns of wheat plants, leading to the formation of slush-like ice instead of a more injurious kind of ice. This implies that the injury is due to a direct effect of the ice crystals, which would, therefore, have to penetrate the cells from the intercellular spaces, or be formed intracellularly.

Although primary direct freezing is, by definition, due to intracellular freezing, this does not explain the injury. It is usually assumed to be a mechanical disruption of the protoplasmic structure. Even when the intracellular freezing is conducted in such a way as to produce small (but microscopically detectable) crystals and to maintain the protoplast intact, the killing is just as complete as when the protoplasm is disrupted by larger intracellular crystals (e.g., in amoeba; Smith *et al.*, 1951). Similarly, injury is not due to a purely physical compression of the protoplasm between the frozen vacuole and cell wall, since intraprotoplasmal freezing has been observed to result in death even when the vacuole does not freeze (Siminovitch and Scarth, 1938). Intracellular freezing injury is, therefore, not due to a general disrupting or crushing effect on protoplasm as a whole, but more likely to a physical impairment of membrane integrity. This hypothesis, however, has not been established experimentally.

Intracellular freezing, as indicated above, may injure plants in nature (1) following a steady-state supercooling or (2) when tissue temperature drops rapidly following a mid-winter warming (sunscald). Except for these relatively rare cases, freezing injury in nature is due to extracellular freezing. All the other theories of freezing injury and resistance have, therefore, been proposed to explain extracellular freezing injury.

B. SECONDARY, FREEZE-INDUCED DEHYDRATION INJURY

1. Solution Effects

Most explanations of freezing injury are based on the increase in concentration of the cell solution on ice formation, resulting from the removal of water as a solvent. These so-called "solution effects" may involve pH changes, increases in salt or ion concentration, close contact of solute molecules, etc.

a. PROTEIN PRECIPITATION THEORY

The first solution-effect theory proposed that the increase in concentration of the cell solution coagulates the protoplasmic proteins and that sugars prevent this coagulation (Lidforss, 1896). In support of this suggestion, Gorke (1906) froze expressed sap from cereal leaves and other plants at temperatures from −4° to −40°C, and observed protein precipitation in the thawed juice. The more resistant the plant to freezing injury, the lower was the temperature required to precipitate the proteins. Adding sugar to the sap before freezing prevented, or at least decreased, the precipitation during subsequent freezing. Gorke, therefore, explained freezing injury as a precipitation of the protoplasmic proteins due to the increased concentration of the cell salts that occurs on freezing, and freezing resistance as a protection of the proteins by the sugars that accumulate on hardening. Gorke's protective effect by sugars was corroborated in the case of grains but not in the case of many other plants (e.g., fruit trees; see Levitt, 1956). Furthermore, as mentioned above, there are many plants that cannot become resistant in spite of high sugar content, so his explanation of freezing resistance is inadequate. It is also incapable of explaining even a small protective effect by plasmolysis in sugar solutions, since the sugars do not enter the cells. Attempts to show a lower salt content in hardy than in nonhardy cells, have failed (see Levitt, 1956). Terumoto (1962), in fact, has shown that lake balls are among the most resistant of algae (they survived −20°C for 26 hr) in spite of their extraordinarily high salt concentration—equivalent to 0.85 *M* NaCl. In the case of marine algae, there was no apparent relationship between the lethal salt concentration at 20°C and the concentration that killed the cells when frozen at −15°C (Terumoto, 1967). Salt injury occurred only in monovalent salt solutions which induced irregularities in the protoplast surface. In divalent salt solutions there was little or no injury and the plasmolyzed cells had a smooth surface. In the case of collards, the hardened plants showed a greater tolerance than the unhardened plants in both NaCl and $CaCl_2$ solutions during either plasmolysis or freezing (Samygin and Matveeva, 1969). This was taken to support Gorke's theory, but, as shown above, hardened plants develop this plasmolysis resistance to nonelectrolytes as well as to salts. Kappen (1969a) has shown that, in spite of their high salt content, the leaves of three halophytic species were able to tolerate freezing at as low as −16° to −20°C in February, with not more than 10% injury, although all were killed in summer by −4° to −7°C. Finally, it has even been possible to increase freezing resistance by inducing the uptake of salts (Sakai, 1961).

Gorke's concept later received support from Ullrich's laboratory, again based on work with grains, as in Gorke's pioneering research (Ullrich and Heber, 1957, 1958, 1961; Heber, 1958a,b, 1959a,b). These investigators thoroughly confirmed the protection by sugars against freezing precipitation of proteins *in vitro* (i.e., in plant juice). They emphasized that in order to be effective *in vivo*, the sugars would have to be in the protoplasm rather than in the vacuole (Heber, 1958a) and suggested that this might explain the lack of correlation between sugars and hardiness in some plants. It was, in fact, possible to show an increased sugar content of the chloroplasts on hardening (Heber, 1959a).

This renewed interest in Gorke's work was perhaps stimulated by Tonzig's (1941) proposal of a formation of mucoprotein (now known as glycoprotein) complexes between the protective sugar and the proteins. Experimental support for Tonzig's proposal was produced by Jeremias (1956), although earlier results had been negative (Levitt, 1954). The positive results of Jeremias have since been disproved by experiments from the same laboratory. The lack of complex formation was shown by dialysis of the plant juice to which sugar had been added. This enabled complete removal of the protective effect (Ullrich and Heber, 1958). Also, more reliable analyses (Heber, 1959b) failed to reveal any relationship between glycoprotein content and freezing resistance, even when sucrose concentration was closely related to hardiness. These investigators then returned to an explanation similar to that given originally by Gortner (1938)—a protection by the formation of uncombined sugar coats around the protein molecules in place of the water coats that are removed by freezing (Ullrich and Heber, 1957, 1958; Heber, 1958a). They suggest a possible formation of H bridges between the OH groups of the carbohydrates and the polar groups of the protein. This interpretation does not agree with direct determinations of hydrogen bond formation between sugars and proteins. They may be formed quite regularly with pentoses and sometimes with hexoses, but not with disaccharides (Giles and McKay, 1962), indicating that only sugar molecules below a limiting size can penetrate the dissolved protein aggregates and form a hydrogen bond complex. Yet the disaccharides and even larger sugars are correlated with freezing resistance. All of the sugars tried by Ullrich and Heber were equally effective on a molar basis, and the maximum effect was attained at about 0.6 *M*. On the other hand, the results are difficult to explain on a simple osmotic basis on account of the pronounced effect of low concentrations and the lack of any increase at concentrations higher than 0.6 *M*.

Heber and his co-workers now appear to have dealt the death-blow to all their earlier confirmation of Gorke's results. They have demonstrated that none of the soluble enzymes tested lose their activities when the plant is killed by freezing (see below). The "protection" of these soluble proteins by sugars in both Gorke's and their experiments is, therefore, an artifact. At least part of the effect of the sugar is to protect the membrane system of the broken chloroplasts. If these membrane systems are allowed to rupture they precipitate much more easily (Uribe and Jagendorf, 1968).

A second revival of Gorke's concept and, in particular, a support for Tonzig's mucoprotein hypothesis, has come from two completely independent laboratories. When frozen at −20°C, the soluble chloroplast protein of tea leaves was precipitated (Sugiyama and Simura, 1967b). The amount precipitated was inversely proportional to the freezing tolerance of the variety, and decreased seasonally in the more resistant varieties from 20% in September to 5-10% in December-February. They also observed an increase in the sugar content of the soluble chloroplast protein of tea leaves from about 12% in September to 20-25% in winter (November-January). The increase paralleled the freezing resistance of the variety. However, the sugar content of the insoluble chloroplast protein was only about 1% and failed to increase. On the basis of Heber's recent results, this would seem to eliminate glycoproteins as a factor in freezing tolerance, since his results point to the insoluble proteins as the ones involved in freezing injury. Furthermore, Sugiyama and Simura (1966) also found their glycoprotein in rice leaves, yet rice is not tolerant of freezing.

A similar support of the Gorke-Tonzig concept has resulted from a newer method of detecting glycoproteins. These results have led Steponkus and Lanphear (1966) and Steponkus (1968) to explain the hardening of *Hedera helix* on the basis of a two-phase process. The first phase occurs in the light and results in an accumulation of sucrose. The second phase proceeds in the dark and is the rate-limiting step. Four hours of the dark reaction are required for every hour of exposure to the light condition. The dark reaction may correspond to Tumanov's second stage of hardening. According to Steponkus (1969a,b) an alteration of protein structure resulting in an increased affinity for sugars corresponds to the second or dark phase. He suggests two types of protein-sugar interactions during cold hardening of *Hedera helix*, leading to (1) covalently linked protein-sugar complexes or glycoproteins and (2) protein-sugar associations via H bonding (less stable than the covalent bonds), as suggested by Heber, and much earlier by Gortner (see above). In opposition to the results of Sugiyama and Simura with tea, both the soluble and the particulate fractions from hardened tissue of *Hedera helix* showed a higher sugar binding capacity than the corresponding protein fractions from unhardened tissue. Furthermore, sucrose showed a higher affinity for the protein fractions than did glucose.

These results from Steponkus' laboratory are directly opposed to those of Heber's group and others (see above). It is again impossible to explain the greatest "protection" of the proteins by the very sugar (sucrose) that shows the lowest ability to H bond or combine covalently with proteins. Furthermore, it is possible that the greater "binding" of the sugar by the proteins from hardened tissue is simply an adsorption phenomenon due to a higher specific surface of the proteins, since no evidence of a covalent or a H bonding has been produced. The fact that the ^{14}C-labeled sugars became attached to proteins which were already supposedly bonded naturally to the sugars in the plant eliminates the possibility of covalent bonding, for this would prevent exchange with the labeled sugars. Until such objections are met, it must still be concluded that the mass of evidence opposes Gorke's protein precipitation theory, and Tonzig's mucoprotein modification of it. Furthermore, no explanation is given for the lack of protection of animal cells (and some algal cells) by sugars against freezing injury. This has been demonstrated even more forcefully by the positive results in the case of animals. Glycoproteins have been found to protect fish against freezing injury (De Vries, 1970). The glycoproteins were unequivocally proved to be present, and it was also proved that the protection they confer is due to freezing avoidance and that the fish are killed as soon as they freeze. Thus, just as in the case of rice, glycoproteins, when present, do not confer freezing tolerance. It, therefore, must be concluded that freezing tolerance is independent of glycoproteins.

Nevertheless, a glycoprotein has recently been isolated from hardened, but not from unhardened, black locust (Brown and Bixby, 1975). Williams (1973) obtained indications that glycopeptides account for 60% of the total osmotic value of hardy Cornus cells and explained its role on the basis of osmotically bound water and nonosmotic effects. Direct measurements of molecular weights and vapor pressures failed to support this conclusion.

b. OTHER SALT CONCENTRATION EFFECTS

On the basis of all the above evidence, the original Lidforss-Gorke concept of a general protein coagulation or precipitation due to the high salt concentration in freeze-dehydrated cells has now been abandoned. Nevertheless, many investigators have produced evidence of other "solution effects." Lovelock (1953) showed that when red blood cells were frozen at different temperatures in 0.16 *M* NaCl containing glycerol in different concentrations, the killing temperature varied inversely with the glycerol concentration. Yet in each case hemolysis occurred at 0.8 *M* NaCl. Meryman (1968) has concluded that injury from extracellular ice in animal cells results from concentration not just of salts but of any extracellular, nonpenetrating solute. That absolute electrolyte concentration is not the cause of injury was shown by exposing the cells to high concentrations of a penetrating electrolyte—NH_4Cl. In opposition to the nonpenetrating NaCl, which injured at 0.8 *M*, the red blood cells survived 4 *M* NH_4Cl. Even some of Lovelock's own evidence does not support a salt injury theory (Levitt, 1958). Nevertheless, Farrant and Woolgar (1970) support Lovelock's concept and most investigators of animal, human, and bacterial cells have accepted "solution effects" as the major cause of freezing injury. It has been proposed, for instance, that the high salt concentration dissociates the enzyme aggregate of cyclic AMP phosphodiesterase and exposes certain groups to hydrolytic cleavage (Sheppard and Tsien, 1974). Opposing results, however, have also been obtained; for instance, the lethal temperature for sea urchin eggs frozen extracellularly was much lower than expected from the lethal salt concentration of the medium (Takahashi and Asahina, 1974).

In the case of plants, Heber and Santarius (1964) did succeed in protecting an insoluble particulate enzyme (ATPase) against inactivation by freezing, when they added sugar before the freeze (see below). They again attempt to explain the protection by H bonding. Again, as they admit, disaccharides are just as effective protectants as monosaccharides, even though disaccharides do not form H-bonded complexes, "presumably because a firmly bound water environment (of the sugar molecule) prevents interaction with protein." In order to make their theory plausible, they therefore conclude that "on freezing, part of this environment is frozen out, and interaction may now become possible." However, a simple fact proves that the "water environment" is not frozen out; for it is the retention of this "water environment" that results in the much lower freezing point of a molar sucrose solution than that of a 1 *M* hexose solution. On the other hand, their evidence that disaccharides are even more effective than monosaccharides in low molar concentrations is in agreement with an osmotic protection due to prevention of chloroplast or lamellar rupture. Later results indicate that sucrose is far more effective than monosaccharides (Heber, 1967), so Heber seems to have eliminated both covalently bound sugars in glycoproteins and H bonding of sugars to proteins, as factors in protein protection. Heber (1968) also prepared chloroplast membranes from hardy spinach which were not inactivated by freezing, yet there was not enough sugar present to protect them. Therefore, the natural tolerance cannot be due to any kind of protein-sugar combination.

Santarius (1969), has confirmed Heber's results with chloroplasts, showing that only photophosphorylation and electron transport were affected by freezing and that a number of soluble enzymes were not inactivated. He also obtained the same results with desiccation as with freezing. He suggests that during dehydration by either freezing or desiccation, the concentration of electrolytes in the remaining solution is responsible for inactivation of the chloroplast membranes. In opposition to this explanation, Heber *et al.* (1973) later suggested that freezing injury may be due to the accumulation of potentially membrane-toxic semipolar substances, such as phenylalanine and phenylpyruvate, which show a much higher toxicity during freezing than inorganic salts.

Heber (1968, 1970) has now succeeded in isolating from hardened spinach leaves first two small proteins (mol. wt. 17,000 and 10,000), then a third, which are not found in nonhardened leaves. They are heat stable and stable against acidification. They are more than twenty to fifty times as effective as sucrose (on a weight basis) in protecting ATPase activity of chloroplast vesicles against inactivation by freezing. Other protein fractions were scarcely or not at all protective. Such results must be interpreted with caution. Both bovine serum albumin and various SH reagents greatly enhance phosphorylation in both red kidney beans and spinach chloroplasts, although having little effect on photoreduction (Howes and Stern, 1969). These substances also protect against inhibition by atabrin, but not against the rapid loss in activity at 0°C. Furthermore, Pullman and Monroy (1963) had previously obtained similar results with mitochondrial ATPase. Both the ATPase and the coupling factor (F_1) in the mitochondria lost their activities on freezing at −55°C (or even at 4°C). In the presence of another mitochondrial protein of low molecular weight, the above two components survive 4 days at −55°C without loss of coupling activity. In spite of this protective effect, the animal tissues from which the mitochondria were obtained (beef heart) are completely intolerant of freezing. It cannot, therefore, be concluded that Heber's proteins are responsible for the freezing tolerance of the spinach leaves.

Similar results have been obtained with the ATPase of *Bacillus megaterium* (Ishida and Mizushima, 1969). This is a membrane-bound enzyme whose activity depends on this binding. When solubilized by dialysis and mild alkaline treatment, its activity is cold labile. On recombination with the alkali-treated membrane in the presence of Ca^{2+} or Mg^{2+} (i.e., by an ionic bond), it is protected against cold inactivation. The general protective effect of proteins against loss in enzymatic activity of other proteins has, in fact, been shown by others. Shikama (1963) found that 0.01% gelatin was much more effective in preventing inactivation on freezing, than were glycerol, glucose, or sucrose.

c. RESISTANCE MECHANISMS

Whatever the mechanism of injury, if due to solution effects, resistance may be induced by either avoidance or tolerance of the increased concentration of the cell solution on freezing. In view of their well-known correlation with freezing resistance, sugars (and other related noninjurious solutes) are the most logical candidates for the avoidance mechanism. Nevertheless, the old concept of bound water has been proposed as the mechanism for both of these kinds of resistance. A kind of tolerance similar to that proposed by Heber is supported by Shikama (1963). He assumes that water adjacent to the proteins forms cubical "icebergs" on the nonpolar regions, thus stabilizing their structure when frozen. Above −75°C, he suggests that they may be converted to the hexagonal form, leading to breakdown of intramolecular H bonds. At these higher freezing temperatures, however, protective substances might stabilize the iceberg structure.

Conversely, on the basis of the amount of unfrozen water in living and killed plants, Tumanov *et al.* (1969) proposed an avoidance mechanism. They concluded that the freezing tolerance of winter wheat depends, to a great extent, on the water-retaining power of the cells. They believe that this is due not only to osmotic forces, but also to the "vital" state of the protoplast.

Weiser (1970) has combined both of these mechanisms in his "vital water" hypothesis—water which is intimately associated with protoplasmic constituents and necessary for life. According to his concept, this water is released at the instant of frost-killing and freezes, accounting for his third exotherm. More recent results by Burke and co-workers (see Chap. 5, c2) have now shown that any exotherms beyond the first one at the freezing point are due to intracellular freezing of ordinary free water in specific cells or tissues following supercooling. Furthermore, the concept of "vital water" has been indisputably put to rest by Johansson and Krull (1970). The amount of water unfrozen (per gram dry matter) was found to be exactly the same in dead as in living wheat plants when frozen either at −2.65° or −9.0°C (Table 9.1). Yet the living plants were uninjured by the freezing in all cases at −2.65°C, and in the hardiest state at −9.0°C. Furthermore, the amount of water left unfrozen at −9.0°C (0.6–0.8 gm/gdm) is only slightly above the value for the amount bound by proteins (0.5 gm/gdm; see Chapter 8). These measure-

TABLE 9.1

Comparison of Unfrozen Water (g/gdm) in Living Intact Plants and in Plants Killed by Chloroform Determined Calorimetrically (Winter Wheat Variety Sammetsvete)[a]

		Unfrozen water (g/gdm)			
		at −2.65°C		at −9.0°C	
Days hardened	Frost-killing temp. (°C)	living	dead	living	dead
0	− 6.5	2.88	2.84	0.60	0.62
10	− 9.0	2.56	2.53	0.81	0.80
20	−11.5	2.27	2.31	0.72	0.74

[a] From Johansson and Krull, 1970.

ments are undoubtedly the most accurate to be found in the literature, due to the painstaking measurements of the constants needed for calculation of ice formation from calorimetric measurements.

Macdowall and Buchanan (1974) have recently concluded from NMR measurements that bound water is slightly higher in hardened than in unhardened wheat plants. More extensive NMR measurements by Burke *et al.* (1974, 1976), however, have failed to detect any difference between the two.

2. Mechanical Effects

a. ILJIN'S THEORY OF MECHANICAL STRESS

On the basis of numerous experiments and observations, Iljin (1933a) proposed his mechanical stress theory of freezing injury and resistance. He was struck by the pronounced cell collapse that occurs on extracellular ice formation, and concluded that the protoplasm of such cells must be subjected to a mechanical stress. If thawing was rapid enough, he observed the cell wall to snap back to its original position, sometimes tearing part of the protoplasmic surface. Due to its lower permeability, the protoplast reabsorbed water less rapidly than the cell wall, and, therefore, showed a temporary "pseudoplasmolysis" during the thawing.

Iljin later went so far as to conclude that the injury occurred only during the thawing, mainly because of his "success" in preventing injury by simply thawing in strongly plasmolyzing solutions. As mentioned above, this was not a true success since he was unable to deplasmolyze these cells successfully. Many phenomena can, however, be explained by Iljin's theory of injury during freezing due to mechanical stress and it was, therefore, widely accepted at least as a working hypothesis. Any factor that reduced the mechanical stress during freezing and thawing would, for instance, confer hardiness. The factors associated with hardiness and the role of avoidance of freeze-dehydration can thus be logically explained by Iljin's theory. The smaller the cell size, the greater the specific surface and, therefore, the less the volume strain per unit surface at any one degree of cell contraction. The greater the sugar concentration, the smaller the water loss and therefore the less the cell contraction and the consequent mechanical stress at any one freezing temperature. An increase in bound water would have the same effect. It would also explain why such factors are of secondary value and therefore not always correlated with hardiness. The primary factor in freezing tolerance would have to be tolerance of freeze-dehydration—the possession of a kind of protoplasm that can be subjected to the mechanical stress without suffering a sufficient strain to produce injury. Siminovitch (Levitt and

Siminovitch, 1940) was, in fact, able to show that protoplasm from hardened plants can survive a greater pressure than protoplasm from unhardened plants. The greater stiffening of the unhardened protoplasm on dehydration (see Chapter 8) would ensure its rupture when subjected to a smaller mechanical stress.

Other observations support Iljin's theory. Winter wheat cells, frozen extracellularly, were severely compressed (Salcheva and Samygin, 1963), leading to death during freezing, and not during thawing. The unhardened cells reached a maximum compression at −7° to −8°C, with several pinched portions. The hardened cells showed maximum compression at −12° to −13°C, but the cells were square with concave walls. The unhardened cells filled out incompletely on thawing and were injured; the hardened smoothed out completely and were uninjured.

Although the above observations lend qualitative support to Iljin's theory, a true test of the theory must be quantitative. According to his concept, the two components of freezing tolerance, which have already been identified as avoidance of dehydration strain and tolerance of dehydration strain (Chapter 7), are really an avoidance of cell contraction and a tolerance of cell contraction, respectively. In the case of wheat leaves, measurement of

TABLE 9.2

Relationship of Cell Contraction on Freezing to Freezing Tolerance[a]

	Frost-killing temperature	Calculated degree of cell contraction at frost-killing temperature
Wheat varieties		
Minhardi	−15	4.9
Lutescens	−15	5.2
Hostianum	−14	5.1
Erythrospermum	−14	5.7
Durable	−14	5.7
Ukrainka	−13	5.1
Zemke	−13	5.9
Schroder	−12	6.2
Pine leaves		
Sept. 23	−10	2.4
Oct. 8	−15	2.5
Nov. 27	−35	2.9
Jan. 1	−40	3.0
Mar. 28	−30	2.8
May 3	−13	2.1

[a] From Levitt, 1956. Adapted from Tumanov and Borodin (1930) and from Pisek (1950).

the second component yielded the opposite of the expected result. On the basis of cryoscopic measurements in a series of wheat varieties, the calculated tolerance of cell contraction on freezing was *inversely* related to the measured freezing tolerance (Table 9.2). Johansson and Krull (1970) and Johansson (1970) obtained a similar inverse relationship during the hardening of wheat and rye varieties, on the basis of direct calorimetric measurements of ice formation (Fig. 9.1). Freezing tolerance in these grains, therefore, cannot be due to tolerance of cell contraction. It was, however, directly proportional to avoidance of cell contraction.

A similar result was earlier obtained with catalpa cells (Scarth and Levitt, 1937). Calculations revealed that 75% of the cell volume was converted to ice at −6°C when unhardened, whereas due to the low water content, the hardened cells could never attain this amount of cell contraction, even if all their water were frozen. This would indicate that the hardened cells owed

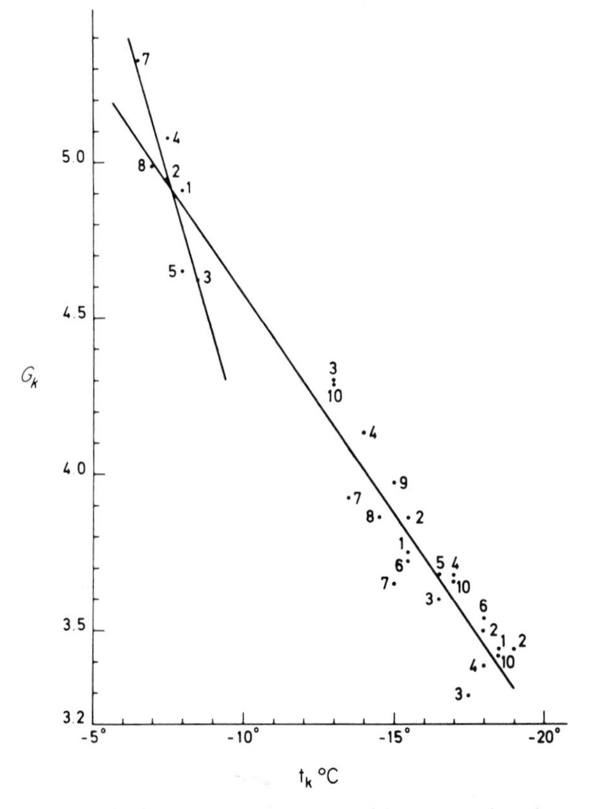

Figure 9.1. Degrees of cell contraction (G_k) in ten different (numbered) rye varieties at the frost-killing temperatures of the plants (t_k°C). (From Johansson, 1970.)

their freezing tolerance to avoidance of cell contraction on freezing. Even the very hardy pine had only half the tolerance of cell contraction possessed by the far less hardy wheats (3.0 at −40°C versus 5.0–6.0 at −12° to −15°C, respectively; Table 9.2). This was again due to the much smaller water content of the pine needles (50–60% versus 85–90% of the fresh weight in wheat leaves), leading to avoidance of cell contraction on freezing. In the pine, however, there was a very small but steady rise in tolerance of cell contraction, paralleling the seasonal increase in freezing tolerance (Table 9.2).

All these data can be explained on the basis of Iljin's theory; they indicate that tolerance of cell contraction is either a negligible or a nonexistent factor in freezing tolerance of pine needles. This does not agree with the many lines of evidence demonstrating a direct relationship between freezing tolerance and tolerance of protoplasmic dehydration (Chapters 7 and 8). These all point to the reverse conclusion—that avoidance of freeze-dehydration (and, therefore, of cell contraction) is the secondary factor and that tolerance of freeze-dehydration is the primary factor (Chapter 7).

Further evidence of this conclusion was obtained in the case of catalpa (Scarth and Levitt, 1937). Although the seasonal hardening led to an enormous increase in avoidance of cell contraction (see above), dehardening occurred in the laboratory without appreciable loss of this avoidance. This can only mean that the loss of freezing tolerance during dehardening was due to a loss of tolerance of freeze-dehydration, without which the marked avoidance of cell contraction was useless.

The explanation of these discrepancies is provided by the above described results with wheat and rye (Table 9.2 and Fig. 9.1). Since there is a straightline inverse relationship between the freezing tolerances of the ten rye varieties and the degree of cell contraction at the freeze-killing point, it must be concluded that the differences in tolerance between these varieties are solely due to avoidance of freeze-induced cell contraction. This conclusion is in agreement with Akerman's (1927) earlier direct relation between freezing tolerance of his wheat varieties and sugar content. If, however, the sugars accumulated even to a minor degree in the protoplasm, this would increase not only the avoidance, but also the protoplasmic tolerance of cell contraction, since it would decrease the freeze-dehydration of the protoplasm at any one freezing temperature, and, therefore, the "stiffening" of the protoplasm on dehydration. Since no such increase in tolerance of freeze-dehydration is associated with freezing tolerance of the rye varieties, this indicates that all the sugars are accumulated in the vacuole.

Other evidence leads to the same conclusion. The freeze-dehydration calculated from the freezing points of the rye varieties is, in each case, identical with the value measured calorimetrically (Chapter 7). This means that the cell is behaving as an ideal osmotic system. The vacuole may resemble such a system, since it consists of a true solution, but the protoplasm is a colloidal system, and, therefore, cannot be expected to approach an ideal osmotic system. Even the vacuole sap would not be expected to act as an ideal osmotic system when largely freeze-dehydrated. As Johansson points out, this ideal behavior can be explained by two mutually compensating factors. The vacuole solutes consist of a mixture of mainly salts and sugars (chiefly sucrose). The freezing-point lowering per mole decreases with the increase in concentration of salts (due to the decreasing dissociation constant) and increases with increase in concentration of sucrose (due to the binding of water). Consequently, as the vacuole is freeze-dehydrated, the net result is that of an ideal osmotic system.

The tolerance of freeze-dehydration does not merely remain constant in these rye varieties. It actually *decreases* with increase in freezing tolerance. On the other hand, it must be emphasized that this decrease occurs only if measured at the lower freeze-killing temperatures of the more tolerant plants. At this lower temperature, the dehydrated protoplasm would be "stiffer" than at the higher temperature even if at the same degree of dehydration; but it must also be more dehydrated, since the above evidence indicates that essentially *all* the increase in sugar content is in the vacuole. The actual protoplasmic strain produced by the mechanical stress (due to cell contraction) must, of course, depend not only on the severity of the stress but also on the protoplasmic resistance to the stress. It is, therefore, this protoplasmic resistance, and not the cellular freeze-dehydration that comprises the second component of freezing tolerance (Fig. 9.2). On this basis, it should be possible to calculate the relative value of the strain produced at the killing point, if the protoplasmic resistance, as well as the degree of cell contraction is known. Unfortunately, the quantitative effects of temperature and dehydration on the protoplasmic resistance are unknown. On the basis of Johansson's results, it can be assumed empirically, that for every 1°C lowering of the freezing point, the protoplasmic resistance to the mechanical stress is decreased by 0.125 of a unit of the degree of cell contraction. Using this quantity as a correction factor, all his values for the ten rye varieties when corrected for a temperature of −5°C, give a resistance (as measured by the cell contraction at the killing point) of 5.1–5.3. If this same correction factor applies to the wheats, then the variety with the lowest freezing tolerance among the eight investigated by Tumanov and Borodin (Schroder) has a 15–20% greater tolerance of freezing dehydration than the variety with the highest freezing tolerance—Minhardi. If these two were crossed, it might, therefore be possible to select a more hardy wheat from the progeny than either parent.

In the case of the pine leaves, on the other hand, the tolerance of freeze-

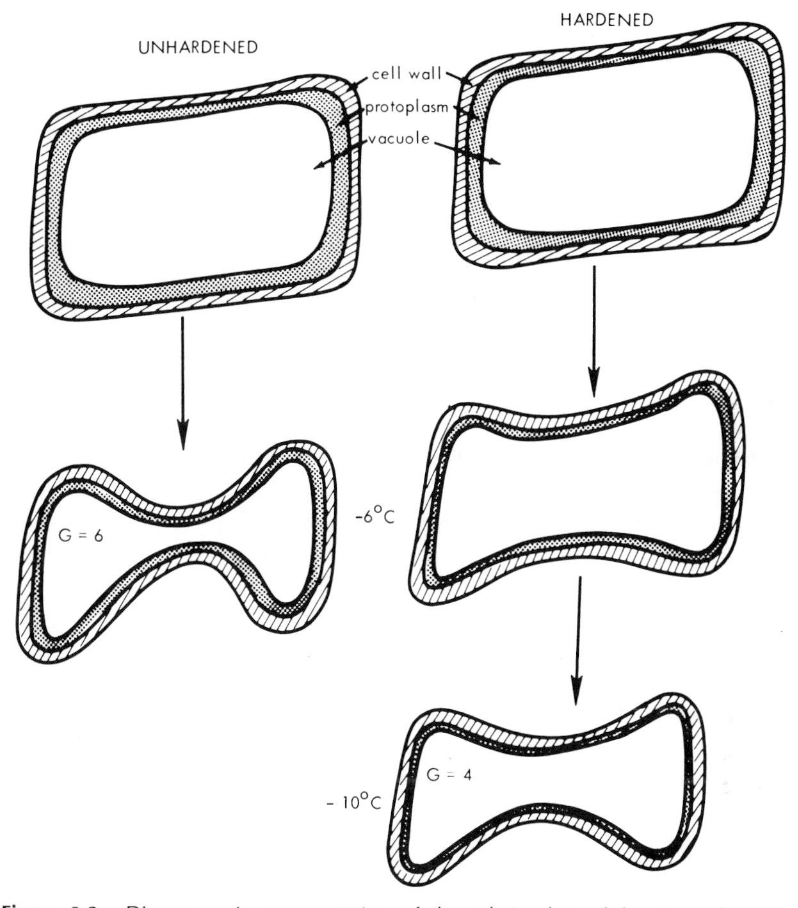

Figure 9.2. Diagrammatic representation of the relationship of freezing tolerance in Johansson's wheat and rye plants to cell contraction and protoplasmic dehydration. Unhardened cell killed at −6°C, when the cell volume is decreased to one-sixth of the original volume, and the protoplasm is dehydrated to one-half of its original volume. Hardened cell killed at − 10°C, although the cell volume is decreased only to one-fourth its original volume. This is because at the lower temperature, the protoplasm is now dehydrated to one-third of its original volume, making it more "brittle" and therefore injured by a smaller mechanical stress due to the smaller degree of cell contraction.

dehydration during January would be not merely 25% greater than in September, as indicated by the respective cell contractions at the killing temperatures (3.0 and 2.4, respectively), but nearly 200% greater (6.75 versus 2.4). It is, of course, not likely that the same correction factor can be applied to the protoplasm of such different plants, but the corrected value is undoubtedly closer to the true value than is the uncorrected one.

It follows from all the above evidence that tolerance of both freeze-dehydration and cell contraction is the primary factor in freezing tolerance, and avoidance of both is a secondary but nevertheless important factor. This is true even in the case of the grains that seemed to indicate the opposite; for all the above measurements were made on the leaves of the grains. It is the crowns, however, that must survive the freezing if the grain seedling is to survive. Measurements on the crowns have revealed that varietal differences are due to tolerance of freeze-dehydration, not to avoidance (see Chapter 7).

In spite of all the supporting evidence, there is also evidence that opposes Iljin's theory. On the basis of Iljin's theory, animal cells should not suffer freezing injury, since they do not possess the stiff, cellulose cell walls responsible for the mechanical stress during cell collapse. Indeed, Asahina (1962a) observed that unfertilized sea urchin eggs become flattened on extracellular freezing, while fertilized eggs contract without loss of their spherical shape. According to Iljin's theory the former would be expected to suffer more injury; but the two kinds of eggs are equally resistant to extracellular freezing injury. Unfortunately, the strongest evidence that the injury is due to the mechanical stress was the prevention of such injury by the use of protective solutions that plasmolyzed the cells and therefore completely eliminated the mechanical stress on the protoplasm by the cell wall. As mentioned above, Iljin greatly overestimated the protective effect of such treatment by accepting plasmolysis without deplasmolysis as evidence of the complete absence of injury. Results (Terumoto, 1962) with algae (known as lake balls) have shown that when frozen in salt solutions the cells plasmolyze, yet injury may or may not occur, depending on the salt and temperature used. The very existence of plasmolysis and deplasmolysis injury and its correlation with freezing injury (see Chapter 8) opposes Iljin's concept. If injury is dependent on the mechanical stress due to cell collapse, there should be no such thing as plasmolysis and deplasmolysis injury, since the cell wall is separated from the protoplasm and unable to subject it to a mechanical stress.

Finally, Siminovitch and his co-workers (see Chapter 8) have demonstrated that extracellular freezing of free plant protoplasts produces protoplast dehydration and contraction in the same way as when enclosed in their cell walls and results in identical injury. Therefore, Iljin's concept of injury due to a mechanical stress induced *in the cell wall* by freeze-dehydration and cell collapse is invalid.

b. MERYMAN'S THEORY OF MINIMAL CRITICAL CELL VOLUME

Meryman also suggests that freezing injury in the absence of intracellular ice is a mechanical effect. Accordingly, the injury due to osmotic dehydration would be basically a damage to the plasma membrane (see Chapter 10). He further proposes (Meryman *et al.*, 1977) that cells generally cannot recover from a greater volume decrease than 40-50% of the unfrozen volume and that although hardening can lower the killing temperature, the volume of the cell at this lower temperature is generally identical to its volume at higher killing temperatures. In other words, Meryman proposes that freezing tolerance is due to avoidance of freeze dehydration and that tolerance of freeze dehydration does not exist. Williams and Williams (1976) in fact, showed that hardy cells of *Cornus florida* possess a greater resistance to dehydrative shrinkage than unhardened cells. Meryman *et al.*, however, accept one exception to this rule—Kharkov wheat. This variety violates the minimum volume rule, according to them by storing in vesicles lipid lost by the membrane during freezing, and returning it to the membrane on thawing.

Meryman's concept is based on measurements (see Table 7.6) indicating that the minimum cell volume is 36% (rather than his 40-50%) and on the results of Johansson and Krull (see Section a), indicating that varietal differences in freezing tolerance of grain varieties is due to avoidance and not tolerance of freeze-dehydration. Here again, however, the quantitative values do not agree with his minimum volume of 40-50%, since some 80-85% of the tissue water was frozen at the freeze-killing temperatures. Since the cell sap behaved as an ideal solution, this means a critical cell volume of only 15-20% of the unfrozen volume. In further opposition to his concept of a single critical volume for hardened and unhardened cells is the above-discussed evidence of tolerance of freeze dehydration, e.g. in hardened pines.

The concept is completely invalidated by both old and new quantitative measurements of the amount of water frozen extracellularly at the freeze-killing temperature (see Chapter 7). (1) Calorimetric measurements showed that the hardening of cabbage seedlings is accompanied by an increased tolerance of freeze-dehydration and, therefore, a tolerance of an increased reduction in cell volume (Levitt, 1939). (2) NMR measurements of ice formation in potato species differing in freezing resistance revealed that all of the differences were completely due to differences in tolerance of freeze dehydration (Chen *et al.*, 1976). (3) Similar results were obtained in the case of wheat varieties when NMR measurements were made on wheat crowns (Gusta *et al.*, 1975) instead of Johansson and Krull's measurements on the leaves. Thus, in three of the four quantitative investigations of the amount of cell water frozen at the freeze-killing temperature, resistance was due to tolerance of freeze dehydration and therefore to a survival of a greater decrease in cell volume. Furthermore, in many if not most cases, the cells were able to survive a shrinkage to far below the 40-50% of the unfrozen cell volume proposed by Meryman. The concept of a single critical cell volume for all cells is, therefore, completely invalid. This has, in fact, been known for decades by the many cell physiologists who have plasmolyzed cells to far below half of their normal volume, deplasmolyzed, and replasmolyzed them without loss of the osmotic properties of the cells or change of their cell sap concentration at normal turgor.

c. MODIFICATIONS OF ILJIN'S THEORY

The Soviet investigators have adopted Iljin's theory of freezing injury, and have repeatedly attempted to explain freezing resistance on its basis. According to Tumanov's (1967) "new hypothesis" (which is really an extension of Iljin's theory), when the plants prepare for winter, the cell contents change from a sol to a gel, and this renders the cell stable against mechanical deformations and dehydration. A slow drop in freezing point of the solution in the gel lattice protects the cells from ice formation. The transformation also makes the cells more inert chemically. The protective substances act not only as antifreezes but also as plasticizers of the gel. In the most freezing-resistant plants he also assumes fundamental changes in the plasmalemma. Apparently, the solidification is supposed to occur only in the vacuole, whereas the cytoplasm becomes "softer." On the other hand, the gel lattice is supposed to form in the cells as a result of synthesis of water-soluble proteins, which presumably accumulate in the protoplasm. More recently, Tumanov (1969) suggested three kinds of resistance: (1) intracellular freezing resistance, (2) dehydration resistance, and (3) resistance to mechanical deformation. This concept is in general agreement with the above conclusions.

An additional kind of mechanical stress has been proposed by Ewart *et al.* (1953). They suggest that the conversion of starch to sugar during hardening is simply a mechanism of removing potentially injurious starch grains, which could damage the protoplasm physically during the dehydration induced by extracellular freezing. In opposition to this hypothesis, Codaccioni and Le Saint (1966) found that the presence of starch in considerable quantity was no obstacle to freezing tolerance in the case of the unhardened layer of cells surrounding the vascular bundle. These cells possessed a natural freezing tolerance. Similarly, plants grown in continuous light had abundant starch in all the parenchyma cells even when hardened.

In view of all this newer information, Iljin's mechanical stress theory of freezing injury and resistance must be modified as follows in order to be valid.

1. If a mechanical stress due to cell collapse produces the injury on freeze dehydration, it must arise in the protoplasm, regardless of the pres-

ence or absence of a cell wall tension. This conclusion is established not only by the existence of freezing injury in free protoplasts (without cell walls) but also by the increased rigidity of the protoplasm of nonhardy cells when dehydrated plasmolytically. One source of this mechanical stress in the protoplasm must be the phase transition of its lipids from the liquid to the solid state which occurs below its phase transition temperature (see Chapter 3) and is enhanced by the freeze-dehydration (Kuiper, 1975).

2. Resistance to the mechanical stress must include both avoidance and tolerance of freeze-dehydration. Avoidance is induced by decreasing the cell collapse and therefore the resulting mechanical stress. Tolerance, or the increase in freeze-dehydration survived by the protoplasm is induced by molecular changes which will be discussed in Chapter 10.

10. Molecular Basis of Freezing Injury and Tolerance

A. EVIDENCE FOR A MOLECULAR BASIS

On the basis of the above analysis of the available information (Chapters 6–9), the effects of freezing on the plant may be summarized as follows:

(a) Freezing injury consists of two main types: (1) primary direct injury due to intracellular freezing, and (2) secondary freeze-dehydration injury due to extracellular freezing. The former is usually ascribed to a direct physical effect of the intracellular ice on the protoplasm. The latter is most logically explained by the mechanical stress and solution effects, both of which increase with the degree of cell dehydration and contraction.

(b) Resistance to the primary direct injury is solely due to avoidance of intracellular freezing. The only freezing tolerance is of extracellular freezing and is due to a combination of avoidance and tolerance of extracellular freezing. These relations are summarized in the scheme below:

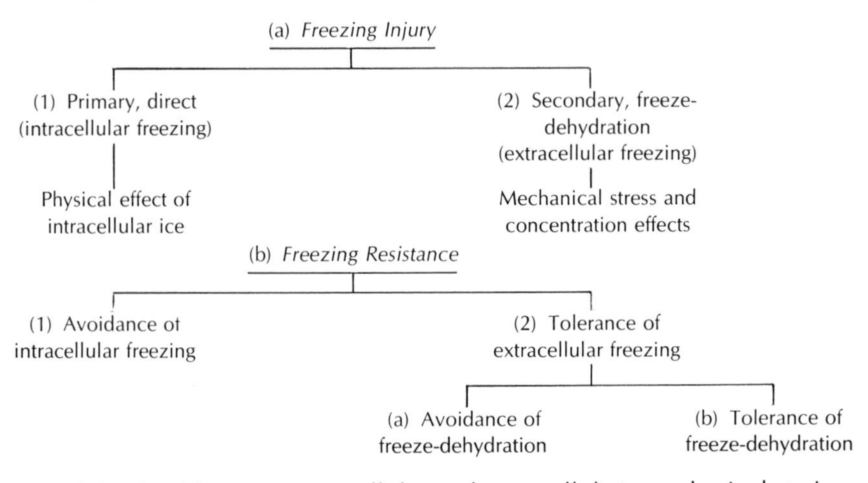

Both kinds of freezing (intracellular and extracellular) are physical strains. The primary freezing stress produces a change in state of the protoplasmic water from liquid to solid; the secondary freeze-induced water stress produces a diffusion of the cell's water to the external ice loci. It might, therefore, be concluded that the injury is also a physical process, and not due to a biochemical, molecular change. In opposition to this conclusion the above two physical strains are elastic and completely reversible. They cannot of themselves be the plastic (irreversible) injurious strain. They must, therefore, produce an indirect plastic strain in the protoplasm which is the immediate cause of the injury. What is the nature of this injurious strain?

1. Primary, Direct Freezing Injury

Protoplasm is so highly structured that it is difficult to conceive of intraprotoplasmal ice formation without some disruption of the organelles, membranes, or simply the groundplasm. If large enough, the expanded crystals must separate the protoplasmic components between which they are formed. If this separation disrupts an essential structure, damage must occur. Consequently, injury due to intracellular freezing has long been explained by the formation of ice crystals in the protoplasm which protrude into the plasma membrane leaving permeable holes upon thawing. This would explain the need for both rapid freezing (inducing a crystal size too small to protrude through the membrane) and rapid thawing (preventing the growth of these small crystals to a sufficient size for membrane damage) for the survival of intracellularly frozen cells. Some evidence does exist of such hypothetical membrane damage. When chloroplast grana were frozen in liquid nitrogen and thawed, centrifuging precipitated a heavy fraction which was much more severely damaged than the light fraction (Uribe and Jagendorf, 1968). Since the light fragments consisted of swollen grana and the heavy fraction did not swell, it was concluded that some of the thylakoid membranes were rendered highly permeable by freezing and were precipitated as the heavy fraction.

If plants are *naturally* frozen intracellularly, the rate of freezing must be slow enough to form microscopically visible ice crystals and therefore membrane disruption and death. The more rapid *artificial* intracellular freezing of animal and human cells for purposes of cryopreservation may permit survival by avoiding the formation of microscopically visible crystals, especially when cryopreservative is present. Even this kind of intracellular freezing is often fatal, and the main areas of damage are at the cell surface (Hunt *et al.*, 1977). Yet this intracellular freezing injury occurs without a change in composition of the membrane proteins or any elution of them (Skrabut *et al.*, 1976). It must, therefore, be due to disruption of the other major component of the membrane, the lipids. This is to be expected from their phase transition to the solid crystalline state, at temperatures that vary with the species, but are always well above the cryopreservation temperature. Even crystals of submicroscopic size would lead to cell expansion, penetrating into the now solid lipid layer. Presumably, those cells that survive this intracellular freezing do so due to an ability to repair the damage on thawing, before it becomes irreversible. Extracellular freezing would prevent this kind of damage due to the spacial separation of the ice from the lipid layer.

2. Secondary Freeze-Dehydration Injury

a. RELATION TO CHEMICAL REACTIONS

It has long been known that the temperature coefficient (Q_{10}) of chemical reactions is much larger than that of physical processes. Therefore, the lowering of the plant's temperature to its freeze-killing point would be expected to slow down all chemical reactions so much more than it slows down the physical process of diffusion, that freeze-dehydration should occur much more rapidly than any accompanying chemical reactions. Since freezing injury can be essentially instantaneous, this would seem to exclude chemical reactions as a possible cause of injury.

Until recently, however, reaction rates and Q_{10}'s in aqueous media have not been followed below 0°C due to the complications introduced by the ice formation. The new technique of *cryoenzymology* has overcome this problem by the use of mixtures of water with antifreezes (ethylene glycol, methanol, etc.). It has been shown that freezing temperatures do not inactivate enzymes in the absence of ice formation (Douzou, 1973, 1977; Douzou *et al.*, 1975; Fink, 1974, 1976; Hazelwood and Dawson, 1976; Kraicscovits and Douzou, 1973; Travers *et al.*, 1975). Special procedures are required to prevent or delay the injurious effects of the antifreeze on proteins (Travers *et al.*, 1975). In this way, a number of enzyme systems (hemoproteins, peroxidases, etc.) have been found to continue their activity down to at least −115°C, though at more decelerated rates than expected. However, the Arrhenius plot for peroxidase activity remained a straight line from 20°C down to the lowest temperature (−65°C) investigated (Maurel *et al.*, 1974, and Maurel and Travers, 1973), showing that the mechanism of peroxidase activity is not modified by the low temperature. Chloroplasts were shown to reduce artificial electron acceptors measurably at −25°C (Cox, 1975). It was possible to observe the reaction even at a temperature of −40°C. Nevertheless, some reactions may be affected indirectly. In wheat leaves, photosynthesis depends on the P_{685} chlorophyll *a* complex at low temperatures, but at higher temperatures the active complex is P_{695} (Devay and Feher, 1975). This is because P_{685} complex is stable at temperatures below the freezing point, P_{695} is not.

In contrast to these expected decreases in rates of enzymatically controlled reactions in the absence of ice formation many (mainly nonenzymatic, bimolecular) reactions are now known to show an anomalous acceleration as a result of freezing (Table 10.1), by as much as 5–1000 times (Grant and Alburn, 1967; Grant, 1969) in spite of the low temperature. Even in the case

TABLE 10.1

Comparative Rates of Nonenzymic Reactions in Frozen and Unfrozen Aqueous Systems[a,b]

			Rate of reaction[c,d]	
Type of reaction	Substrate	Catalyst	Unfrozen	Frozen
Spontaneous hydrolysis	Acetic anhydride	None	30X (+5°C)	X (−10°C)
	β-Propiolactone	None	Considerable	0 (5 hr at −10°)
Acid-catalyzed hydrolysis	Acetic anhydride	HCl	X (5°)	3–27X (−10°)
Base-catalyzed hydrolysis	Acetic anhydride	Acetate	X (5°)	2.7X (−10°)
Imidazole-catalyzed hydrolysis	β-Propiolactone	Imidazole	X (5°)	7X (−10°)
	Penicillin G (pH 7.7)	Imidazole	X (0°)	18X (−8°); 16X (−18°); 5X (−28°); 1X (−78°)
Acid-catalyzed chemical dehydration	5-Hydro-6-hydroxy-deoxyuridine	HCl	X (30°) 0 (22°)	12X (−10°) Rapid (−20°)
Oxidation	Ascorbic acid (pH 5.5)	None	X (1°)	3X (−11°)
	Ascorbic acid (pH 5.5)	$CuCl_2$	X (1°)	3.5X (−11°)
Reduction	Potassium ferricyanide	KCN	Stable above freezing	Complete conversion to ferrocyanide in: 7 hr (−12°); 2 hr (−25°); 107 sec (−78°)
Catalyzed decomposition of peroxide	Hydrogen peroxide (pH 7.2)	$FeCl_3$	X (1°)	13–28X (−11°)
	Hydrogen peroxide (pH 7.2)	$CuCl_2$	Stable (1°)	Quite rapid (−11°)
Hydroxyaminolysis	Amino acid methyl esters (pH 7.2–7.7)	None	X (1°)	1.7–5.5X (−18°)
	Amides (pH 7.0)	Buffer	X (0°)	1.3–7X (−18°)

[a] Substrate concentration ranged from 0.0001 to 0.02 *M*.
[b] From Fennema, 1966.
[c] X = the lower rate of the two.
[d] 0 = undetectable.

of enzymatically controlled reactions, there are some exceptions to the above described deceleration. Thus, small increases were observed in frozen systems relative to otherwise identical supercooled systems (e.g., at −4°C) in the rate of hydrolysis of sucrose by invertase (Tong and Pincock, 1969). Even nonenzymatic reactions do, however, occur in frozen cells, e.g., the dechlorination of DDT in frozen avian blood stored at −20°C for 9 weeks, supposedly due to a redox reaction (Echobichon and Saschenbrecker, 1967).

More than one factor may be involved in such anomalous accelerations. In the case of enzymatically controlled reactions, the decrease in catalytic velocity with lowering of temperature (expected on thermodynamic grounds) may be partially and sometimes fully offset by an increase in enzyme-substrate affinity (Somero and Hochachka, 1969); but this cannot explain an actual net acceleration. Four other explanations have been proposed (Lund *et al.*, 1969), two of which have received the most attention. According to one interpretation (Grant, 1966) it is due to the ice crystals acting as catalysts. This could explain the greater injury due to intracellular than to extracellular freezing, since only in the former do the ice crystals contact the reactants within the protoplasm.

The second-order kinetics, with up to 1000 times more rapid rates than in supercooled liquid solutions, led Pincock and Kiovsky (1966) to conclude that it is a concentration effect. In the case of freeze-dehydration injury, only the latter explanation can apply, since ice forms extracellularly, and the dehydrated cell contents have no contact with the crystals. On the basis of Pincock and Kiovsky's interpretation, at least some of the chemical reactions that occur in the protoplasm may be expected to accelerate as a result of freeze-dehydration. Obviously, not all the reactions will be equally affected, and the net result may be some completely new reactions.

This concentration effect, however, is possible only within a relatively small temperature range below the freezing point, for essentially all of the plant's freezable water is frozen at about −30°C. In agreement with this conclusion, Araki (1977a,b) observed a degradation of mitochondrial phospholipids in rabbit liver frozen slowly at temperatures above −25°C. Maximum degradation occurred at −15°C, and none occurred when frozen rapidly to −196°C. This maximum may indicate that below −15°C the small amount of solvent left in the liquid state may limit the reaction rate due to the decreased diffusion.

Since chemical reactions are shown to continue in the absence of freezing at the lowest freezing temperatures tested, and since the rates must increase when the concentration of the cell solution is increased by freeze dehydration, indirect chemical strains must accompany the primary freezing strain.

They must, therefore, be considered in any attempt to explain freezing injury.

b. RELATION TO ENZYME INACTIVATION

Many attempts have been made to detect specific enzymes responsible for freezing injury. Ullrich and Heber (1961) found that five soluble enzymes which were not sedimented by centrifuging, were neither coagulated nor inactivated by freezing. Even some of the particulate enzymes showed no change in activity after freezing. McCown *et al.* (1969b) were also unable to detect any decrease in enzyme activity in carnation tissue killed by freezing. In fact freezing is used as a routine method for isolating them (Rhodes and Stewart, 1974), to preserve them (Judel, 1975), or even to activate them (North, 1973; Sheppard and Tsien, 1974). Rapid freezing of tissues from 12 unrelated species of higher plants in liquid N_2 was used to render the cells permeable to substrates. The activities and properties of the enzymes were quite comparable to those obtained by conventional extraction (Rhodes and Stewart, 1974). Unfortunately, this kind of freezing is intracellular rather than extracellular.

Nevertheless, some enzymes are inactivated by freezing, for instance chloroplast ATPase as shown by Heber (see Sec. B,1). In the case of seeds, some proteins have been precipitated by freezing and have been called cryoproteins (Ghetie and Buzila, 1964a). These were apparently concentrated in the cotyledons and were bound to 2–10% lipid (Ghetie and Buzila, 1964b). They were, perhaps, storage proteins and therefore presumably not present in other leaves, and not active as enzymes. Enzymes may also be cryoinactivated. Mitochondrial MDH (malic dehydrogenase) is unstable when frozen (Nakanish *et al.*, 1969) and pectin esterase from both coleus and bean was denatured by freezing and thawing (Lamotte *et al.*, 1969). In the case of mouse liver, the succinate cytochrome *c* reductase complex was particularly sensitive to and suitable for evaluating freeze-thaw injury (Fishbein and Stowell, 1968). Hanafusa (1967) detected marked decreases in enzymatic activity and conformational parameters of fibrous proteins (myosin A and B; H- and L-meromyosin) as a result of freezing (at −30°, −79°, or −196°C), but no change in the conformational parameters of globular proteins (G actin and catalase) in spite of the decrease in enzymatic activity. He concludes that freeze-thawing causes a partial unfolding of the helical structure in the fibrous proteins (25% in myosin A), but not in the globular proteins. Other enzymes are also inactivated by freezing (Anderson and Nath, 1975; Darbyshire, 1975; Whittam and Rosano, 1973), or at least undergo a conformational change (Yu and Jo, 1973). Others are reversibly inactivated by cold in the absence of substrate (Huang and Cabib, 1973). In the case of mitochondria from rabbit liver, the effect is highly specific. α-Ketoglutarate dehydrogenase was inactivated by freezing although other enzymes involved in the oxidation of α-ketoglutarate and succinate were essentially unaffected (Araki, 1977a).

In summary, all the work with plants indicates that in spite of *in vitro* inactivation of some soluble enzymes by freezing, *in vivo* freezing fails to inactivate them. It is, of course, conceivable that a key enzyme may yet be found in plants that is inactivated by *in vivo* freezing. In view of the mass of negative evidence this seems unlikely. Furthermore, enzyme inactivation can injure a plant only by a disturbance in its metabolism. Such a disturbance would occur very slowly at the low temperature inducing freezing injury.

Whether an enzyme or some other molecule is involved in freezing injury, the specific cell structure initially damaged must first be identified.

B. MEMBRANE DAMAGE AS THE INITIAL INJURIOUS STRAIN

1. Evidence That Freezing Injury Is Due to Membrane Damage

(a) The first sign of injury in a thawed, leaf-bearing plant is the water-soaked and consequent flaccid appearance of the leaves, due to infiltration of the intercellular spaces with liquid. The cells have, therefore, lost either their semi-permeability or their ability to reabsorb the solutes in the intercellular fluid. Both of these losses are dependent on membrane properties. They may, of course, be the result, rather than the cause of injury. There are, however, several lines of evidence that point to the semi-permeable cell membrane as the locus of the initial freezing injury.

(b) Maximov (1912) protected cells against freezing injury by use of nonpenetrating solutes. Since these protective solutes do not penetrate the membrane and are immediately effective, Maximov concluded that they must exert their effect on the surface of the protoplasm and that freezing injury must be due to plasma membrane injury.

(c) The increase in cell permeability which accompanies hardening (see Chapter 8) must be due to a change in the plasma membrane, since this structure controls the permeability of the protoplast. It may be postulated that the protoplasmic structure altered by hardening is likely to be the one damaged by freezing in the unhardened plant.

(d) The prevention of ectoplasmic stiffening on dehydration of the protoplasm is characteristic of the hardened state (see Chapter 8). The ectoplast, or outer layer of the protoplast, includes the plasma membrane.

(e) The inactivation of membrane-associated ATPase by freezing of chloroplast thylakoids (Heber, 1968; Heber *et al.*, 1973) is primarily due to membrane destruction. Although in these experiments rapid freezing at −20°C was used, Steponkus and co-workers (1977) showed that slow freezing of isolated thylakoids also damaged the light-induced proton uptake. The plasma membrane-bound ATPase of *Hedera helix* was also freezing labile, and protected by sucrose during freezing (Steponkus and Wiest, 1973).

(f) Direct observations of electron micrographs have revealed freeze-induced membrane damage. Examination of freeze-substituted and freeze-etched tomato fruit indicated that membranes are the most sensitive cell components to freezing, the tonoplast being more sensitive than the plasmalemma (Mohr and Stein, 1969). This technique, however, undoubtedly involves intracellular freezing. Even though frozen slowly, however, at −4.4 and −6.7°C, citrus cells show disruption of essentially all the membranes (plasmalemma, tonoplast, chloroplast, mitochondrial), as well as disorganization of the cell contents (Young and Yelenosky, 1973). Meristematic cells of wheat roots show damage to the mitochondrial membranes as a result of freezing (Gazeau, 1974). The membranes of microorganisms and the membrane-bound enzymes also suffer structural changes as a result of freezing and thawing (Aithal *et al.*, 1975). The first sites showing damage in frozen leukocytes are the plasma membrane and the lysosomes (Crowley *et al.*, 1973). Even the nuclear membrane has been reported damaged by freezing at −10°C for 2 days in the coleorhiza of roots of a wheat-witch grass hybrid (Petrovskaya-Baranova, 1974).

From the above observations, it may be concluded that all the cell membranes may be injured by freezing, although some may be more susceptible than others. Since no difference could be detected in the respiratory functions of mitochondria isolated from lethally and nonlethally frozen rye coleoptile cells, mitochondrial membranes must be less freezing sensitive than the plasmalemma (Singh *et al.*, 1977b). In support of this conclusion, freeze-injured potato leaf cells revealed protoplasm damage though the mitochondria and chloroplasts appeared normal (Palta and Li, 1978).

(g) Combining electron microscopy with biochemical measurements, Garber and Steponkus (1976a,b) concluded that freezing damaged the light-induced proton uptake of spinach thylakoids due to a preferential release of CF_1 (coupling factor) from their outer surface and a concomitant loss of Ca^+-ATPase activity. Three lesions were actually identified: (1) loss of plastocyanin, (2) loss of coupling factor, and (3) loss of semipermeability.

The freeze inactivation of the thylakoid membranes was accelerated in the presence of NaCl, and the addition of sugars protected the thylakoids. The cryoprotective effect was not due to a colligative mechanism, but to a protection of the membrane site of injury, apparently by maintaining the CF particles on the membrane, and therefore protecting the Ca^{2+}-ATPase activity. Similarly, the NaCl-induced freezing damage was not a solution effect since different degrees of damage were associated with the same concentration of NaCl in the frozen state. The damage was again associated with inactivation of the membrane-bound CF_1.

Whether or not freezing injury in nature is due to the above kinds of membrane damage is still an unanswered question. Steponkus *et al.* (1977) point out that the plasma membrane is more likely to be the locus of injury than the chloroplast thylakoid, and unlike the latter it possesses no photophosphorylating system. Furthermore, the damage to the thylakoids when frozen *in vitro* may be due to ice pressure, which does not exist in the frozen plant. In support of this objection, the damaged thylakoids are converted from the tubular to the vesicular form, indicating compression.

The effect of cold acclimation on the chloroplast thylakoids is perhaps more conclusive evidence that the cell membranes are the locus of freezing injury. Electron micrographs revealed a decrease in particle concentration on the inner fracture face to one-half that in thylakoids from nonacclimated plants. The particles were of only one size (±140A) in the thylakoids from acclimated plants, but were of two sizes (±100 and ±165 Å) in those from nonacclimated plants. Furthermore, sucrose afforded greater protection of proton uptake by thylakoids from acclimated tissue than by those of nonacclimated tissue. Steponkus *et al.* (1977) therefore suggested that acclimation alters the membrane in a way which renders the CF_1 particles less susceptible to release.

All the above observations, unfortunately, need to be reconfirmed, for the electron micrographs may be interpreted in a completely different manner due to the techniques used (Paul Armond, personal communication).

(h) The conductivity method (see Chapter 7) measures freezing injury by the amount of electrolyte efflux from the tissue immediately after thawing. For over 40 years, it has been the most successful objective method of measuring freezing injury. The method has usually been tacitly assumed to measure the percentage of cells killed and therefore made freely permeable to the cell solutes, which then all diffuse out. Recent results (Palta *et al.*, 1977a) have shown that this interpretation is incorrect. Onion bulbs frozen at a temperature from which they later fully recovered (−5°C) showed increased conductivity immediately after thawing. Therefore, freezing must have damaged the cells temporarily, though *none* were killed. Others frozen at a temperature (−11°C) which led to death of some of the tissues a few days later, had no dead cells immediately after thawing. These bulbs also showed a higher conductivity than those frozen at −5°C. The conductivity increase, therefore, *predicted* the later death of the cells. Yet there was no other observable injury to the cells when examined immediately after thawing. They plasmolyzed normally, showed normal cytoplasmic streaming, and the optical properties of the protoplasm were indistinguishable from those of unfrozen cells.

The conductivity method, therefore, measured not the percentage of onion cells killed, but the degree of injury to the still living cells. Since the conductivity measures the net loss of electrolytes from the cells, a loss which is normally prevented by the external cell membrane (the plasmalemma) and since no other cell injury was initially observable, damage to the outer cell membrane of the onion must have been the initial freezing injury.

What is the nature of this membrane damage? The most logical explanation is a loss of semi-permeability or in other words, an increase in cell permeability. Direct measurements, however, with tritiated water, showed that the permeability of the onion cells to water was unchanged by freezing at temperatures that resulted in a pronounced increase in conductivity. It was also shown that essentially all the electrolyte efflux was accounted for by the concentration of K^+ salt present and that the osmolality of the cells was decreased by an amount approximately equal to this loss of K^+ salt. If this efflux is not due to an increase in cell permeability, the only other explanation is damage to the actual uptake mechanism—the ion pump. The normal living cell is not in a state of equilibrium, but in a steady state. The concentration of cell solutes in general, and of K^+ salts in particular, remains constant only if the rate of influx equals the rate of efflux, yielding a net change of zero. The much higher concentration of the cell sap than that of its environment is maintained by an active influx via a cell pump, which balances the slow passive efflux due to diffusion. Any damage to the pump decreases the influx and therefore leads to a net efflux of cell solutes. Since the pumps occur in the membrane, the above-described membrane damage induced by freezing is best explained by a damage to the ion pumps, and perhaps mainly the K^+ pump. In agreement with the above results and the proposed interpretation, freeze-drying of *E. coli* leads to a membrane dysfunction, apparently to a permease, making it leaky to K^+ (Israeli *et al.*, 1974). The damage can be partially repaired on incubation in nutrient medium.

In opposition to these results, Le Saint-Quervel (1977) obtained a very good correlation between conductivity values and cabbage cell survival determined by vital staining with neutral red immediately after thawing. But she thawed the frozen leaf discs by immersing them in 50 ml distilled water at room temperature. This rapid thawing has long been known to increase freezing injury (see Chapter 6). The above described slow post-thawing injury (p. 256) was presumably converted to an immediate, rapid-thaw injury. In agreement with this interpretation and with Palta's results, earlier observations (Levitt, 1957a) demonstrated that cabbage cells were able to stain vitally after thawing even when the freezing injury led to death 1–2 days later.

(i) Relation of membrane damage to salt concentration. The K^+ efflux that was shown to occur even at a mild noninjurious freezing temperature (−5°C) may be explained by the salt concentration theory only if the moderate (approximately 5×) increase in salt concentration of the cell can be shown to produce this effect. The post-thawing increase in K^+ efflux, however, can be a salt concentration effect only if the K^+ concentration *in the intercellular spaces* becomes high enough to damage the plasma membrane. The initial loss of cell solutes, as shown above, is not fatal as long as the cells can reabsorb the ions slowly during and after thawing, or as long as the proposed damage to the ion pump is reparable after thawing. But the salt concentration outside the cell may conceivably convert the temporary, reparable damage to an irreparable one. The high external concentration of K^+ ions, for instance, might replace the Ca^{2+} in the membrane, leading to a loss of semipermeability (in other words a damaging increase in permeability) of the cell, or of ion pump activity since Ca^{2+} enhances ion pump activity (the Viets effect). In favor of this interpretation, freeze inactivation of thylakoid membrane ATPase was accelerated in the presence of NaCl (Garber and Steponkus, 1976a), and mitochondrial ATPase was similarly affected (Bruni *et al.*, 1977). The salt concentration would, therefore, operate by increasing the membrane damage. The logical conclusion then, is that the external salt concentration injures by damaging the cell membrane but that the first stage of membrane damage, which releases the external salt solution, precedes this salt injury. In the case of rat erythrocytes, for instance, once the membrane is damaged, a high external salt concentration can produce the same effect on an enzyme activity as freezing (Sheppard and Tsien, 1974). It is obvious, therefore, that the salt concentration hypothesis and the membrane concept of freezing injury are equivalent. The salt concentration, however, is only one component of the membrane concept, which also takes into account the increased concentration of other substances such as the proteins (see Sec. 2,a,ii). In support of the primary nature of the membrane damage and the secondary nature of the salt concentration, the lethal freezing temperature for sea urchin eggs was lower than expected from the lethal salt concentration in the medium (Takahashi and Asahina, 1974).

2. Membrane Component Damaged

Since cell membranes consist basically of lipids and proteins, membrane damage must be due to either lipid or protein changes, or both. The initial damaging event, therefore, cannot involve other molecules such as nucleic acids. If, as indicated above, the ion pumps are the seat of this initial injury, proteins (ATPases) must be involved. This does not, however, exclude the lipids since they are also required for the operation of the ion pump. What, then, are the effects of freezing low temperatures on proteins and lipids?

a. DIRECT LOW TEMPERATURE EFFECTS

Although it is known that low temperature per se is incapable of inducing freezing injury, the molecular changes that occur at freezing low temperatures in the absence of freezing must be considered as possible first steps leading to the freeze-induced injurious change.

i. Lipid Changes. Lipids are essential components of the cell membranes and any change in them could conceivably injure the cell by leading to a loss in semi-permeability. Furthermore, they are the only substances in the plant that, like water, are present in the liquid state at normal temperatures and may be converted to the solid state at lower temperatures. This "phase transition" from the liquid crystalline to the solid (gel) state has already been indicated to play a role in chilling injury (see Chapter 3) and must be considered as a possible factor in freezing injury.

A phase change in the membrane lipids from the liquid crystalline to the solid, gel state may conceivably play a role in freezing injury in two ways. (1) It markedly lowers the permeability of the membrane (about 3 times; see Chap. 3), and therefore may lead to intracellular freezing injury by preventing a rapid enough rate of exosmosis of water to insure extracellular freezing. In the case of erythrocytes, for instance, calculations show that more than 95% of the cell water is transferable during freezing, yet measurements indicated that at least 20% was retained (Levin and Cravalho, 1976). This was explained by a decreased membrane conductivity at subzero temperatures. (2) Even if no intracellular freezing occurred, the solidified membrane would be more likely to "fracture" under the mechanical stress of cell contraction that accompanies extracellular freezing. A quantitative evaluation of these two possible sources of freezing injury indicates that they would be a danger only if the phase transition occurred at temperatures no lower than −2 to −4°C (Levitt, 1978). This is because the great majority of frozen plant cells would have up to 50–75% of their freezable water frozen extracellularly by the time they attain a temperature of −2° to −4°C. The small remaining amount of freezable water would be able to exosmose as rapidly as the cell is normally cooled in nature, even in the case of cells with low permeability, which therefore would not freeze intracellularly. Similarly, the major component of the cell contraction would have occurred by this time, and the loss of additional water to the extracellular ice loci could cause only slight fractures which would be readily reversible on thawing. The logical conclusion is that the phase transition temperature is likely to be an important factor in freezing injury only in the case of very sensitive plants. Any plant that is tolerant of freezing temperatures of −2° to −4°C is probably injured at lower freezing temperatures by molecular changes other than the lipid phase transition.

ii. Protein Changes. Proteins may undergo three kinds of changes at low temperatures in the absence of freezing. (1) A loss of quaternary structure may lead to dissociation of a large protein molecule into smaller monomeric subunits, with (reversible) loss in enzymatic activity (Markert, 1965). (2) A loss of tertiary structure may occur due to unfolding (denaturation) of the molecule or subunit (Brandts, 1967). (3) The unfolded molecules may aggregate by forming physical bonds between the newly exposed sensitive chemical groups that were previously buried within the folded molecule. These physical bonds may be "hydrophobic" (actually a repulsion by the more polar groups) hydrogen, or electrostatic in nature. The unfolding and aggregation have been illustrated as follows (Brandts, 1967):

$$N \underset{\text{normal } T}{\overset{\text{low } T}{\rightleftharpoons}} D \xrightarrow{\text{low } T} A$$

where N is the native, D is the denatured, and A is the aggregated protein molecules.

The unfolding (denaturation) of proteins at low temperatures may also be followed by an aggregation due to chemical (covalent) bond formation. The best known is the formation of intermolecular disulfide (SS) bonds between the unfolded molecules. Unlike the physical bonds, which are reversible, the chemical bonds are thermodynamically irreversible and, therefore, more likely to be injurious.

b. FREEZING EFFECTS

i. Effects on Lipids. These direct effects of freezing temperatures (in the absence of freezing) on physical or chemical reactions in the cell cannot, by themselves, explain freezing injury which always fails to occur in the supercooled cell. What effects can the freezing process have, by itself, on these reactions?

As shown above, freezing can accelerate chemical reactions, due presumably to the increased concentration on freeze-dehydration. Unfortunately, there is little or no experimental evidence as to the effects of freezing *per se* on lipids. The lipid molecules are unlikely to be altered by the freezing of the cell's water, since lipids are essentially hydrophobic and, therefore, possess little or no freezable water. Even ice crystals formed outsid but in contact with, the surface lipid of cells do not penetrate the layer Chapter 6). Nevertheless, lipids extracted from dogwood collapse c Langmuir trough at a value close to the freezing energy required to des the cells (Williams and Ranasastry, 1973). Besides this physical effect, results indicate that many reactions, including the oxidation of ascorbic (Table 10.1), are accelerated by freezing. Since the peroxidation of lipids been shown to damage cell membranes in the presence of ascorbic (Bidlack and Tappel, 1973) and since it is the phospholipids with uns rated fatty acids that are selectively degraded, this damaging peroxidatio the membrane lipids may conceivably be accelerated by freezing, parti larly since unsaturation of the fatty acids is increased by the low temperat before exposure to freezing (see above). This possibility has not yet be investigated directly. However, O_2 has been observed to injure freeze-dri bacteria (Israeli *et al.*, 1974).

The formation of singlet molecular oxygen in illuminated chloroplasts was proposed as the explanation for the lipid peroxidation observed in light-irradiated chloroplast fragments (Takahama and Nishimura, 1975). The superoxide anion radical was suggested as the precursor of the singlet oxygen. The degradation of phosphatidylcholine to phosphatidic acid is associated with freezing injury in poplar and black locust (Yoshida and Sakai, 1974; Yoshida, 1978), indicating a direct chemical effect of freezing on the lipids. Similarly, whether at −20°C, in dry ice or in liquid N_2, freezing increased the level of phosphatidic acid from 4.7 to 50% of the total in immature cotyledons of soybeans and diminished the levels of PC, PE, and *N*-acyl PE from 54.1 to 6.6% (Wilson and Rinne, 1976). They, therefore, suggested that enzymic destruction of the phospholipids occurred during freezing and thawing. Degradation of PL was also observed in rabbit liver mitochondria when frozen slowly above −25°C, though not when frozen rapidly at −196°C (Araki, 1977b). Maximum degradation resulted from freezing at −15°C. In poplar and black locust, the enzyme phospholipase-D was found both in the particulate and soluble cell fractions (Yoshida, 1978). Although it was associated with several membranes, the highest specific activity was in the microsome fraction. In opposition to this suggestion, lipase hydrolysis was found to decrease from 5.4% at −2°C to 4.3% at −12°C (Parducci and Fennema, 1978). On this basis, the lower the freezing temperature, the less the injury caused by lipid hydrolysis.

ii. Effects on Proteins. In spite of these indications that freezing injury may be related to the effects of freezing on the lipids, other lines of evidence point to the proteins as the molecules affected. Thus the stiffening of the protoplast surface shown above (Chapter 8) on plasmolytic dehydration must be due to the proteins and not the lipids, because (a) the experiments were performed at room temperature and therefore above the phase transition temperature of the membrane lipids, and (b) the oil drop method would dissolve the surface lipids, yet it showed the stiffening of the membrane surface. The possible effects of freezing on proteins have already been indicated by the low temperature effects in the absence of freezing (see Sec. 2,a,ii):

$$N \underset{\text{normal } T}{\overset{\text{low } T}{\rightleftharpoons}} D \xrightarrow{\text{low } T} A$$

It is obvious that the denaturation by itself cannot cause the damage, since it is simply a physical unfolding which is readily reversible on warming. Thus, the enzyme RuDPC from *Euglena gracilis* undergoes a conformational change when there is a change in temperature, the low temperature forms being more hydrophobic externally, the high temperature form being more reactive enzymatically (Wildner and Henkel, 1977). Activity is restored in the low temperature form by incubating at 50°C. The irreversible change is the aggregation. This aggregation is unlikely to occur in the absence of freezing, due to the combined effect of the low temperature (slowing down any reactions) and the protection of the reactive groups by the water molecules. On the basis of Pincock's interpretation (see Sec. A,2,a), this aggregation would result from the increased concentration on freeze-dehydration and, therefore, the closer approach of the denatured molecules to each other. Interaction between the adjacent molecules would also be favored by the denaturation, since this would permit contact between chemical groups previously protected from each other inside the folds of the native molecules. The interactions could be physical, due to the formation of intermolecular "hydrophobic," hydrogen, electrostatic bonds, or chemical due to covalent bonding.

C. THE SH HYPOTHESIS OF FREEZING INJURY

1. The Concept

When they are sufficiently close, chemical combination occurs between the S atoms of adjacent protein molecules, (1) by oxidation of 2 SH groups, or (2) by SH $\rightleftharpoons$ SS interchange, as follows:

$$2\ RSH + \tfrac{1}{2}\ O_2 \rightleftharpoons RSSR + H_2O \qquad (10.1)$$

$$\begin{array}{l} R_1{-}S \\ \;|\quad\;\; | \\ R_2{-}S \end{array} + R_3SH \rightleftharpoons HSR_1R_2SSR_3 \qquad (10.2)$$

On the basis of these reactions and a series of experimental results (see Chapter 8), a SH theory of freezing injury and resistance was first proposed (Levitt, 1962). The freeze-induced injury due to SS bonding between protein molecules was postulated to occur as in Fig. 10.1 The theory was later modified (Levitt and Dear,1970) to conform to the above principles of denaturation followed by aggregation, and may be summarized as follows.

(a) Low temperature denatures proteins reversibly, unmasking reactive SH groups.

(b) Freeze-dehydration removes cell water, producing cell contraction which applies stress on the protoplasmic proteins, activating their SS bonds, and decreasing the distance between the reversibly denatured protein molecules.

(c) Intermolecular bonding due to SH oxidation or SH $\rightleftharpoons$ SS or SS $\rightleftharpoons$SS interchange, aggregates the proteins irreversibly, killing the cell.

It must be emphasized that the SH hypothesis does not exclude the possibility that bonds other than SS bridges may induce aggregation. H bonds, however, are not likely to be involved due to their strong bonding to water molecules. Hydrophobic bonds are not likely to occur, because they are weakened by the low temperature, and still further by freezing, since this removes the water which is the real cause of hydrophobic bonding by repelling the hydrophobic groups. However, the results of Goodin and Levitt (1970) suggest that, on thawing, hydrophobic groups may lead to aggregation of the proteins before reimbibition of the thaw-water. When embedded in a hydrophobic environment, electrostatic and H bonds are stronger (Koshland and Kirtley, 1966), and, therefore, may help to strengthen these weak bonds. The strongest of all are the SS bonds, and these may confer an irreversibility to otherwise reversible bonding.

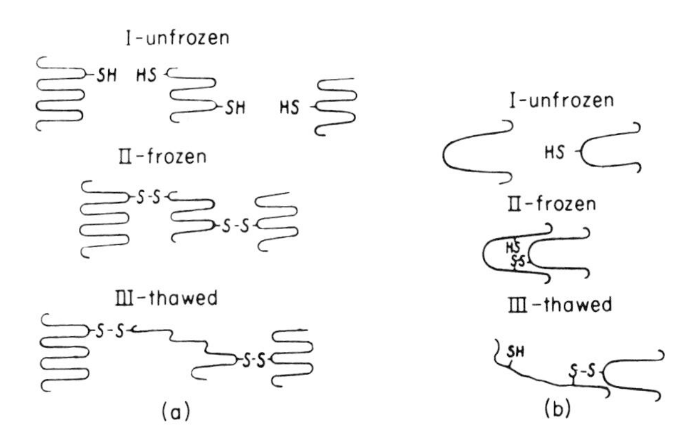

Figure 10.1. Postulated mechanisms of protein unfolding due to intermolecular SS formation during freezing (from Levitt, 1962).

2. Supporting Evidence

a. CORRELATIONS WITH INJURY AND TOLERANCE

The SH⇌SS hypothesis was based on the following correlations. (i) An increase in protein SS was found to accompany freezing injury; no such increase occurred when the freezing failed to injure (Levitt, 1967c). More recent measurements have revealed that these early SS values are probably exaggerated, since they involved unfolding of the molecule and, therefore, included the previously masked SH groups. The relative relationship has, however, been confirmed qualitatively by labeling the proteins with *p*-[^{14}C]chloromercuribenzoate and separating them by disc electrophoresis (Morton, 1969). This evidence indicates that only a small fraction of the SH groups of the soluble proteins is converted to intermolecular SS bonds on freezing. Reliable measurements on the insoluble proteins have not yet been obtained. (ii) The SH content of the homogenate of a plant (and of the proteins in it) was proportional to freezing tolerance, and this was due to a greater resistance of the tolerant plants to SH oxidation during homogenization (Chapter 8). (iii) An increase in activity of the GSH oxidizing system also paralleled the increase in freezing tolerance, due at least partly to an increase in ascorbic acid. This system could conceivably protect the protein SH groups by scavenging free oxygen, H_2O_2, or other oxidizing agents.

b. TESTS WITH MODEL SYSTEMS

Following the above correlations, the hypothesis was tested with pure proteins as model systems, in order to determine whether or not freeze-dehydration actually can induce intermolecular SS bonding of protein molecules. Many pertinent results are already available in the literature on proteins. Table 10.2 lists a number of enzymes inactivated by freezing. They are all SH proteins. Among these, lipoyl dehydrogenase is a particularly clear case. In the reduced (SH) form it is inactivated by freezing; in the oxidized (SS) form it is not inactivated by freezing (Massey *et al.*, 1962). A relationship between ATPase inactivation on freezing and SH content has been found in the case of purified myofibrillar proteins of chicken pectoralis (Khan *et al.*, 1968). When frozen at −30°C, the proteins showed a loss of SH content, solubility, ATPase activity, and water-binding capacity. Blocking the SH groups with PCMB (*p*-chloromercuribenzoate) also inactivated the ATPase, but failed to alter its solubility or water-binding capacity. Oxidation of the SH groups to SS by H_2O_2 had the same effect as freezing. The logical explanation is that freezing induced an aggregation of the proteins by intermolecular SS formation, since ATPase is an SH enzyme (Kuokol *et al.*, 1967), requiring thiol for activity (Heber, 1967). It has, of course, long been

TABLE 10.2

SH-Containing Enzymes Inactivated by Freezing[a]

Enzyme	Molecular weight	SH (groups/ molecule)	SS (groups/ molecule)	Protectant against cryoinactivation
1. Lactic dehydrogenase	170,000	14		Glutathione, mercaptoethanol
2. Triosephosphate dehydrogenase	100,000	11-12	0	Mercaptoethanol
3. Glutamic dehydrogenase	1,000,000	90-120	0	Mercaptoethanol
4. Lipoyl dehydrogenase	100,000	8-12	2	Oxidation: SH → SS
5. Catalase	248,000	(Altered by SH reagents		
6. Myosin	594,000	45		
7. 17β-Hydroxy steroid dehydrogenase		(Inactivated by SH reagents)		
8. Succinate dehydrogenase	200,000	(Inactivated by SH reagents)		
9. Phosphoglucomutase	74,000	7.5		

[a] From Levitt, 1966a.

known that the extraction of active enzymes is often dependent on the addition of thiols to the extraction medium (Anderson and Rowan, 1967). Obviously, any tying up of the SH group associated with the active site (e.g., by intermolecular SS formation) would certainly inactivate the ATPase. In the case of three dehydrogenases and one transferase, it has even been possible to prevent their inactivation during freezing, by the addition of a small amount of thiol (Table 10.2).

It is questionable, however, whether the above experiments involve a true freeze-dehydration, as it occurs in the case of living cells. Protein solutions frozen in test tubes may undergo freeze-dehydration if frozen at high temperatures (e.g., −5°C). The ice then forms first adjacent to the colder tube wall, and as more and more ice crystallizes externally, the solution gradually becomes more and more concentrated in the warmer center of the test tube (Goodin and Levitt, 1970). The freezing of protein solutions, however, is usually performed at much lower temperatures, and the ice crystals probably form rapidly, trapping the protein molecules between them. In the experiments described above with the myofibrillar proteins, which present the most clear-cut evidence of intermolecular SS formation on freezing, freeze-dehydration may have occurred as a secondary process; for although frozen at −30°C, they were subsequently stored at −5°C for 10 weeks, allowing the larger ice crystals to grow at the expense of the smaller crystals, and presumably leaving pockets of concentrated protein between the large crystals.

A true test of the hypothesis obviously requires a model protein system that can be freeze-dehydrated in the same way as the protoplasm of extracellularly frozen cells. This can be achieved with any protein that forms an aqueous gel, by inoculating the slightly supercooled gel surface with an ice crystal. If the temperature is maintained slightly below the freezing point, the ice forms slowly externally to the gel. When this ice sheet is removed, the remaining unfrozen gel is correspondingly thinner due to the freeze-dehydration. Gelatin itself contains no SH groups; but a thiolated form of gelatin, called Thiogel is available. Unlike ordinary gelatin, the melting point of the gel. When the gel was freeze-dehydrated (i.e., frozen extragellu-bond formation can, therefore, be followed simply by measuring the melting point of the gel. When the gel was freeze-dehydrated (i.e., frozen extracellularly) its melting point rose much more rapidly than when kept unfrozen at 1°C (Fig. 10.2). That this rise in melting point was, indeed, due to intermolecular SS bonding was proved by adding GSH, which reduced the SS groups and returned the melting point to its original value (Fig. 10.3). The hypothesis, therefore, holds for the model system—freeze-dehydration does accelerate intermolecular SS formation. That the dehydration, rather than the freezing *per se*, is the initiating factor was established by dehydrating the Thiogel osmotically at 1°C. The intermolecular SS bonding was accelerated even more than by freeze-dehydration (Table 10.3).

The model system can also be used as a test of the tolerance mechanism. Freezing tolerance has two components: avoidance of freeze-dehydration and tolerance of freeze-dehydration. The first of these is easily tested with the model system. The addition of solutes to the gel confers avoidance by

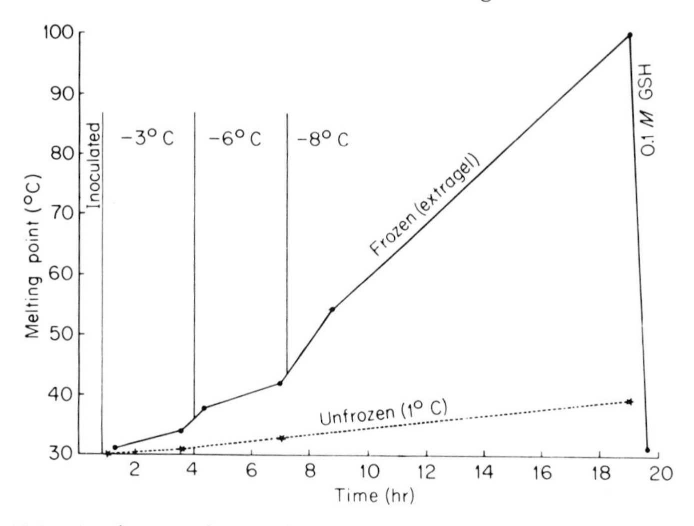

Figure 10.2. Acceleration of intermolecular SS formation (shown by rise in melting point) in Thiogel on freezing (from Levitt, 1965).

TABLE 10.3

Effect of Dehydration by PVP Solution on Melting Point of Thiogel[a,b]

	0.1 *M* Buffer		0.1 *M* Buffer containing 36% PVP	
Time (hr)	Thiogel dimensions (mm)	Melting point (°C)	Thiogel dimensions (mm)	Melting point (°C)
0		29		29
1	11 × 11	30	8 × 8	100
4½	10 × 12	34		
7½		39		
12½		90–100		
20	½ hr in 0.1 *M* GSH			31

[a] Thiogel squares stirred continuously in 0.1 *M* phosphate buffer (pH 7.3) and in similar buffer containing 36% polyvinylpyrrolidone.
[b] From Levitt, 1965.

decreasing the degree of freeze-dehydration. It also retarded the intermolecular SS bonding on freeze-dehydration (Fig. 10.3).

Thiogel, however, is a denatured protein and, therefore, cannot strictly test the SH hypothesis. A native protein, bovine serum albumin (BSA) was, therefore, investigated. Although it does not form a gel, it can be freeze-dehydrated in a dialyzing sac by inoculating the external surface of the sac (Goodin and Levitt, 1970) or more simply by inoculating a solution in a test tube at a temperature just below its freezing point, allowing the freezing to occur from outside inward. Native BSA contains 17 SS groups per molecule and no SH groups. Freeze-dehydration of this native protein failed to induce aggregation detectable by precipitation on centrifuging. When, however, the protein was denatured in 6 *M* urea, and the SS groups were reduced by addition of thiol, the subsequently purified protein aggregated (precipitated) completely on freeze-dehydration. Measurements revealed the formation of SS groups and the aggregate was completely resolubilized in the presence of a thiol and urea, although not by urea alone.

If the SH groups of the denatured and reduced BSA were tied up by combination with a SH reagent (N-ethylmaleimide or NEM), freeze-dehydration in some cases failed to precipitate the protein (Goodin and Levitt, 1970). In other cases, precipitation did occur, particularly if the protein was warmed before rehydration. This was apparently due to intermolecular apolar bonding, since it was reversed by 6 *M* urea. This may, perhaps, explain postthawing injury in plants warmed rapidly after thawing. However, the NEM added seventeen extra hydrophobic groups per protein molecule, and, therefore, these results are not necessarily typical of an unaltered protein.

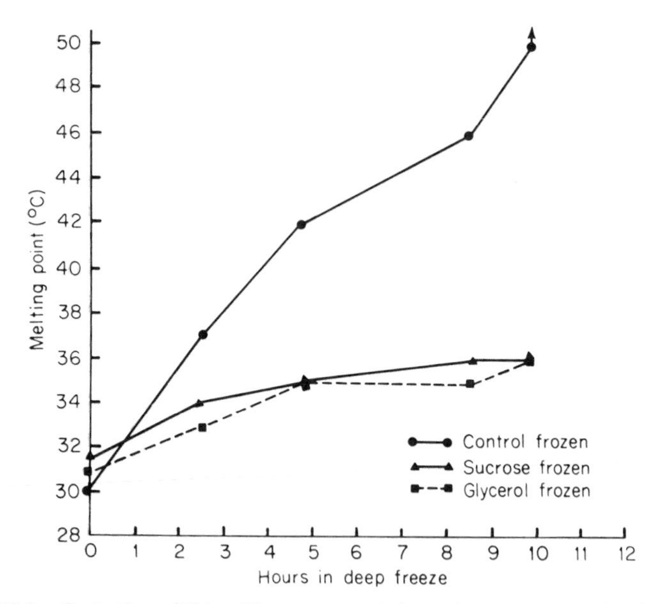

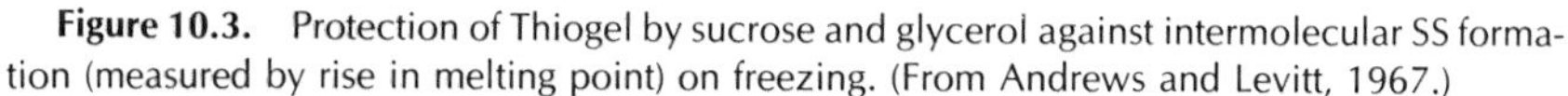

Figure 10.3. Protection of Thiogel by sucrose and glycerol against intermolecular SS formation (measured by rise in melting point) on freezing. (From Andrews and Levitt, 1967.)

The results with BSA support the SH hypothesis; for the native SS protein did not aggregate on freeze-dehydration, but the denatured, SH protein did. This agrees with the above results with lipoyl dehydrogenase (Massey *et al.*, 1962).

The two model systems clearly establish one part of the SH hypothesis. Freeze-dehydration of a SH protein does induce intermolecular SS bonding. In both cases, however, this involved a denatured protein. Unfortunately, no test has yet been made on a native SH protein.

A further observation with a model protein may help to explain the effect of the mechanical stress due to cell contraction. When the fibrous protein keratin is subjected to tension, the mechanical stress greatly facilitates the formation of new SS bonds between previously existing ones (Feughelmann, 1966). In fact, the temperature at which this exchange reaction occurs is lowered from 100° to 20°C by the tension. Subjection of the plant cell to a similar mechanical stress due to freeze-dehydration may be expected to produce the same effect.

c. TESTS WITH LIVING CELLS

One part of the SH hypothesis is now firmly established. Denatured SH proteins are induced to form intermolecular SS bonds by freeze-dehydration. It has also been shown that SS groups form in cells injured by freeze-dehydration. It remains to be proved that this is the cause and not the result of freezing injury. Furthermore, perhaps the plant proteins do not denature reversibly before freezing, as they do in the case of Brandts' model proteins. He has, in fact, demonstrated that some proteins denature at so low a temperature that their solutions must be maintained in the unfrozen state by the addition of an antifreeze, in order to induce reversible denaturation (Brandts *et al.*, 1970). This is undoubtedly why many enzymes are neither denatured nor inactivated by freezing. On the contrary, many of them are preserved over long periods of time in the frozen state. Furthermore, even enzymes that are inactivated by freezing in the pure state will not necessarily be inactivated *in vivo*.

Attempts to prevent disulfide injury were first made by adding GSH to sections of tissue before freezing (Levitt and Hasman, 1964). No change in freezing tolerance occurred. Unfortunately, it was not known whether or not the GSH entered the cells. Furthermore, even if it did enter, it is known that some applied thiols may be immediately oxidized on entering the living cell, others may not (Eldjarn, 1965). It was, therefore, decided to use a thiol that (1) is known to penetrate the cell readily, and (2) is not a normal constituent of cells and, therefore, is not likely to be rapidly oxidized by the cell's enzymes. This second attempt succeeded in altering the freezing tolerance of the cells, but in the unexpected direction. Sections of unhardened tissue are protected against freezing injury by freezing in sugar solutions (see Chapter 6). When treated with the thiol mercaptoethanol, this protection is removed (Krull and Levitt, 1972). Similarly, SH reagents which tie up SH groups, actually lowered the freezing tolerance of cabbage cells (Levitt, 1969a). The first of these negative results cannot be fully understood until the nature of the sugar protection is known (see below). They are explainable, however, on the basis of known SH reactions. Too low a concentration of small-molecule thiols to keep the protein SH groups reduced may serve to trigger SH $\rightleftharpoons$ SS interchange reactions, thus increasing the very process against which it is being used to protect the plant. Too high a concentration may split essential intramolecular SS bonds. Similarly, at least one of the SH reagents that lowered freezing tolerance (PCMB), may induce a conformational change in the protein (Sugiyama and Akazawa, 1967).

Negative results have also been obtained with chloroplast thylakoids (Heber and Santarius 1964; see Chapter 8). This negative result is not surprising. No amount of freeze dehydration can bring the membrane proteins closer to each other, since they are imbedded in nonremovable lipids. Membrane proteins can aggregate only with soluble proteins of the cytoplasm adjacent to them (see Sec. 3,b). Since the thylakoid membranes were separated from the soluble proteins before freezing, aggregation via SS or any other bonds was impossible. Thus, although the above prediction from the SH hypothesis is not supported by the tests to date, the negative results may conceivably be due to uncontrolled complications.

Some positive results have been obtained with living cells. A natural increase in cell SH, due perhaps to GSH accumulation, occurs on fertilization of sea urchin eggs. This is accompanied by a rise in freezing tolerance (Asahina and Tanno, 1963). Similarly, a small increase in freezing tolerance of rat uterus tissue has been obtained by adding thiol (Wirth *et al.*, 1970).

More recently Takahashi and Asahina, (1977) also obtained negative results with sea urchin eggs. There was no change in SH content accompanying freeze-killing. On the basis of their results, the only possible role of SS bonds would be by SH⇌SS interchange rather than SH oxidation to SS bonds. This possibility is supported by isolation of a protein from spinach leaves which catalyzes ^{35}S exchange with S of CSH (Schmidt, 1977).

Spraying with 0.1 *M* solution of CSH increased the freezing tolerance of bark tissue of stems of *Cornus stolonifera*, from a killing temperature of about −5°C before spraying, to −11°C 24 hr after spraying. No increases were detected 36 or 48 hr after treatment (Li and Weiser, 1973). This time factor may explain the above negative results, for it emphasizes the transient nature of the protection. Molecular SS formation by application of thiols, must be balanced against the complexity of the interactions, which is indicated by the recent discovery of a new protein involved in oxidation-reduction reactions via its SH groups. Thioredoxin is a chloroplast protein (but also found in *E. coli*) that is reduced by ferredoxin in the presence of another newly discovered enzyme: ferredoxin-thioredoxin reductase (Wolosiuk and Buchanan, 1977). The reaction is apparently regulated by oxidation and reduction of glutathione in the presence of GSH peroxidase and GSSG reductase respectively. It is obvious, therefore, that any freeze-induced change in SH and SS groups (at least in green cells) must depend on interactions with GSH and the above four enzymes. Chloroplast thylakoids, for instance, may be lacking some of these stromal components and, therefore, unable to duplicate *in vitro* the effect assumed to occur *in vivo*.

It is interesting to note that a biological stress (application of Vaccineurin III) produces a decreased content of GSH in rat liver within 30 min, at the same time as GSSG and other SS compounds increase, and that after 24 hr this change is reversed (Harisch and Schole, 1974).

Both GSH and GSH reductase occur in chloroplasts of spinach. The GSH is believed to stabilize the enzymes of the Calvin cycle and keep the ascorbic acid reduced (Foyer and Halliwell, 1976; Jablonski and Anderson, 1978). The GSH reductase is supposed to regulate the GSH: GSSG ratio, and the resulting high concentration of GSH to protect sensitive SH groups and to regulate enzyme activity (Schaedle and Bassham, 1977; Jablonski and Anderson, 1978; Halliwell and Foyer, 1978). However, the reduction of GSSG to GSH was accompanied by evolution of O_2 (Jablonski and Anderson, 1978) and therefore opposed the suggestion (Schaedle and Bassham, 1977) that the enzyme protects against high concentrations of O_2. The enzyme has been purified (Halliwell and Foyer, 1978). The content of glutathione and

GSSG-reductase varies seasonally in pine needles, with increased levels during winter. It has, therefore, been proposed that they play an important role in the winter hardiness of the leaves of evergreens (Esterbauer and Grill, 1978).

3. Relation of SH and SS to Membranes

a. SUSCEPTIBILITY OF MEMBRANE PROTEINS TO SS BONDING

Several lines of evidence indicate that membrane proteins are particularly susceptible to intermolecular SS bonding.

1. Membrane proteins are high in SH content. Although the plasma membrane proteins of plants have not been investigated, those of animals and microorganisms have. The proteins of erythrocyte membranes are higher in SH content than the model system Thiogel (Table 10.4) and therefore may be expected to form intermolecular SS bonds even more readily than Thiogel on freezing. Similarly, six water-soluble, globular proteins were found on the surface of *Paramecium aurelia*, and were remarkably high in cystine (more than 10%; Reisner *et al.*, 1969). Mitochondrial structural (i.e., membrane) proteins have extraordinarily high SH contents. By use of radioactive maleimides, which localize the reactive SH groups in the mitochondrial membrane of rat liver, radioactivity was found in the basic, structural proteins and not significantly in the oligomycin insensitive ATPase protein (Zimmer, 1970). In the case of the only plant membranes investigated—the chloroplast structural proteins—the published results range from values below those for Thiogel to values well above them (Table 10.4).

2. Tests with SH reagents have localized the SH groups in membranes, for instance on the surface of intact Ehrlich ascites tumor cells, human blood platelets, and lymphocytes (Mehrishi and Grasetti, 1969). In the membranes of red blood cells, 7% of the SH groups are readily reactive with *N*-ethylmaleimide, up to 25% with chlormerodrin, and the remaining 75% only with $HgCl_2$ (Rothstein and Weed, 1963). The deleterious effect of Cu^{2+} on *Chlorella pyrenoidosa* is due to binding to the cytoplasmic membrane (Steeman-Nielsen *et al.*, 1969). As mentioned above, Cu^{2+} can combine with SH groups. Similarly, contractile elements in the ectoplast (the outer protoplasmic layer) of *Mougeotia* are inhibited by PCMB, and the inhibition is reversed completely by cysteine (Schönbohm, 1969).

b. PERMEABILITY EFFECTS OF SH REAGENTS

It has long been known that both Cu^{2+} and Hg^{2+} inhibit the entry of glycerol into red blood cells. Since the amount of Hg^{2+} required to alter the

TABLE 10.4

Total Potential SH Groups (SH + 2SS) in Membrane Proteins[a]

Protein	SH groups/ 10,000 mol. wt.	Reference (see Levitt and Dear, 1970)
Thiogel	0.6–0.8	
Chloroplast structural		
Chlorella pyrenoidosa	1.2	Weber, 1962
Allium porrum	0.3	Weber, 1962
Antirrhinum majus	0.32	Weber, 1962
Spinacea oleracea	0.44	Weber, 1962
	1.2	Criddle, 1966
Beta vulgaris	0.5–0.8	Bailey *et al.*, 1966
Mitochondrial structural		
Neurospora	2.0–2.5	Woodward and Munkres, 1966
Yeast	5.0	Woodward and Munkres, 1966
Beef heart	4.6	Criddle *et al.*, 1962
Beef heart	2.5	Woodward and Munkres, 1966
Beef heart	1.0	Lenaz *et al.*, 1968
Erythrocyte membrane	1.0	Morgan and Hanahan, 1966
Erythrocyte membrane	1.0	Mazia and Ruby, 1968
Liver membrane (Eigen)	0.0	Neville, 1969

[a] From Levitt and Dear, 1970.

membrane properties is of the same order of magnitude as the estimated SH content, Webb (1966) suggests that the change in permeability is due to reaction of the Hg with SH groups (forming S-Hg-S bridges) in and around the membrane pores, thus impeding the passage of substances across the membrane. A similar effect could be produced by SS bridges between adjacent membrane proteins. Consequently, a conversion of SS to SH groups in the membrane proteins could conceivably account for the increase in permeability. In agreement with this conclusion, some reagents that combine with SH groups lower the cell permeability (Levitt, 1971). Epidermal cells of red cabbage were less permeable to urea after treatment at low temperatures 0°–3°C) with four SH reagents and more permeable after treatment with one (Table 10.5). Both changes were reversed by treatment with a thiol (ME), or after longer periods in the absence of thiol or other SH reagents. Similar results were obtained with strawberry leaves (Younis, 1969). Phenylmercuric acetate poisons many enzymatic systems by reactions with SH and SS groups. It initially increases membrane permeability of plant cells, but later causes denaturation of the proteins (Waisel *et al.*, 1969). Similar results have been obtained with human red cells, and to a lesser degree other biological membranes, (Naccache and Sha'afi, 1974). PCMBS and DTNB (1 m*M*) inhibited the rate of transport of water and small hydrophilic (but not hydrophobic) electrolytes across the membrane. This effect was reversed by

TABLE 10.5

Relative Permeability of Red Cabbage Epidermal Cells to Urea after Treatment at 1°–3°C with the Given SH Reagent[a,b]

SH reagent	Concentration (molar)	Time in solution (hr)	No. of trials	Relative permeability
β-mercaptoethanol (ME)	10^{-2}	1–24	11	0.97 ± 0.15
Dithiothreitol (DTT)	10^{-2}	5–40	4	0.60 ± 0.17
Iodoacetamide (IA)	10^{-4}	7–40	11	0.66 ± 0.10
Methylmercuric iodide (MMI)	10^{-5}	9–24	8	0.59 ± 0.15
N-Ethylmaleimide (NEM)	10^{-5}	12–18	3	0.50 ± 0.10
p-Chloromercuribenzoate (PCMB)	10^{-4}	9–40	11	3.7 ± 1.5

[a] Urea concentration 0.45–0.75 *M*. Treatment solution: 0.1 *M* 9:1 NaCl + $CaCl_2$ with or without (control) the given partial concentration of SH reagent.

$$\text{Relative permeability} = \frac{\text{(time for deplasmolysis in urea after treatment in control solution)}}{\text{(time for deplasmolysis in urea after treatment for same time in solution with SH reagent)}}$$

[b] From Levitt, 1972a.

CSH. NEM and IAA had no significant effect. They conclude that there are two populations of membrane-bound SH groups, which differ in location and participate in control of water transfer—one closer to the outer surface and readily accessible to PCMBS, the other probably located in the membrane interior.

There is some evidence that SS bridges may also form between soluble proteins and adjacent membrane proteins. Such mixed aggregates are apparently responsible for the loss of semipermeability and the consequent injury in abnormal red blood cells containing Heinz bodies (Jacob *et al.*, 1968). Labeling of SH groups with [^{14}C]*p*-chloromercuribenzoate has led to evidence of a similar mixed aggregation of proteins as a result of freezing injury to cabbage leaves (Morton, 1969). Direct evidence of SS formation in cell membranes as a result of radiation injury has been obtained in the case of isolated erythrocyte ghosts, i.e., the membranes (Sutherland and Pihl, 1968; see Chapter 8, Vol. 2).

c. EFFECT OF SH REAGENTS ON ACTIVE UPTAKE

These affect the *active* uptake by red blood cells (Webb, 1966). Other cells and tissues have shown similar effects. A protein involved in active transport of lactose was located in the membrane of *E. coli*, and has a SH group essential for transport (Yariv *et al.*, 1969). Two mercaptans (2, 3-dimercaptopropanol and 2-mercaptoethanol) at low concentrations (10–100 μM) inhibit and at higher concentrations (above 1 mM) activate the membrane enzyme (Na^+ + K^+)-ATPase (Bader *et al.*, 1970). Leaf disks of *Saccharum officinarum*, which were first allowed to accumulate large quantities of sugar, secreted about one-third of it into the external medium when treated with 10^{-4} M of the SH reagent iodoacetate (Schoolar and Edelman, 1970). No other respiratory inhibitors induced secretion, indicating a selective effect of the SH reagent IA on the cell membranes. Similarly, the SH reagent NEM interferes with the active uptake of amino acids by cells of *Escherichia coli* (Janick *et al.*, 1977).

d. POSSIBLE DENATURATION OF MEMBRANE PROTEINS

Membrane lipoproteins contain more hydrophobic groups than do the soluble lipoproteins, which in turn have more than soluble (nonlipo) proteins (Hatch and Bruce, 1968). Since the low temperature-induced unfolding of proteins is due to a weakening of the hydrophobic bonds, the membrane proteins will undergo greater denaturation at low temperatures, and will therefore be more readily aggregated on freezing. These membrane lipoproteins have a marked tendency to self-aggregate after removal of the lipids (Hatch and Bruce, 1968). Since both low temperature and dehydration weaken hydrophobic bonds, if these lipoproteins are linked by hydrophobic bonds, they may become separated from the lipids on freezing, leading to irreversible protein aggregation, loss of ion pump activity and finally of semipermeability, and death.

All these facts favor the conclusion that extracellular freezing injury is initiated in the plasma membrane, and that the change occurs in the membrane proteins due to intermolecular SS formation, causing coaggregation of proteins. How can this freezing injury arise?

D. MECHANISM OF FREEZE-INDUCED MEMBRANE DAMAGE

On the basis of all the above evidence, the following explanation of the mechanism of freezing injury is proposed.

1. Cells essentially without any freezing tolerance. These are cells whose membrane lipids undergo a phase transition from the liquid crystalline to the solid-gel state at temperatures above or insufficiently below their freezing point. If frozen rapidly, they freeze intracellularly due to the decreased permeability of the gelled lipid membrane and are killed due to the puncturing of their plasma membranes by the intracellular ice crystals and the consequent loss of semipermeability. If frozen slowly enough, they may

freeze extracellularly, and the consequent cell collapse fractures the now solid lipid membrane, causing an efflux of cell solutes through the fractures and therefore death of the cell.

2. Cells with varying degrees of extracellular freezing tolerance.

(a) Extracellular freezing produces a freeze-dehydration.

(b) The resulting cell collapse produces a mechanical tension on the cell surface and therefore on the surface membranes (plasmalemma and tonoplast).

(c) This tension pulls the membrane molecules apart at the point of weakest intermolecular bonds—the hydrophobic bonds between the lipids, which are further weakened by the low temperature.

(d) The technique of freeze-fracturing has established the locus of their separation—between the two lipid layers of the membrane. In the case of freeze-dehydration, however, there is no knife to produce a freeze-fracture, so the strain will occur as a slippage between the two lipid layers. This slippage per se would not damage the integrity of the lipid layers and, therefore, would not alter the permeability of the membrane. It would, however, displace those membrane proteins which project into both lipid layers. The membrane would also, however, be subjected to a linear strain—a stretch. Since the inner lipid layer is on the convex side of the collapsed cell surface, it would be subject to the greater stretch, and the membrane protein would, therefore, be displaced into this layer. If the combined slippage and stretch are sufficiently pronounced, the protein molecule may protrude or even be extruded into the cytosol.

(e) At least some of these membrane proteins must be components of the ion pumps. Therefore, their dislodgement would impair the ion uptake mechanism. The lipid displacement within the membrane would be spontaneously reversible on thawing. Even the protein dislodgement could be reversed, though more slowly, if the protein is not completely dislodged into the cytosol. This reversibility could lead to repair of the damage to the ion uptake mechanism. During the freezing, however, the freeze-dehydration would concentrate the cell contents, bringing the cytosol proteins and other solutes into close contact with the dislodged membrane proteins. This could lead to protein aggregation, via the low temperature-induced unfolding (denaturation), followed by intermolecular, probably SS, bonding.

This concept would explain why extracellular freezing should depress the active uptake mechanism without damaging the semipermeability of the membrane. Even though small "holes" arise in the membrane due to dislodgement of the membrane proteins, these would be simultaneously filled wity cytoplasmic components. On thawing, only the most hydrophobic components of the cytoplasm (perhaps the aggregated membrane proteins) would remain in the membrane, and its semipermeability would be unimpaired. The membrane proteins, however, would be aggregated with cytosol protein and, therefore, unable to function as ion pumps even if reconstituted into the membrane. The degree of damage to the ion pumps, and therefore to the cell, would depend on the proportion dislodged from the membrane and aggregated.

(f) In some plants and under some extreme conditions, a loss of semipermeability may conceivably occur. If, for instance, the linear stretch of the membrane produces a sufficiently large separation between the lipid molecules of the inner layer, contact may be made between the unsaturated bonds within the layer and the enzyme systems of the cytosol, leading to peroxidation of the lipids and loss of membrane semipermeability.

E. FREEZING RESISTANCE AND THE HARDENING MECHANISM

1. Problems in Identification of Resistance Factors

The search for the factors associated with stress resistance is often futile. This is because of the usually ignored law of limiting factors. Due to this law, each factor can be demonstrated to be correlated with freezing resistance only if it happens to be limiting. Many of the apparently contradictory results of different investigators attempting to identify a hardiness factor are due to the operation of this law. In the case of some varieties of wheat, for instance, sugar content alone accounts for the differences in freezing resistance. These varieties fail to show a correlation between other factors (e.g. permeability, lipid unsaturation, etc.) and freezing resistance. Conversely, other wheat varieties show no correlation between sugar content and freezing resistance, and the two may, in some cases, be actually inversely correlated.

The following analysis attempts to avoid this problem by considering the theoretical possibilities first and comparing the expected relations with the experimental results.

2. Kinds of Resistance

Of the six possible kinds of freezing resistance (see Fig. 7.3), the avoidance of freezing temperatures and the total absence of free water play no role in the resistance of whole vegetative plants exposed above the snow cover. The remaining four tolerance mechanisms are all important and must be developed in hardy plants to a sufficient degree during the hardening process. Furthermore, to be effective, they must develop in a certain order. This explains the previously puzzling fact that some well-established hardiness factors may be developed without conferring any resistance on the plant. Potato tubers, for instance, can be maintained essentially indefinitely at 0° to 5°C without suffering injury. Two metabolic changes occur at this temperature that are characteristic of the hardening process: a conversion of starch to sugar and a synthesis of proteins from amino acids (Levitt, 1954). Nevertheless, these changes fail to increase the freezing resistance of the tubers, and they are killed by the slightest freeze. A primary hardiness factor is apparently missing from the potato tuber, and therefore, the above secondary factors are essentially ineffective. On the other hand, not all the resistance adaptations must be developed in the same plant or tissue. Freezing avoidance due to supercooling, for instance, if developed to a sufficient degree, eliminates the need for the remaining three kinds of resistance, and vice versa. Since supercooling is the decisive adaptation only in buds of some woody plants and in their wood parenchyma and since the possible adaptations leading to its development are still little understood (see Chapter 7), only the remaining three kinds of resistance will be considered.

3. Kinds of Resistance-Adaptations

a. AVOIDANCE OF INTRACELLULAR FREEZING

The plant must first develop adequate avoidance of intracellular freezing before tolerance of extracellular freezing can do any good. This requires (1) the maintenance of its membrane lipids in the normal liquid-crystalline phase at, and slightly below, the freezing point of the plant and (2) a sufficiently high permeability of the cell to water when in the normal liquid-crystalline phase. During the hardening period, it must therefore, (a) lower the phase transition temperature of its membrane lipids below its freezing point unless it already possesses this property constitutively, and (b) it must increase its permeability to water, again unless its constitutive permeability is sufficient. Even in the case of very hardy plants (e.g. fruit trees), specific organs such as flowers and fruit may never develop these properties and therefore are completely devoid of freezing resistance. The only known method used by the plant to lower the phase transition temperature of its membrane lipids is by an increase in the unsaturation of the fatty acids. Other methods may conceivably exist but have not yet been detected in plants. This increase in unsaturation may also increase the permeability of the membrane when in the normal, liquid-crystalline state (Stein, 1967) and therefore may account for at least some of the increase observed during hardening (see Chapter 8). However, the high permeability may also depend on the SH content of the membrane, since decreasing it has been shown to lower cell permeability (see Sec. C,3,b). Some plants possess avoidance of intracellular freezing constitutively and do not show this increase in unsaturation or in fluidity of the membrane lipids during hardening. Black locust, for instance, in the unhardened state during summer is already able to survive freezing at any temperature above −10°C (Singh *et al.*, 1977a).

b. AVOIDANCE OF FREEZE-DEHYDRATION

Once the plant has developed adequate avoidance of intracellular freezing, it must then develop sufficient tolerance of extracellular freezing to survive the freezing temperatures of its environment. It may do this during the hardening period by accumulating soluble substances which therefore decrease the amount of freeze-dehydration at any one freezing temperature. The substances accumulated during hardening are sugars or sugar alcohols.

c. TOLERANCE OF FREEZE-DEHYDRATION

Although moderate degrees of resistance can be developed during hardening, due to increases in the above two adaptations, higher degrees of resistance require adequate tolerance of freeze-dehydration. The changes during hardening, which lead to the development of this adaptation are least understood and, therefore, require the development of a working hypothesis. According to the above-discussed concept of freeze-induced membrane damage, the following adaptations may be proposed.

i. Avoidance of Membrane Separation. Siminovitch *et al.*, (1968) and Singh *et al.*, (1975) have established this mechanism. They found that total lipids increased during fall hardening by only 20–40%, but that phospholipids increased more than 100%. The increase in the polar phospholipids was, therefore, at least partially at the expense of the nonpolar or neutral lipids. This increase corresponded to the increase in membrane content of the cells since the organelles, and the cytoplasm as a whole also showed a 100% increase. On the other hand, starved cells increased markedly in freezing tolerance and in phospholipid content without any increase in organelles or cytoplasm. On the basis of modern concepts of membrane structure, the lipid content of a membrane is fixed (e.g. as a bimolecular leaflet), and therefore, no increase per unit membrane area would be expected. The only possible way in which an increase in membrane lipids could occur, as Siminovitch *et al*. suggest, is by folding of the protoplast surface. If this occurred, it would, of course, be expected to prevent the tension on the protoplast surface and the resulting membrane separation during freeze-induced cell collapse. It should be pointed out that any such infolding of the membrane must include an increase in membrane proteins as well as phospholipids.

ii. Tolerance of Membrane Separation. This is the least understood of all the adaptations and therefore requires the most speculation. Although the precise nature of the membrane damage is still not established, evidence of the existence of a tolerance of such membrane damage has been presented above—by the loss of the ability of some otherwise normal cells to reabsorb the solutes that leaked into the intercellular spaces as a result of freezing injury, and by their ability to repair this damage.

1. In the case of cells whose ion pump activity may be damaged by freezing, one adaptation might be the accumulation during hardening of a reserve of membrane proteins capable of acting as ion pumps. This would favor, by mass action, their incorporation into the membrane during freezing or thawing, preventing or repairing the loss of membrane-controlled ion uptake.

2. The aggregation of these ion-pump proteins on freeze-dehydration must be prevented. Two mechanisms are conceivable: (a) prevention of protein unfolding (denaturation) and (b) prevention of intermolecular bond formation. The first could be accomplished by a strengthening of the *intra*-molecular bonds, for instance, by the replacement of hydrophobic bonds with hydrophilic bonds. This cannot occur in the membrane proteins, since their insertion into the lipid layer requires a high degree of hydrophobicity. This conclusion is supported by the complete absence of any difference between hardy and nonhardy plants in the degree of hydrophobicity of their insoluble proteins (Chou and Levitt, 1972). It may conceivably occur, however, in the soluble proteins, preventing them from unfolding and aggregating with membrane proteins. In support of this conclusion, hardening has been found to increase the hydrophily and correspondingly to decrease the hydrophobicity of the soluble Fraction I protein in cabbage (Shomer-Ilan and Waisel, 1975), as well as an unidentified soluble protein in wheat (Rochat and Therrien, 1975b). In support of the second conclusion, is the evidence of an increased reducing capacity, which opposes intermolecular SS bonding.

3. Although no change in permeability was detected in the described cases of apparent ion-pump damage, other cells exposed to more severe freezing definitely did lose their property of semi-permeability and were dead on thawing. This must involve damage to the membrane lipids, due perhaps to peroxidation of the unsaturated bonds exposed during membrane separation. The adaptation required to prevent this kind of injury would be the avoidance of lipid peroxidation. This again would require the development of a high reduction capacity. This capacity may depend on a relative increase in activity of an enzyme controlling the reduction reaction. It has been shown, for instance, that gluthathione peroxidase inhibits lipid peroxidation in rat liver microsomes (McCay *et al.*, 1976). The rearrangement of peroxidase isoenzymes is suggested to be an adaptive mechanism against low-temperature stress (Petrova and Mishustina, 1976). The enzyme SOD has also been suggested as an essential defense in aerobic organisms against O_2 toxicity (Fridovitch, 1975; Dagley, 1975). Tocopherol and ubiquinones

also protect unsaturated PL of biological membranes against destructive peroxidation (Maggio *et al.*, 1977).

All the above six adaptations leading to the four kinds of freezing resistance are tabulated in Table 10.6.

4. Degrees of Freezing Resistance (Hardiness)

On the basis of these six resistance adaptations, the different degrees of freezing resistance can be explained as follows:

a. TENDER PLANTS (COMPLETELY UNHARDY)

The membrane lipids of these plants possess a phase transition temperature above their freezing point and the plants therefore have no avoidance of intracellular freezing. These plants always freeze intracellularly and, therefore, are killed by the "first touch of frost."

b. SLIGHTLY HARDY PLANTS

The phase transition temperature of their membrane lipids is slightly below their freezing point, due in some cases at least, to increased unsaturation of the lipids. They, therefore, possess avoidance of intracellular freezing and are able to freeze extracellularly. Due to the absence of other hardiness factors (e.g. lack of accumulation of sugars, etc.) they possess low avoidance and tolerance of freeze dehydration. They are able to survive freezing at no lower than about −5°C.

c. MODERATELY HARDY

Besides a low membrane lipid phase transition temperature, they develop a higher cell sap concentration during hardening due to the accumulation of sugars or other solutes. They, therefore, possess marked avoidance of both intracellular freezing and extracellular freeze-dehydration but little tolerance of freeze-dehydration. They are able to survive about −5 to −10°C.

d. VERY HARDY

In addition to these adaptations already described, these plants must develop one or more of the remaining four adaptations, although there is insufficient evidence to be sure of the order. Ion pump inactivation precedes loss of semipermeability. The plant may conceivably prevent this inactivation by protein changes—accumulation of membrane proteins and increase in hydrophily or SH masking of soluble proteins, leading to avoidance of ion pump inactivation. Besides a marked avoidance of both intracellular freezing and

TABLE 10.6

Adaptations Required for the Four Different Kinds of Freezing Resistance in Extracellularly Frozen Plants

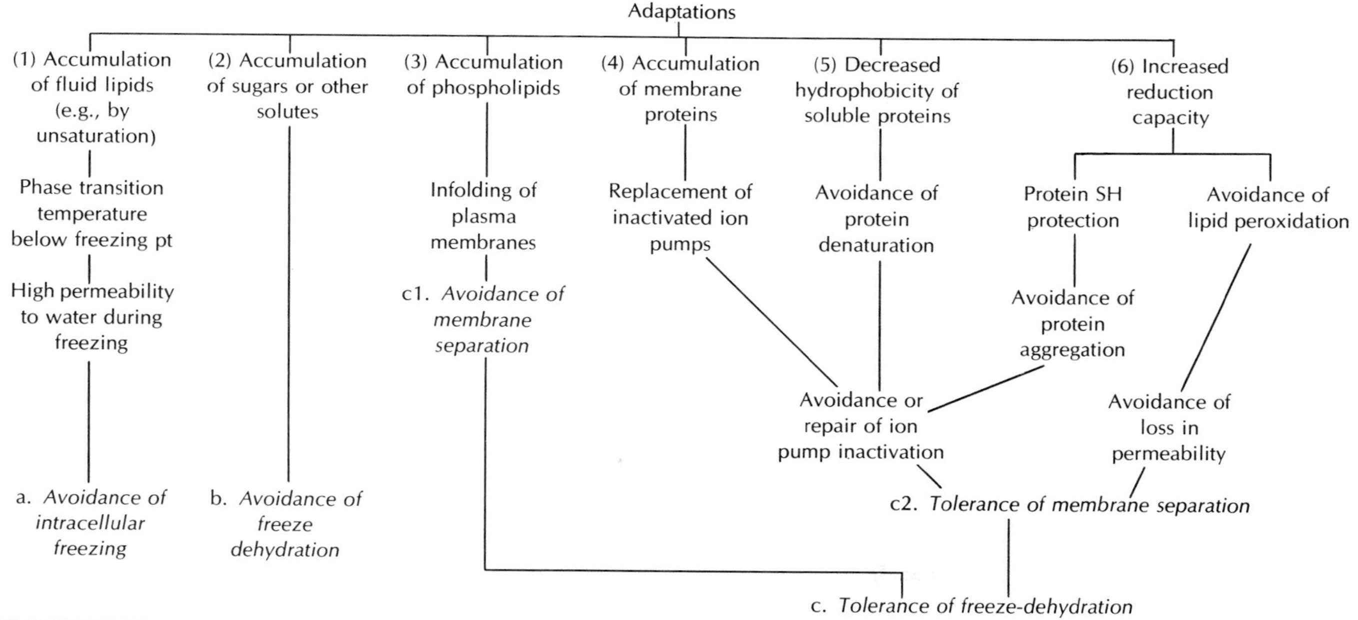

Adaptations

(1) Accumulation of fluid lipids (e.g., by unsaturation)	(2) Accumulation of sugars or other solutes	(3) Accumulation of phospholipids	(4) Accumulation of membrane proteins	(5) Decreased hydrophobicity of soluble proteins	(6) Increased reduction capacity	
Phase transition temperature below freezing pt		Infolding of plasma membranes	Replacement of inactivated ion pumps	Avoidance of protein denaturation	Protein SH protection	Avoidance of lipid peroxidation
High permeability to water during freezing		c1. *Avoidance of membrane separation*			Avoidance of protein aggregation	
			Avoidance or repair of ion pump inactivation			Avoidance of loss in permeability
a. *Avoidance of intracellular freezing*	b. *Avoidance of freeze dehydration*		c2. *Tolerance of membrane separation*			
			c. *Tolerance of freeze-dehydration*			

extracellular freeze-dehydration, these plants would also possess a well-developed tolerance of freeze-dehydration. They would be able to survive −10 to −20°C.

e. EXTREMELY HARDY

These plants must possess a more nearly complete avoidance of ion pump inactivation due to the prevention of membrane protein aggregation (e.g. by SS bonding). This requires the sixth adaptation-increased reducing capacity. This factor also leads to avoidance of loss of semipermeability by prevention of lipid peroxidation. Plants in this group therefore develop a greater tolerance of freeze-dehydration due to an increased tolerance of membrane separation. Alternatively, the third adaptation may be adopted; an accumulation of phospholipids and membrane proteins may confer avoidance of membrane separation due to an infolding of the membrane. These plants are able to survive −20° to −196°C or lower.

It must be emphasized that all these adaptations are quantitative, and therefore, all gradations in freezing resistance exist. Furthermore, it is not intended to imply that the adaptations develop in rigid succession one at a time. In fact, the metabolic processes basic to these adaptations indicate that they probably all develop simultaneously, or at least overlap in development (see Sec. F). Furthermore, some of these adaptations are perhaps alternatives. The increase in phospholipids, for instance, according to the explanation of Singh *et al.* (1977a), may conceivably eliminate or at least decrease the requirements of protein accumulation and increase in hydrophily.

5. Hardening Stages

This concept of the different degrees of freezing resistance also throws some light on the concept of "hardening stages." There has been a tendency to classify these postulated stages on the basis of the environmental factors—the photoperiod stage, the low temperature stage, etc. This classification is of little use, partly because not all plants require a specific photoperiod for hardening, and some that harden on exposure to a specific photoperiod may achieve the same degree of hardening in the absence of the favorable photoperiod by exposure to a water stress (Chen *et al.*, 1977). It would be far more meaningful to classify the hardening stages on the basis of the hardiness factor involved; for instance (1) the membrane liquefying stage (2) the sugar accumulation stage, etc. Unfortunately, as mentioned before, these factors probably do not develop in a rigid order. Their development may overlap, or may even occur simultaneously.

F. METABOLIC CONTROL OF THE HARDENING PROCESS

Low temperature (in the absence of freezing) may affect enzyme activity, and therefore cell metabolism, in five ways.

1. A direct kinetic effect—a decrease in reaction rate, the amount depending on the E_{act} of the specific enzyme.
2. Inactivation of membrane enzymes due to the phase transition of the associated membrane lipids.
3. Inactivation of soluble enzymes due to weakening of hydrophobic bonds, leading to dissociation or unfolding of the enzyme protein.
4. Inductive formation of inhibitors or stimulators of specific enzymes—e.g. growth inhibitors or promoters.
5. A change in enzyme concentration.

Hardy plants must first be able to avoid enzyme inactivation by (2) and (3). Second, their enzymes must have E_{act} favorable for the development of freezing resistance. Finally, (4) and (5) must also be favorable for development of freezing resistance.

1. Direct Kinetic Effect

a. EFFECT ON GROWTH AND ASSIMILATION

The first effect of the low temperature is kinetic—a slow down of all the metabolic processes. This retardation will not be uniform for all reactions because of differences in the temperature (or Arrhenius) coefficients. Some reactions may even be stopped altogether, due to lack of substrate (Selwyn, 1966). The net result may be a marked change in the relative quantities of the different substances present in the plant. Thus, when young plants are raised at normal growing temperatures (e.g., 25°C) nearly all of the photosynthetic products are used up in the production of new growth, either as the raw materials or in the energy-supplying respiratory process. A small amount accumulates as sugars or starch. When the temperature is dropped to 5°C, growth nearly stops and photosynthesis continues at a decreased rate. In hardy plants, the decrease in utilization of photosynthetic products due to the nearly complete growth stoppage is greater than the decrease in rate of photosynthesis. As a result, the net accumulation of these products is greater than at 25°C, and the concentration of sugars in the plant increases (Fig. 10.4).

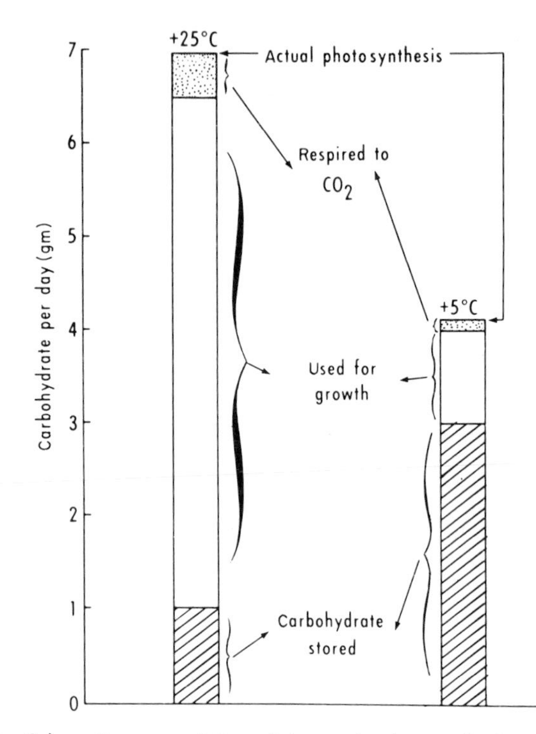

Figure 10.4. Schematic representation of changes in photosynthesis, growth, and carbohydrate storage due to low (hardening) temperatures. Relative photosynthesis and respiration values based on Andersson (1941). (From Levitt, 1967b.)

b. EFFECT ON PHOTOREACTIONS RELATIVE TO ASSIMILATION

Although the net accumulation of carbohydrates at 5°C is greater than at 25°C, the actual rate of photosynthesis definitely and markedly decreases. This is to be expected, since the process is measured by the rate of CO_2 assimilation to carbohydrates, consisting of a series of ordinary dark chemical reactions which have relatively high temperature coefficients. Furthermore, since the carbohydrates are primarily in the soluble form in hardened plants, their accumulation may be sufficient to inhibit carbon assimilation. The "light reactions" of photosynthesis are not usually measured, and since they are more closely related to the true photochemical reactions of photosynthesis, they occur much more rapidly and they must, therefore, have low temperature coefficients. In other words, they must take place at nearly the same rate at 5° as at 25°C. As direct evidence of this, photophosphorylation has been found to occur in spinach at a good rate not only at 0° but even at −10°C (Hall, 1963), whereas oxidative phosphorylation stopped completely at −2°C. At 25°C, the ATP and NADPH produced photosynthetically are undoubtedly nearly all used up by the rapid CO_2 assimilation. At 5°C, due to the markedly decreased CO_2 assimilation, only a small part is used up photosynthetically, and the major part is therefore available for other metabolic processes (Fig. 10.5).

Even at normal temperatures, a brief red irradiation caused an immediate rise of NADPH level and an immediate drop of $NADP^+$ level in the coleoptilar node of etiolated *Avena* seedlings (Fujii and Kondo, 1969). According to Arnon (1969), if $NADP^+$ turnover ceases and NADPH accumulates, only cyclic photophosphorylation can operate, supporting protein synthesis which requires only the ATP. The accumulated carbohydrate, ATP, and NADPH may, therefore, support an increased synthesis of RNA, proteins, and phospholipids at the low, hardening temperatures. In the case of winter perennials, the accumulation of these substances is even greater because growth inhibitors accumulate in late summer and early fall. As a result, all growth stops and the plant becomes dormant. Dormancy is a state of growth inactivity, but not a state of metabolic inactivity. Dormant seeds of *Avena fatua*, for instance, are capable of synthesizing protein at a rate comparable to that of nondormant seeds (Chen and Varner, 1970). Due to this continued

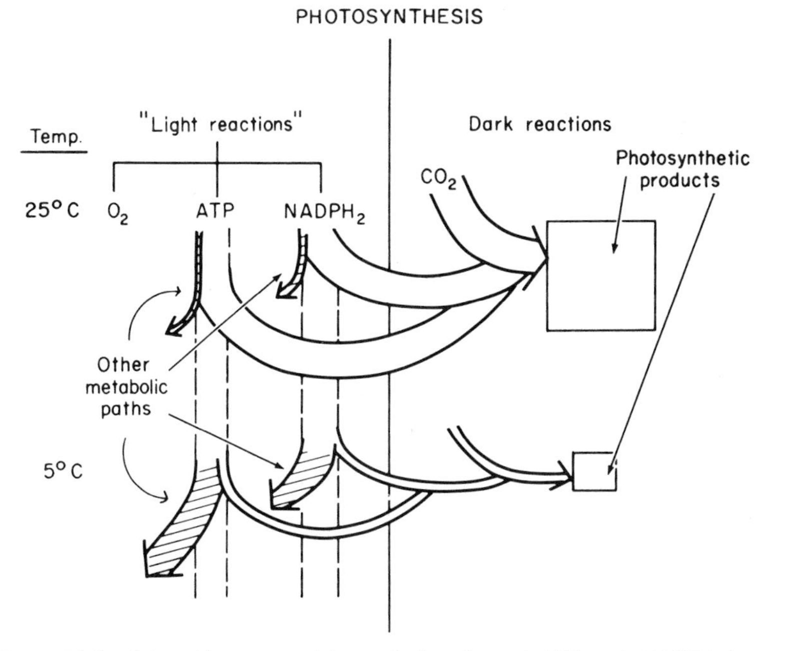

Figure 10.5. Schematic representation of diversion of ATP and NADPH to non-photosynthetic pathways at 5°C. (From Levitt, 1967b.)

active metabolism in winter perennials, without any utilization in growth, substances accumulate in the fall to an even greater degree than in the case of the winter annuals which do continue to grow, although at a decreased rate. The winter perennials, therefore, attain an even higher degree of freezing tolerance than do the winter annuals.

A basic result of the kinetic changes at hardening temperatures is, therefore, an accumulation of reducing power, produced photosynthetically (by photoelectron transport) but not used up in carbon assimilation. Although this prediction is based on the above theoretical considerations, it is supported experimentally. (1) Kuraishi *et al.* (1968) demonstrated an increased ratio of $NADPH/NADP^+$ at hardening temperatures. (2) Ascorbic acid in the reduced form has been shown to accumulate during hardening (see Chapter 8). (3) An artificial difference in SH content, between hardy and less hardy tissues, arises due to more rapid oxidation in the less hardy tissues during the first few minutes after homogenization (Schmuetz, 1969; Chapter 8). This difference in oxidation rate between the hardy and less hardy plants cannot be detected later at room temperature. Benson *et al.* (1949) long ago showed that the photosynthetic reducing potential is available for carbon assimilation for a few minutes after the light is turned off, but that after longer periods in the dark, photosynthetic carbon reduction cannot occur. (4) The importance of oxidation as a factor in freezing injury is supported by some experiments with cucumber seedlings. When grown in air, they were killed by a single night at −2°C. When grown in 2% O_2 + 98% argon they were not killed until the temperature dropped 5°–8°C below this value (Siegel *et al.*, 1969). An atmosphere of nitrogen and CO_2 was similarly able to protect plants of *Haworthia* at temperatures of −30°C. (5) Low temperature treatment of grape leaves and shoots sharply changed the intensity of the redox processes, especially in the frost resistant cultivar (Sherer *et al.*, 1972).

In the case of photosynthesizing leaves, the ultimate in reducing power would be expected in those most hardy evergreen plants whose net carbon assimilation decreases to zero during winter (see Chapter 8); photosynthesis produces both reducing (NADPH) and oxidizing (O_2) substances. In these plants, therefore, the NADPH accumulates due to the decreased carbon assimilation but the released oxygen is immediately reduced to water due to an equal rate of respiration. The direct relationship between respiration rate and freezing tolerance (see Chapter 8) is, therefore, readily understood.

The internal xylem cells do not contain chlorophyll and are less freezing tolerant than the cortical cells. Presumably even they undergo an increase in reduction potential because they are separated from the air by the surrounding chlorophyll-containing cortical cells. There are, however, nonphotosynthesizing tissues not surrounded by chlorophyll-containing cells, that are fully freezing-resistant, e.g., buds and the cambium. These are meristematic and such tissues have high reduction intensities (Van Fleet, 1954). It must also be remembered that some buds possess freezing avoidance due to undercooling (Chapter 7) and, therefore, may not have as high a freezing tolerance as the photosynthesizing tissues.

How is this reduction capacity maintained (a) at very low temperatures when presumably photosynthesis ceases and (b) at night? Many investigators have succeeded in measuring photosynthesis in hardy plants at temperatures below freezing. The extreme has been measured in very hardy lichens below −20°C (Lange, 1965). If this has any survival value it cannot be due to the negligible accumulation of carbohydrates at these low temperatures, but could be due to the maintenance of the high reduction capacity. It must be realized, of course, that oxidation and reduction reactions always take place simultaneously. The hardy plant must permit the harmless oxidation of carbohydrate reserves, while preventing the harmful oxidation of lipids and protein SH. For example, the Russian investigators (Tumanov and Trunova, 1963) have developed methods of hardening wheat plants artificially in complete darkness and therefore, of course, in the absence of photosynthesis. They do this by feeding sugars to the seedlings via their roots, at low temperatures and over about a 2-week-period. This is done only after the plants have begun to harden, and therefore presumably have already undergone the first protoplasmic changes. Furthermore they have found that not all sugars work. Only those that are metabolized by the plant give good results. It is, perhaps, the oxidation of these sugars that prevents the oxidation of lipids or protein SH.

The accumulation of NADPH is apparently insufficient by itself to induce hardening, for it confers chilling tolerance, on peas, but no freezing tolerance (see above). It is presumably necessary to have a complete reduction series, including reduced ascorbic acid, the GSH oxidation-reduction system, etc. This may, perhaps, explain the decrease in freezing tolerance induced by SH reagents. The first effect of these may be to inactivate the reducing system by combining with GSH or some other component.

What role can this increase in reducing power play in the hardening process? It could favor the accumulation of substances less oxidized than carbohydrates, such as the nucleic acids, proteins, and phospholipids. This may explain the relatively small accumulation of sugars in some of the most hardy plants (e.g. conifers).

On the basis of the SH hypothesis, this could lead to a reduction of SS to SH in some proteins, or it could protect the protein SH groups by maintaining them in the reduced state. The first of these effects, if it involved membrane proteins, could account for the observed increase in permeability on hardening. The second, protective effect could prevent the aggregation of proteins by intermolecular SS bonds.

Experimental evidence of such a change has been obtained by Asahi (1964). Spinach chloroplasts, indeed, reduce protein disulfides in the light, but not in the dark. NADPH was not, however, able to reduce the protein in the dark. Asahi concluded from these and other results, that reduction was due to the photosynthetic electron transport system, but did not involve NADPH. This does, of course, support the general concept of hardening described above, although suggesting the replacement of NADPH by some other reducing substance formed photosynthetically. More recent results point to thioredoxin as the substance. It is involved in the reduction processes of photosynthesis, probably in conjunction with GSH, GSSG, GSH peroxidase, and GSH reductase (Wolosiuk and Buchanan, 1977). In the case of the photosynthesizing bacterium, *Chromatium*, direct evidence has been produced of this dependence of protein SH on photosynthesis (Hudock *et al.*, 1965). Triosephosphate dehydrogenase prepared from cells grown in the light had 4.2 SH groups per mole enzyme, whereas the same enzyme extracted from nonphotosynthesizing cells grown on organic medium in the dark had only 2.4 SH groups. During the early hardening of plants, an increase in protein SH has, indeed, been repeatedly observed. Although this is an oxidation artifact, in at least some cases (see Chapter 8), it is still conceivable that a real change in this direction may occur in the living cell, at least in the case of certain enzymes. Unsaturation may, perhaps arise in conjunction with the photoelectron transport reactions at hardening temperatures. The increase in reducing power may also prevent the formation of lipid peroxides from these unsaturated fatty acids, a reaction which is probably highly injurious to membranes (Christophersen, 1969). The danger of peroxidation is suggested by the strong acceleration of H_2O_2 decomposition and ascorbic acid oxidation in frozen solutions in the presence of Cu and Fe salts (Grant, 1969). The net result of the increase in reducing power would, on this basis, be to maintain the membrane lipids in a state of optimum fluidity and, therefore, with maximum ability to protect and repair membrane separation.

The above metabolic changes during hardening are summarized in Table 10.7. They are capable of explaining five of the six adaptations listed in Table 10.6. The proposed decrease in hydrophobicity of the soluble proteins is the only one unexplained, and also the one with the least experimental support.

Thus, plants with enzymes that have quantitatively suitable E_{act} will develop metabolic changes at low (hardening) temperatures that lead to hardening. Just how the low temperature leads to (4) and (5) is unknown. The increase in protein (and, therefore, enzyme) concentration may be a general one. More likely, the formation of growth inhibitors or decrease in growth

TABLE 10.7

Metabolic Changes During Hardening, Leading to the Adaptations Listed in Table 10.5

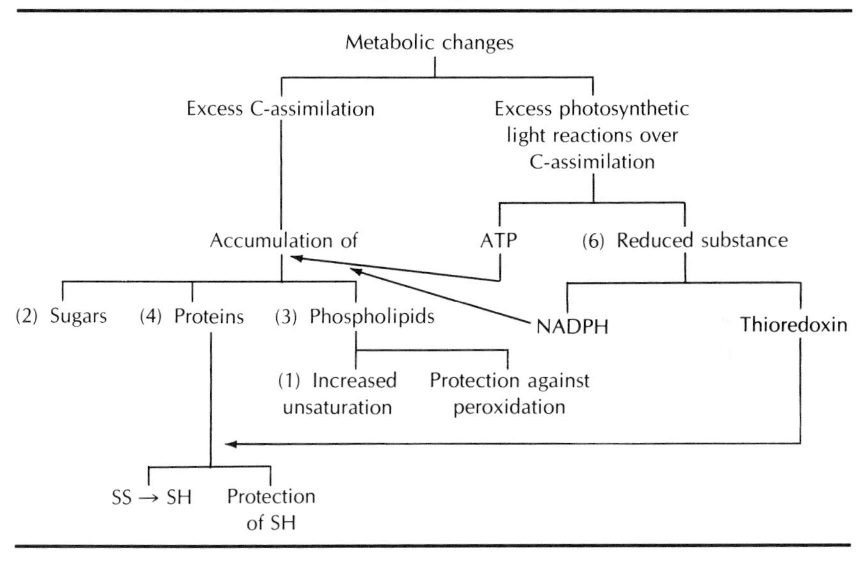

stimulators will prevent the formation of enzymes involved in growth while permitting the accumulation of photosynthetic enzymes.

Other environmental conditions favoring hardening (short photoperiod, dehydration, etc.) must produce similar effects on the metabolism of the plant leading to increased freezing resistance. The short day, for instance, presumably decreases the activity or concentration of enzymes involved in the growth of perennials, relative to those involved in C assimilation.

The inability of some plants to harden adequately, in the above manner when exposed to hardening conditions is not fully understood. In the case of tender, chilling sensitive plants the explanation is clear. They are actually injured at hardening temperatures (and therefore cannot harden) due to the phase transition of their membrane lipids (see Chapter 3). Others (e.g. potato) are chilling resistant yet unable to harden for reasons which have not yet been determined. The differences must depend in some way on differences in metabolism on exposure to hardening temperatures. Perhaps this is due to enzyme inactivation at hardening temperatures due to some factor other than the phase transition of the membrane lipids. This may conceivably be determined by a high degree of hydrophobicity of the enzyme protein, leading to unfolding (denaturation) and therefore to reversible inactivation. The evidence for this, however, is as yet minimal (see Sec E3 C2).

Bibliography

(Chapters 4–10)

Ahring, R. M., and Irving, R. M. (1969). A laboratory method for determining cold hardiness in bermudagrass. *Cynodon dactylon* (L.). Pers. *Crop. Sci.* **9**, 615–618

Aithal, H. N., Kalra, V. K., and Brodie, A. F. (1975). Alteration of *Mycobacterium phlei* membrane structure by freezing and thawing: Reversal by heating. *Arch. Biochem. Biophys.* **168,** 122–132.

Akamatsu, Y. (1974). Osmotic stabilization of unsaturated fatty acid auxotrophs of *Escherichia coli. J. Biochem.* **76,** 553–561.

Akerman, A. (1927). "Studien über den Kältetod und die Kälteresistenz der Pflanzen," pp. 1–232. Berlingska Boktryck. Lund, Sweden.

Albrecht, R. M., Orndorff, G. R., and MacKenzie, A. P. (1973). Survival of certain microorganisms subjected to rapid and very rapid freezing on membrane filters. *Cryobiology,* **10,** 233–239.

Alexandrov, V. Ya., and Shukhtina, G. G. (1964). The state of the protoplasm of plant cells in winter. A criticism of the theory of P. A. Henkel and E. Z. Oknina "the separation of protoplasm." *In* "Cytological Aspects of Adaptation of Plants to the Environmental Factors" (V. Ya. Alexandrov, Ed.). Acad. Sci. USSR. Moscow. Pp. 136–154.

Algera, L. (1936). Concerning the influence of temperature treatment on the carbohydrate metabolism, the respiration, and the morphological development of the tulip. I–III. *Kon. Ned. Akad. Wetensch. Amsterdam Proc.* **39,** 1–29.

Amirshahi, M. C., and Patterson, F. L. (1956). Development of a standard artificial freezing technique for evaluating cold resistance of oats. *Agron. J.* **48,** 181–184.

Anderson, J. O., and Nath, J. (1975). The effects of freeze-preservation on some pollen enzymes: I. Freeze-thaw stresses. Cryobiol. **12**, 160–168.

Anderson, J. W., and Rowan, K. S. (1967). Extraction of soluble leaf enzymes with thiols and other reducing agents. *Phytochemistry* **6,** 1047–1056.

Anderson, L. E., Nehrlich, S. C., and Champigny, M. L. (1978). Light modulation of enzyme activity. Activation of the light effect mediators by reduction and modulation of enzyme activity by thiol-disulfide exchange? *Plant Physiol.* **61,** 601–605.

Anderson, G. (1944). "Gas Change and Frost Hardening Studies in Winter Cereals," 163 pp. Håkan Ohlssons Boktryckeri, Lund, Sweden.

Andrews, C. J., and Pomeroy, M. K. (1977). Mitochondrial activity and ethanol accumulation in ice-encased winter cereal seedlings. *Plant Physiol.* **59,** 1174–1178.

Andrews, C. J., and Pomeroy, M. K. (1978). The effect of anaerobic metabolite on survival and ultrastructure of winter wheat in relation to ice encasement. *Plant Physiol.* **61,** Supple.: 17.

Andrews, C. J., Pomeroy, M. K., and de La Roche, I. A. (1974a). Changes in cold hardiness of overwintering winter wheat. *Can. J. Plant Sci.* **54,** 9–15.

Andrews, C. J., Pomeroy, M. K., and de La Roche, I. A. (1974b). The influence of light and diurnal freezing temperature on the cold hardiness of winter wheat seedlings. *Can. J. Bot.* **52,** 2539-2546.

Andrews, J. E., and Roberts, D. W. A. (1961). Association between ascorbic acid concentration and cold hardening in young winter wheat seedlings. *Can. J. Bot.* **39,** 503-512.

Andrews, J. E., Horricks, J. S., and Roberts, D. W. A. (1960). Interrelationships between plant age, root-rot infection, and cold hardiness in winter wheat. *Can. J. Bot.* **38,** 601-611.

Andrews, S. and Levitt, J. (1967). The effect of cryoprotective agents on intermolecular SS formation during freezing of Thiogel. *Cryobiology* **4,** 85-89.

Angell, C. A., and Sare, E. J. (1970). Vitreous water: identification and characterization. *Science* **168,** 281.

Angelo, E., Iverson, V. E., Brierley, W. G., and Landon, R. H. (1939). Studies on some factors relating to hardiness in the strawberry. *Minn. Agr. Exp. Sta. Tech. Bull.* **135,** 1-36.

Antoniani, C., and Lanzani, G. A. (1963). Influence of germination in the cold on the peroxidases of *Triticum vulgare* embryos. *Vitalstoffe Zivilisationskrankh.* **8,** 39-40.

Aoki, K. (1950). Analysis of the freezing process of living organisms. I. The relation between the shape of the freezing curve and the mode of freezing in plant tissues. *Low Temp. Sci.* **3,** 219-227.

Apelt, A. (1907). Neue Untersuchungen über den Kältetod der Kartoffel. *Beitr. Biol. Pflanz.* **9,** 215-262.

Araki, T. (1977a). Freezing injury in mitochondiral membranes. I. Susceptible components in the oxidation systems of frozen and thawed rabbit liver mitochondria. *Cryobiology,* **14,** 144-150.

Araki, T. (1977b). Freezing injury in mitochondrial membranes. II. Degradation of phospholipid in rabbit liver mitochondria during freezing and storage at low temperatures. *Cryobiology,* **14,** 151-159.

Arndt, K. (1974). The influence of temperature on the vitamin C content of Brussels sprouts. *Angew Bot.* **48,** 125-136.

Arnon, D. I. (1969). Regulation of ferredoxin-dependent photosynthetic activity of isolated chloroplasts. *Abstr. Int. Bot. Congr. Seattle,* p. 5.

Arntzen, C. J. (1978). Dynamic structural features of chloroplast lamellae. *In* "Current Topics in Bioenergetics." (D. R. Sanadi and L. P. Vernon, eds.) Academic Press, New York, vol. 8 pp. 112-160.

Arny, D. C., Lindow, S. E., and Upper, C. D. (1976). Frost sensitivity of *Zea mays* increased by application of *Pseudomonas syringae. Nature* **262,** 282-284.

Aronsson, A. (1975). Influence of photo- and thermoperiod on the initial stages of frost hardening and dehardening of phytotron-grown seedlings of Scots pine and Norway spruce. *Stud. Forest. Suec.* **128,** 1-20.

Aronsson, A., and Eliasson, L. (1970). Frost hardiness in Scots pine. 1. Conditions for test on hardy plant tissues and for evaluation of injuries by conductivity measurements. *Stud. Forest Suec.* **77,** 1-30.

Aronsson, A., Ingestad, T., and Loof, Lars-Goran. (1976). Carbohydrate metabolism and frost hardiness in pine and spruce seedlings grown at different photoperiods and thermoperiods. *Physiol. Plant.* **36,** 127-132.

Asahi, T. (1964). Sulfur metabolism in higher plants. IV. Mchanism of sulfate reduction in chloroplasts. *Biochim. Biophys. Acta* **82,** 58-66.

Asahina, E. (1954). A process of injury induced by the formation of spring frost on potato sprout. *Low Temp. Sci. Ser.* **B11,** 13-21.

Asahina, E. (1956). The freezing process of plant cell. *Contrib. Inst. Low Temp. Sci. Hokkaido Univ.* **10,** 83-126.

Asahina, E. (1958). On a probable freezing process of molluscan cells enabling them to survive at a super-low temperature. *Low Temp. Sci. Ser.* **B16**, 65-75.

Asahina, E. (1961). Intracellular freezing and frost resistance in egg-cells of the sea urchin. *Nature (London)* **191,** 1263-1265.

Asahina, E. (1962a). Frost injury in living cells. *Nature (London)* **1976,** 445-446.

Asahina, E. (1962b). A mechanism to prevent the seeding of intracellular ice from outside in freezing living cells. *Low Temp. Sci. Ser.* **B20,** 45-56.

Asahina, E. (ed.). (1967). Freezing injury in egg cells of the sea urchin. "Cellular Injury and Resistance in Freezing Organisms," pp. 211-230. Inst. Low. Temp. Sci. Hokkaido Univ., Sapporo, Japan.

Asahina, E., and Takahashi, T. (1978). Freezing tolerance in embryos and spermatozoa of the sea urchin. *Cryobiology,* **15,** 122-127.

Asahina, E., and Tanno, K. (1963). A remarkably rapid increase of frost resistance in fertilized egg cells of the sea urchin. *Exp. Cell Res.* **31,** 223-225.

Asahina, E., Aoki, K., and Shinozaki, J. (1954). The freezing process of frost-hardy caterpillars. *Bull. Entomol. Res.* **45,** 329-339.

Asahina, E., Shimada, K., and Hisada, Y. (1970). A stable state of frozen protoplasm with invisible intracellular ice crystals obtained by rapid cooling. *Exp. Cell. Res.* **59,** 349-358.

Augusten, H. (1963). Der Einfluss der Temperatur auf einige Stoffwechselgrossen bei Bulbillen und Wurzelknollen von *Ficaria verna* Huds. *Beitr. Biol. Pflanz.* **38,** 421-424.

Babenko, V. I., and Gevorkyan, A. M. (1967). Accumulation of oligosaccharides and their significance for the low temperature hardening of cultivated cereals. *Fiziol. Rast.* **14,** 727-736.

Babenko, V. I., and Ruzhitskaya, I. P., (1973). Characteristics of the dynamics of organic acids in winter wheat during hardening and freezing. *Dokl. Vses Ord. Lenina Akad. S-Kh Nauk Im. V. I. Lenina.* **7,** 10-12.

Bader, H., Wilkes, A. B., and Jean, D. H. (1970). The effect of hydroxylamine mercaptans, divalent metals and chelators on ($Na^+ + K^+$)—ATPase. A possible control mechanism. *Biochim. Biophys. Acta* **198,** 583-593.

Baig, M. N., and Tranquillini, W. (1976). Studies on upper timberline: morphology and anatomy of Norway spruce and stone pine needles from various habitat conditions. *Can. J. Bot.* **54,** 1622-1632.

Bajaj, Y. P. S. (1976). Regeneration of plants from cell suspensions frozen at −20, −70, and −196°C. *Physiol. Plant.* **37,** 263-268.

Baker, B. S. and Jung, G. A. (1970b). Response of four perennial grasses to high temperature stress. *Proc. 11th Int. Grassland. Congr.* 499-502.

Baker, B. S., and Jung, G. A. (1972). Growth and metabolic changes occuring in orchard grass during temperature acclimation. *Bot. Gaz.* **133,** 120-126.

Baker, B. S., and Jung, G. A. (1968). Effect of environmental conditions on the growth of four perennial grasses. I. Response to controlled temperature. *Agron. J.* **60,** 155-158.

Bakumenko, N. I. (1974). Effect of frost on respiration metabolism in wheat seedlings. *Fiziol. Rast.* **21,** 135-140.

Bamberg, S., Schwartz, W., and Tranquillini, W. (1967). Influence of day length on the photosynthetic capacity of stone pine (*Pinus cembra L.*). *Ecology* **48,** 264-269.

Bank, H. (1973). Visualization of freezing damage: II. Structural alterations during warming. *Cryobiology,* **10,** 157-170.

Bannier, L. J. and Steponkus, P. L. (1976). Cold acclimation of *Chrysanthemum* callus cultures. *J. Am. Soc. Hortic. Sci.* **10,** 409-412.

Barashkova, E. A., and Udovenko, G. V. (1973). Metabolic changes associated with a temperature drop in winter wheat varieties which vary by frost resistance. *Fiziol. Biokhim. Kul't Rast.* **5,** 19-25.

Barnett, J. R. (1975). Seasonal variation of organelle numbers in sections of fusiform cambium cells of *Pinus radiata D. Don. N. Z. J. Bot.* **13,** 325-332.

Barr, M. L., and Potter, L. D. (1974). Chlorophylls and carotenoids in aspen bark *Populus tremuloides. Southwest Nat.* **19,** 147-154.

Bartetzko, H. (1909). Untersuchungen über das Erfrieren von Schimmelpilzen. *Jahrb. Wiss. Bot.* **47,** 57-98.

Bartholic, J. F. (1972). Thin layer foam for plant freeze protection. *Proc. Fla. State Hortic. Soc.* **85,** 299-302.

Bates C. G. (1923). Physiological requirements of Rocky Mountain trees. *J. Agr. Res.* **24,** 97-164.

Bauer, H., Huter, M., and Larcher, W. (1969). Der Einfluss und die Nachwirkung von Hitze-und Kältestress auf den CO_2 Gaswechsel von Tanne und Ahorn. *Ber. Deut. Bot. Ges.* **82,** 65-70.

Bauer, M., Chaplin, C. E., Schneider, G. W., Barfield, B. J., and White, G. M. (1976). Effects of evaporative cooling during dormancy on "Redhaven" peach wood and fruit bud hardiness. *J. Am. Soc. Hortic. Sci.* **101,** 452-454.

Baust, J. G. (1973). Mechanisms of cryoprotection in freezing tolerant animal systems. *Cryobiology* **10,** 197-205.

Beattie, D. J., and Flint, H. L. (1973). Effect of K level on frost hardiness of stems of Forsythia x intermedia *Zab Lynwood. J. Am. Soc. Hortic. Sci.* **98,** 539-541.

Becquerel, P. (1907). Recherches sur la vie latente des grains. *Ann. Sci. Nat. Ser.* **9,** 193-311.

Becquerel, P. (1932). L'anhydrobiose des tubercules des Renoncules dans l'azote liquide. *C.R. Acad. Sci. Paris* **194,** 1974-1976.

Becquerel, P. (1949). L'action du froid sur la cellule végétale. *Botaniste* **34,** 57-74.

Becquerel, P. (1954). La cryosynérèse cytonucléoplasmique jusqu' aux confins du zero absolu, son role pour la vegétation polaire et la conservation de la vie. *8th Congr. Int. Bot. Rapports et Communications.* **11,** 269-270.

Belt, Helena, (1972). The effect of chlorocholine chloride on cold-resistance of different cultivars of winter rape. *Zesz. Nauk. Uniw. Mikolaja. Kopernika Toruniu Nauki. Mat-Przyr Biol.* **14,** 229-237.

Belzile, L., Paquin, R., and Therrien, H. P. (1973). The effects of (2-chloroethyl) trimethylammonium chloride on the development and frost resistance of 'Dover' winter barley. *Can. J. Plant Sci.* **53,** 31-36.

Benko, B., and Pillar, J. (1965). Relation of sugars to the frost resistance of apple trees. *Ved. Pr. Vyzk. Ustavu Rastlinnej Vyroby Piestanoch.* **3,** 291-298.

Benson, A. A., Calvin, M., Haas, V. A., Aronoff, S., Hall, A. G., Bassham, J. A., and Weigl, J. W. (1949). C^{14} in Photosynthesis. *In* "Photosynthesis in Plants" (J. Franck and W. E. Loomis, eds.), pp. 381-402. Iowa State College Press, Ames, Iowa.

Bergmann, H. (1973). Influence of 5°C treatment on the metabolism of purine and pyrimidine derivatives and nucleic acids in etiolated and green seedlings of *Hordeum vulgare L. Biol. Zentralbl.* **92,** 443-453.

Bervaes, J. C. A. M., and Kylin, A. (1972). Long and short term development of frost hardiness in Pinus sylvestris and heavy particle adenosine triphosphatase. *Physiol. Plant.* **27,** 178-181.

Bervaes, J. C. A. M., Kuiper, P. J. C., and Kylin, A. (1972). Conversion of digalactosyl diglyceride (extra long chain conjugates) into monogalactosyl diglyceride of pine needle chloroplasts upon dehardening. *Physiol. Plant.* **27,** 231-235.

Bervaes, J. C. A. M., Ketchie, D. O., and Kuiper, J. C. (1977). Kinetics of freezing damage in apple bark and pine needles. *Physiol. Plant.* **40,** 35–38.

Bidlack, W. R., and Tappel, A. L. (1973). Damage to microsomal membrane by lipid peroxidation. *Lipids,* **8,** 177–182.

Biebl, R. (1967b). Temperaturresistenz tropischer Urwaldmoose. *Flora (Jena)* **157,** 25–30.

Biebl, R. (1967c). Kurtztag-Einflüsse auf arktische Pflanzen während der arktischen Langtage. *Planta* **75,** 77–84.

Biel, E. R., Havens, A. V., and Sprague, M. A. (1955). Some extreme temperature fluctuations experienced by living plant tissue during winter in New Jersey. *Bull. Amer. Meteorol. Soc.* **36,** 159–162.

Billings, W. D., and Mooney, H. A. (1968). The ecology of arctic and alpine plants. *Biol. Rev. Cambridge Phil. Soc.* **43,** 481–529.

Bird, C. J., and McLachlan, J. (1974). Cold-hardiness of zygotes and embryos of Fucus (Phaeophyceae, Fucales). *Phycologia,* **13,** 215–226.

Bittenbender, H. C. and Howell, Jr., G. S. (1974). Adaptation of the Spearman-Karber method for estimating the T_{50} of cold stressed flower buds. *J. Am. Soc. Hortic. Sci.* **99,** 187–190.

Bittenbender, H. C. and Howell Jr., G. S. (1975). Interactions of temperature and moisture content on spring de-acclimation of flower buds of highbush blueberry. *Can. J. Sci.* **55,** 447–452.

Bixby, J. A., and Brown, G. N. (1975). Ribosomal changes during induction of cold hardiness in black locust seedlings. *Plant Physiol.* **56,** 617–621.

Björkman, O., Nobs, M. A., and Hiesey, W. M. (1970). Growth, photosynthetic, and biochemical responses of contrasting *Mimulus* clones to light intensity and temperature. *Carnegie Inst. Washington Yearb.* **68,** 614–620.

Blazich, F. A., Evert, D. R., and Bee, D. E. (1974). Comparison of three methods of measuring winter hardiness of internodal stem sections of Forsythia intermedia *Lynwood. J. Am. Soc. Hortic. Sci.* **99,** 211–214.

Bobart, J. (1684). "Philosophical Transactions and Collections to the End of the year 1700," (Abridged and disposed under general heads) vol. 2, 4th ed., pp. 155–160, 1731.

Bogdanov, P. (1935). Photoperiodism in species of woody plants. Preliminary contribution. *Exp. Sta. Rec.* **73,** 22.

Bolduc, R. J. Cherry, J. H., and Blair, B. O. (1970). Increase in indoleacetic acid oxidase activity of winter wheat by cold treatment and gibberellic acid. *Plant Physiol.* **45,** 461–464.

Bondarenko, V. I., Artyukh, A. D., and Makarova, A. Ya. (1975). Frost-resistance and productivity of winter wheat as a function of conditions of moistening. *Fiziol. Biokhim. Kul't Rast.* **7,** 398–403.

Bondarenko, V. N., Tkalich, I. D. and Artyukh, A. D. (1973). Effect of soil moisture conditions on frost resistance, winter-hardiness and yield of winter wheat. *Dokl. Vses. Akad. S-Kh. Nauk. Lm. V. I. Lenina.* **11,** 12–14.

Borzakivs'ka, I. V. (1965). Changes in the pigment system of the bark of some trees in winter. *Ukr. Bot. Zh.* **22,** 19–25.

Borzakivs'ka, I. V., and Motruk, V. V. (1969). Phosphorus metabolism in woody plant seedlings in connection with their winter hardiness resulting from the effect of variable temperatures on the seeds. *Ukr. Bot. Zh.* **26,** 67–72.

Boussiba, S., Rikin, A., and Richmond, A. E. (1975). The role of abscisic acid in cross-adaptation of tobacco plants. *Plant Physiol.* **56,** 337–339.

Boutron, P., and Kaufmann, A. (1978). Stability of the amorphous state in the system water-glycerol-dimethylsulfoxide. *Cryobiology,* **15,** 93–108.

Bouwkamp, J. C., and Honma, S. (1969). The inheritance of frost resistance and flowering response in broccoli (*Brassica oleracea* yar. *italica*). *Euphytica* **18,** 395-397.

Brach, E. J., Fejer, S. O., and Wilner, J. (1976). Diurnal rhythm in electric impedance in Juniper. *Phyton. Rev. Int. Bot. Exp.* **34,** 81-82.

Brandts, J. F. (1967). *In* "Thermobiology" (A. H. Rose, ed.), pp. 25-72. Academic Press, New York.

Brandts, J. F., Fu, J., and Nordin, J. H. (1970). The low temperature denaturation of chymotrypsinogen in aqueous solution and in frozen aqueous solution. *In:* "The Frozen Cell" (G. E. W. Wolstenholme and M. O'Connor, eds.), pp. 189-208. Ciba Foundation Symposium, J. & A. Churchill, London.

Bregvadze, M. A. (1974). Leaf respiration of some evergreen plants in relation to their winter-hardiness. *Soobshch. Akad. Nauk. Gruz SSSR.* **75,** 437-440.

Bregvadze, M. A., Sharashidze, N. M., and Tarkashvili, D. V. (1975). The dynamics of the activity of the endogenous regulators of growth in the leaves of Cinnamomum japonicum and Cinnamomum camphora in relation to their frost resistance. *Soobshch Akad. Nauk. Gruz. SSSR.* **79,** 153-156.

Brierley, W. G., and Landon R. H. (1946). A study of cold resistance of the roots of the Latham red raspberry. *Proc. Amer. Soc. Hort. Sci.* **47,** 215-218.

Brierley, W. G., and Landon, R. H., (1954). Effects of dehardening and rehardening treatments upon cold resistance and injury of Latham raspbery canes. *Proc. Amer. Soc. Hort. Sci.* **63,** 173-178.

Brown, G. N., and Bixby, J. A. (1973a). Ribonuclease activity during induction of cold hardiness in mimosa epicotyl and hypocotyl tissues. *Cryobiology,* **10,** 152-156.

Brown, G. N., and Bixby, J. A. (1973b). Quantitative and qualitative changes in total protein and soluble protein during induction of cold hardiness in black locust stem tissues. *Cryobiology* **10,** 529-530.

Brown, G. N., and Bixby, J. A. (1975). Soluble and insoluble protein patterns during induction of freezing tolerance in black locust seedlings. *Physiol. Plant.* **34,** 187-191.

Brown, G. N. and Bixby, J. A. (1976). Relationship between black locust seedling age and induction of cold hardiness. *For Sci.* **22,** 208-210.

Brown, G. N., Bixby, J. A., Melcarek, P. K., Hinckley, T. M., and Rogers, R. (1977). Xylem pressure potential and chlorophyll fluorescence as indicators of freezing survival in black locust and western hemlock seedlings. *Cryobiology,* **14,** 94-99.

Brown, J. H., Bula, R. J., and Low, P. F. (1970). Physical properties of cytoplasmic protein-water extracts from roots of hardy and nonhardy *Medicago sativa* ecotypes. *Cryobiology* **6,** 309-314.

Brown, H. T., and Escombe, F. (1897). The influence of very low temperatures on the germinative power of seeds. *Proc. Roy. Soc.* **62,** 160-165.

Brown, M. S. and Reuter, F. W. (1974). Freezing of nonwoody plant tissues: III. Videotape micrography and the correlation between individual cellular freezing events and temperature changes in the surrounding tissue. *Cryobiology.* **11,** 185-191.

Brown, M. S., Pereira, E. S. B., and Finkle, B. J. (1974). Freezing of nonwoody plant tissues: II. Cell damage and the fine structure of freezing curves. *Plant Physiol.* **53,** 709-711.

Bruni, A., Frigeri, L., and Bigon, E. (1977). Cold lability of membrane bound F-ATPase. *Biochim. Biophys. Acta.* **462,** 323-332.

Buchanan, D. W., Biggs, R. H., and Bartholic, J. F. (1974). Cold hardiness of peach and nectarine trees growing at 29-30°N latitude. *J. Am. Soc. Hortic. Sci.* **99,** 256-259.

Bugaevsky, M. F. (1939a). Contribution to the study of causes of death in root crops subjected to low temperatures. *C. R. Acad. Sci. URSS* **25,** 527-530.

Buhlert, H. (1906). Untersuchungen über das Auswintern des Getreides. *Landwirt. Jahrb.* **35,** 837–887.

Bula, R. J., and Smith, D. (1954). Cold resistance and chemical composition in overwintering alfalfa, red clover and sweet clover. *Agron J.* **446,** 397–401.

Bull, H. B., and Breese, K. (1968). Protein hydration I. Binding sites. *Arch. Biochem. Biophys.* **128,** 488–496.

Burke, M. J., Bryant, R. G., and Weiser, C. J. (1974). Nuclear magnetic resonance of water in cold acclimating red osier dogwood stem. *Plant Physiol.* **54,** 392–398.

Burke, M. S., Gusta, L. V., Quamme, H. A., Weiser, C. J., and Li, P. H. (1976). Freezing and injury in plants. *Annu. Rev. Plant Physiol.* **27,** 507–528.

Bylinska, E. (1975). The relationship of winter transpiration of selected species from the genera *Viburnum, Populus,* and *Lonicera* to the northern boundary of their geographical distribution. *Monogr. Bot.* **50,** 5–59.

Campbell, R. K., and Sorensen, F. C. (1973). Cold-acclimation in seedling Douglas-fir related to phenology and provenance. *Ecology,* **54,** 1148–1151.

Calder, F. W., Canham, W. D., and Fensom, D. S. (1973). Some effects of Alar-85 (N-dimethylamino succinamic acid) on the physiology of alfalfa and Ladino clover. *Can. J. Plant Sci.* **53,** 269–278.

Candolle, C. de. (1895). Sur la vie latente des graines. *Bibl. Univ. Arch. Sci. Phys. Nat.* (3) **33,** 497–512.

Carmichael, J. W. (1962). Viability of mold cultures stored at −20°C. *Mycologia* **54,** 432–436.

Caspary, R. (1854). Auffallende Eisbildung auf Pflanzen. *Bot. Zeitung* **12,** 665–674, 681–690, 697–706.

Caspary, R. (1857). Bewirkt die Sonne Risse in Rinde und Holze der Bäume? *Bot. Zeitung* **15,** 153–156, 329–335, 345–350, 361–371.

Chabot, J. F., and Chabot, B. F. (1975). Developmental and seasonal patterns of mesophyll ultrastructure in *Abies balsamea. Can. J. Bot.* **53,** 295–304.

Chambers, R., and Hale, H. P. (1932). The formation of ice in protoplasm. *Proc. Roy. Soc.* **B110,** 337–352.

Chandler, W. H. (1913). The killing of plant tissue by low temperature. *Mo. Agr. Expt. Sta. Res. Bull.* No. 8, 171 pp.

Cheban, A. I. (1968). Content and accumulation of nucleic acids in buds of wintering grape plant eyes of different stages on its seasonal development. *Fiziol. Rast.* **15,** 329–335.

Chen, P., Li, P. H., and Weiser, C. J. (1975). Induction of frosthardiness in red-osier dogwood stems by water stress. *Hortscience,* **10,** 372–374.

Chen, P. and Li, P. H. (1976). Effects of photoperiod, temperature, and certain growth regulators on frost hardiness of *Solanum* species. *Bot. Gaz.* **137,** 105–109.

Chen, P. M. and Li, P. H. (1977). Induction of frost hardiness in stem cortical tissues of *Cornus stolonifera* Michx. by water stress. II. Biochemical changes. *Plant Physiol.* **59,** 240–243.

Chen, P. M., Burke, M. J., and Li, P. H. (1976). The frost hardiness of several *Solanum* species in relation to the freezing of water, melting point depression, and tissue water content. *Bot. Gaz.* **137,** 313–317.

Chen, P. M., Li, P. H., and Burke, M. J. (1977). Induction of frost hardiness in stem cortical tissues of *Cornus stolonifera* Michx. by water stress. I. Unfrozen water in cortical tissues and water status in plants and soil. *Plant. Physiol* **59,** 236–239.

Chen, S. S. C., and Varner, J. E. (1970). Respiration and protein synthesis in dormant and after-ripened seeds of *Avena fatua. Plant Physiol.* **46,** 108–112.

Chian, L. C., and Wu, S. H. (1965). Cytological studies on the cold resistance of plants. The changes of the intercellular materials of wheat in the overwintering period. *Acta Bot. Sinica* **13,** 196–207.

Chien, Ling-Cheng, Ching, Yu-Hsaing, and Chang, Pau-Tien (1973). Cytological studies on the cold resistance of plants: Ultastructural changes of the wheat cells in the winter period. *Acta Bot. Sin.* **15,** 22-36.

Ching, Te. May, and Slabaugh, W. H. (1966). X-ray diffraction analysis of ice crystals in coniferous pollen. *Cryobiology* **2,** 321-327.

Chomel, M. (1710). Sur les arbres morts par la gelée de 1709. "Histoire de l'Academie Royal des Sciences" (Avec les memoires, etc.)., pp. 59-61.

Chou, J. C. -L. and Levitt, J. (1972). The hydrophobicity of proteins from hardy (freeze resistant) and nonhardy species of grains. *Cryobiology,* **9,** 266-270.

Christersson, L. (1973). The effect of inorganic nutrients on water economy and hardiness of conifers: I. The effect of varying potassium, calcium and magnesium levels on water content, transpiration rate and the initial phase of development of frost hardiness of Pinus silvestris L. seedlings. *Stud. Forest. Suec.* **103,** 1-26.

Christersson, L. (1975a). Frost hardiness development in rapid- and slow-growing Norway spruce seedlings. *Can. J. Forest Res.* **5,** 340-343.

Christersson, L. (1975b). Frost hardiness development in Pinus silvestris L. seedlings at different levels of potassium and calcium fertilization. *Can. J. Forest Res.* **5,** 738-740.

Christophersen, B. O. (1969). Reduction of linolenic acid hydroperoxide by a glutathione peroxidase. *Biochim. Biophys. Acta* **176,** 463-470.

Chrominski, A., Belt, H., and Michniewicz, M. (1969). Effect of (2-chloroethyl)trimethylammonium chloride (CCC) on frost resistance, yield and seed quality indexes of winter rape. *Rocz. Nauk. Roln. Ser.* **A95,** 191-197.

Chuvashina, N. P. (1962). Using cytochemical methods in studies of the winter hardiness of apple trees. *Tr. Tsentr. Genet. Lab. Im. I. V. Michurina* **8,** 67-72.

Civinskij, V. (1934). Capacity of cotton to withstand cold. *C. R. Acad. Sci. URSS* **1,** 149-150.

Clements, H. F. (1938). Mechanisms of freezing resistance in the needles of *Pinus ponderosa* and *Pseudotsuga mucronata. Res. Stud. State Coll. Wash.* **6,** 3-45.

Cochran, P. H., and Berntsen, C. M. (1973). Tolerance of lodgepole and ponderosa pine seedlings to low night temperatures. *Forest. Sci.* **19,** 272-280.

Codaccioni, M. (1968). Réactions aux brusques abaisements thermiques des plantes du *Mentha viridis* cultivés in vitro. *C. R. Acad. Sci. (Paris)* **276,** 1499-1502.

Codaccioni, M., and Le Saint-Quervel, A. M. (1966). Endurcissement artificiel au gel chez le Chou de Milan. Role eventuel de l'amidon et sa localisation dans les parties aeriennes. *C. R. Acad. Sci. (Paris)* **263,** 1837-1840.

Coffman, F. A. (1955). Results from uniform winter hardiness nurseries of oats for the five years 1947-1951, inclusive. *Agron. J.* **47,** 54-57.

Cohn, F. and David, G. (1871). Wirkung der Kälte auf Pflanzenzellen. *Naturforscher* **39,** 316.

Coleman, E. A., Bula, R. J., and Davis, R. L. (1966). Electrophoretic and immunological comparisons of soluble root proteins of *Medicago sativa L.* genotypes in the cold hardened and nonhardened condition. *Plant Physiol.* **41,** 1681-1685.

Collison, R. C., and Harlan, J. D. (1934). Winter injury of Baldwin apple trees and its relation to previous tree performances and nutritional treatment. *N. Y. Agr. Exp. Sta. Geneva Tech. Bull.* **647,** 1-13.

Connor, K. W., and Ashwood-Smith, M. J. (1973a). Ineffectiveness of dimethyl sulfone as a cryoprotective agent. *Cryobiology,* **10,** 87-88.

Connor, W., and Ashwood-Smith, M. J. (1973b). Cryoprotection of mammalian cells in tissue culture with polymers; possible mechanism. *Cryobiology* **10.** 488-496.

Constaninescu, E. (1933). Weitere Beiträge zur Physiologie der Kälteresistenz bei Wintergetreide. *Planta* **21,** 304-323.

Cook, T. W., and Duff, D. T. (1976). Effects of K fertilization on freezing tolerance and carbohydrate content of *Festuca arundinacea* Schreb: maintained as turf. *Agron. J.* **68,** 116–119.
Cooper, W. C. (1959). Cold hardiness in citrus as related to dormancy. *Proc. Fl. State Hort. Soc.* **72,** 61–66.
Cooper, W. C., Gorton, B. S., and Taylor, S. D. (1954). Freezing tests with small trees and detached leaves of grapefruit. *Proc. Amer. Soc. Hort. Sci.* **63,** 167–172.
Cooper, W. C., Paynado, A., and Otey, G. (1955). Effects of plant regulators on dormancy, coldhardiness and leaf form of grapefruit trees. *Proc. Amer. Soc. Hort. Sci.* **66,** 100–101.
Corns, G. and Schwerdtfeger, G. (1954). Improvement in low temperature resistance of sugar beet and garden beet seedlings with sodium TCA and Dalapon. *Can. J. Agr. Sci.* **34,** 639–641.
Coville, F. V. (1920). The influence of cold in stimulating the growth of plants. *J. Agr. Res.* **20,** 151–160.
Cox, R. P. (1975). The reduction of artificial electron acceptors at subzero temperatures by chloroplasts suspended in fluid media. *Biochim. Biophys. Acta* **387,** 588–598.
Cox, W., and Levitt, J. (1969). Direct relation between growth and frost hardening in cabbage leaves. *Plant Physiol.* **44,** 923–928.
Cox, W. and Levitt, J. (1972). An improved, leaf-disk method for determining the freeze-killing temperature of leaves. *Cryobiology,* **9,** 249–254.
Cox, W. and Levitt, J. (1976). Interrelations between environmental factors and freezing resistance of cabbage leaves. *Plant Physiol.* **57,** 553–558.
Craig, D. L., Gass, D. A., and Fensom, D. S. (1970). Red raspberry growth related to electrical impedance studies. *Can J. Plant Sci.* **50,** 59–66.
Craker, L. E., Gusta, L. V., and Weiser, C. J. (1969). Soluble proteins and cold hardiness of two woody species. *Can. J. Plant. Sci.* **49,** 279–286.
Crowley, J. P., Rene, A., and Valeri, C. R. (1973). The morphology of leukocytes freeze-preserved in dimethylsulfoxide. *Cryobiology* **10,** 524–525.
Cunningham, J. L. (1973). Preservation of rust fungi in liquid nitrogen. *Cryobiology,* **10,** 361–363.
Czapek, F. (1901). Der Kohlenhydrat-Stoffwechsel der Laubblätter in Winter. *Ber. Deut. Bot. Ges.* **19,** 120–127.
Dagley, S. (1975). Microbial degradation of organic compounds in the biosphere. *Am. Sci.* **63,** 681–689.
Dalmer, M. (1895). Über Eisbildung in Pflanzen mit Rucksicht auf die anatomische Beschaffenheit derselben. *Flora (Jena)* **80,** 436–444.
Dantuma, G., and Andrews, J. E. (1960). Differential response of certain barley and wheat varieties to hardening and freezing during sprouting. *Can. J. Bot.* **38,** 133–151.
Darbyshire, B. (1975). The results of freezing and dehydration of horseradish peroxidase. *Cryobiology,* **12,** 276–281.
Dart, R. K., and Stretton, R. J. (1976). Effect of temperature on fatty acid composition in *Aspergillus. Microbios.* **3,** 31–34.
Davis, D. L. and Finkner, V. C. (1973). Influence of temperature on sterol biosynthesis in Triticum aestivum. *Plant Physiol.* **52,** 324–326.
Davis, D. L., and Gilbert, W. B. (1970). Winter hardiness and changes in soluble protein fractions of bermudagrass. *Crop Sci.* **10,** 7–8.
Day, W. R., and Peace, T. R. (1937). The influence of certain accessory factors on frost injury to forest trees. II. Temperature conditions before freezing. III. Time factors. *Forestry* **11,** 13–29.
De Jong, D. W., Olson, A. C., Hawker, K. M., and Jansen, E. F. (1968). Effect of cultivation

temperature on peroxidase isozymes of tobacco plant cells grown in suspension. *Plant Physiol.* **43,** 841-844.

De la Roche, I. A., Keller, W. A., Singh, J., and Siminovitch, D. (1977). Isolation of protoplasts from unhardened and hardened tissues of winter rye and wheat. *Can. J. Bot.* **55,** 1181-1185.

De la Roche, I. A., Andrews, C. J., Pomeroy, M. K., Weinberger, P. and Kates, M. (1972). Lipid changes in winter wheat seedlings (*Triticum aestivum*) at temperatures inducing cold hardiness. *Can. J. Bot.* **50,** 2401-2409.

De la Roche, I. A., Andrews, C. J., and Kates, M. (1973). Changes in phospholipid composition of a winter wheat cultivar during germination at 2 C and 24 C. *Plant Physiol.* **51,** 468-473.

De la Roche, I. A., Pomeroy, M. K., and Andrews, C. J. (1975). Changes in fatty acid composition in wheat cultivars of contrasting hardiness. *Cryobiology,* **12,** 506-512.

De Long, W. A., Beaumont, J. H., and Willaman, J. J. (1930). Respiration of apple twigs in relation to winter hardiness. *Plant Physiol.* **5,** 509-534.

De Silva, N. S., Weinberger, P., Kates, M., and de la Roche, I. A. (1975). Comparative changes in hardiness and lipid composition in two near-isogenic lines of wheat (spring and winter) grown at 2°C and 24°C. *Can J. Bot.* **53,** 1899-1905.

De Vries, A. L. (1970). Cold resistance in fishes in relation to protective glycoproteins. *Cryobiology* **6,** 585.

Dear, J. (1973). A rapid degradation of starch at hardening temperatures. *Cryobiology,* **10,** 78-81.

Decheva, R., Kosseva, D., and Zolotovich, G. (1972). A study on the carbohydrate exchange during the annual developmental cycle of certain oil-bearing rose plants. *Rastenievd. Nauki.* **9,** 11-20.

Dereuddre, J. (1978). Effets de divers types de refroidissements sur la teneur en eau et sur la résistance au gel des bourgeons de rameaux d' Épicés en vie ralenties. *Physiol. Vég.* **16,** 469-489.

Dévay, M. (1965). The biochemical processes of vernalization. III. The changes of ascorbic acid oxidizing capacity in the course of vernalization. *Acta Agron. Acad. Sci. Hung.* **14,** 93-97.

Devay, M. and Feher, M. T. (1975). Temperature-induced changes in the pattern of chlorophyll complexes in wheat leaves. *Biochem. Physiol. Pflanz.* **168,** 561-566.

Devay, M. and Paldi, E. (1977). Cold-induced tRNA synthesis in wheat cultivars during the hardening period. *Plant Sci. Lett.* **8,** 191-195.

Dexter, S. T. (1932). Studies of the hardiness of plants: a modification of the Newton pressure method for small samples. *Plant Physiol.* **7,** 721-726.

Dexter, S. T. (1933). Effect of several environmental factors on the hardening of plants. *Plant Physiol.* **8,** 123-129.

Dexter, S. T., (1935). Growth, organic nitrogen fractions and buffer capacity in relation to hardiness of plants. *Plant Physiol.* **10,** 149-158.

Dexter, S. T., Tottingham, W. E., and Graber, L. F. (1932). Investigations of hardiness of plants by measurement of electrical conductivity. *Plant Physiol.* **7,** 63-78.

Diller, K. R. (1975). Intracellular freezing: effect of extracellular supercooling. *Cryobiology,* **12,** 480-485.

Dogramadzhyan, A. D. and Marutyan, S. A. (1972). Metabolic characteristics of vine shoots in relation to frosthardiness. *Biol. Zh. Arm.* **25,** 51-60.

Dolnicki, A. and Piskornik, Z. (1971). Studies on the frost resistance of wheat: VII. Effect of growth regulators on peroxidase activity. *Acta Agrar. Silvestria Ser. Agrar.* **11,** 105-129.

Dorsey, M. J. (1934). Ice formation in the fruit bud of the peach. *Proc. Amer. Soc. Hort. Sci.* **31,** 22–27.

Dorsey, M. J., and Strausbaugh, P. D. (1923). Winter injury to plum during dormancy. *Bot. Gaz. (Chicago),* **76,** 113–142.

Dougall, D. K., and Wetherell, D. F. (1974). Storage of wild carrot cultures in the frozen state. *Cryobiology,* **11,** 410–415.

Doughty, C. C., and Hemerick, G. A. (1975). Impedance as a measurement of blueberry bud hardiness. *J. Am. Soc. Hortic. Sci.* **100,** 115–118.

Douzou, P. (1973). Enzymology at sub-zero temperatures. Mol. *Cell Biochem.* **1,** 15–27.

Douzou, P. (1977). "Cryobiochemistry: An Introduction." Academic Press, New York.

Douzou, P., Hui Bon Hoa, G., and Petsko, G. A. (1975). Protein crystallography at subzero temperatures: Lysozymesubstrate complexes in cooled mixed solvents. *J. Mol. Biol.* **96,** 367–380.

Doyle, J., and Clinch, P. (1927). Seasonal changes in conifer leaves, with special reference to enzymes and starch formation. *Proc. Roy. Irish Acad. Sect.* **B37,** 373–414.

Draper, S. R. (1975). Amino acid changes associated with the development of cold hardiness in perennial ryegrass. *J. Sci. Food Agric.* **26,** 1171–1176.

Du Hamel, H. L., and de Buffon, G. L. L. (1740). Observations des différents effets que produisent sur les Végétaux les grandes gelées d'Hiver et les petites gelées du Printemps. 1737. *Mem. Math. Phys. Acad. Roy. Sci. (Paris),* pp. 273–298.

Durzan, D. J. (1968a). Nitrogen metabolism of *Picea glauca.* I. Seasonal changes of free amino acids in buds, shoot apices, and leaves and the metabolism of uniformly labelled ^{14}C-L-arginine by buds during the onset of dormancy. *Can. J. Bot.* **46,** 909–919.

Durzan, D. J. (1968b). Nitrogen metabolism of *Picea glauca.* II. Diurnal changes of free amino acids, amides, and guanidine compounds in roots, buds, and leaves during the onset of dormancy of white spruce saplings. *Can. J. Bot.* **46,** 921–928.

Durzan, D. J. (1969). Nitrogen metabolism of *Picea glauca.* IV. Metabolism of uniformly labelled ^{14}C-L-arginine (carbamyl-^{14}C)-L citrulline, and (1, 2, 3, 4-^{14}C)-γ-guanidinobutyric acid during diurnal changes in the soluble and protein nitrogen associated with the onset of expansion of spruce buds. *Can. J. Biochem.* **47,** 771–783.

Echobichon, D. J., and Saschenbrecker, P. W. (1967). Dechlorination of DDT in frozen blood. *Science* **156,** 663–665.

Eggert, R. (1944). Cambium temperatures of peach and apple trees in winter. *Proc. Amer. Soc. Hort. Sci.* **45,** 33–36.

Eldjarn, L. (1965). Some biochemical effects of S-containing protective agents and the development of suitable SH/SS systems for the *in vitro* studies of such effects. *Progr. Biochem. Pharmacol.* **1,** 173–185.

Emmert, F. H., and Howlett, S. (1953). Electrolytic determination of the resistance of fifty-five apple varieties to low temperatures. *Proc. Amer. Soc. Hort. Sci.* **62,** 311–318.

Esterbauer, H., and Grill, D. (1978). Seasonal variation of glutathione and glutathione reductase in needles of *Picea abies. Plant Physiol.* **61,** 119–121.

Ewart, M. H., Siminovitch, D., and Briggs, D. R. (1953). Studies on the chemistry of the living bark of the black locust tree in relation to frost hardiness. VI. Amylase and phosphorylase systems of the bark tissues. *Plant Physiol.* **28,** 629–644.

Farkas, T., Deri-Hadlaczky, E., and Belea, A. (1975). Effect of temperature upon linolenic acid level in wheat and rye seedlings. *Lipids,* **10,** 331–334.

Farrant, J., Walter, C. A., Lee, H., and McGann, L. E. (1977). Use of two-step cooling procedures to examine factors influencing cell survival following freezing and thawing. *Cryobiology,* **14,** 273–276.

Farrant, J., and Woolgar, A. E. (1970). Possible relationships between the physical properties of solutions and cell damage during freezing. *In* "The Frozen Cell" (G. E. Wolstenholme and M. O'Connor, eds.), pp. 97-130. Churchill, London.

Faw, W. F., and Jung, G. A. (1972). Electrophoretic protein patterns in relation to low temperature tolerance and growth regulation of alfalfa. *Cryobiology,* **9,** 548-555.

Faw, W. F., Shih, S. C., and Jung, G. A. (1976). Extractant influence on the relationship between extractable proteins and cold tolerance of alfalfa. *Plant Physiol.* **57,** 720-723.

Fedorova, A. I., Gagulaeva, A. P., and Molokova, N. I. (1976). Study of abscisic acid in the Siberian larch and its growth rate. *Fiziol. Rast.* **23,** 80-87.

Fejer, S. O. (1976). Combining ability and correlations of winter survival, electrical impedance and morphology in juvenile apple trees. *Can. J. Plant. Sci.* **56,** 303-309.

Fennema, O. (1966). An over-all view of low temperature food preservation. *Cryobiology* **3,** 197-213.

Feughelmann, M. (1966). Sulphydryl-disulfide interchange and the stability of keratin structure. *Nature (London)* **211,** 1259-1260.

Field, C. P. (1939). Low temperature injury to fruit blossom. I. On the damage caused to fruit blossom by varying degrees of cold. *Ann. Rep. (26th year). 1938. East Malling Res. Sta.,* pp. 127-138.

Fink, A. L. (1974). The trypsin-catalyzed hydrolysis of *N*-α-benzyloxycarbonyl-L-Lysine p-nitrophenyl ester in dimethylsulfoxide at sub-zero temperatures. *J. Biol. Chem.* **249,** 5027-5032.

Fink, A. L. (1976). Cryoenzymology: the use of sub-zero temperatures and fluid solutions in the study of enzyme mechanisms. *J. Theor. Biol.* **61,** 419-445.

Fishbein, W. N., and Stowell, R. E. (1968). Studies on the mechanism of freezing damage to mouse liver using a mitochondrial enzyme assay. I. Temporal localization of the injury phase during slow freezing. *Cryobiology* **4,** 283-289.

Fishbein, W. N., and Winkert, J. W. (1977). Parameters of biological freezing damage in simple solutions: catalase. 1. The characteristic pattern of intracellular freezing damage exhibited in a membraneless system. *Cryobiology* **14,** 389-398.

Fishbein, W. N., and Winkert, J. W. (1978). Parameters of biological freezing damage in simple solutions: catalase. II. Demonstration of an optimum recovery cooling rate curve in a membraneless system. *Cryobiology* **15,** 168-177.

Foote, K. C., and Schaedle, M. (1976a). Physiological characteristics of photosynthesis and respiration in stems of *Populus tremuloides Michx. Plant Physiol.* **58,** 91-94.

Foote, K. C., and Schaedle, M. (1976b). Diurnal and seasonal patterns of photosynthesis and respiration by stems of *Populus tremuloides. Michx. Plant Physiol.* **58,** 651-655.

Foury, F., Boutry, M., and Goffeau, A. (1977). Efflux of potassium induced by Dio-9, a plasma membrane ATPase inhibitor in the yeast *Schizosaccharomyces pombe. J. Biol. Chem.* **252,** 4577-4583.

Fowler, D. B., Siminovitch, D., and Pomeroy, M. K. (1973). Evaluation of an artificial test for frost hardiness in wheat. *Can. J. Plant Sci.* **53,** 53-59.

Foyer, C. H., and Halliwell, B. (1976). The presence of glutathione and glutathione reductase in chloroplasts: a proposed role in ascorbic acid metabolism. *Planta* **133,** 21-25.

Fridovich, I. (1975). Oxygen: Boone and Bane. *Am. Scientist.* **63,** 54-59.

Friedman, K. J. (1977). Role of lipids in the *Neurospora crassa* membrane. I. Influence of fatty acid composition on membrane lipid phase transitions. *J. Memb. Biol.* **32,** 33-47.

Frischenschlager, B. (1937). Versüche uber die Keimstimmung an einigen Gemüsearten. *Gartenbauwissenschaft* **11,** 159-166.

Fuchigami, L. H., Weiser, C. J., and Richardson, D. G. (1973). The influence of sugars on growth and cold acclimation of excised stems of red-osier dogwood. *J. Am. Soc. Hort. Sci.* **98,** 444-447.

Fuchs, W. H. (1935). Worauf beruht die Erhöhung der Kälteresistenz durch reichliche Kaliernährung. *Ernaehr. Pflanze* **31,** 233-234.

Fujil, T., and Kondo, N. (1969). Changes in the levels of nicotinamide nucleotides and in activities of NADP-dependent dehydrogenases after a brief illumination of red light. *Develop. Growth Differ.* **11,** 40-45.

Furoya, N. (1974). The chemical properties of lignin of ray paranchyma cells of *Quercus crispula. Bull. Gov. Exp. Stn.* (Toyko) **263,** 35-42.

Futrell, M. C., Lyles, W. E., and Pilgrim, A. J. (1962). Ascorbic acid and cold hardiness in oats. *Plant Physiol. Suppl.* **37,** 70.

Garber, M. P., and Steponkus, P. L. (1973b). Alterations in chloroplast membranes during cold acclimation and freezing. *Cryobiology,* **10,** 532.

Garber, M. P., and Steponkus, P. L. (1973a). Alterations in chloroplast membranes during cold acclimation and freezing. II. Alterations during cold acclimation. *Cryobiology,* **10,** 531.

Garber, M. P., and Steponkus, P. L. (1976a). Alterations in chloroplast thylakoids during an in vitro freeze-thaw cycle. *Plant Physiol.* **57,** 673-680.

Garber, M. P., and Steponkus, P. L. (1976b). Alterations in chloroplast thylakoids during cold acclimation. *Plant Physiol.* **57,** 681-686.

Gaskin, M. H. (1959). Effects of certain plant growth regulators upon cold hardiness of lychees. *Proc. Fl. State Hortic. Soc.* **72,** 353-356.

Gassner, G., and Grimme, C. (1913). Beiträge zur Frage der Frosthärte der Getreidepflanzen. *Ber. Deut. Bot. Ges.* **31,** 507-516.

Gazeau, C.-M. (1974). Influence du glycerol sur le maintien des ultrastructures mitochondriales dans les tissus méristematiques des racines de Blé, soumis à des congélations intenses (−196°C). *C. R. Acad. Sci. (Paris).* **278,** 2429-2432.

Genevès, L. (1955). Recherches sur les effets cytologiques du froid. *Rev. Cytol. Biol. Veg.* **16,** 1-207.

Genevès, L. (1957). Sur le role des ecailles dans la résistance au froid des bourgeons de Marronier: *Aesculus Hippocastanus. C. R. Acad. Sci. (Paris)* **244,** 2083-2085.

Genkel, P. A., and Oknina, E. Z. (1954). Diagnosis of plant frost resistance by the depth of the rest of the tissues and cells. *Akad. Nauk. SSSR Timiriazev Inst. Plant Physiol. Moscow,* pp. 1-34.

Genkel, P. A., and Zhivukhnia, G. M. (1959). The process of protoplasm isolation as the second phase of hardening of winter wheat. *Dokl. Bot. Sci.* **127,** 216-219.

George, M. F., and Burke, M. J. (1976). The occurrence of deep supercooling in cold hardy plants. *Curr. Adv. Plant Sci.* **22,** 349-360.

George, M. F., and Burke, M. J. (1977a). Cold hardiness and deep supercooling in xylem of shagbark hickory. *Plant Physiol* **59,** 319-325.

George, M. F., and Burke, M. J. (1977b). Supercooling in overwintering azalea flower buds. Additional freezing parameters. *Plant Physiol.* **59,** 326-328.

George, M. F., Burke, M. J., and Weiser, C. J. (1974). Supercooling in overwintering azalea flower buds. *Plant Physiol.* **54,** 29-35.

George, M. F., Burke, M. J., Pellett, H. M., and Johnson, A. G. (1974). Low temperature exotherms and woody plant distribution. *Hortscience,* **9,** 519-522.

Gerloff, E. D., Richardson, T., and Stahmann, M. A. (1966). Changes in fatty acids of alfalfa roots during cold hardening. *Plant Physiol.* **41,** 1280-1284.

Gerloff, E. D., Stahmann, M. A., and Smith, D. (1967). Soluble proteins in alfalfa roots as related to cold hardiness. *Plant Physiol.* **42,** 895-899.

Geslin, H. (1939). La lutte contre les gelées et les seuils de résistance des principales cultures fruitières. *Ann. Epiphyt. Phytogenet.* **5,** 7-16.

Getman, F. H., and Daniels, F. (1937). "Outlines of Theoretical Chemistry." John Wiley, New York.

Ghetie, V., and Buzila, L. (1964a). Cryoprecipitation in the cytoplasmic fluids of some seeds. *Rev. Roum. Biochim.* **1,** 41-49.

Gheties, V., and Buzila, L. (1964b). Localization of cryoproteins in seeds. *Rev. Roum. Biochim.* **1,** 293-301.

Giles, C. H., and McKay, R. B. (1962). Studies in hydrogen bond formation. XI. Reactions between a variety of carbohydrates and proteins in aqueous solutions. *J. Biol. Chem.* **237,** 3288-3392.

Giosan, N., Biucan, D., and Lupas, V. (1962). Contribution to the study of the absorption of micronutrients by winter wheat at low temperatures. *Stud. Cercet. Agron.* **12,** 33-36.

Gleir, J. H., and Caruso, J. L. (1973). Low-temperature induction of starch degradation in roots of a biennial weed. *Cryobiology,* **10,** 328-330.

Glerum, C. (1973). Annual trends in frost hardiness and electrical impedance for seven coniferous species. *Can. J. Plant. Sci.* **53,** 881-889.

Godman, R. M. (1959). Winter sunscald of yellow birch. *J. Forest.* **57,** 368-369.

Göppert, H. R. (1830). "Über die Wärme-Entwickelung in den Pflanzen, deren Gefrieren und die Schützmittel gegen dasselbe." Max and Comp. Berlin.

Göppert, H. R. (1883). "Über das Gefrieren, Erfrieren der Pflanzen und Schutzmittel dagegen." "Altes und Neues," pp. 1-87. Enke, Stuttgart.

Goetz, A., and Goetz, S. S. (1938). Vitrification and crystallization of protophyta at low temperatures. *Proc. Amer. Phil. Soc.* **97,** 361-388.

Golodriga, P. Ya., and Osipov, A. V. (1972). An express method and instrument for diagnosing frost resistance in plants. *Fiziol. Biokhim. Kul't Rast.* **4,** 650-655.

Gonda, K., and Koga, S. (1973). Low-temperature thermograms of Saccharomyces cerevisiae. *J. Gen. Appl. Microbiol.* **19,** 393-396.

Goodale, G. L. (1885). "Gray's Botanical Textbook," vol. 2: Physiological Botany. Ivison, Blakeman, Taylor, New York.

Goodin, R. (1969). On the cryoaggregation of bovine serum albumin. Thesis, Univ. of Missouri, Columbia, Missouri.

Goodin, R., and Levitt, J. (1970). Cryoaggregation of bovine serum albumin. *Cryobiology* **6,** 333-338.

Gorke, H. (1906). Über chemische Vorgänge beim Erfrieren der Pflanzen. *Landwirtsch. Vers. Sta.* **65,** 149-160.

Gortner, R. A. (1938). "Outlines of Biochemistry." John Wiley, New York.

Goryshina, T. K., and Kovaleva, J. A. (1967). Seasonal temperature adaptations of prevernal, remoral ephemeroids. *Bot. Zh. (Leningrad)* **52,** 629-640.

Goujon, C., Maia, N., and Doussinault, G. (1968). Frost resistance in wheat. II. Reactions at the coleoptile stage studied under artificial conditions. *Ann. Amelior Plant,* **18,** 49-57.

Grahle, A. (1933). Vergleichende Untersuchungen über strukturelle und osmotische Eigenschaften der Nadeln verschiedener Pinus-Arten. *Jahrb. Wiss. Bot.* **78,** 203–294,

Grandfield, C. O. (1943). Food reserves and their translocation to the crown buds as related to cold and drought resistance in alfalfa. *J. Agr. Res.* **67,** 33-47.

Granhall, I. (1943). Genetical and physiological studies in interspecific wheat crosses. *Hereditas* **29,** 269-380.

Grant, N. H. (1966). The biological role of ice. *Discovery* **27,** (No. 8), 26-30.

Grant, N. H. (1969). Biochemical reactions in essentially nonaqueous systems. *Cryobiology* **6,** 182-187.

Grant, N. H., and Alburn, H. E. (1967). Reactions in frozen systems. VI. Ice as a possible model for biological structured-water systems. *Arch. Biochem. Biophys.* **118,** 292-296.

Greathouse, G. A., and Stuart, N. W. (1937). Enzyme activity in cold hardened and unhardened red clover. *Plant Physiol.* **12,** 685-702.

Green, D. G., and Ratzlaff, C. D. (1975). An apparent relationship of soluble sugars with hardiness in winter wheat varieties. *Can. J. Bot.* **53,** 2198-2201.

Green, D. G., Ferguson, W. S., and Warder, F. G. (1970). Effects of decenylsuccinic acid on ^{32}P uptake and translocation by barley and winter wheat. *Plant Physiol.* **45,** 1-3.

Greenham, C. G., and Daday, H. (1957). Electrical determination of cold hardiness in *Trifolium repens L.* and *Medicago sativa* L. *Nature (London)* **180,** 541-543.

Greenham, C. G., and Daday, H. (1960). Further studies on the determination of cold hardiness in *Trifolium repens L.* and *Medicago sativa L. Aust. J. Agr. Res.* **11,** 1-15.

Grenier, G., Mazliak, P., Tremolieres, A., and Willemot, C. (1973). Cold influence on fatty acid synthesis in the roots of two alfalfa varieties, one cold-resistant and the other less resistant. *Physiol. Vég.* **11,** 253-265.

Grenier, G., and Willemot, C. (1974). Lipid changes in roots of frost hardy and less hardy alfalfa varieties under hardening conditions. *Cryobiology,* **11,** 324-331.

Grenier, G., Hope, H. J., Willemot, C., and Therrien, H.-P. (1975). Sodium-1, 2-^{14}C acetate incorporation in roots of frost-hardy and less hardy alfalfa varieties under hardening conditions. *Plant Physiol.* **55,** 906-912.

Grenier, G., and Willemot, C. (1975). Lipid phosphorus content and ^{33}Pi incorporation in roots of alfalfa varieties during frost hardening. *Can. J. Bot.* **53,** 1473-1477.

Griggs, W. H., Harris, R. W., and Iwakiri, B. T. (1956). The effectiveness of growth-regulators in reducing fruit loss of Bartlett pears caused by freezing temperatures. *Proc. Amer. Soc. Hort. Sci.* **67,** 95-101.

Grzesiuk, S., Login, A., Rejowski, A., and Sojka, E. (1974a). Frost hardiness on wheats and its relation to the quality of proteins in the tillering nodes. *Hodowla Rosl. Aklim. Nasienn.* **18,** 1-10.

Grzesiuk, S., Login, A., Rejowski, A., and Sojka, E. (1974b). Frost hardiness of rye and its relation to the quality of proteins. *Hodowla Rosl. Aklim. Nasienn.* **18,** 11-18.

Gunar, I. I., and Sileva, M. N. (1954). Variation of the sugar content of winter wheats during the hardening-off process. *Fiziol. Rast.* **1,** 141-145.

Gusta, L. V., Burke, M. J., and Kapoor, A. C. (1975). Determination of unfrozen water in winter cereals at sub-freezing temperatures. *Plant Physiol.* **56,** 707-709.

Gusta, L. V., and Fowler, D. B. (1976). Effects of temperature on dehardening and rehardening of winter cereals. *Can. J. Plant Sci.* **56,** 673-678.

Haberlandt, F. (1875). "Wissenschaftlich-Praktische Untersuchungen auf dem Gebiete des Pflanzenbaues." Herausg, v. F. Haberlandt.

Haest, G. W. M., Kamp, D., Plasa, G., and Deuticke, B. (1977). Intra-and intermolecular cross-linking of membrane proteins in intact erythrocytes and ghosts by SH-oxidizing agents. *Biochim. Biophys. Acta.* **469,** 226-230.

Hall, D. O. (1963). Photosynthetic phosphorylation above and below 0°C. *Diss. Abstr.* **24,** 4962-4963.

Hall, T. C., McLeester, R. C., McCown, B. H., and Beck, G. E. (1969). Enzyme changes during acclimation. *Cryobiology* **6,** 263.

Halliwell, B., and Foyer, C. H. (1978). Properties and physiological function of a glutathione reductase purified from spinach leaves by affinity chromatography. *Planta.* **139,** 9-17.

Hamilton, D. F. (1973). Factors influencing dehardening and rehardening of Forsythia x intermedia stems. *J. Am. Soc. Hort. Sci.* **98,** 221-223.

Hanafusa, N. (1967). Denaturation of enzyme protein by freeze-thawing. p. 33. *In* "Cellular Injury and Resistance in Freezing Organisms" (E. Asahina, ed.). Inst. Low Temp. Sci., Hokkaido, Japan.

Hanafusa, N. (1974). Role of protective additives on freeze-drying of enzyme protein. Low Temp. Sci. Ser. B. Biol. Sci. **32,** 1-8.
Hansen, D. H., and Klikoff, L. G. (1972). Water stress in krummholz, Wasatch Mountains, Utah. *Bot. Gaz.* **133,** 392-394.
Harisch, G., and Schole, J. (1974). Der Glutathionstatus der Rattenleber in Abhängigkeit vom Lebensalter und von akuter Belastung. Z. Naturforsch. **29c,** 261-266.
Harper, J. E., and Paulsen, G. M. (1967). Changes in reduction and assimilation of nitrogen during the growth cycle of winter wheat. *Crop Sci.* **7,** 205-209.
Harrington, L. D., Pellett, N. E., and Bee, D. E. (1977). Influence of daminozide on growth and cold acclimation of Taxus roots. *Hortscience* **12,** 257-258.
Harris, R. E. (1973). Relative hardiness of strawberry cultivars at three times of the winter. *Can. J. Plant Sci.* **53,** 147-152.
Hart, M. L. T., and Van DerMolen, W. H. (1972). Winter killing of grasses in Icelandic bay-fields. *Res. Inst. Nedri. as. Hveragerdi Icel. Rep.* **5,** 1-14.
Hartgerink, A. P., and Mayo, J. M. (1976). Controlled-environment studies on net assimilation and water relation of *Dryas intergrifolia. Can. J. Bot.* **54,** 1884-1895.
Harvey, R. B. (1918). Hardening process in plants and developments from frost injury. *J. Agr. Res.* **15,** 83-112.
Harvey, R. B. (1922). Varietal differences in the resistance of cabbage and lettuce to low temperatures. *Ecology* **3,** 134-139.
Harvey, R. B. (1923). Relation of the color of bark to the temperature of the cambium in winter. *Ecology* **4,** 391-394.
Harvey, R. B. (1930). Length of exposure to low temperature as a factor in the hardening process in tree seedlings. *J. Forest.* **28,** 50-53.
Hatakeyama, I. (1957). Relation between the freezing point and the cold hardiness of plant tissue. *Physiol. Ecol. (Jap.)* **7,** 89-97.
Hatakeyama, I. (1960). The relation between growth and cold hardiness of leaves of *Camellia sinensis. Biol. J. Nara Women's Univ.* **10,** 65-69.
Hatakeyama, I. (1961). Studies on the freezing of living and dead tissues of plants, with special reference to the colloidally bound water in living state. *Mem. Coll. Sci. Kyoto Imp. Univ. Ser.* **B28,** 401-429.
Hatano, S. (1978). Studies on frost hardiness in *Chlorella ellipsoidea:* effects of antimetabolites, surfactants, hormones, and sugars on the hardening process in the light and dark. *In* "Plant Cold Hardiness and Freezing Stress." (P. H. Li and A. Sakai, eds.), pp. 175-196. Academic Press, New York.
Hatano, S., Sadakane, H., Tutumi, M., and Natanabe, T. (1976a). Studies on frost hardiness in *Chlorella ellipsoidea:* I. Development of frost hardiness of *Chlorella ellipsoidea* in synchronous culture. *Plant Cell. Physiol.* **17,** 451-458.
Hatano, S., Sadakane, H., Tutumi, M., and Watanabe, T. (1976b). Studies on frost hardiness in *Chlorella ellipsoidea:* II. Effects of inhibitors of RNA and protein synthesis and surfactants on the process of hardening. *Plant Cell Physiol.* **17,** 643-651.
Hatch, F. T., and Bruce, A. L. (1968). Amino-acid composition of soluble and membranous lipoproteins. *Nature (London)* **218,** 1166-1168.
Havis, J. R. (1976). Root hardiness of woody ornamentals. *Hortscience.* **11,** 385-386.
Hayden, R. I., Moyse, C. A., Calder, F. W., Crawford, D. P., and Fensom, D. S. (1969). Electrical impedance studies on potato and alfalfa tissue. *J. Exp. Bot.* **20,** 177-200.
Hayes, H. K., and Aamodt, O. S. (1927). Inheritance of winter hardiness and growth habit in crosses of Marquis with Minhardi and Minturki wheats. *J. Agr. Res.* **35,** 223-236.
Hazlewood, G. P., and Dawson, R. M. C. (1976). A phospholipid deacylating system of bacteria active in a frozen medium. *Biochem. J.* **153,** 49-53.

Heber, U. (1958a). Ursachen der Frostresistenz bei Winterweizen. I. Die Bedeutung der Zucker für die Frostresistenz. *Planta* **52,** 144-172.

Heber, U. (1958b). Ursachen der Frostresistenz bei Winterweizen. II. Die Bedeutung von Aminosäuren und Peptiden für die Frostresistenz. *Planta* **52,** 431-446.

Heber, U. (1959a). Beziehungen zwischen der Grosse von Chloroplasten und ihrem Gehalt an löslichen Eiweissen und Zuckern im Zusammenhang mit dem Frostresistenz Problem. *Protoplasma* **51,** 284-298.

Heber, U. (1959b). Ursachen der Frostresistenz bei Winterweizen. III. Die Bedeutung von Proteinen fur die Frostresistenz. *Planta* **54,** 34-67.

Heber, U. (1967). Freezing injury and uncoupling of phosphorylation from electron transport in chloroplasts. *Plant Physiol.* **42,** 1343-1350.

Heber, U. (1968). Freezing injury in relation to loss of enzyme activities and protection against freezing. *Cryobiol.* **5,** 188-201.

Heber, U. (1970). Proteins capable of protecting chloroplast membranes against freezing. *In* "The Frozen Cell" (G. E. W. Wolstenholme and M. O'Connor, eds.), pp. 175-188. J. and A. Churchill, London.

Heber, U., and Santarius, K. A. (1964). Loss of adenosine triphosphate synthesis caused by freezing and its relationship to frost hardiness problems. *Plant Physiol.* **39,** 712-719.

Heber, U., Tyankova, L., and Santarius, K. A. (1973). Effects of freezing on biological membranes *in vivo* and *in vitro*. *Biochim. Biophys. Acta.* **291,** 23-37.

Hedlund, T. (1917). Über die Möglichkeit von der Ausbildung des Weizens im Herbst auf die Winterfestigkeit der verschiedenen Sorten zu schliessen. *Bot. Centr.* **135,** 222-224.

Heinrichs, D. H. (1959). Germination of alfalfa varieties in solutions of varying osmotic pressure and relationship to winter hardiness. *Can. J. Plant Sci.* **39,** 384-394.

Hendershott, C. H. (1962a). The influence of maleic hydrazide on citrus trees and fruits. *Proc. Amer. Soc. Hort. Sci.* **80,** 241-246.

Hendershott, C. H. (1962b). The responses of orange trees and fruits to freezing temperatures. *Proc. Amer. Soc. Hort. Sci.* **80,** 247-254.

Hermbstädt, S. F. (1808). Über die Fähigkeit der lebenden Pflanzen im Winter Wärme zu erzeugen. *Ges. Naturforsch. Freunde Berlin zweiter Jahrgang,* pp. 316-319.

Hewett, E. W. (1976). Seasonal variations of cold hardiness in apricots. *N. Z. J. Agric. Res.* **19,** 353-358.

Highkin, H. R. (1967). Effect of temperature on formation of peroxidase enzymes. *Plant Physiol. Supp.* **42,** 5-16.

Hildreth, A. C. (1926). Determination of hardiness in apple varieties and the relation of some factors to cold resistance. *Minn. Agr. Exp. Sta. Tech. Bull.,* No. 42.

Hillard, J. H. (1975). Eragrostis curvula: Influence of low temperatures on starch accumulation, amylolytic activity and growth. *Crop Sci.* **15,** 293-294.

Hirtz, Rolf-Dieter and Menke, W. (1973). Not freezing water in the lamellar system of chloroplasts. *Z. Naturforsch. Teil. C. Biochem. Biophys. Biol. Virol.* **28,** 230.

Hoffman, H. (1857). "Witterung und Wachsthum, Oder Grundzüge der Pflanzenklimatologie," pp. 312-334. Leipzig.

Holm-Hansen, O. (1967). Factors affecting the viability of lyophilized algae. *Cryobiology* **4,** 17-23.

Holubowica, T., and Boe, A. A. (1969). Development of cold hardiness in apple seedlings treated with gibberellic acid. *J. Amer. Soc. Hort. Sci.* **94,** 661-664.

Holzl, J., and Bancher, E. (1968). Untersuchungen zur Frage des Uberlebens Pflanzlicher Epidermen in flüssigen Stickstoff und Freon. *Z. Pflanzenphysiol.* **58,** 310-326.

Hopkins, E. F. (1924). Relation of low temperatures to respiration and carbohydrate changes in potato tubers. *Bot. Gaz. (Chicago)* **78,** 311-325.

Hoshino, M., Oizumi, H., and Okubo, T. (1972). Studies on the assimilation and translocation of $^{14}CO_2$ in ladino clover: VII. Effect of temperature, especially low temperature on the translocation of ^{14}C-assimilation. *Proc. Crop Sci. Soc. Jpn.* **41,** 509-513.

Howell, G. S., and Stockhouse, S. S. (1973). The effect of defoliation time on acclimation and dehardening in tart cherry (*Prunus cerasus* L). *J. Am. Soc. Hortic. Sci.* **98,** 132–136.

Howes, C. D., and Stern, A. I. (1969). Photophosphorylation during chloroplast development in Red Kidney bean. I. Characterization of the mature system and the effect of BSA and sulfhydryl reagents. *Plant Physiol.* **44,** 1515-1522.

Huang, K. P., and Cabib, E. (1973). Yeast glycogen synthetase in the glucose-6-phosphate independent form: a case of cold lability without major changes in molecular size. *Biochim. Biophys. Acta.* **302,** 240-248.

Hudock, G. A., Mellin, D. B., and Fuller, R. C. (1965). Alternative forms of triosephosphate dehydrogenase in Chromatium. *Science* **150,** 776-777.

Huner, N. P. A., and Macdowall, F. M. (1976). Chloroplastic proteins of wheat and rye grown at warm and cold-hardening temperatures. *Can. J. Biochem.* **54,** 848-853.

Hunt, C. J., Beadle, D. J., and Harris, L. W. (1977). An ultrastructural study of the recovery of Chinese hamster ovary cells after freezing and thawing. *Cryobiology,* **14,** 135-143.

Hwang, Shuh-Wei, and Horneland, W. (1965). Survival of alga cultures after freezing by controlled and uncontrolled cooling. *Cryobiology* **1,** 305-311.

Hwang, Shuh-Wei, and Howells, A. (1968). Investigation of ultra-low temperatures for fungal cultures. II. Cryo-protection afforded by glycerol and dimethyl sulfoxide to 8 selected fungal cultures. *Mycologia* **60,** 622-626.

Hylmö, B. (1943). Disackaridbildning hos viktål vid kall väderlek. *Medd. N:R* **17** *Från Statens Trädgårdsförsök.,* pp. 1-37.

Iljin, W. S. (1933a). Über Absterben der Pflanzengewebe durch Austrocknung und über ihre Bewahrung vor dem Trockentode. *Protoplasma* **19,** 414-442.

Iljin, W. S. (1933b). Über den Kältetod der Pflanzen und seine Ursachen. *Protoplasma* **20,** 105-124.

Iljin, W. S. (1934). The point of death of plants at low temperatures. *Bull. Ass. Russe Rech. Sci. Prague Sect. Sci. Nat. Math.* **1** (6) (No. 4), 135-160.

Iljin, W. S. (1935a). The relation of cell sap concentration to cold resistance in plants. *Bull. Ass. Russe Rech. Sci. Prague, Sect. Sci. Nat. Math.* **13** 3(8), 33-55.

Irias, J. J., Olmsted, M. R., and Utter, M. F. (1969). Pyruvate carboxylase: Reversible inactivation by cold. *Biochemistry* **8,** 5136-5148.

Irmscher, E. (1912). Über die Resistenz der Laubmoose gegen Austrocknung und Kälte. *Jahrb. Wiss. Bot.* **50,** 387-449.

Irving, R. M. (1969a). Characterization and role of an endogenous inhibitor in the induction of cold hardiness in *Acer negundo. Plant Physiol.* **44,** 801-805.

Irving, R. M. (1969b). Influence of growth retardants on development and loss of hardiness of *Acer negundo. J. Amer. Soc. Hort. Sci.* **94,** 419-422.

Irving, R. M., and Lanphear, F. O. (1967a). Environmental control of cold hardiness in woody plants. *Plant Physiol.* **42,** 1191-1196.

Irving, R. M., and Lanphear, F. O. (1967b). Dehardening and the dormant condition in *Acer* and *Viburnum. Proc. Amer. Soc. Hort. Sci.* **91,** 699–705.

Irving, R. M., and Lanphear, F. O. (1968). Regulation of cold hardiness in *Acer negundo. Plant Physiol.* **43,** 9-13.

Isaac, W. E. (1933). Some observations and experiments on the drought resistance of *Pelvetia canaliculata. Ann. Bot.* **47,** 343-348.

Ishida, M., and Mizushima, S. (1969). The membrane ATPase of *Bacillus megaterium.* II. Purification of membrane ATPase and their recombination with membrane. *J. Biochem.* **66,** 133-138.

Israeli, E., Giberman, E., and Kohn, A. (1974). Membrane malfunctions in freeze-dried *Escherichia coli*. *Cryobiology* **11,** 473–477.

Ivanov, S. M. (1939). Activity of growth process-principal factor in frost resistance of citrus plants. *C. R. Acad. Sci. URSS* **22,** 277–281.

Iwanoff, L. (1924). Über die Transpiration der Holzgewächse im Winter. I. *Ber. Deut. Bot. Ges.* **42,** 44–49. II. pp. 210–218.

Jacob, H. S., Brain, M. C., Dacil, J. V., Carrell, R. W., and Lehmann, H. (1968). Abnormal haem binding and globin SH group blockade in unstable haemoglobins. *Nature (London)* **218,** 1214–1217.

Jablonski, P. P., and Anderson, J. W. (1978). Light-dependent reduction of oxidized glutathione by ruptured chloroplasts. *Plant Physiol.* **61,** 221–225.

Janick, P. A., Grunwald, G. B., and Wood, J. M. (1977). The effects of *N*-ethylmaleimide on active amino acid transport in *Escherichia coli*. *Biochim. Biophys. Acta.* **464,** 328–337.

Janssen, D. W., and Busta, F. F. (1973). Repair of injury in *Salmonella anatum* cells after freezing and thawing in milk. *Cryobiology* **10,** 386–392.

Jennings, D. L., Carmichael, E., and Costin, J. J. (1972). Variation in the time of acclimation of raspberry canes in Scotland and Ireland and its significance for hardiness. *Hortic. Res.* **12,** 187–200.

Jennings, D. L., and Carmichael, E. (1975). Some physiological changes occurring in overwintering of raspberry plants in Scotland. *Hortic. Res.* **14,** 103–108.

Jenny, J. (1953). Trials of measuring winter temperatures of apple and pear buds. *Rev. Viticult. Arboricult.* **9,** 26–27.

Jensen, I. J. (1925). Winter wheat studies in Montana with special reference to winter killing. *J. Amer. Soc. Agron.* **17,** 630–631.

Jeremias, K. (1956). Zur Physiologie der Frosthärtung (Unter besonderer Berucksichtigung von Winterweizen). *Planta* **47,** 81–104.

Johansson, N.-O. (1970). Ice formation and frost hardiness in some agricultural plants. *Nat. Swed. Inst. Plant Protection Contrib.* **14,** (132), 364–382.

Johansson, N.-O., and Krull, E. (1970). Ice formation, cell contraction, and frost killing of wheat plants. *Nat. Swed. Inst. Plant Protection Contrib.* **14**(131), 343–362.

Johansson, N. -O., and Torsell, B. (1956). Field trials with a portable refrigerator. *Acta Agr. Scand.* **6,** 81–99.

Johansson, N.-O., Albertson, C. E., and Månsson, T. (1955). Undersökningar över höstetets härdning och avhärdning. *Sver. Utsaedesfoeren. Tidsk.*, pp. 82–96.

Johnston, W. J., and Dickens, R. (1976). Centipedegrass cold tolerance as affected by environmental factors. *Agron. J.* **68,** 83–85.

Jonassen, G. W. (1973). A study of the lethal effects of low temperature on swede rootlets. *Brassica napus rapifera Metzg. Sinsk. Meld. Nor. Landbrukshoegsk.* **52,** 1–17.

Judel, G. K. (1975). Influence of pH and temperature on the activity of phenol oxidase and peroxidase from sunflower plants. *Biochem. Physiol. Pflanz.* **167,** 243–252.

Jung, G. A. (1962). Effect of uracil, thiouracil, and guanine on cold resistance and nitrogen metabolism of alfalfa. *Plant Physiol.* **37,** 768–774.

Jung, G. A., and Smith, D. (1960). Influence of extended storage at constant low temperature on cold resistance and carbohydrate reserves of alfalfa and medium red clover. *Plant Physiol.* **35,** 123–125.

Jung, G. A., and Smith, D. (1962). Trends of cold resistance and chemical changes over winter in the roots and crowns of alfalfa and medium red clover. I. Changes in certain nitrogen and carbohydrate fractions. *Agron. J.* **53,** 359–364.

Jung, G. A., Shih, S. C., and Shelton, D. C. (1967a). Seasonal changes in soluble protein, nucleic acids, and tissue pH related to cold hardiness of alfalfa. *Cryobiology* **4,** 11–16.

Jung, G. A., Shih, S. C., and Shelton, D. C. (1967b). Influence of purines and pyrimidines on cold hardiness of plants. III. Associated changes in soluble protein and nucleic acid content and tissue pH (*Medicago sativa* varieties "Vernal" and "Caliverde"). *Plant Physiol.* **42,** 1653-1657.

Kacperska-Palacz, A., Blaziak, M., and Wcislinska, B. (1969). The effect of growth retardants CCC and B-9 on certain factors related to cold acclimation of plants. *Bot. Gaz. (Chicago)* **130,** 213-221.

Kacperska-Palacz, A., and Wcislinska, B. (1972). Electrophoretic pattern of soluble proteins in the rape leaves in relation to frost hardiness. *Physiol. Veg.* **10,** 19-25.

Kacperska-Palacz, A. and Egierszdorff, S. (1972). Effects of cold hardening and CCC treatment on hydration and frost and desiccation hardiness of plant tissues. *Bot. Gaz.* **133,** 355-360.

Kacperska-Palacz, A., Debska, Z., and Jakubowska, A. (1975). The phytochrome involvement in the frost hardening process of rape seedlings. *Bot. Gaz.* **136,** 137-140.

Kacperska-Palacz, A., Dlugokecka, E., Breitenwald, J., and Wcislinska, B. (1977a). Physiological mechanisms of frost tolerance: possible role of protein in plant adaptation to cold. *Biol. Plant.* **19,** 10-17.

Kacperska-Palacz, A., Jasinska, M., Sobczyk, E. A., and Wcislinska, B. (1977b). Physiological mechanisms of frost tolerance: subcellular localization and some physical-chemical properties of protein fractions accumulated under cold treatment. *Biol. Plant.* **19,** 18-26.

Kainmüller, C. (1975). Temperatur-resistenz von Hochgebergspflanzen. Osterreich. Akad. Wiss. Sitz. Math.-naturwiss. Kl. (Jahrg. 1975), pp. 67-75.

Karcher, H. (1931). Über die Kälteresistenz einiger Pilze und Algen. *Planta* **14,** 515-516.

Kakhana, B. M., and Arasimovich, V. V. (1973). Conversions of fructosans in artichoke tubers depending on storage temperature. *Izv. Akad. Nauk. Mold. SSR. Ser. Biol. Khim. Nauk.* **3,** 24-29.

Kaku, S. (1973). High ice nucleating ability in plant leaves. *Plant Cell Physiol.* **14,** 1035-1038.

Kaku, S. (1975). Analysis of freezing temperature distribution in plants. *Cryobiology* **12,** 154-159.

Kaku, S., and Iwaya, M. (1978). Low temperature exotherms in xylems of evergreen and deciduous broad-leaved trees in Japan with reference to freezing resistance and distribution range. *In* "Plant Cold Hardiness and Freezing Stress." (P. H. Li and A. Sakai, eds.), Academic Press, New York. pp. 227-239.

Kallinin, F. L., and Stasevskaya, N. P. (1972). Nucleic acids and nuclear proteins of winter wheat meristems during growth inhibition by lowered temperatures. *Fiziol. Biokhim. Kul't Rast.* **4,** 599-601.

Kantser, A. N. (1972). Dynamics of lignin content and frost-resistance of woody plants. *Fiziol. Biokhim. Kul't. Rast.* **4,** 92-95.

Kanwisher, J. (1957). Freezing and drying in intertidal algae. *Biol. Bull.* **113,** 275-285.

Kappen, L. (1969a). Frost resistance of native halophytes in relation to their salt, sugar, and water content in summer and winter. *Flora Abt.* **B158,** 232-260.

Kappen, L. and Maier, M. (1973). The importance of some non-volatile carbonic acids for the freezing tolerance of the halophyte Halimione portulacoides under different levels of sodium chloride. Oecologia (*Berlin*) **12,** 241-250.

Karimov, Kh. Kh., Cherner, R. I., and Rakhmonov, A. (1974). Potential photosynthesis and content of plastid pigments in winter-vegetating forage crops. *Izv. Akad. Nauk. Tadzh. SSR* Otd. *Biol. Nauk.* **3,** 34-41.

Karmanenko, N. M. (1972). Energy efficiency of respiration and phosphorus metabolism in winter wheat varieties with different winter hardiness. *Fiziol. Rast.* **19,** 807-812.

Keller, T. (1973). CO_2 exchange of bark of deciduous species in winter. *Photosynthetica* **7,** 320-324.

Kenefick, D. G., and Swanson, C. R. (1963). Mitochondrial activity in cold acclimated winter barley. *Crop Sci.* **3,** 202-205.

Kentzer, T. (1967). Further investigations concerning the influence of chlorocholine chloride (CCC) on the frost resistance of tomato plants. *Rocz. Nauk. Roln. Ser.* **A93,** 511-522.

Kerner von Marilaun, A. (1894). "The Natural History of Plants," vol. 1, part 2 (translated by F. W. Oliver), pp. 539-557. Holt, New York.

Kessler, W. (1935). Über die inneren Ursachen der Kälteresistenz der Pflanzen. *Planta* **24,** 312-352.

Kessler, W., and Ruhland, W. (1938). Weitere Untersuchungen über die inneren Ursachen der Kälteresistenz. *Planta* **28,** 159-204.

Kessler, W., and Ruhland, W. (1942). Über die inneren Ursachen der Kälteresistenz der Pflanzen. *Forschungsdienst* **16,** 345-351.

Ketchie, D. O., and Burts, W. D. (1973). The relation of lipids to cold acclimation in "Red Delicious" apple tress. *Cryobiology* **10,** 529.

Ketchie, D. O., Murren, C., and Weeks, T. E. (1973). The effect of cryoprotectants on fruit trees. *Cryobiology* **10,** 533.

Khan, A. W., Davidkova, E., and van den Berg, L. (1968). On cryodenaturation of chicken myofibrillar proteins. *Cryobiology* **7,** 184-188.

Khisamutdinova, V. I., Vasil'eva, I. M., and Estrina, R. I. (1972). Respiration intensity and the effect of 2,4-DNP on it in winter wheat during autumn hardening. *Fiziol. Biokhim. Kul't. Rast.* **4,** 46-51.

Khisamutdinova, V. I., Vasil'eva, I. M., Kuz'mina, G. G., and Vershinin, A. A. (1975). Effect of chlorocholine chloride on energy exchange and the state of water in winter wheat during hardening. *Fiziol. Rast.* **22,** 1048-1054.

Khokhlova, L. P., Denisova, G. A., Rykalova, N. I., and Khazhina, R. G. (1974). Study of chemical composition, state of water and some functional characteristics of winter wheat mitochondria. *Fiziol. Rast.* **21,** 252-259.

Khokhlova, L. P., Eliseeva, N. S., Stupishina, E. A., Bondar, I. G., and Suleimanov, I. G. (1975). Effect of autumn hardening on electrophoretic properites and protein structure of mitochondria in winter wheat. *Fiziol. Rast.* **22,** 831–837.

Kiesselbach, T. A., and Ratcliff, J. A. (1918). Freezing injury of seed corn. *Nebr. Agr. Exp. Sta. Bull.* **163,** 1-16.

Kimball, S. L., and Salisbury, F. B. (1973). Ultrastructural changes of plants exposed to low temperatures. *Am. J. Bot.* **66,** 1028-1033.

Kitaura, K. (1967). Supercooling and ice formation in mulberry trees. *In* "Cellular Injury and Resistance in Freezing Organisms" (E. Asahina, ed.), pp. 143-156. Inst. Low Temp. Sci., Hokkaido, Japan.

Klages, K. H. (1926). Relation of soil moisture content to resistance of wheat seedlings to low temperatures. *J. Amer. Soc. Agron.* **18,** 184-193.

Kneen, E., and Blish, M. J. (1941). Carbohydrate metabolism and winter hardiness of wheat. *J. Agr. Res.* **62,** 1-26.

Knowlton, H. E., and Dorsey, M. J. (1927). A study of the hardiness of the fruit buds of the peach. *West Virginia Exp. Sta. Bull.* **211,** 1-28.

Kohn, H. (1959). Experimenteller Beitrag zur Kenntnis der Frostresistenz von Rinde, Winterknospen und Bluten verschiedener Apfelsorten. *Gartenbauwissenschaft* **24,** 314–329.

Kohn, H., and Levitt, J. (1965). Frost hardiness studies on cabbage grown under controlled conditions. *Plant Physiol.* **40,** 476–480.

Kohn, H., and Levitt, J. (1966). Interrelations between photoperiod, frost hardiness and sulfhydryl groups in cabbage. *Plant Physiol.* **41,** 792–796.

Kohn, H., Waisel, Y., and Levitt, J. (1963). Sulfhydryls—a new factor in frost resistance. V. Direct measurements on proteins and the nature of the change in SH during vernalization of wheat. *Protoplasma* **57,** 556–568.

Kol, E. (1969). Die Binnegewasser. Vol. XXIV. Kryobiologie. Biologie und Limnologie des Schnees und Eises. I. Kryovegetation. 216 p. Schweizerbat'she, Verlagsbuchhandlung (Nagele U. Obermiller), Stuttgart, Germany.

Kol, E. (1975). Cryobiological researches in the High Tatra: I. Acta. Bot. *Acad. Sci. Hung.* **21,** 61–75.

Kolomycev. G. G. (1936). Winter hardiness and earliness of wheats. *C. R. Acad. Sci. URSS* **12,** 351–356.

Kolosha, O. I., and Reshetnikova, T. P. (1967). The effect of the temperature on the content of macroenergetic phosphorus in the tillering nodes and on the frost resistance of winter wheat. *Rast. Ustoich. Rast.* **SB3,** 188–193.

Kolosha, O. I., and Kostenko, I. I. (1973). The water regime, enzymatic activity and frost-resistance of winter wheat. *Dokl. Vses. Ord. Lenina. Akad. S-Kh. Nauk. Im. V. I. Lenina.* **3,** 11–14.

Konovalov, I. N. (1955). Experiments on increasing the frost resistance of stock and cabbage by the action of extracts of winter-resisting plants. *Dokl. Akad. Nauk. SSSR* **101,** 767–770.

Koshland, D. E., and Kirtley, M. E. (1966). Protein structure in relation to cell dynamics and differentiation. *In* "Major problems in Developmental Biology" (M. Locke, ed.), pp. 217–250. Academic Press, New York.

Kraicsovits, F. and Douzou, P. (1973). Chymotrypsin-catalyzed hydrolysis at subzero temperatures: Stabilization of ES complexes. *Biochimie* **55,** 1007–1010.

Kramer, P. J. (1937). Photoperiodic stimulation of growth by artificial light as a cause of winter killing. *Plant Physiol.* **12,** 881–883.

Krasavtsev. O. A. (1961). Acclimatization of arboreal plants to extremely low temperatures. *Izv. Akad. Nauk. SSSR Ser. Biol.* **2,** 228–232.

Krasavtsev, O. A. (1962). Fluorescence of woody plant cells in the frozen state. *Sov. Plant Physiol.* **9,** 282–288.

Krasavtsev, O. A. (1968). Amount of nonfrozen water in woody plants at different temperatures. *Fiziol. Rast.* **15,** 225–235.

Krasavtsev, O. A. (1969). Effect of prolonged frost on trees. *Fiziol. Rast.* **16,** 228–236.

Krasavtsev, O. A. (1973). Effect of initial negative temperatures in frost-hardening of plants. *Fiziol. Rast.* **20,** 24–31.

Krasnuk, M., Jung, G. A., and Witham, F. H. (1976). Electrophoretic studies of several dehydrogenases in relation to the cold tolerance of alfalfa. *Cryobiology* **13,** 375–393.

Krasnuk, M., Jung, G. A., and Witham, F. H. (1975). Electrophoretic studies of the relationship of peroxidases, polyphenol oxidase, and indoleacetic acid oxidase to cold tolerance of alfalfa. *Cryobiology,* **12,** 62–80.

Kretschmer, G., and Beyer, B. (1970). Einfluss von CCC auf die Winterhärte von Weizen. *Albrecht Thaer Arch.* **14,** 93–104.

Krull, E., and Levitt, J. (1972). Reversal by mercaptoethanol of protective effect of solutes against frost injury of red cabbage. *Physiol. Plant.* **27,** 259–261.

Kruuv, J., Lepock, J. R., and Keith, A. D. (1978). The effect of fluidity of membrane lipids on freeze-thaw survival of yeast. *Cryobiology,* **15,** 73–79.

Kuiper, P. J. C. (1967). Surface-active chemicals as regulators of plant growth, membrane permeability and resistance to freezing. *Mededeel. Landbouwhoogesh. Wageningen* **67,** 1–23.

Kuiper, P. J. C. (1969). Surface-active chemicals membrane permeability and resistance to freezing. *Proc. 1st Int. Citrus Symp.* **2,** 593–595.

Kuiper, P. J. C. (1972). Temperature response of ATPase of bean roots as related to growth temperature and to lipid requirement of the ATPase. *Physiol. Plant.* **26,** 200–205.

Kuiper, P. J. C. (1975). Water structure in membranes. Symp. "State of intracellular water and its biological importance." XIIth Internat. Bot. Cong. Leningrad.

Kuksa, I. N. (1939). The effect of mineral nutrition on winter hardiness and yield of winter wheat. *Himiz. Soc. Zemled.* **1,** 70–79. (*Herb. Abstr.* **9,** 635, 1939).

Kunisch, H. (1880). Über die tödliche Einwirkung niederer Temperaturen auf die Pflanzen. Inaug. Diss. 55 pp., Breslau.

Kuntz, I. D., Jr., Brassfield, T. S., Law, G. D., and Purcell, G. V. (1969). Hydration of macromolecules. *Science* **163,** 1329–1331.

Kuokol, J., Dugger, W. M., Jr., and Palmer, R. L. (1967). Inhibitory effect of peroxyacetyl nitrate on cyclic photophosphorylation by chloroplasts from "Black Valentine" bean leaves. *Plant Physiol.* **42,** 1419–1422.

Kuraishi, S., Arai, N., Ushijinia, T., and Tazaki, T. (1968). Oxidized and reduced nicotinamide adenine dinucleotide phosphate levels of plants hardened and unhardened against chilling injury. *Plant Physiol.* **43,** 238–242.

Kull, U. (1973). Temperature independent accumulation of raffinose in barks of Populus. *Ber. Dtsch. Bot. Ges.* **86,** 499–503.

Kylin, H. (1917). Über die Kälteresistenz der Meeresalgen. *Ber. Deut. Bot. Ges.* **35,** 370–384.

Laine, J., Roxby, N., and Coukell, M. B. (1975). A simple method of storing cellular slime mold amaebae. *Can. J. Microbiol.* **21,** 959–962.

Lamotte, C. E., Gochnauer, C., LaMotte, L. R., Mathur, J. R., and Davies, L. L. R. (1969). Pectin esterase in relation to leaf abscission in Coleus and Phaseolus. *Plant Physiol.* **44,** 21–26.

Landi, Renzo. (1974). The crown-freezing technique in the wheat breeding for cold area. *Genet. Agrar.* **28,** 381–391.

Lange, O. L. (1965). Der CO_2 Gaswechsel von Flechten bei tiefen Temperaturen. *Planta* **64,** 1–19.

Lange, O. L. (1962). Die Photosynthese der Flechten bei tiefen Temperaturen und nach Frostperioden. *Ber. Deut. Bot. Ges.* **75,** 351–352.

Lange, O. L., and Metzner, H. (1965). Lichtabhängiger Kohlenstoff-Einbau in Flechten bei tiefen Temperaturen. *Naturwissenschaften* **52,** 191–192.

Langlet, O. (1934). Om vriationen hos tallen (*P. silvestris* L.) och dess samband med klimatet. *Sv. Skogsvardsf. Tidskr.* **32,** 87–110.

Lapins, K. (1962). Artificial freezing as a routine test of cold hardiness of young apple seedlings. *Proc. Amer. Soc. Hort. Sci.* **81,** 26–34.

Larcher, W. (1954). Die Kälteresistenz mediterraner Immergrüner und ihre Beeinflussbarkeit. *Planta* **44,** 607–635.

Larcher, W. (1957). Frosttrocknis an der Waldgrenze und in der alpinen Zwergstrauchheide auf dem Patscherkofel bei Innsbruck. *Veroeff. Ferdinandeum Innsbruck* **37,** 49–81.

Larcher, W. (1959). Das Assimilationsvermögen von *Quercus ilex* und *Olea europea* im Winter. *Ber. Deut. Bot. Ges.* **72,** (18).

Larcher, W. (1968). Die Temperaturresistenz als Konstitutionsmerkmal der Pflanzen. *In* "Klimaresistenz Photosynthese und Stoffproduktion" (H. Polster, ed.), pp. 7–21. Deut. Akad. Landwirtsch., Berlin.

Larcher. W. (1969). Zunahme des Frostabhärtungsvermögen von *Quercus ilex* im Laufe der Individualentwicklung. *Planta* **88,** 130–135.

Larcher, W., and Eggarter, H. (1960). Anwendung des Triphenyltetrazoliumchlorids zur Beurteilung von Frostschäden in verschiedenen Achsengeweben bei *Pinus*-Arten, und Jahresgang der Resistenz. *Protoplasma* **41,** 595–619.

Larcher, W., *et al.* (1969). Anwendung und Zuverlössigkeit der Tetrazoliummethode zur Feststellung von Schäden in pflanzlichen Gewebe. *Mikroskopie* **25,** 207–218.
Larcher, W. (1975). Pflanzenökologische Beobachtungen in der Paramostufe der venezelanischen Anden. Österreich. Akad. Wiss. Jahrg. 1975, Sitz. math-Naturwiss Klasse: 194–213.
Lavee, S., and Galston, A. W. (1968). Hormonal control of peroxidase activity in cultured Pelargonium pith. *Amer. J. Bot.* **55,** 890–893.
Lawaczeck, R., Kainosho, M., and Chan, S. I. (1976). The formation and annealing of structural defects in lipid bilayer vesicles. *Biochim. Biophys. Acta* **443,** 313–330.
Lawrence, T., Cooper, J. P., and Breese, E. L. (1973). Cold tolerance and winter hardiness in Lolium perenne: II. Influence of light and temperature during growth and hardening. *J. Agric. Sci.* **80,** 341–348.
Layne, R. E. C. (1963). Effect of vacuum drying, freeze-drying and storage environment on the viability of pea pollen. *Crop. Sci.* **3,** 433–436.
Leddet, C. (1974). Action de certains oses et polyols sur la résistance au gel des tissus de *Topinambour maintenus* en survie. *C. R. Acad. Sci. (Paris)* **278,** 2131–2134.
Leddet, C., and Schaeverbeke, J. (1975). Action de la proline sur la résistance au gel des tissus de *Topinambour maintenus* en survie. *Acad. Sci. (Paris).* **280,** 2849–2852.
Leibo, S. P., Mazur, P., and Jackowski, S. C. (1974). Factors affecting survival of mouse embryos during freezing and thawing. *Exp. Cell. Res.* **89,** 79–88.
Leibo, S. (1976). Freezing damage of bovine erythrocytes: simulation using glycerol concentration changes at subzero temperatures. *Cryobiology* **13,** 587–598.
Le Saint (Quervel), A. -M. (1956). Quelques expériences sur la résistance au gel et la surfusion de jeunes plantes étiolées placées a −4°C. *Rev. Gen. Bot.* **63,** 514–523.
Le Saint, A. -M. (1957). Mise en évidence d'une chute de la perméabilité aux gas des tissus de pomme, pendant le gel, a partir de la cessation de la surfusion. *Rev. Gen. Bot.* **64,** 334–338.
Le Saint, A. -M. (1958). Comparaison de la résistance au froid de jeunes plantes de pois étiolées ou chlorophylliennes. *Rev. Gen. Bot.* **65,** 471–477.
Le Saint (-Quervel), A.-M. (1960). Études des variations comparées des acides aminés libres et des glucides solubles, au cours de l'acquisition et de la perte de l'aptitude à résister au gel chez le chou de Milan. *C. R. Acad. Sci. (Paris)* **251,** 1403–1405.
Le Saint, A. -M. (1966). Observations physiologiques sur le gel et l'endurcissement au gel chez le chou de Milan. Thèses preésenté à la faculté des sciences de l'Université de Paris, 93 pp.
Le Saint, A.-M. (1969a). Comparison of free protein and soluble carbohydrate levels in relation to the unequal sensitivity to freezing of the savoy cabbage cv. "Pontoise." *C. R. Acad. Sci. (Paris)* **D268,** 310–313.
Le Saint (-Quervel), A. -M. (1969b). Corrélations entre la résistance au gel, l'eclairement et la teneur en proline libre chez le chou de Milan Cult. Pontoise. *C. R. Acad. Sci. (Paris)* **269,** 1423–1426.
Le Saint, A. -M., and Catesson, A. M. (1966). Variations simultanées des teneurs en eau, en sucre solubles, en acide aminés et de la pression osmotique dans la phloeme et la cambium de Sycamore pendant les périodes de repos apparent et de reprise de la croissance. *C. R. Acad. Sci. (Paris)* **263,** 1463–1466.
Le Saint, A. M., and Frotte, M. (1972). Variations of soluble carbohydrates and starch in the Milan cabbage in the course of becoming inured to freezing and its transmission from one part to another of the same plant. C. R. *Acad. Sci. Ser. D.* **274,** 1035–1037.
Le Saint-Quervel, A.-M. (1977). Recherche et utilization d'une méthode d'évaluation de la résistance au gel de fragments de feuilles. *C. R. Acad. Sci (Paris).* **284,** 41–44.

Levin, R. L., and Cravalho, E. G. (1976). A membrane model describing the effect of temperature on the water conductivity of erythrocyte membranes at subzero temperatures. *Cryobiology,* **13,** 415–429.

Levitt, J. (1939). The relation of cabbage hardiness to bound water, unfrozen water, and cell contraction when frozen. *Plant Physiol.* **14,** 93–112.

Levitt, J. (1941). "Frost Killing and Hardiness of Plants," 211 pp. Burgess, Minneapolis, Minnesota.

Levitt, J. (1954). Investigations of the cytoplasmic particulates and proteins of potato tubers. III. Protein synthesis during the breaking of the rest period. *Physiol. Plant.* **7,** 597–601.

Levitt, J. (1956). "The Hardiness of Plants." 278 pp. Academic Press, New York.

Levitt, J. (1957a). The moment of frost injury. *Protoplasma* **48,** 289–302.

Levitt, J. (1957b). The role of cell sap concentration in frost hardiness. *Plant Physiol.* **32,** 237–239.

Levitt, J. (1958). Frost, drought, and heat resistance. *Protoplasmatologia* **6,** 87 pp.

Levitt, J. (1959a). Effects of artificial increases in sugar content on frost hardiness. *Plant Physiol.* **34,** 401–402.

Levitt, J. (1959b). Bound water and frost hardiness. *Plant Physiol.* **33,** 674–677.

Levitt, J. (1962). A sulfhydryl-disulfide hypothesis of frost injury and resistance in plants. *J. Theoret. Biol.* **3,** 355–391.

Levitt, J. (1965). Thiogel—a model system for demonstrating intermolecular disulfide bond formation on freezing. *Cryobiology* **1,** 312–316.

Levitt, J. (1966a). Cryochemistry of plant tissue: protein interactions. *Cryobiology* **3,** 243–251.

Levitt, J. (1966b). Winter hardiness in plants. *In* "Cryobiology" (H. T. Meryman, ed.), pp. 495–563. Academic Press, New York.

Levitt, J. (1967a). Status of the sulfhydryl hypothesis of freezing injury and resistance. *In* "Molecular Mechanisms of Temperature Adaptation." (C. Ladd Prosser, ed.), pp. 41–51. Amer. Assoc. Adv. Sci., Washington, D.C.

Levitt, J. (1967b). The mechanism of hardening on the basis of the SH $\leftrightarrows$ SS hypothesis of freezing injury. *In* "Cellular Injury and Resistance in Freezing Organisms." (E. Asahina, ed.), pp. 51–61. Inst. Low Temp. Sci. Hokkaido Univ., Sapporo, Japan.

Levitt, J. (1967c). The role of SH and SS in the resistance of cells to high and low temperatures. *In* "The Cell and Environmental Temperature" (A. S. Troshin, ed.), pp. 269–274. Pergamon, Oxford.

Levitt, J. (1969a). The effect of sulfhydryl reagents on freezing resistance of hardened and unhardened cabbage cells. *Cryobiology* **5,** 278–280.

Levitt, J., (1972a). The effect of sulfhydryl reagents on cell permeability. pp. 107–111 in: Plant response to climate factors. Proceedings of the Uppsala Symposium. Paris, Unesco (Ecology and Conservation 5).

Levitt, J. (1972b). "Responses of plants to environmental stresses." 1st ed. Academic Press, New York.

Levitt, J. (1978). An overview of freezing injury and survival, and its interrelationship to other stresses. *In* "Plant Cold Hardiness and Freezing Stress." (P. H. Li and A. Sakai, eds.). Academic Press, New York. pp. 3–15.

Levitt, J., and Dear, J. (1970). The role of membrane proteins in freezing injury and resistance. *In* "The Frozen Cell." (G. E. W. Wolstenholme and M. O'Connor, eds.), pp. 148–174. Ciba Foundation Symp., J. & A. Churchill, London.

Levitt, J., and Hasman, M. (1964). Mechanism of protection by non-penetrating and non-toxic solutes against freezing injury to plant cells. *Plant Physiol.* **39,** 409–412.

Levit, J., and Scarth, G. W. (1936). Frost-hardening studies with living cells. I. Osmotic and bound water changes in relation to frost resistance and the seasonal cycle. *Can. J. Res.* **C14,** 267-284.

Levitt, J., and Siminovitch. D. (1940). The relation between frost resistance and the physical state of protoplasm. I. The protoplasm as a whole. *Can. J. Res.* **C18,** 550-561.

Levitt, J., Sullivan, C. Y., and Krull, E. (1960). Some problems in drought resistance. *Bull. Res. Counc. Isr.* **8D,** 173-179.

Levitt, J., Sullivan, C. Y., Johansson, N-O., and Pettit, R. M. (1961). Sulfhydryls—a new factor in frost resistance. Changes in SH content during frost hardening. *Plant Physiol.* **36,** 611-616.

Levitt, J., Sullivan, C. Y., and Johansson, N-O. (1962). Sulfhydryls—a new factor in frost resistance. III. Relation of SH increase during hardening to protein, glutathione, and glutathione oxidizing activity. *Plant Physiol.* **37,** 266-271.

Lewis, F. J., and Tuttle, G. M. (1920). Osmotic properties of some plant cells at low temperatures. *Ann. Bot.* **34,** 405-416.

Lewis, F. J., and Tuttle, G. M. (1923). On the phenomena attending seasonal changes in the organisation in leaf cells of *Picea canadensis.* (Mill.) B. S. P. *New Phytol.* **22,** 225-232.

Li, P. H. and Palta, J. P. (1978). Frost hardening and freezing stress in tuber-bearing *Solanum* species. *In* Plant Cold Hardiness and Freezing Stress. (P. H. Li and A. Sakai, eds.), pp. 49-71. Academic Press, New York.

Li, P. H., and Weiser, C. J. (1967). Evaluation of extraction and assay methods for nucleic acids from red osier dogwood and RNA, DNA, and protein changes during cold acclimation. *Proc. Amer. Soc. Hort. Sci.* **91,** 716-727.

Li, P. H., and Weiser, C. J. (1969a). Metabolism of nucleic acids in one-year-old apple twig during cold hardening and dehardening. *Plant Cell Physiol.* **10,** 21-30.

Li, P. H., and Weiser, C. J. (1969b). Influence of photoperiod and temperature on potato foliage protein and 4S RNA. *Plant Cell Physiol.* **10,** 929-934.

Li, P. H., and Weiser, C. J. (1973). Short term increases in the cold tolerance of red osier dogwood stems induced by application of cysteine. *Plant Physiol.* **52,** 685-687.

Lidforss, B. (1896). Zur Physiologie und Biologie der wintergrünen Flora. *Bot. Centr.* **68,** 33-44.

Lidforss, B. (1907). Die wintergrüne Flora. *Lunds Univ. Arsskr. Afd.* **2,** 1-76.

Lindley, J. (1842). Observations upon the effects produced on plants by the frost which occurred in England in the winter of 1837-38. (Read in 1838). *Trans. Hort. Soc. (London)* **3,** 225-315.

Lindow, S. W., Arny, D. C., Upper, C. D., and Brachet, W. R. (1978). The role of bacterial ice nuclei in frost injury to sensitve plants. "Plant Cold Hardiness and Freezing Stress." (P. H. Li and A. Sakai eds.), pp. 249-263. Academic Press, New York.

Ling, G. N. (1968). The physical state of water in biological systems. *Food Technol.* **22,** 1254-1258.

Lin Wu, Pei-Hsing. (1974). Effects of temperature on the metabolic pattern of incorporation of ^{14}C by the moss Dicranum scoparium incubated with acetate-2-^{14}C. *Ohio J. Sci.* **74,** 200-208.

Lipman, C. B. (1936a). The tolerance of liquid air temperatures by dry moss protonema. *Bull. Torrey Bot. Club* **63,** 515-518.

Lipman, C. B. (1936b). Normal viability of seeds and bacterial spores after exposure to temperatures near the absolute zero. *Plant Physiol.* **11,** 201-205.

Lipman, C. B. (1937). Tolerance of liquid air temperatures by spore-free and very young cultures of fungi and bacteria growing on agar media. *Bull Torrey Bot. Club* **64,** 537-546.

Lipman, C. B., and Lewis, G. N. (1934). Tolerance of liquid-air temperatures by seeds of higher plants for sixty days. *Plant Physiol.* **9,** 392-394.

Livne, A. (1968). Membrane lipids as site for freezing injury. *Isr. J. Chem.* 152p.

Lockett, M. C., and Luyet, B. J. (1951). Survival of frozen seeds of various water contents. *Biodynamica* **7**(134), 67-76.

Lona, F. (1962). Resistenza al freddo in relazione all'effecto delle gibberelline, antigibberelline e delle radiazioni morfogenetische. *Ateneo Parmense* **33,** (Suppl. 6), 209-213.

Lona, F., Squarza, A., Bocchi, A., and Cantoni, G. (1956). Le reazione al freddo di piante erbacee in rapporto all'azione di sostanze stimolanti ed inibenti l'attivita plastimatica. *Pubb. Chim. Biol. Med. Ist. Carlo Erba Ric. Ter.* **2,** 473-494.

Lovelock, J. E. (1953). The mechanism of the protective action of glycerol against haemolysis by freezing and thawing. *Biochim. Biophys. Acta* **11,** 28-36.

Lucas, J. W. (1954). Subcooling and ice nucleation in lemons. *Plant Physiol.* **29,** 245-251.

Lund, D. B., Fennema, O., and Powrie, W. D. (1969). Enzymic and acid hydrolysis of sucrose as influenced by freezing. *J. Food. Sci.* **34,** 378-382.

Lundegårdh, H. (1914). Einige Bedingungen der Bildung und Auflösung der Stärke. *Jahrb. Wiss. Bot.* **53,** 421-463.

Lundquist, V., and Pellet, H. (1976). Preliminary survey of cold hardiness levels of several bulbous ornamental plant species. *Hortscience* **11,** 161–162.

Lusena, C. V., and Cook, W. H. (1953). Ice propagation in systems of biological interest. I. Effect of membranes and solutes in a model cell system. *Arch. Biochem. Biophys.* **46,** 232-240.

Luyet, B. J. (1937). The vitrification of organic colloids of protoplasm. *Biodynamica* **29,** 1-14.

Luyet, B. J. (1940). "Life and Death at Low Temperatures," Monogr. No. 1. Biodynamica, Normandy, Missouri.

Luyet, B. J. (1951). Survival of cells, tissues, and organisms after ultrarapid freezing. *In* "Freezing and Drying" (R. J. C. Harris, ed.), pp. 3-23. Institute of Biology, London.

Luyet, B. J., and Galcs, G. (1940). The effect of the rate of cooling on the freezing point of living tissues. *Biodynamica* **3**(65), 157-169.

Luyet, B. J., and Gehenio, P. M. (1937). The double freezing point of living tissues. *Biodynamica* **1**(30), 1-23.

Luyet, B. J., and Gehenico, P. M. (1938). The lower limit of vital temperatures, a review. *Biodynamica* **33,** 1-92.

Luyet, B. J., and Grell, Sister M. (1936). A study with the ultracentrifuge of the mechanism of death in frozen cells. *Biodynamica* **23,** 1-16.

Luyet, B. J., and Hodapp, E. L. (1938). On the effect of mechanical shocks on the congelation of subcooled plant tissues. *Protoplasma* **30,** 254-257.

Lybeck, B. R. (1959). Winter freezing in relation to the rise of sap in tall trees. *Plant Physiol.* **34,** 482-486.

McCay, P. B., Gibson, D. D., Fong, K. L., and Hornbrook, K. K. (1976). Effect of glutathione peroxidase activity on lipid peroxidation in biological membranes. *Biochim. Biophys. Acta.* **431,** 459–468.

McCoan, B. H., Beck, G. E., and Hall, T. C. (1969a). The hardening of three clones of Dianthus and the corresponding complement of peroxidase isoenzymes. *J. Amer. Soc. Hort. Sci.* **94,** 691-693.

McCown, B. H., Hall, T. C., and Beck, G. E. (1969b). Plant leaf and stem proteins. II. Isozymes and environmental change. *Plant Physiol.* **44,** 210-216.

McCown, B. H., McLeester, R. C., Beck, G. E., and Hall, T. C. (1969c). Environment-induced changes in peroxidase zymograms in the stems of deciduous and evergreen plants. *Cryobiology* **5,** 410-412.

MacDonald, M. A., Fensom, D. S., and Taylor, A. R. A. (1974). Electrical impedance in Ascophyllum nodosum and Fucus vesiculosus in relation to cooling, freezing, and desiccation. *J. Phycol.* **10,** 462-469.

MacDowall, F. D. H. (1974). Growth kinetics of Marquis wheat: VI. Genetic dependence and winter hardening. *Can. J. Bot.* **52,** 151-157.

MacDowall, F. D. H., and Buchanan, G. W. (1974). Estimation of the water of hydration in wintering wheat leaves by proton magnetic resonance. *Can. J. Biochem.* **52,** 652-654.

Macfayden, A. (1900). On the influence of the temperature of liquid air on bacteria. *Proc. Roy. Soc.* **66,** 180-182, 339-340.

McGann, L. E., Kruuv, J., Frim, J., and Frey, H. E., (1974). Factors affecting the repair of sublethal freeze-thaw damage in mammalian cells. II. The effect of ouabain. *Cryobiology* **11,** 332-339.

McGann, L. E., Kruuv, J., Frim, J., and Frey, H. E. (1975). Factors affecting the repair of sublethal freeze-thaw damage in mammalian cells. I. Suboptimal temperature and hypoxia. *Crybiology,* **12,** 530-539.

McGrath, J. J., Cravalho, E. G., and Huggins, C. E. (1975). An experimental comparison of intracellular ice formation and freeze-thaw survival of HeLa s-3 cells. *Cryobiology,* **12,** 540-550.

McCrath, M. S., and Daggett, P.-M. (1977). Cryopreservation of flagellar mutants of *Chlamydomonas reinhardtii. Can. J. Bot.* **55,** 1794-1796.

McGuire, J. J., and Flint, H. L. (1962). Effects of temperature and light on frost hardiness of conifers. *Proc. Amer. Soc. Hort. Sci.* **80,** 630-635.

McKenzie, J. S., Weiser, C. J., and Li, P. H. (1974c). Changes in water relations of Cornus stolonifera during cold acclimation. *J. Am. Soc. Hortic. Sci.* **99,** 223-228.

McKenzie, J. S., Weiser, C. J., and Li, P. H. (1974a). Effects of red and far red light on the initiation of cold acclimation in *Cornus stolonifera* Michx. *Plant Physiol.* **53,** 783-789.

McKenzie, J. S., Weiser, C. J., Stadelmann, E. J., and Burke, M. J. (1974b). Water permeability and cold hardiness of cortex cells in *Cornus stolonifera* Michx.: A preliminary report. *Plant Physiol.* **54,** 173-176.

McLeester, R. C., Weiser, C. J., and Hall, T. C. (1968). Seasonal variations in freezing curves of stem sections of *Cornus stolonifera* Michx. *Plant Cell Physiol.* **9,** 807-817.

McLeester, R. C., Weiser, C. J., and Hall, T. C. (1969). Multiple freezing points as a test for viability of plant stems in the determination of frost hardiness. *Plant Physiol.* **44,** 37-44.

Magistad, O. G., and Truog, E. (1925). Influence of fertilizers in protecting corn against freezing. *J. Amer. Soc. Agron.* **17,** 517-526.

Mair, B. (1968). A gradient in cold resistance of ash bud sequences. *Planta* **82,** 164-169.

Manis, R. E., and Knight, R. J., Jr. (1967). Avocado germ plasm evaluation: Technique used in screening for cold tolerance. *Proc. Fla. State Hort. Soc.* **80,** 387-391.

Marcellos, H., and Single, W. V. (1975). Temperatures in wheat during radiation frost. *Aust. J. Exp. Agric. Anim. Husb.* **15,** 818-822.

Maggio, B., Diplock, A. T., and Lucy, J. A. (1977). Interactions of tocopherol and ubiquinones with monolayers of phospholipids. *Biochem. J.* **161,** 111-121.

Mark, A. F. (1975). Photosynthesis and dark respiration in three alpine snow tussocks. (*Chionochloa* spp.) under controlled environments. *N. Z. J. Bot.* **13,** 93-122.

Mark, J. J. (1936). The relation of reserves to cold resistance in alfalfa. *Iowa Agr. Expt. Sta. Res. Bull.* **208,** 305-335.

Markert, C. L. (1965). Lactate dehydrogenase isozymes: Dissociation and recombination of subunits. *Science* **140,** 1329-1330.

Markova, L. E. (1973). Change in carbohydrate content in some plants vegetating during the mesothermal period of the year. *Uzb. Biol. Zh.* **17,** 38-40.

Markowski, A., Myczkowski, J., and Lebek, J. (1962). Preliminary investigations on changes in nitrogen compounds of wheat embryos in the course of germination under various temperature conditions. *Bull. Acad. Pol. Sci. Cl. V* **10,** 145-150.

Marlangeon, R. C. (1969). Effects of gibberellic acid and other drugs on the permanence of green autumn foliage and the induction of resistance to frost in peach flowers. *Phyton* **26,** 107-111.

Marlangeon, R. C. (1968). Attempts to induce chemical resistance to frost on grape vines under outdoor conditions. *Phyton. Rev. Int. Bot. Exp.* **25,** 53-60.

Marshall, D. C. (1961). The freezing of plant tissue. *Aust. J. Biol. Sci.* **14,** 368-390.

Martin, J. F. (1932). The cold resistance of Pacific Coast spring wheats at various stages of growth as determined by artificial refrigeration. *J. Amer. Soc. Agron.* **24,** 871-880.

Martincic, A., Gams, M., Vogelnik, K., Batic, F., and Vrhovsek, D. (1975). Net-photosynthetic activity of the holly, *Ilex aquifolium* L. in winter conditions. *Biol. Vestn.* **23,** 45-52.

Martsolf, J. D., Ritter, C. M., and Hatch, A. H. (1975). Effect of white latex paint on temperature of stone fruit tree trunks in winter. *J. Am. Soc. Hortic. Sci.* **100,** 122-129.

Marutyan, S. A., Dogramadzhyan, A. D., and Abadzhyan, R. A. (1972). Effect of low temperatures on protein fractions of grape shoots. *Dokl. Akad. Nauk. SSSR. Ser. Biol.* **204,** 1010-1012.

Massey, V., Hofmann, T., and Palmer, G. (1962). The relation of function and structure in lipoyl dehydrogenase. *J. Biol. Chem.* **237,** 3820-3828.

Mathias, E. L., Bennett, O. L., and Lundberg, P. E. (1973). Effect of rates of nitrogen on yield, nitrogen use and winter survival of Midland bermudagrass *Cynodon dactylon L. Pers.* in Appalachia. *Agron. J.* **65,** 67-68.

Maurel, P. and Travers, F. (1973). Effet des basses températures sur la cinétique d'une réaction peroxidasique. *Acad. Sci. (Paris)* **276,** 3383-3386.

Maurel, P., Travers, F., and Douzou, P. (1974). Spectroscopic determinations of enzyme-catalyzed reactions at subzero temperatures. *Analyt. Biochem.* **57,** 555- 563.

Maximov, N. A. (1908). Zur Frag über das Erfrieren der Pflanzen. *J. Botan. Ed. Sect. Bot. Soc. Imp. Nat. St. Petersburg,* 32-46. (*Bot. Centr.* **110,** 597-598.)

Maximov, N. A. (1912). Chemische Schutzmittel der Pflanzen gegen Erfrieren. *Ber. Deut. Bot. Ges.* **30,** 52-65, 293-305, 504-516.

Maximov, N. A. (1914). Experimentelle und kritische Untersuchungen über das Gefrieren und Erfrieren der Pflanzen. *Jahrb. Wiss. Bot.* **53,** 327-420.

Maximov, N. A. (1929a). Internal factors of frost and drought resistance in plants. *Protoplasma* **7,** 259-291.

Mazelis, M., and Fowden, L. (1969). Conversion of ornithine into proline by enzymes from germinating peanut cotyledons. *Phytochemistry* **8,** 801-809.

Mazur, P. (1960). Physical factors implicated in the death of microorganisms at subzero temperatures. *Ann. N. Y. Acad. Sci.* **85,** (Art. 2), 610-629.

Mazur, P. (1961). Physical and temporal factors involved in the death of yeast at subzero temperatures. *Biophys. J.* **1,** 247-264.

Mazur, P. (1963). Kinetics of water ions from cells at subzero temperatures and the likelihood of intracellular freezing. *J. Gen. Physiol.* **47,** 347-369.

Mazur, P. (1970). Cryobiology: the freezing of biological systems. *Science* **168**, 939-949.

Mazur, P., and Schmidt, J. J. (1968). Interactions of cooling velocity, temperature, and warming velocity on the survival of frozen and thawed yeast. *Cryobiology* **5,** 1-17.

Mazur, P., Rhian, M. A., and Mahlandt, B. G. (1957). Survival of *Pasturella Tulariensis* in sugar solutions after cooling and warming at subzero temperatures. *J. Bacteriol.* **73,** 394-397.

Mazur, P. (1977). The role of intracellular freezing in the death of cells cooled at supraoptimal rates. *Cryobiology* **14,** 251-272.

Merishi, J. N., and Grassetti, D. R. (1969). Sulfhydryl groups on the surface of intact Ehrlich ascites tumor cells, human blood platelets, and lymphocytes. Nature (*London*) **224,** 563-564.

Meindl, T. (1934). Weitere Beiträge zur protoplasmatischen Anatomie des Helodea-Blattes. *Protoplasma* **21,** 362-393.

Meryman, H. T. (1966a). "Cryobiology," 775 pp. Academic Press, New York.

Meryman, H. T. (1966b). Review of biological freezing. *In* "Cryobiology" (H. T. Meryman, ed.), Academic Press, New York. pp. 3-114.

Meryman, H. T. (1967). The relationship between dehydration and freezing injury in the human erythrocyte. *In* "Cellular Injury and Resistance in Freezing Organisms" (E. Asahina, ed.), pp. 231-244. Inst. Low Temp. Sci. Sapporo, Hokkaido.

Meryman, H. T. (1968). Modified model for the mechanism of freezing injury in erythrocytes. *Nature (London)* **218,** 333-336.

Meryman, H. T. (1970). The exceeding of a minimum tolerable cell volume in hypertonic suspension as a cause of freezing injury. *In* "The Frozen Cell" (G. E. W. Wolstenholme and M. O'Connor, eds.), pp. 51-67. Ciba Found. Symp., J. and A. Churchill, London.

Meryman, H. T., Williams, R. J., and Douglas, M. St. J. (1977). Freezing injury from "solution effects" and its prevention by natural or artificial cryoprotection. *Cryobiology* **14,** 287-302.

Meyer, B. S. (1932). Further studies on cold resistance in evergreens, with special reference to the possible role of bound water. *Bot. Gaz. (Chicago)* **94,** 297-321.

Meyer, H. W., and Winkelmann, H. (1969). Die Gefrierätzung und die Struktur biologischer Membranen. *Protoplasma* **68,** 253-270.

Mia, A. J. (1972). Fine structure of the ray parenchyma cells in Populus tremuloides in relation to senescence and seasonal changes. *Tex. J. Sci.* **24,** 245-260.

Michaelis, P. (1934). Okologische Studien an der alpinen Baumgrenze. IV. Zur Kenntnis des winterlichen Wasserhaushaltes. *Jahrb. Wiss. Bot.* **80,** 169-247.

Michel-Durand, E. (1919). Variation des substances hydrocarbonées dans les feuilles. *Rev. Gen. Bot.* **31,** 145-156, 196-204.

Miller, A. A. (1969). Glass transition temperature of water. *Science* **163,** 1325-1326.

Miller, L. K. (1969). Freezing tolerance in an adult insect. *Science* **166,** 105-106.

Miller, R. H., and Mazur, P. (1976). Survival of frozen thawed human red cells as a function of cooling and warming velocities. *Cryobiology,* **13,** 404-414.

Miller, R. W., de la Roche, I., and Pomeroy, M. K. (1974). Structural and functional responses of wheat mitochondrial membranes to growth at low temperatures. *Plant Physiol.* **53,** 426-433.

Milner, H. W., and Hiesey, W. M. (1964). Photosynthesis in climatic races of Mimulus (Scrophulariaceae). I. Effect of light intensity and temperature on rate. *Plant Physiol.* **39,** 208-213.

Mininberg, S. Ya, and Shumik, S. A. (1972). Respiration intensity and catalase activity in grape varieties differing in frost-resistance under various nutritive conditions. *Fiziol. Biokhim. Kul't. Rast.* **4,** 614-618.

Mitra, S. K. (1921). Seasonal changes and translocation of carbohydrate materials in fruit spurs and two-year-old seedlings of apple. *Ohio J. Sci.* **21,** 89-90.

Modlibowska, I. (1968). Effects of some growth regulators on frost damage. *Cryobiology* **5,** 175-187.

Modlibowska, I., and Rogers, W. S. (1955). Freezing of plant tissues under the microscope. *J. Exp. Bot.* **6,** 384-391.

Mohr, W. P., and Stein, M. (1969). Effect of different freeze-thaw regimes on ice formation and ultrastructural changes in tomato fruit parenchyma tissue. *Cryobiology* **6,** 15-31.

Molisch, H. (1897). "Untersuchungen über das Erfrieren der Pflanzen," pp. 1-73. Fischer, Jena.

Moor, H. (1960). Reaktionsweisen der Pflanzen auf Kälteeinflüsse. *Z. Schweiz. Forstv.* **30** (Festschrift. Prof. Frey-Wyssling), 211-222.

Moretti, A. (1953). Physiological effects of winter treatments of chemicals upon grapevine. *Riv. Fruitticolt. Viticolt. Orticolt.* **15,** 2-25.

Morren, C. (1838). Observations anatomiques sur la congélation des organes des végétaux. *Bull. Acad. Roy. Sci. Belles-lett. Bruxelles* **5,** 65-66, 93-111.

Morris, G. J. (1976). The cryopreservation of *Chlorella:* 1. Interactions of rate of cooling, protective additive and warming rate. *Arch. Microbiol.* **107,** 57-62.

Morris, G. J. (1976). The cryopreservation of *Chlorella:* 2. Effect of growth temperature on freezing tolerance. *Arch. Microbiol.* **107,** 309-312.

Morris, I., and Farrell, K. (1971). Photosynthetic rates, gross patterns of carbon dioxide assimilation and activities of ribulose diphosphate carboxylase in marine algae grown at different temperatures. *Physiol. Plant.* **25,** 372-377.

Morris, J. Y., and Tranquillini, W. (1969). Uber den Einfluss des osmotischen Potentiales des Wurzelsubstrates auf die photosynthese von *Pinus contorta*-Samlingen im Wechsel der Jahreszeiten. *Flora Abt.* **B158,** 277-287.

Morton, W. (1969). Effects of freezing and hardening on the sulfhydryl groups of protein fractions from cabbage leaves. *Plant Physiol.* **44,** 168-172.

Moschkov, B. S. (1935). Photoperiodismus und Frosthärte ausdauernder Gewächse. *Planta* **23,** 774-803.

Moser, W. (1969). Die Photosyntheseleistung von Nivalpflanzen. *Ber. Deut. Bot. Ges.* **82,** 63-64.

Müller-Thurgau, H. (1880). Über das Gefrieren und Erfrieren der Pflanzen. *Landwirtsch. Jahrb.* **9,** 133-189.

Müller-Thurgau, H. (1882). Über Zuckeranhäufung in Pflanzentheilen in Folge niederer Temperatur. *Landwirtsch. Jahrb.* **11,** 751-828.

Müller-Thurgau, H. (1886). Über das Gefrieren und Erfrieren der Pflanzen. II. Theile. *Landwirtsch. Jahrb.* **15,** 453-610.

Musich, V. N. (1968). The content and composition of sugars in winter wheat plants on hardening. *Ref. Zh. Biol.* No. 36111.

Myers, J. S., and Jakoby, W. B. (1975). Glycerol as an agent eliciting small conformational changes in alcohol dehydrogenase. *J. Biol. Chem.* **250,** 3785-3789.

Mytsyk, L. P. (1972). Cold resistance of *Lolium perenne* L. main shoots in the first year of vegetation in fall seeding. *Ekologiya* **3,** 92–94.

Naccache, P., and Sha'afi, R. I. (1974). Effect of PCMBS on water transfer across biological membranes. *J. Cell Physiol.* **8,** 449-456.

Nag, K. K., and Street, H. E. (1975a). Freeze preservation of cultured plant cells: I. The pretreatment phase. *Physiol. Plant.* **34,** 254-260.

Nag, K. K., and Street, H. E. (1975b). Freeze preservation of cultured plant cells: II. The freezing and thawing phases. *Physiol. Plant.* **34,** 261-265.

Nägeli, C. (1861). Über die Wirkung des Frostes auf die Pflanzenzellen. *Sitzber. Math. Phys. Kl. Bayer. Akad. Wiss. Munchen* pp. 264-271.

Nakanish, M., Wilson, A. C., Nolan, R. A., Gorman, G. C., and Bailey, G. S. (1969). Phenoxyethanol: Protein preservative for taxonomists. *Science* **163,** 681-683.

Nath, J. and Anderson, J. O. (1975). Effect of freezing and freeze-drying on the viability and storage of *Lilium longiflorum* L. and *Zea mays* L. pollen. *Cryobiology,* **12,** 18–88.

Nath, J., and Gonda, S. R. (1975). Effects of freezing and thawing on glycerol mutants of *Escherichia coli. Cryobiology* **12,** 321-327.

Nei, T., Araki, T., and Matsusaka, T. (1967). The mechanism of cellular injury by freezing in microorganisms. *In* "Cellular Injury and Resistance in Freezing Organisms" (E. Asahina, ed.), pp. 157-170. Inst. Low Temp. Sci., Hokkaido.

Neilson, R. E., Ludlow, M. M., and Jarvis, P. G. (1972). Photosynthesis in Sitka spruce (*Picea sitchensis* (Bong.) Carr.): II. Response to temperature. *J. Appl. Ecol.* **9,** 721–745.

Newman, E. I., and Kramer, P. J. (1966). Effect of decenylsuccinic acid on the permeability and growth of bean roots. *Plant Physiol.* **41,** 606–609.

Newton, R., Brown, W. R., and Anderson, J. A. (1931). Chemical changes in nitrogen fractions of plant juices on exposure to frost. *Can. J. Res.* **5,** 327–332.

Noble, A., and Lowe, K. F. (1974). Alcohol-soluble carbohydrates in various tropical and temperate pasture species. *Trop. Grassl.* **8,** 179–188.

North, M. J. (1973). Cold-induced increase of glycerol kinase in *Neurospora crassa. FEBS. Lett.* **35,** 67–70.

Novikov, V. A. (1928). Cold resistance of plants. II. *J. Exp. Landwirtsch. Südosten Eur-Russlands* **6,** 71–100.

Nyuppieva, K. A. (1973). Short-term effect of negative temperatures on the content and the state of pigments in the leaves of potato plants with different resistance to low temperatures. *Fiziol. Rast.* **20,** 17–23.

Nyuppieva, K. A., Khein, K. Y., and Osipova, O. P. (1972). Effect of negative temperatures on the photosyntheitc apparatus of potato species with different frost resistances. *Fiziol. Rast.* **19,** 258–264.

Ogolevets, I. V. (1976a). Hardening of isolated callus tissue of woody plants with different frost resistances. *Fiziol. Rast.* **23,** 115–119.

Ogolevets, I. V. (1976b). Study of hardening of isolated callus tissue of trees with different frost-resistance. *Fiziol. Rast.* **23,** 139–145.

Okanenko, A. A. (1974). Dynamics of fats in annual shoots of apple varieties differing in frost-resistance. *Fiziol. Biokhim. Kul't Rast.* **6,** 90–94.

Oknina, E. Z., and Markovich, A. A. (1951). Means of increasing resistance to cold in *Rosa gallica. Izv. Akad. Nauk. SSSR Ser. Biol.,* pp. 107–114.

Olien, C. R. (1961). A method of studying stresses occurring in plant tissue during freezing. *Crop Sci.* **1,** 26–28.

Olien, C. R. (1965). Interference of cereal polymers and related compounds with freezing. *Cryobiology* **2,** 47–54.

Olien, C. R. (1967). Freezing stresses and survival. *Annu. Rev. Plant Physiol.* **18,** 387–408.

Olien, C. R. (1973). Thermodynamic components of freezing stress. *J. Theor. Biol.* **39,** 201–210.

Olien, C. R. (1974). Energies of freezing and frost desiccation. *Plant Physiol.* **53,** 764–767.

Olien, C. R., and Su-En Chao. (1973). Liquid water content of cell walls in frozen tissues evaluated by electrophoresis of indicators. *Crop Sci.* **13,** 674–676.

Olien, C. R., Marchetti, B. L., and Chomyn, E. V. (1968). Ice structure in hardened winter barley. *Mich. Agr. Exp. Sta. Quart. Bull.* **50,** 440–448.

Omran, A. O., Atkins, I. M., and Gilmore, E. C. Jr. (1968). Heritability of cold hardiness in flax. *Crop Sci.* **8,** 716–719.

Onoda, N. (1937). "Mikroskopische Beobachtungen über das Gefrieren einiger Pflanzenzellen in flüssigem Paraffin." Botan. Inst. der Kaiserlichen Univ. zu Kyoto. *Bot. and Zool* **5,** 1845–2188.

Orii, Y., and Iizuka, T. (1975). Change in effective pH of salt solutions on freezing, as evidenced by altered reactivities of heme towards carbonyl reagents. *J. Biochem.* **77,** 1123–1126.

Ormrod, D. P., and Layne, R. E. (1974). Temperature and photoperiod effects on cold hardiness of peach scion-rootstock combinations. *Hortscience* **9,** 451–453.

Ostaplyuk, E. D. K. (1967). A physiological characterization of the frost resistance of winter barley. *Rost. Ustoich. Rast.* **3,** 203–209.

Otsuka, K. (1971). Survival of pollen cells at super-low temperatures. *Low Temp. Sci. Ser. B. Biol. Sci.* **29,** 107–111.

Otsuka, K. (1972). The ultrastructure of mulberry cortical parenchyma cells related to the change of.the freezing resistance in spring. *Low Temp. Sci. Ser. B. Biol. Sci.* **30,** 33–44.

Overton, E. (1899). Beobachtungen und Versuche über das Auftreten von rothem Zellsaft bei Pflanzen. *Jahrb. Wiss. Bot.* **33,** 171-177.

Paldi, E., and Dévay, M. (1977). Characteristics of the rRNA synthesis taking place at low temperatures in wheat cultivars with varying degrees of frost hardiness. *Phytochem.* **16,** 177-179.

Palta, J. P., Levitt, J., and Stadelmann, E. J. (1977a). Freezing injury in onion bulbs. I. Evaluation of the conductivity method and analysis of ion and sugar efflux from injured cells. *Plant Physiol.* **60,** 393-397.

Palta, J. P., Levitt, J., and Stadelmann, E. J. (1977b). Freezing injury in onion bulbs. II. Post-thawing injury or recovery. *Plant Physiol.* **60,** 398-401.

Palta, J. P., Levitt, J., Stadelmann, E. J., and Burke, M. J. (1977c). Dehydration of onion cells: A comparison of freezing vs. desiccation and living vs. dead cells. *Physiol. Plant.* **41,** 273-279.

Palta, J. P., and Li, P. H. (1978). Cell membrane properties in relation to freezing injury. *In* "Plant Cold Hardiness and Freezing Stress." (P. H. Li and A. Sakai, eds.), pp. 93-115. Academic Press, New York.

Pankratova, S. I., and Khokhlova, L. P. (1977). Dynamics of the phospholipid content in tillering nodes of winter wheat during autumn hardening. *Fiziol. Biokhim. Kul't Rast.* **9,** 129-135.

Pantanelli, E. (1918). Sur la resistanza delle piante al freddo. *Atti reale Accad. Ital. Mem. Cl. Sci. Fis. Mat. Nat.* **27,** 126-130, 148-153. (*Biol. Abstr.* **2,** 1135, 1919).

Papahadjopoulos, D., Vail, W. J., and Moscarello, M. (1975). Interaction of a purified hydrophobic protein from myelin with phospholipid membranes: Studies on ultra- structure, phase transitions and permeability. *J. Membr. Biol.* **22,** 143-164.

Paquin, R., Belzile, L., Willemot, C., and St.-Pierre, J.-C. (1976). Effects of some growth retardants and gibberellic acid on the resistance to cold of alfalfa. *Medicago sativa. Can. J. Plant Sci.* **56,** 79-86.

Parducci, L. G., and Fennema, O. (1978). Rate and extent of enzymatic lipolysis at subfreezing temperatures. *Cryobiology* **15,** 199-204.

Parish, G. R. (1974). Seasonal variation in the membrane structure of differentiating shoot cambial-zone cells demonstrated by freeze-etching. *Cytobiologie* **9,** 131–143.
131-143.

Parker, J. (1951). Moisture retention in leaves of conifers of the Northern Rocky Mountains. *Bot. Gaz. (Chicago)* **113,** 210-216.

Parker, J. (1953). Some applications and limitations of tetrazolium chloride. *Science* **118,** 77-79.

Parker, J. (1956). Drought resistance in woody plants. *Bot. Rev.* **22,** 241-289.

Parker, J. (1958). Sol-gel transitions in the living cells of conifers and their relation to resistance to cold. *Nature (London)* **182,** 1815.

Parker, J. (1959a). Seasonal variations in sugars of conifers with some observations on cold resistance. *Forest Sci.* **5,** 56-63.

Parker, J. (1959b). Seasonal changes in white pine leaves; a comparison of cold resistance and free-sugar fluctuations. *Bot. Gaz. (Chicago)* **121,** 46-50.

Parker, J. (1960). Survival of woody plants at extremely low temperatures. *Nature (London)* **187,** 1133.

Parker, J. (1962). Relationships among cold hardiness, water-soluble protein, anthocyanins, and free sugars in *Hedera helix* L. *Plant Physiol.* **37,** 809-813.

Parkes, A. S. (1964). Cryobiology. *Cryobiology* **1,** 3.

Parsons, L. R. (1978). Water relations, stomatal behavior, and root conductivity of red osier dogwood during acclimation to freezing temperatures. *Plant Physiol.* **62,** 64-70.

Pauli, A. W., and Mitchell, H. L. (1960). Changes in certain nitrogenous constituents of winter wheat as related to cold hardiness. *Plant Physiol.* **35,** 539-542.

Paulsen, G. M. (1968). Effect of photoperiod and temperature on cold hardening in winter wheat. *Crop. Sci.* **8,** 29-32.

Pellett, N. E. (1973). Influence of nitrogen and phosphorus fertility on cold acclimation of roots and stems of two container-grown woody plant species. *J. Am. Soc. Hortic. Sci.* **98,** 82-86.

Pellett, N. E., and White, D. B. (1969). Relationship of seasonal tissue changes to cold acclimation of *Juniperus chinensis* Hetzi. *J. Amer. Soc. Hort. Sci.* **94,** 460-462.

Peltier, G. L., and Kiesselbach, T. A. (1934). The comparative cold resistance of spring small grains. *J. Amer. Soc. Agron.* **26,** 681-686.

Perkins, H. J., and Andrews, J. E. (1960). The effects of uptake of certain sugars and amino acids on the cold hardiness of young wheat plants. *Naturwissenschaften* **24,** 608-609.

Perry, T. O., and Hellmers, H. (1973). Effects of abscisic acid on growth and dormancy of two races of red maple. *Bot. Gaz.* **134,** 283-289.

Petit-Thouars, A. du. (1817). "Le verger Francais ou traité général de la culture des arbres fruitiers, etc," pp. 6-45. Paris.

Petrie, A. H. K., and Arthur, J. I. (1943). Physiological ontogeny in the tobacco plant. The effect of varying water supply on the drifts in dry weight and leaf area and on various components of the leaves. *Austr. J. Exp. Biol. Med. Sci.* **21,** 191-200.

Petrova, O. V., and Mishustina, P. S. (1976). Peroxidase isozymes in corn leaves at lower temperatures. *Fiziol. Biokhim. Kul't. Rast.* **8,** 174-177.

Petrovskaya-Baranova, T. P. (1972). Histochemical study of chloroplasts of winter wheat during overwintering. *Byull. Gl. Bot. Sad. Akad. Nauk. SSSR.* **85,** 71-75.

Petrovskaya-Baranova, T. P. (1974). Lysis of the nucleus and nuclear envelope in frozen cells of the coleorhiza of grasses. *Fiziol. Rast.* **21,** 1248-1251.

Pfeiffer, M. (1933). Frostuntersuchungen an Fichtentrieben. *Tharandter Forstl. Jahrb.* **84,** 664-695.

Pieniazek, J., and Wisniewska, J. (1954). The properties of the protoplasm in the tissues of one-year old fruit-tree shoots in the course of different phenophases. *Bull. Acad. Pol. Sci. Cl. 2* **2,** 149-152.

Pikush, G. R. (1974). Effect of chlorocholine chloride on formation of frost-resistance in winter wheat. *Fiziol. Biokhim. Kul't Rast.* **6,** 54-60.

Pincock, R. E., and Kiovsky, T. E. (1966). Reactions in frozen solutions. XI. The reaction of ethylene chlorohydrin with hydroxyl ion in ice. *J. Amer. Chem. Soc.* **88,** 4455-4459.

Pisek, A. (1950). Frosthärte und Zusammensetzung des Zellsaftes bei *Rhododendron ferrungineum, Pinus cembra* und *Picea excelsa. Protoplasma* **39,** 129-146.

Pisek, A. (1953). Wie schutzen sich die Alpenpflanzen gegen Frost? *Umschau* Heft, 21.

Pisek, A. (1958). Versuche zur Frostresistenzprüfung von Reinde, Winterknospen und Blüten einiger Arten von Obsthölzern. *Gartenbauwissenschaft* **23,** 54-74.

Pisek, A. (1962). Frostresistenz von Bäumen. *Meded. Inst. Vered. Tuenbougew, Wageningen* **182,** 74-79.

Pisek, A., and Kemnitzer, R. (1967). Der Einfluss von Frost auf die Photosynthese der Weisstanne (*Abies alba* Mill). *Flora (Jena)* **157,** 315-326.

Pisek, A., Sohm, H., and Cartellieri, E. (1935). Untersuchungen über osmotischen Wert und Wassergehalt von Pflanzen und Planzengesellschaften der aplinen Stufe. *Beitr. Bot. Centr.* **52,** 634-675.

Pogosyan, K. S., and Sakai, A. (1972a). Effect of the thawing rate on survival of grapes. *Fiziol. Rast.* **19,** 1204-1210.

Pogosyan, K. S., and Sakai, A. (1972b). Effect of negative temperature conditions on frost resistance of grape plants. *Biol. Zh. Arm.* **25,** 49-55.

Pojarkova, H. A. (1924). Winterruhe, Reservestoffe, und Kälteresistenz bei Holzpflanzen. *Ber. Deut. Bot. Ges.* **42,** 420-429.

Polishchak, L. K., Dibrova, L. S., Zablotskaya, K. M., and Lapchik, V. F. (1968). The importance of oxidation-reduction processes in the frost resistance of plants. *Rost. Ustoich. Rast.* **SB4,** 122-129.

Pomerleau, R., and Ray, R. G. (1957). Occurrence and effects of summer frost in a conifer plantation. *Forest Res. Division Tech. Note (Ottawa, Can.)* **51,** 1-15.

Pomeroy, M. K. (1976). Swelling and contraction of mitochondria from cold-hardened and nonhardened wheat and rye seedlings. *Plant Physiol.* **57,** 469-473.

Pomeroy, M. K. (1977). Ultrastructural changes during swelling and contraction of mitochondria from cold-hardened and non-hardened winter wheat. *Plant Physiol.* **59,** 250-255.

Pomeroy, K., and Andrews, C. J. (1975). Effect of temperature on respiration of mitochondria and shoot segments from cold-hardened and nonhardened wheat and rye seedlings. *Plant Physiol.* **56,** 703-706.

Pomeroy, M. K., and Andrews, C. J. (1978). Metabolic and ultrastructural changes in winter wheat during ice encasement under field conditions. *Plant Physiol.* **61,** 806-811.

Pomeroy, M. K., Andrews, C. J., and Fedak, G. (1975). Cold hardening and dehardening responses in winter wheat and winter barley. *Can. J. Plant Sci.* **55,** 529-536.

Pribor, D. B. (1975). Biological interactions between cell membranes and glycerol or DMSO. *Cryobiology* **12,** 309-320.

Prillieux, E. (1869). Sur la formation de glacons a l'intérieur des plantes. *Ann. Sci. Nat. Paris* (Ser. 5) **12,** 125-134.

Prillieux, E. (1872). Coloration en bleu des fleurs de quelques orchidées sous l'influence de la gelée. *Bull. Soc. Bot. Fr.* **19,** 152-155.

Prilutskii, A. V., Doskoch, Ya. E., and Tarusov, B. N. (1974). Characteristics of spontaneous ultra-weak chemoluminescence and some morphological and physiological indices of frost resistance in mulberry trees as affected by the thermoperiod. *S-Kh. Biol.* **9,** 397-404.

Proebsting, E. L., Jr. (1959). Cold hardiness of Elberta peach fruit buds during four winters. *Proc. Amer. Soc. Hort. Sci.* **74,** 144-153.

Proebsting, E. L. Jr., and Mills, H. H. (1973). Bloom delay and frost survival in ethephon-treated sweet cherry. *Hortscience* **8,** 46-47.

Proebsting, E. L. Jr, and Mills, H. H. (1974). Time of gibberellin application determines hardiness response of "Bing" cherry buds and wood. *J. Am. Soc. Hort. Sci.* **99,** 464-466.

Protsenko, D. F., and Rubanyuk, E. A. (1967). Amino acid metabolism in winter rye and wheat during the overwintering period. *Rost. Ustoich. Rast. Sb.* **3,** 161-169.

Puempel, B., Goebl, F., and Tranquillini, W. (1975). Growth mycorrhiza and frost resistance of *Picea abies* seedlings following fertilization with different levels of nitrogen. *Eur. J. For. Pathol.* **5,** 83-97.

Pugh, E. L., and Kates, M. (1975). Characterization of a membrane-bound phospholipid desaturase system of *Candida lipolytica*. *Biochim. Biophys. Acta.* **380,** 442-453.

Pullman, M. E., and Monroy, G. C. (1963). A naturally occurring inhibitor of mitochondrial adenosine triphosphatase. *J. Biol. Chem.* **238,** 3762-3769.

Pushkar, N. S., Shenberg, M. G., and Oboznaya, E. I. (1976). On the mechanism of cyroprotection by polyethylene oxide. *Cryobiology* **13,** 142-146.

Quamme, H. A. (1976). Relationship of the low temperature exotherm to apple and pear production in North America. *Can. J. Plant Sci.* **56,** 493-500.

Quamme, H., Weiser, C. J., and Stushnoff, C. (1973). The mechanism of freezing injury in xylem of winter apple twigs. *Plant Physiol.* **51,** 273-277.

Quamme, H. A., Layne, R. E. C., Jackson, H. O., and Spearman, G. A. (1975). An improved exotherm method for measuring cold hardiness of peach flower buds. *Hortscience.* **10,** 521-523.

Quatrano, R. S. (1968). Freeze preservation of cultured flax cells utilizing dimethyl sulfoxide. *Plant Physiol.* **43,** 2057-2061.

Quisenberry, K. S., and Bayles, B. B. (1939). Growth habit of some winter wheat varieties and its relation to winter hardiness and earliness. *J. Amer. Soc. Agron.* **31,** 785-789.

Rachie, K. O., and Schmid, A. R. (1955). Winter hardiness of birdsfoot trefoil strains and varieties. *Agron. J.* **47,** 155-157.

Radzievsky, G. B., and Shekhtman, Ya L. (1955). The application of roentgeno-structural analysis for the study of ice formation in plant grains. *Dokl. Akad. Nauk. SSSR* **101,** 1051-1053.

Raese, J. T., Williams, M. W., and Billingsley, H. D. (1977). Sorbitol and other carbohydrates in dormant apple shoots as influenced by controlled temperatures. *Cryobiology,* **14,** 373-378.

Ragan, P., and Nylund, R. E. (1977). Influence of N-P-K fertilizers on low temperature tolerance of cabbage seedlings. *Hortscience,* **12,** 320-321.

Rajashekar, C., and Burke, M. J. (1978). The occurrence of deep undercooling in the genera *Pyrus, Prunus,* and *Rosa:* a preliminary report. *In* "Plant Cold Hardiness and Freezing Stress" (P. H. Li and A. Sakai eds.), pp. 213-225. Academic Press, New York.

Rasmussen, D. H., Macaulay, M. N., and MacKenzie, A. P. (1975). Supercooling and nucleation in single cells. *Cryobiology,* **12,** 328-339.

Regehr, D. L., and Bazzaz, F. A. (1977). Low temperature photosynthesis in successional winter annuals. Ecology, **57,** 1297-1303.

Rehfeldt, G. E. (1977). Growth and cold hardiness of intervarietal hybrids of Douglas-fir. *Theor. Appl. Genet.* **50,** 3-16.

Reid, D. M., Pharis, R. P., and Roberts, D. W. A. (1974). Effects of four temperature regimens on the gibberellin content of winter wheat cv. *Kharkov. Physiol. Plant.* **30,** 53-57.

Rein, R. (1908). Untersuchungen über den Kältetod der Pflanzen. *Z. Naturforsch.* **80,** 1-38.

Reisner, A. H., Rowe, J., and Sleigh, R. W. (1969). Concerning the tertiary structure of the soluble surface proteins of Paramecium. *Biochemistry* **8,** 4637-4644.

Reum, J. A. (1835). "Pflanzenphysiologie, oder das Leben Wachsen und Verhalten der Pflanzen," pp. 168-169. Arnoldische Buchhandlung. Dresden u. Leipzig.

Rey, L. (1961). Automatic regulation of the freeze-drying of complex systems. *Biodynamica* **8,** 241-260.

Rhodes, D., and Stewart, G. R. (1974). A procedure for the *in vivo* determination of enzyme activity in higher plant tissue. *Planta* **118,** 133-144.

Richter, H. (1968a). Die Gefrierresistenz glyzerinbehandelter Campanulazellen. *Protoplasma* **66,** 63-78.

Richter, H. (1968b). Die Reaktion hochpermeable Pflanzenzellen auf drei Gefrierschutzstoffe (Glyzerin, Aethylenglukal, Dimethylsulfoxid). *Protoplasma* **65,** 155-166.

Riddle, W. A., and Pugach, S. (1976). Cold hardiness in the scorpion, *Paruroctonus aquilonalis. Cryobiology* **13,** 248-253.

Rieth, A. (1966). Zur Kenntnis der Lebensbedingungen von *Porphyridium cruentum* (Ag.) Naeg. VI. Wachstum in einum Rhythmus von Frost und Wärmeperioden. *Biol. Zentralbl.* **85,** 569-578.

Rikin, A., Waldman, M., Richmond, A. E., and Dovrat, A. (1975). Hormonal regulation of morphogenesis and cold-resistance: I. Modifications by abscisic acid and by gibberellic acid in alfalfa *Medicago sativa* L. seedlings. *J. Exp. Bot.* **26,** 175–183.

Rimpau, R. H. (1958). Untersuchungen über die Wirkung von kritischer Photoperiode und Vernalisation auf die Kältresistenz von *Triticum aestivum* L. *Z. Pflanzenzuecht* **40,** 275–318.

Riov, J., and Brown, G. N. (1976). Comparative studies of activity and properties of ferredoxin-NADP reductase during cold hardening of wheat. *Can. J. Bot.* **54,** 1896-1902.

Riov, J., and Brown, G. N. (1978). Properties of chloroplast membrane-bound ferredoxin-NADP$^+$ reductase during cold hardening of wheat. No indication of qualitative membrane changes during cold hardening. *Cryobiology* **15,** 80-86.

Rivera, V., and Corneli, E. (1931). Rassenga die casi fitopathologici osservati nel 1929 (danni da freddo e da crittogame). *Riv. Patol. Veg.* **21,** 65-100.

Robbins, M. L., and Whitwood, W. N. (1973). Deep-cold treatment of seeds: Effect on germination and on callus production from excised cotyledons. *Hortic. Res.* **13,** 137-141.

Roberts, D. W. A. (1967). The temperature coefficient of invertase from the leaves of cold-hardened and cold-susceptible wheat plants. *Can. J. Bot.* **43,** 1347-1357.

Roberts, D. W. A. (1969a). A comparison of the peroxidase isozymes of wheat plants grown at 6°C and 20°C. *Can. J. Bot.* **47,** 263-265.

Roberts, D. W. A. (1969b). Some possible roles for isozymic substitutions during cold hardening in plants. *Int. Rev. Cytol.* **26,** 303-328.

Roberts, D. W. A. (1973). A survey of the multiple forms of invertase in the leaves of winter wheat, *Triticum aestivum.* L. Emend Thell. ssp. *vulgare. Biochim. Biophys. Acta.* **321,** 220-227.

Roberts, D. W. A. (1975). The invertase completment of cold-hardy and cold-sensitive wheat leaves. *Can. J. Bot.* **53,** 1333-1337.

Roberts, D. W. A., and Grant, M. N. (1968). Changes in cold hardiness accompanying development in winter wheat. *Can. J. Plant Sci.* **48,** 369-376.

Roberts, R. H. (1922). The development and winter injury of cherry blossom buds. *Wisc. Agr. Expt. Sta. Res. Bull.,* No. 52.

Robinson, D. M. (1973). Repair of freezing injury in mammalian cells grown in serial culture. *Cryobiology,* **10,** 413-420.

Robinson, T. W. (1957). The phraeatophyte problem. *Symp. Phraeatophytes,* pp. 1-12. Rep. Southwest Reg. Meeting Amer. Geophys. Union, Sacramento, California.

Robson, M. G., and Jewiss, O. R. (1968). A comparison of British and North African varieties of tall fescue (*Festuca arundinacea.*) II. Growth during winter and survival at low temperatures. *J. Appl. Ecol.* **5,** 179-190.

Rochat, E., and Therrien, H. P. (1975a). Ultramicroscopic study of the cytologic modifications in winter wheat during the hardening process to cold weather. *Can. J. Bot.* **53,** 536-543.

Rochat, E., and Therrien, H. P. (1975b). Étude des protéines des blés resistant, Kharkov et sensible, Selkirk, au cours de l'endurcissement au froid. I. Protéines solubles. *Can. J. Bot.* **53,** 2411-2416.

Rochat, E., and Therrien, H. P. (1975c). Protein study of the resistant wheats *Kharkov* and sensitive wheat *Selkirk* during hardening to cold. II. Soluble proteins and proteins of the chloroplasts and membranes. *Can. J. Bot.* **53,** 2417-2424.

Rochat, E., and Therrien, H. P. (1976c). Study of amino acids in relation to resistance to cold in Kharkov and Kent winter wheats. *Nat. Can.* **103,** 517-525.

Rochat, E., and Therrien, H. P. (1976a). Metabolism of ribonucleic acids of winter wheat *Triticum aestivum* L. during hardening at low temperatures. *Nat. Can.* **103,** 441-450.

Rochat, E., and Therrien, H. P. (1976b). Effects of antimetabolites and of some exogenous substances on hardening under cold conditions of winter wheat *Triticum aestivum* L. *Nat. Can.* **103,** 451-456.

Rodionov, V. S., Nyuppieva, K. A., and Zakharova, L. S. (1973). Effect of low temperature on concentration of galacto-and phospholipids in potato leaves. *Fiziol. Rast.* **20,** 525-531.
Rogers, W. S. (1954). Some aspects of spring frost damage to fruit and its control. *J. Roy. Hort. Soc.* **79,** 29-36.
Rogers, W. S., Modlibowska, I., Ruxton, J. P., and Slater, C. H. W. (1954). Low temperature injury to fruit blossom. IV. Further experimetns on water-sprinkling as an anti-frost measure. *J. Hort. Sci.* **29,** 126-141.
Romanova, L. N. (1967). The physiological bases for the resistance to cold of winter crops. *Ref. Zh. Biol.* 6G132.
Rosa, J. T. (1921). Investigation on the hardening process in vegetable plants. *Mo. Agr. Expt. Sta. Res. Bull.,* No. 48.
Rothstein, A., and Weed, R. I. (1963). The functional significance of sulfhydryl groups in the cell membrane. AEC Res. Div. Rep. UR-633, p. 35.
Rottenberg, W. (1968). Die Standardisierung von Frostresistenz Untersuchungen angewandt an Aussenepidermiszellen von *Allium cepa* L. *Protoplasma* **65,** 37-48.
Rouschal, E. (1939). Zur Ökologie der Macchien: I. Der sommerliche Wasserhaushalt der Macchienpflanzen. *Jahrb. Wiss. Bot.* **87,** 436-523.
Rowley, J. A., Tunnicliffe, C. G., and Taylor, A. O. (1975). Freezing sensitivity of leaf tissue of C_4 grasses. *Aust. J. Plant Physiol.* **2,** 447-451.
Rudorf, W. (1938). Keimstimmung und Photoperiode in ihrer Bedeutung fur die Kälteresistenz. *Zuechter* **10,** 238-246.
Rutherford, P. P., and Weston, E. W. (1968). Carbohydrate changes during cold storage of some inulin-containing roots and tubers. *Phytochemistry* **7,** 175-180.
Sachs, J. (1860). Krystallbildungen bei dem Gefrieren und Veränderung der Zelhäute bei dem Aufthauen saftiger Pflanzentheile, mitgetheilt von W. Hofmeister. *Ber. Verhandl. Sächs. Akad. Wiss. Leipzig. Math. Phys. Kl.* **12,** 1-50.
Sagisaka, S. (1972). Decrease in the metabolic activity in poplar twigs to a fatal level during storage in frozen state. *Low. Temp. Sci. Ser. B. Biol. Sci.* **30,** 15-21.
Sagisaka, S. (1974). Effect of low temperature on amino acid metabolism in wintering poplar: Arginine–glutamine relationships. *Plant Physiol.* **53,** 319–322.
Sagisaka, S. (1974). Transition of metabolisms in living poplar bark from growing to wintering stages and vice versa: Changes in glucose 6-phosphate and 6-phosphogluconate dehydrogenase activities and in the levels of sugar phosphates. *Plant Physiol.* **54,** 544-549.
Sakai, A. (1955a). The relationship between the process of development and the frost hardiness of the mulberry tree. *Low Temp. Sci. Ser.* **B13,** 21-31.
Sakai, A. (1955b). The seasonal changes of the hardiness and the physiological state of the cortical parenchyma cells of mulberry tree. *Low Temp. Sci. Ser.* **B13,** 33-41.
Sakai, A. (1956a). The effect of temperature on the maintenance of the frost hardiness. *Low Temp. Sci. Ser.* **B14,** 1-6.
Sakai, A. (1956b). The effect of temperature on the hardening of plants. *Low Temp. Sci. Ser.* **B14,** 7-15.
Sakai, A. (1957). The effect of maleic hydrazide upon the frost hardiness of twig of mulberry tree. *J. Sericult. Sci. Jap.* **26,** 13-20.
Sakai, A. (1958). Survival of plant tissue at super-low temperature. II. *Low Temp. Sci. Ser.* **B16,** 41-53.
Sakai, A. (1960a). The frost hardening process of woody plant. VII. Seasonal variations in sugars. *Low Temp. Sci. Ser.* **B18,** 1-14.
Sakai, A. (1960b). The frost hardening process of woody plant. VIII. Relation of polyhydric alcohol to frost hardiness. *Low Temp. Sci. Ser.* **B18,** 15-22.

Sakai, A. (1961). Effect of polyhydric alcohols to frost hardiness in plants. *Nature (London)* **189,** 416–417.

Sakai, A. (1962). Studies on the frost-hardiness of woody plants. I. The causal relation between sugar content and frost-hardiness. *Contr. Inst. Low Temp. Sci. Ser.* **B11,** 1–40.

Sakai, A. (1966). Temperature fluctuation in wintering trees. *Physiol. Plant.* **19,** 105–114.

Sakai, A. (1967). Mechanism of frost damage on basal stems in young trees. *Low Temp. Sci. Ser.* **B25,** 45–57.

Sakai, A. (1968). Survival of plant tissue at super-low temperatures. VII. Methods for maintaining viability of less hardy plant cells at super-low temperatures. *Low Temp. Sci. Ser.* **B26,** 1–11.

Sakai, A. (1970a). Freezing resistance in willows from different climates. *Ecology* **51,** 485–491.

Sakai, A. (1970b). Mechanism of desiccation damage of conifers wintering in soil-frozen areas. *Ecology* **51,** 657–664.

Sakai, A., and Yoshida, S. (1968a). The role of sugar and related compounds in variations of freezing resistance. *Cryobiology* **5,** 160–174.

Sakai, A., and Yoshida, S. (1968b). Protective actions of various compounds against freezing injury in plant cells. *Low Temp. Sci. Ser.* **B26,** 13–21.

Sakai, A. (1972). Freezing resistance of trees in North America. *Low Temp. Sci. Ser. B. Biol. Sci.* **30,** 77–89.

Sakai, A. (1973). Characteristics of winter hardiness in extremely hardy twigs of woody plants. *Plant Cell Physiol.* **14,** 1–9.

Sakai, A. (1974). Characteristics of winter hardiness in extremely hardy twigs. *Fiziol. Rast.* **21,** 141–147.

Sakai, A., and Otsuka, K. (1972). A method for maintaining the viability of less hardy plant cells after immersion in liquid nitrogen. *Plant Cell Physiol.* **13,** 1129–1133.

Sakai, A., and Sugawara, Y. (1973). Survival of poplar callus at super-low temperatures after cold acclimation. *Plant Cell Physiol.* **14,** 1201–1204.

Sakharova, A. S., and Yakupov, N. A. (1969). A morphophysiological method for predicting the winter-hardiness of woody plants. From *Ref. Zh. Biol.* No. 5G136.

Salcheva, G., and Samygin, G. (1963). A microscopic study of freezing of the tissues of winter wheat. *Fiziol. Rast.* **10,** 65–72.

Salt, R. W. (1950). Time as a factor in the freezing of undercooled insects. *Can. J. Res.* **D28,** 285–291.

Salt, R. W. (1955). Extent of ice formation in frozen tissues and a new method for its measurement. *Can J. Zool.* **33,** 391–403.

Salt, R. W. (1957). Natural occurrence of glycerol in insects and its relation to their ability to survive freezing. *Can. Entomol.* **89,** 491–494.

Salt, R. W. (1958). Role of glycerol in producing abnormally low supercooling and freezing points in an insect, *Bracon cephi* (Gahan). *Nature (London)* **181,** 1281.

Salt, R. W. (1961). Principles of insect cold-hardiness. *Ann. Rev. Entomol.* **6,** 55–74.

Salt, R. W. (1962). Intracellular freezing in insects. *Nature (London)* **193,** 1207–1208.

Salt, R. W., and Kaku, S. (1967). Ice nucleation and propagation in spruce needles. *Can. J. Bot.* **45,** 1335–1346.

Saltykovskij, M. I., and Saprygina, E. S. (1935). The frost-resistance of winter cereals at different stages of development. *C. R. Acad. Sci. URSS* **4,** 99–103.

Samygin, G. A., and Matveeva, N. M. (1967). Protective effect of glycerine and other substances which easily penetrate protoplasts during the freezing of plant cells. *Fiziol. Rast.* **14,** 1048–1056.

Samygin, G. A., and Matveeva, N. M. (1969). Protective action of salt solutions during freezing of plant cells. *Fiziol. Rast.* **15,** 552–560.

Santarius, K. A. (1969). Der Einfluss von Elecktrolyten auf Chloroplasten beim Gefrieren und Trocknen. *Planta* **89,** 23-46.

Santarius, K. A. (1973). Freezing: The effect of eutectic crystallization on biological membranes. *Biochim. Biophys. Acta.* **291,** 38-50.

Saprygina, E. S. (1935). Frost resistance of spring wheats. (On the effect of length of the "light" stage on the hardiness of wheats). *C. R. Acad. Sci. URSS* **3,** 325-328.

Sarbhoy, A. K., Ghosh, S. K., Lal, S. P., and Lall, G. (1975). Investigation on the preservation of fungi by the lyophilization technique. *Indian Phytopathol.* **27,** 361-363.

Sarhan, F., and D'Aoust, M. J. (1975). RNA synthesis in spring and winter wheat during cold acclimation. *Physiol. Plant.* **35,** 62-65.

Sarkisova, M. M., and Chailakhyan, M. Kh. (1974). Effect of growth regulators on budding and frost resistance of the apricot. *Biol. Zh. Arm.* **27,** 3-9.

Savitskii, I. L. (1976). Photosynthetic activity in apple tree organs. *Fiziol. Biokhim. Kul't Rast.* **8,** 53-56.

Sawada, S. I., and Miyachi, S. (1974a). Effects of growth temperature on photosynthetic carbon metabolism in green plants: 1. Photosynthetic activities of various plants acclimatized to varied temperatures. *Plant Cell Physiol.* **15,** 111-120.

Sawada, S. I., and Miyachi, S. (1974b). Effects of growth temperature on photosynthetic carbon metabolism in green plants. II. Photosynthetic $^{14}CO_2$-incorporation in plants acclimatized to varied temperatures. *Plant Cell Physiol.* **15,** 225-238.

Sawada, S. I., Matsushima, H., and Miyachi, S. (1974). Effects of growth temperature on photosynthetic carbon metabolism in green plants: III. Differences in structure, photosynthetic activities and activities of ribulose diphosphate carboxylase and glycolate oxidase in leaves of wheat grown under varied temperatures. *Plant Cell Physiol.* **15,** 239-248.

Scarth, G. W. (1936). The yearly cycle in the physiology of trees. *Trans. Roy. Can. Soc. Sect. 5* **30,** 1-10.

Scarth, G. W. (1941). Dehydration injury and resistance. *Plant Physiol.* **16,** 171-179.

Scarth, G. W., and Levitt, J. (1937). The trost-hardening mechanism of plant cells. *Plant Physiol.* **12,** 51-78.

Scarth, G. W., and Lloyd, F. E. (1930). "Elementary Course in General Physiology." Wiley, New York.

Schacht, H. (1857). "Lehrbuch der Anatomie und Physiologie der Gewächse," pp. 525-529. Berlin.

Schaedle, M., and Bassham, J. A. (1977). Chloroplast glutathione reductase. *Plant Physiol.* **59,** 1011-1012.

Schaffnit, E. (1910). Studien über der Einfluss neiderer Temperaturen auf die pflanzliche Zell. *Mitt. Kaiser-Wilhelm Inst. Landwirtsch. Bromberg* **3,** 93-144.

Schander, R., and Schaffnit, E. (1919). Untersuchungen über das Auswintern des Getreides. *Landwirtsch. Jahrb.* **52,** 1-66.

Scheffer, F., and Lorenz, H. (1968). Pool-Aminosäuren während des Wachstums und der Entwicklung einiger Weizensorten. I. Pool-Aminosaüren in keimenden Samen, in Blättern mehrerer Entwicklungsstadien sowie im wachsenden und reifenden Korn. *Phytochemistry* **7,** 1279-1288.

Scheumann, W. (1968). Die Dynamik der Frostresistenz und ihre Bestimmung an Gehözen im Massentest. *In* "Klimaresistenz Photosynthese und Stoffproduktion." (H. Polster, ed.), pp. 45-54. Deut. Akad. Landwirtsch. Berlin.

Scheumann, W., and Börtitz, S. (1965). Studien zur Physiologie der Frosthärtung kei Koniferen. *Biol. Zentralbl.* **84,** 489-500.

Schlösser, L. (1936). Frosthärte und Polyploidie. *Zuechter* **8,** 75-80.

Schmalz, H. (1957). Untersuchungen über den Einfluss von photoperiodischer Induktion und Vernalisation auf die Winterfestigkeit von Winterweizen. *Z. Pflanzenzucht.* **38,** 147-180.

Schmalz, H. (1958). Die generative Entwicklung von Winterweizensorten mit unterschiedlicher Winterfestigkeit bei Fruhjahrsaussaat nach Vernalisation mit Temperaturen unter-und oberhalb des Gefrierpunktes. *Zuechter* **28,** 193-203.

Schmidt, A. (1977). Protein catalyzed isotopic exchange reaction between cysteine and sulfide in spinach leaves. *Z. Naturforsch.* **32c,** 219-225.

Schmuetz, W. (1962). Weitere Untersuchungen über die Beziehung zwischen Sulfhydryl-Gehalt und Winterfestigkeit von 15 Weizensorten. *Z. Acker Pflanzenbau* **115,** 1-11.

Schmuetz. W. (1969). Zu Fragen der physiologischen Resistenz bei Getreide. *Vortr. Pflanzenzuechter* **12,** 105-122.

Schmuetz, W., Sullivan, C. Y., and Levitt, J. (1961). Sulfhydryls—A new factor in frost resistance. II. Relation between sulfhydryls and relative resistance of fifteen wheat varieties. *Plant Physiol.* **36,** 617-620.

Schölm, H. E. (1968). Untersuchungen zur Hitze-und Frostresistenz einheimscher Susswasseralgen. *Protoplasma* **65,** 97-118.

Schönbohm, E. (1969). Inhibition of light-induced movement of Mougeotia chloroplasts by *p*-chloromercuribenzoate: Investigations on the mechanics of chloroplast movement. *Z. Pflanzenphysiol.* **61,** 250-260.

Scholander, P. F., Flagg, W., Hock, R. J., and Irving, L. (1953). Studies on the physiology of frozen plants and animals in the Arctic. *J. Cell Comp. Physiol.* **42,** 1-56.

Schoolar, A. I., and Edelman, J. (1970). Production and secretion of sucrose by sugarcane leaf tissue. *J. Exp. Bot.* **21,** 49-57.

Schübler, G. (1827). Beobachtungen über die Temperatur der Vegetabilien und einige damit verwandte Gegenstände. *Ann. Phys. Chem.* **10,** 581-592.

Schulz, E. D., Mooney, H. A., and Dunn, E. L. (1967). Wintertime photosynthesis of bristle-cone pine (*Pinus aristata*) in the White Mountains of California. *Ecology* **48,** 1044-1047.

Schumacher, E. (1875). II. Beiträge zur Morphologie und Biologie der Hefe. *Sitzber. Akad. Wiss. Wien Math. Naturwiss. Kl. Abt. I* **70,** 157-188.

Schwarz, W. (1968). Der Einfluss der Temperatur und Tageslänge auf die Frosthärte der Zirbe. *In* "Klimaresistenz Photosynthese und Stoffproduktion" (H. Polster, ed.), pp. 55-63. Deut. Akad. Landwirtsch. Berlin.

Seibert, M. (1976). Shoot initiation from carnation shoot apices frozen to −196°C. *Science.* **191,** 1178-1179.

Seibert, M., and Wetherbee, P. J. (1977). Increased survival and differentiation of frozen herbaceous plant organ cultures through cold treatment. *Plant Physiol.* **59,** 1043-1046.

Selwyn, M. J. (1966). Temperature and photosynthesis. II. A mechanism for the effects of temperature on carbon dioxide fixation. *Biochim. Biophys. Acta* **126,** 214-224.

Senebier, J. (1800). "Physiologie Végétale," Vol. 3, pp. 282-304. Paschoud, Genève.

Senn, G. (1922). Untersuchungen über die Physiologie der Alpenpflanzen. *Verhandl. Schweiz. Naturforsch. Ges.* **103,** 154-168.

Senser, M., Schoetz, F., and Beck, E. (1975). Seasonal changes in structure and function of spruce chloroplasts. *Planta* **126,** 1-10.

Sergeeva, K. A. (1968). The phosphorus metabolism of woody plants in relation to their winter hardiness. *Tr. Inst. Ekol. Rast. Zhivotn. Ural. Fil. Akad. Nauk. SSSR* **62,** 30-36.

Sergeev, L. I., Sergeeva, K. A., and Kandarova, I. V. (1959). The appearance of starch in the generative buds of woody plants in winter. *Byull. Glavnogo Bot. Sada Akad. Nauk. SSSR* **35,** 70-75.

Sestakov, V. E. (1936). Frost resistance of winter crops during the light stage. *C. R. Acad. Sci. URSS* **3,** 395-398.

Sestakov, V. E., and Sergeev, L. I. (1937). Changes in frost resistance and in the properties of cell protoplasm in winter wheat during the photo-stage. *Bot. Zh. SSR* **22,** 351-363.

Sestakov, V. E., and Smirnova, A. D. (1936). Temperature hardening and the differentiation of the embryonic spike in winter wheats during the light stage of development. *C. R. Acad. Sci. URSS* **3,** 399-403.

Sharashidze, N. M. (1972). Water regimen of some leaf-bearing evergreens in connection with their overwintering at the Batumi Botanical Gardens. *Izv. Batum Bot. Sad. Akad. Nauk. Gruz. SSR.* **17,** 56-70.

Sharon, N. (1977). Lectins. *Sci. Am.* **236,** 108-119.

Shaw, R. H. (1954). Leaf and air temperatures under freezing conditions. *Plant Physiol.* **29,** 102-104.

Shearman, L. L., Olien, C. R., Marchetti, B. L., and Everson, E. H. (1973). Characterization of freezing inhibitors from winter wheat cultivars. *Crop. Sci.* **13,** 514-519.

Sheppard, H., and Tsien, W. H. (1974). The effect of slow freezing and hypertonic NaCl on the hydrolytic activity of rat erythrocyte cyclic AMP phosphodiesterase and its sensitivity to the inhibitor, D, L-4-(3-butoxy-4-methoxybenzyl)-2-imidazolidinone. *Biochim. Biophys. Acta* **341,** 489-496.

Sherer, V. A., Marieva, G. M., Krasyuk, S. E., and Keleberda, M. I. (1972). Redox regime in leaves and shoots of grape varieties with different degrees of frost resistance. *Fiziol. Biokhim. Kul't Rast.* **4,** 499-503.

Sherman, J. K. (1962). Questionable protection by intracellular glycerol during freezing and thawing. *J. Cell. Comp. Physiol.* **61,** 67-83.

Sherman, J. K., and Liu, K. C. (1976). Relation of ice formation to ultrastructural cryoinjury and cryoprotection of rough endoplasmic reticulum. *Cryobiology,* **13,** 599-608.

Shih, S. C., Jung, G. A., and Shelton, D. C. (1967). Effects of temperature and photoperiod on metabolic changes in alfalfa in relation to cold hardiness. *Crop Sci.* **7,** 385-389.

Shikama, K. (1963). Denaturation of catalase and myosin by freezing and thawing. *Sci. Rep. Tohoku Univ. Ser. 4* **29,** 91-106.

Shimada, K., and Asahina, E. (1975). Visualization of intracellular ice crystals formed in very rapidly frozen cells at −27°C,. *Cryobiology* **12,** 209-218.

Shiomi, N., and Hori, S. (1972). Some characteristics of rapidly labeled nucleic acids in barley coleoptiles grown at cold temperature. *Soil Sci. Plant Nutr.* **18,** 93-96.

Shiroya, T., Lister, G. R., Slankis, V., Krotkov, G., and Nelson, C. D. (1966). Seasonal changes in respiration, photosynthesis and translocation of the ^{14}C labelled products of photosynthesis in young *Pinus strobus* L. plants. *Ann. Bot.* **30,** 81-90.

Shmatok, I. D. (1958). Seasonal dynamics of ascorbic acid in leaves of plants under polar conditions. *Fiziol. Rast.* **5,** 341-344.

Shmelev, I. K. (1935). Frost resistance of fruit trees. *Bull. Appl. Bot. Genet. Plant Breeding (Leningrad) Ser.* 3 No. 6, pp. 263-277.

Shmueli, E. (1953). Irrigation studies in the Jordan Valley. I. Physiological activity of the banana in relation to soil moisture. *Bull. Res. Counc. Isr.* **3,** 228-247.

Shmueli, E. (1960). Chilling and frost damage in banana leaves. *Bull. Res. Counc. Isr. Sect. Bot.* **8D,** 225-238.

Shomer-Ilan, A., and Waisel, Y. (1975). Cold hardiness of plants: Correlation with changes in electrophoretic mobility, composition of amino acids and average hydrophobicity of fraction-I-protein. *Physiol. Plant.* **34,** 90-96.

Shpota, V. I., and Bochkareva, E. V. (1974). Conditions for hardening and frost resistance of winter Cruciferae plants. *Fiziol. Rast.* **21,** 833-836.

Siegel, S. M., Speitel, T., and Stoecker, R. (1969). Life in earth extreme environments: a study of cryobiotic potentialities. *Cryobiology* **6,** 160-181.

Simatos, D., Faure, M., Bonjour, E., and Couach, M. (1975). The physical state of water at low temperatures in plasma with different water contents as studied by differential thermal analysis and differential scanning calorimetry. *Cryobiology* **12,** 202–208.

Siminovitch, D. (1963). Evidence from increase in ribonucleic acid and protein synthesis in autumn for increase in protoplasm during the frost-hardening of black locust bark cells. *Can. J. Bot.* **41,** 1301-1308.

Siminovitch, D., and Briggs, D. R. (1949). The chemistry of the living bark of the black locust tree in relation to frost hardiness. I. Seasonal variations in protein content. *Arch. Biochem. Biophys.* **23,** 8-17.

Siminovitch, D., and Briggs, D. R. (1953a). Studies on the chemistry of the living bark of the black locust in relation to its frost hardiness. III. The validity of plasmolysis and desiccation tests for determining the frost hardiness of bark tissue. *Plant Physiol.* **28,** 15-34.

Siminovitch, D., and Briggs, D. R. (1953b). Studies on the chemistry of the living bark of the black locust tree in relation to frost hardiness. IV. Effects of ringing on translocation, protein synthesis and the development of hardiness. *Plant Physiol.* **28,** 177–200.

Siminovitch, D., and Briggs, D. R. (1954). Studies on the chemistry of the living bark of the black locust in relation to its frost hardiness. VII. A possible direct effect of starch on the susceptibility of plants to freezing injury. *Plant Physiol.* **29,** 331-337.

Siminovitch, D., and Levitt, J. (1941). The relation between frost resistance and the physical state of protoplasm. II. *Can. J. Res.* **C19,** 9-20.

Siminovitch, D., and Scarth, G. W. (1938). A study of the mechanism of frost injury to plants. *Can. J. Res.* **C16,** 467–481.

Siminovitch, D., Wilson, C. M., and Briggs, D. R. (1953). Studies on the chemistry of the living bark of the black locust in relation to its frost hardiness. V. Seasonal transformation and variations in the carbohydrates: starch-sucrose interconversions. *Plant Physiol.* **29,** 383–400.

Siminovitch, D., Therrien, H., Wilner, J., and Gfeller, F. (1962). The release of amino acids and other ninhydrin-reacting substances from plant cells after injury by freezing; a sensitive criterion for the estimation of frost injury in plant tissues. *Can. J. Bot.* **40,** 1267-1269.

Siminovitch, D., Rhéaume, B., and Sachar, R. (1967a). Seasonal increase in protoplasm and metabolic capacity in the cells during adaptation to freezing. *In* "Molecular Mechanisms of Temperature Adaptation" (C. L. Prosser, ed.) pp. 3–40. Pub. No. 84. Amer. Assoc. Adv. Sci. Washington, D.C.

Siminovitch, D., Gfeller, F., and Rhéaume, B. (1967b). *In* "Cellular Injury and Resistance in Freezing Organisms" (E. Asahina, ed.), pp. 93–118. Inst. Low Temp. Sci. Hokkaido Univ., Sapporo, Japan.

Siminovitch, D., Rhéaume, B., Pomeroy, K., and Lepage, M. (1968). Phospholipid, protein, and nucleic acid increases in protoplasm and membrane structures associated with development of extreme freezing resistance in black locust tree cells. *Cryobiology* **5,** 202-225.

Singh, J., de la Roche, I. A., and Siminovitch, D. (1975). Membrane augmentation in freezing tolerance of plant cells. *Nature* **257,** 669–670.

Singh, J., de la Roche, I. A., and Siminovitch, D. (1977a). Differential scanning calorimeter analyses of membrane lipids isolated from hardened and unhardened black locust bark and from winter rye seedlings. *Cryobiology* **14,** 620–624.

Singh, J., de la Roche, I., and Siminovitch, D. (1977b). Relative insensitivity of mitochondria in hardened and nonhardened rye coleoptile cells to freezing *in situ*. *Plant Physiol.* **60,** 713-715.

Siminovitch, D., Singh, J., and de la Roche, I. A. (1975). Studies on membranes in plant cells resistant to extreme freezing. I. Augmentation of phospholipids and membrane substance without changes in unsaturation of fatty acids during hardening by black locust bark. *Cryobiology* **12,** 144–153.

Siminovitch, D., Singh, J., and de la Roche, I. A. (1978). Freezing behavior of free protoplasts of winter rye. *Cryobiology* **15,** 205–213.

Simon, E. W., and Wiebe, H. H. (1975). Leakage during inhibition, resistance to damage at low temperature and the water content of peas. *New Phytol.* **74,** 407–412.

Simione, Jr., F. P., and Daggett, P.-M. (1977). Recovery of a marine dinoflagellate following controlled and uncontrolled freezing. *Cryobiology* **14,** 362–366.

Skrabut, E. M., Crowley, J. P., Catsimpoolas, N., and Valeri, C. R. (1976). The effect of cryogenic storage on human erythrocyte membrane proteins as determined by polyacrylamide-gel electrophoresis. *Cryobiology* **13,** 395–403.

Slatyer, R. O. (1976). Water deficits in timberline trees in the Snowy Mountains of south-eastern Australia. *Oecologia* **24,** 357–366.

Slavnyi, P. S., and Musienko, N. N. (1972). Formation of wheat root system and its frost resistance in dependence on nutrition. *Fiziol. Biokhim. Kul't Rast.* **4,** 68–73.

Slosarek, M., Sourek, J., and Mikova, Z. (1976). Results of longterm preservation of mycobacteria by means of freeze-drying. *Cryobiology* **13,** 218–224.

Smirnova, I. S. (1959). The problem of the winter resistance of the root systems of apple and pear trees in a nursery. *Trudy Plodoovshch. Inst. I. V. Michurina* **10,** 43–50.

Smith, A. M. (1915). The respiration of partly dried plant organs. *Brit. Ass. Advan. Sci. Rep.*, p. 725.

Smith, A. P. (1974). Bud temperature in relation to nyctinastic leaf movement in an Andean giant rosette plant. *Biotropica,* **6,** 263–266.

Smith, A. U., Polge, C., and Smiles, J. (1951). Microscopic observation of living cells during freezing and thawing. *J. Roy. Microsc. Soc.* **71,** 186–195.

Smith, D. (1949). Differential survival of ladino and common white clover encased in ice. *Agron. J.* **41,** 230–234.

Smith, D. (1952). The survival of winter-hardened legumes encased in ice. *Agron. J.* **44,** 469–473.

Smith, D. (1968a). Carbohydrates in grasses: IV. Influence of temperature on the sugar and fructosan composition of Timothy (*Phleum pratense*) plant parts at anthesis. *Crop Sci.* **8,** 331–334.

Smith, D. (1968b). Varietal chemical differences associated with freezing resistance in forage plants. *Cryobiology* **5,** 148–159.

Smith, T. J. (1942). Responses of biennial sweet clover to moisture, temperature, and length of day. *J. Amer. Soc. Agron.* **34,** 865–876.

Smolenska, G., and Kuiper, P. J. C. (1977). Effect of low temperature upon lipid and fatty acid composition of roots and leaves of winter rape plants. *Physiol. Plant.* **41,** 29–35.

Smol'skaya, E. M. (1964). Changes with age and season in the chlorophyll content of the shoots of woody plants. *In* "The Effect of Soil Conditions on the Growth of Woody Plants," pp. 100–114. Nauka Tekhnika, Minsk. From *Ref. Zh. Biol.* 1964. 23G3 (Translation).

Snell, K. (1932). Die Beschleunigung der Keimung bei der Kartoffelknolle. *Ber. Deut. Bot. Ges.* **52A,** 146–161.

Snope, A. J., and Ellison, J. H. (1963). Freeze-drying improves preservation of pollen. *N. J. Agr.* **45,** 8–9.

Somero, G. N., and Hochachka, P. W. (1969). Isoenzymes and short-term temperature compensation in poikilotherms: Activation of lactate dehydrogenase isoenzymes by temperature decreases. *Nature (London)* **223,** 194–195.

Sorenson, F. C., and Ferrell, W. K. (1973). Photosynthesis and growth of Douglas-fir seedlings when grown in different environments. *Can. J. Bot.* **51,** 1689–1698.

Souzu, H. (1973). The phospholipid degradation and cellular death caused by freeze-thawing or freeze-drying of yeast. *Cryobiology* **10,** 427–431.

Sprague, M. A. (1955). The influence rate of cooling and winter cover on the winter survival of ladino clover and alfalfa. *Plant Physiol.* **30,** 447–451.

Sprague, V. G., and Graber, L. F. (1940). Physiological factors operative in ice sheet injury of alfalfa. *Plant Physiol.* **15,** 661–673.

Sprague, V. G., and Graber, L. F. (1943). Ice sheet injury to alfalfa. *J. Amer. Soc. Agron.* **35,** 881–894.

Srivastava, G. C., and Fowden, L. (1972). The effect of growth temperature on enzyme and amino-acid levels in wheat plants. *J. Exp. Bot.* **23,** 921–929.

Steemann-Nielsen, E., and Jorgensen, C. K. (1968). The adaptation of plankton algae. I. General part. *Physiol. Plant.* **21,** 401–413.

Steeman-Nielsen, E., Kamp-Nielsen, L., and Wuim-Andersen, S. (1969). The effect of deleterious concentrations of copper on the photosynthesis of *Chlorella pyrenoidosa. Physiol. Plant.* **22,** 1121–1133.

Stein, W. D. (1967). "The Movement of Molecules across Cell Membranes." Academic Press, New York.

Steiner, M. (1933). Zum Chemismus der osmotischen Jahresschwankungen einiger immergrüner Holzgewächse. *Jahrb. Wiss. Bot.* **78,** 564–622.

Steinhübel, G., and Halas, L. (1969). Seasonal trends in the rates of dry matter production in the evergreen and winter green broadleaf woody plants. *Photosynthetica* **3,** 244–254.

Stembridge, G. D., and Larue, J. H. (1969). The effect of potassium gibberellate on flower bud development in the Redskin peach. *J. Amer. Soc. Hort. Sci.* **94,** 492–495.

Steponkus, P. L. (1968). Cold acclimation of *Hedera helix*—a two-step process. *Cryobiology* **4,** 276–277.

Steponkus, P. L. (1969a). Protein-sugar interactions during cold acclimation. *Cryobiology* **6,** 285.

Steponkus, P. L. (1969b). Protein sugar interactions during cold acclimation. *Abstr. 11th Int. Bot. Congr. Seattle, Wash.,* p. 209.

Steponkus, P. L., and Lanphear, F. O. (1966). The role of light in cold acclimation of woody plants. *Proc. 17th Int. Hort. Congr.* **1,** 93.

Steponkus, P. L., and Lanphear, F. O. (1967a). Light stimulation of cold acclimation: Production of a translocatable promoter. *Plant Physiol.* **42,** 1673–1679.

Steponkus, P. L., and Lanphear, F. O. (1967b). Factors influencing artificial cold acclimation and artificial freezing of *Hedera Helix* "Thorndale." *Proc. Amer. Soc. Hort. Sci.* **91,** 735–641.

Steponkus, P. L., and Lanphear, F. O. (1968a). The role of light in cold acclimation of *Hedera Helix* var "Thorndale." *Plant Physiol.* **43,** 151–156.

Steponkus, P. L., and Lanphear, F. O. (1968b). The relationship of carbohydrates to cold acclimation of *Hedera helix* L. v. Thorndale. *Physiol. Plant.* **2,** 777–791.

Steponkus, P. L., and Wiest, S. C. (1973). Freezing injury of plant plasma membranes. *Cryobiology* **10,** 532.

Steponkus, P. L., Garber, M. P., Myers, S. P., and Lineberger, R. D. (1977). Effects of cold acclimation and freezing on structure and function of chloroplast thylakoids. *Cryobiology* **14,** 303–321.

Stergios, B. G., and Howell, G. S., Jr. (1973). Evaluation of viability tests for cold stressed plants. *J. Am. Soc. Hort. Sci.* **98,** 325–330.

Stewart, I., and Leonard, C. D. (1960). Increased winter hardiness in citrus from maleic hydrazide sprays. *Proc. Amer. Soc. Hort. Sci.* **75,** 253–256.

Stoller, E. W. (1977). Differential cold tolerance of quackgrass and Johnsongrass rhizomes. *Weed Sci.* **25,** 348-351.

Stoller, E. W., and Weber, E. J. (1975). Differential cold tolerance, starch, sugar, protein, and lipid of yellow and purple nutsedge tubers. *Plant Physiol.* **55,** 859-863.

Stout, D. G., Steponkus, P. L., and Cotts, R. M. (1973). Cold acclimation effects on the water diffusion permeability of plant cell membranes. *Cryobiology* **10,** 530.

Straib, W. (1946). Beiträge zur Frosthärte des Weizens. *Zuechter* **17,** 1-12.

Strömer, M. (1749). Gedanken über die Ursache warum die Bäume bei starkem Winter erfrieren, wobei die Möglichkeit solchem vorzubeugen erwiesen wird. *Der Kgl. Schwed. Akad. Wiss. Abhandl. Nat. Haushalt. Mech. Jahre 1739 und 1740 (Aus dem Schwed. übersetzt)* **1,** 116-121.

Stuber, C. W., and Levings, C. S., III (1969). Auxin induction and repression of peroxidase isozymes in oats (*Avena sativa* L). *Crop Sci.* **9,** 415-416.

Stuckey, I. H., and Curtis, O. F. (1938). Ice formation and the death of plant cells by freezing. *Plant Physiol.* **13,** 815-833.

Stushnoff, C., and Junttila, O. (1978). Resistance to low temperature injury in hydrated lettuce seed by supercooling. *In* "Plant Cold Hardiness and Freezing Stress" (P. H. Li and A. Sakai eds.), pp. 241-247. Academic Press, New York.

Sugawara, Y., and Sakai, A. (1974). Survival of suspension-cultured sycamore cells cooled to the temperature of liquid nitrogen. *Plant Physiol.* **54,** 722-724.

Sugawara, Y., and Sakai, A. (1976). Cold acclimation processes in cultured plant cells: I. Effect of 2, 4-D on the cold acclimation of callus-cultured tuber cells of Jerusalem artichoke, *Helianthus tuberosus* L. *Low Temp. Sci.* **34,** 1-8.

Sugawara, Y., and Sakai, A. (1978). Cold acclimation of callus cultures of Jerusalem artichoke. *In* "Plant Cold Hardiness and Freezing Stress" (P. H. Li and A. Sakai, eds.), pp. 197-210. Academic Press, New York.

Sugiyama, T., and Akazawa, T. (1967). Structure and function of chloroplast proteins. I. Subunit structure of wheat Fraction I protein. *J. Biochem. (Toyko)* **62,** 474-482.

Sugiyama, N., and Simura, T. (1966). Studies on varietal differentiation of the frost resistance in the tea plant. II. Artificial hardening and dehardening. *Jap. J. Breed.* **16,** 165-173.

Sugiyama, N., and Simura, T. (1967a). Studies on varietal differentiation of the frost resistance of the tea plant. III. With special emphasis on the relation between their frost resistance and chloroplast soluble protein. *Jap. J. Breed.* **17,** 37-42.

Sugiyama, N., and Simura, T. (1967b). Studies on the varietal differentiation of frost resistance of the tea plant. IV. The effects of sugar level combined with protein in chloroplasts on the frost resistance. *Jap. J. Breed.* **17,** 292-296.

Sugiyama, N., and Simura, T. (1968a). Studies on the varietal differentiation of frost resistance of the tea plant. V. The histochemical observations of the variation of glucose content in the stems of the tea plant during frost hardening. *Jap. J. Breed.* **18,** 37-41.

Sugiyama, N., and Simura, T. (1968b). Studies on the varietal differentiation of frost resistance of the tea plant. VI. The distribution of sugars in cells during frost hardening in the tea plant with special reference to histochemical observations by using ^{14}C sucrose. *Jap. J. Breed.* **18,** 27-32.

Sulakadze, T. S. (1961). Growth substances and frost resistance in plants. *Izv. Akad. Nauk. SSSR Ser. Biol.* **4,** 551-560.

Sulakadze, T. S., and Rapava, L. P. (1973). The significance of natural growth regulators to the wintering of winter-vegetative plants. *Soobshch. Akad. Nauk. Gruz. SSR.* **70,** 173-176.

Sun, C. N. (1958). The survival of excised pea seedlings after drying and freezing in liquid nitrogen. *Bot. Gaz. (Chicago)* **119,** 234-236.

Sundbom, E., and Bjorn, L. O. (1977). Phytoluminography: Imaging plants by delayed light emission. *Physiol. Plant.* **40,** 39–41.

Suneson, C. A., and Peltier, G. L. (1934). Cold resistance adjustments of field hardened winter wheats as determined by artificial freezing. *J. Amer. Soc. Agron.* **26,** 50–58.

Suneson, C. A., and Peltier, G. L. (1938). Effect of weather variants on field hardening of winter wheat. *J. Amer. Soc. Agron.* **30,** 769–778.

Sutherland, R. M., and Pihl, A. (1968). Repair of radiation damage to erythrocyte membranes: The reduction of radiation-induced disulfide groups. *Radiat. Res.* **34,** 300–314.

Svec, L. V., and Hodges, H. F. (1973). Respiratory activity in barley seedlings during cold hardening in controlled and natural environments. *Can. J. Plant Sci.* **53,** 457–463.

Swanson, C. R., and Adams, M. W. (1959). Some metabolic responses of alfalfa seedlings to freezing. *Plant Physiol.* **34,** 372–376.

Swartz, H. M. (1971). Effect of oxygen on freezing damage. II. Physical-chemical effects. *Cryobiology* **8,** 255–264.

Sycheva, Z. F., Balagurova, N. I., and Vasyukova, V. A. (1975). Effect of chlorocholine chloride on the respiration and frost resistance of potato plants. *Fiziol. Rast.* **22,** 176–180.

Sycheva, Z. F., Drozdov, S. N., Vasyukova, V. A., and Bystrova, Z. A. (1972). Metabolism of phosphoric compounds in potato leaves during frosts. *Biol. Nauki.* **15,** 83–87.

Takahama, U., and Nishimura, M. (1975). Formation of singlet molecular oxygen in illuminated chloroplasts. Effects on photoinactivation and lipid peroxidation. *Plant Cell Physiol.* **16,** 737–748.

Takahashi, T., and Asahina, E. (1974). Injury by extracellular freezing in the egg cells of sea urchins, a preliminary report. *Low Temp. Sci. Ser. B.* **32,** 9–18.

Takahashi, T., and Asahina, E. (1977). Protein-bound SH groups in frozen thawed egg cells of the sea urchin. *Cryobiology* **14,** 367–372.

Takehara, I., and Asahina, E. (1960a). Glycerol in the overwintering prepupa of slug moth, a preliminary note. *Low Temp. Sci. Ser. B* **18,** 51–56.

Takehara, I., and Asahina, E. (1960b). Frost resistance and glycerol content in overwintering insects. *Low Temp. Sci. Ser. B* **18,** 57–65.

Takehara, I., and Asahina, E. (1961). Glycerol in a slug caterpillar. I. Glycerol formation, diapause, and frost resistance in insect reared at various graded temperatures. *Low Temp. Sci. Ser. B* **19,** 29–36.

Terumoto, I. (1957a). The relation between the activity of phosphorylase and the frost hardiness in table beet. *Low Temp. Sci. Ser. B* **15,** 31–38.

Terumoto, I. (1957b). The frost hardiness of onion. *Low Temp. Sci. Ser. B* **15,** 39–44.

Terumoto, J. (1959). Frost resistance in plant cells after immersion in a solution of various inorganic salts. *Low Temp. Sci. Ser. B* **17,** 9–19.

Terumoto, I. (1960a). Ice masses in root tissue of table beet. *Low Temp. Sci. Ser. B* **18,** 39–42.

Terumoto, I. (1960b). Effect of protective agents against freezing injury in lake ball. *Low Temp. Sci. Ser. B* **18,** 43–50.

Terumoto, I. (1961). Frost resistance in a marine alga *Enteromorpha intestinalis* (L.) Link. *Low Temp. Sci. Ser. B* **19,** 23–28.

Terumoto, I. (1962). Frost resistance in a fresh alga, *Aegagropila sauteri* (Nees) Kütz. I. *Low Temp. Sci. Ser. B* **20,** 1–24.

Terumoto, I. (1967). Frost resistance in algae cells. *In* "Cellular Injury and Resistance in Freezing Organisms" (E. Asahina, ed.), pp. 191–210. Inst. Low Temp. Sci., University of Hokkaido, Sapporo, Japan.

Thiselton-Dyer, W. (1899). The influence of the temperature of liquid hydrogen on the germinative power of seeds. *Proc. Roy. Soc.* **65,** 361–368.

Thomas, D. A., and Barber, H. N. (1974). Studies on leaf characteristics of a cline of Eucalyptus urnigera from Mount Wellington, Tasmania: I. Water repellency and the freezing of leaves. *Aust. J. Bot.* **22,** 501-512.

Thomas, W. D., and Lazenby, A. (1968). Growth cabinet studies into cold-tolerance of *Festuca arundinacea* populations. II. Response to pretreatment conditioning and to number and deviation of low temperature periods. *J. Agr. Sci.* **30,** 347-353.

Thompson, K. F., and Taylor, J. P. (1968). Chemical composition and cold hardiness of the pith in marrow-stem Kale, *J. Agr. Sci.* **30,** 375-381.

Thomson, A. J. (1974). The effect of autumn management on winter damage and subsequent spring production of six varieties of Lolium perenne grown at Cambridge. *J. Br. Grassl. Soc.* **29,** 275-284.

Thomson, L. W., and Zalik, S. (1973). Lipids in rye seedlings in relation to vernalization. *Plant Physiol.* **52,** 268-273.

Thouin, A. (1806). Observations sur l'effet des gelées précoces qui ont eu lieu les 18, 19, 20 vendémiaire an XIV (11, 12, et 13 Octobre 1805). *Ann. Museum Hist. Nat. tome septieme,* pp. 85-114.

Thren, R. (1934). Jahreszeitliche Schwankungen des osmotischen Wertes verschiedener ökologischer Typen in der Umgebung von Heidelberg. Mit einen Beitrag zur Methodik der Pressaftuntersuchung. *Z. Bot.* **26,** 448-526.

Tiffney, W. N. Jr. (1972). Snow cover and the Diapensia lapponica habitat in the White Mountains, New Hampshire. *Rhodora* **74,** 358-377.

Till, O. (1956). Über die Frosthärte von Pflanzen sommergrüner Laubwälder. *Flora (Jena)* **143,** 499-542.

Timmis, R., and Worrall, J. (1974). Translocation of dehardening and bud-break promoters in climatically "split" Douglas-fir. *Can. J. For. Res.* **4,** 229-237.

Timmis, R., and Worrall, J. (1975). Environmental control of cold acclimation in Douglas fir during germination, active growth and rest. *Can. J. For. Res.* **5,** 464-477.

Timmis, R., and Tanaka, Y. (1976). Effects of container density and plant water stress on growth and cold hardiness of Douglas-fir seedlings. *For. Sci.* **22,** 167-172.

Timofejeva, M. (1935). Frost resistance of winter cereals in connection with phasic development and hardening of plants. *C. R. Acad. Sci. URSS* **1,** 64-67.

Todd, G. W., and Levitt, J. (1951). Bound water in *Aspergillus niger. Plant Physiol.* **26,** 331-336.

Toman, F. R., and Mitchell, H. L. (1968). Soluble proteins of winter wheat crown tissues, and their relationship to cold hardiness. *Phytochemistry* **7,** 365-373.

Tong, Min-Min, and Pincock, R. E. (1969). Denaturation and reactivity of invertase in frozen solutions. *Biochemistry* **8,** 908-913.

Tonzig, S. (1941). "I Muco-proteidi e la vita della cellula vegatale." Libreria Universitaria di G. Randi, Padova.

Torssell, B. (1959). Hardiness and survival of winter rape and winter turnip rape. *Växtodling* **15,** 1-168.

Torssell, B., and Hellstrom, N. (1955). Investigations on oil turnips and oil rape. IV. Estimation of plant status. *Acta Agr. Scand.* **5,** 31-38.

Towill, L., and Mazur, P. (1975). Studies on the reduction of 2, 3, 5-triphenyltretrazolium chloride as a viability assay for plant tissue cultures. *Can. J. Bot.* **53,** 1097-1102.

Towill, L. E., and Mazur, P. (1976). Osmotic shrinkage as a factor in freezing injury in plant tissue cultures. *Plant Physiol.* **57,** 290-296.

Towill, L. E., Mazur, P., and Howland, G. P. (1973). Plasmolysis injury as a factor in freezing injury in plant tissue culture. *Cryobiology,* **10,** 532.

Tranquillini, W. (1957). Standortsklima, Wasserbilanz und CO_2 Gaswechsel junger Zirben (*Pinus cembra* L.) an der alpinen Waldgrenze. *Planta* **49,** 612-661.

Tranquillini, W. (1958). Die Frosthärte der Zirbe unter besonderer Berücksichtigung autochthoner und aus Forstgärten stammender Jungpflanzen. *Forstwiss. Zentralbl.* **77,** 65-128.

Tranquillini, W. (1974). Influence of altitude and duration of the vegetation period on the cuticular transpiration capacity of spruce seedlings in winter. *Ber. Dtsch. Bot. Ges.* **87,** 175-184.

Travers, F., Douzou, P., Pederson, T., and Gunsalus, I. (1975). Ternary solvents to investigate proteins at subzero temperatures. *Biochimie.* **57,** 43-48.

Tremolières, A., Jacques, R., and Mazliak, P. (1973). Regulation by light of linolenic acid production in young pea leaves: Action spectrum, influence of light intensity, role of the phytochrome. *Physiol. Veg.* **11,** 239-251.

Trunova, T. I. (1963). The significance of different forms of sugars in increasing the frost resistance of the coleoptiles of winter wheat. *Fiziol. Rast.* **10,** 495-499.

Trunova, T. I. (1968). Changes in the content of acid soluble phosphorus-containing substances during hardening of winter wheat. *Fiziol. Rast.* **15,** 103-109.

Trunova, T. I. (1969). Effect of 2, 4-dinitrophenol on the frost resistance of Ulyanovka variety winter wheat. *Fiziol. Rast.* **16,** 237-240.

Trunova, T. I. (1975). Physiological properties of winter cereals hardened against frosts: A review. *S-Kh. Biol.* **10,** 694-702.

Trunova, T. I., and Zvereva, G. N. (1974). Protective effect of sugars during freezing of chloroplast suspensions. *Fiziol. Rast.* **21,** 1000-1006.

Tsaregorodtseva, S. O., and Novitskaya, Yu. E. (1973). The state of pigments in the buds of conifers during the winter-spring period. *Fiziol. Rast.* **20,** 1052-1056.

Tsuru, S. (1973). Preservation of marine and fresh water algae by means of freezing and freeze-drying. *Cryobiology* **10,** 445-452.

Tumanov, I. I. (1930). Welken und Dürreresistenz. *Arch. Pflanzenbau* **3,** 389-419.

Tumanov, I. I. (1931). Das abhärten winterannuèller Pflanzen gegen niedrige Temperaturen. *Phytopathol. Z.* **3,** 303-334.

Tumanov, I. I. (1967). Physiological mechanisms of frost resistance of plants. *Fiziol. Rast.* **14,** 520-539.

Tumanov, I. I. (1969). Physiology of plants not killed by frost. *Izv. Akad. Nauk. SSSR Ser. Biol.* **4,** 469-480.

Tumanov, I. I., and Borodin, I. N. (1930). Untersuchungen über die Kälteresistenz von Winterkulturen durch direktes Gefrieren und indirekt Methoden. *Phytopathol. Z.* **1,** 575-604.

Tumanov, I. I., and Khvalin, N. N. (1967). Causes of weak frost resistance of fruit tree roots. *Fiziol. Rast.* **14,** 908-918.

Tumanov, I. I., and Trunova, T. I. (1957). Hardening tissues of winter plants with sugar absorbed from the external solution. *Fiziol. Rast.* **4,** 379-388.

Tumanov, I. I., and Trunova, T. I. (1958). The effect of growth processes on the capacity for hardening. *Fiziol. Rast.* **5,** 108-117.

Tumanov, I. I., and Trunova, T. I. (1963). First phase of frost hardening of winter plants in the dark on sugar solutions. *Fiziol. Rast.* **10,** 140-149.

Tumanov, I. I., Krasavtsev, O, A., and Khvalin, N. N. (1959). Promotion of resistance of birch and black currant to $-253°$ temperatures by frost hardening. *C. R. Acad. Sci. URSS* **127,** 1301-1303.

Tumanov, I. I., Krasavtsev, O. A., and Trunova, T. I. (1969). Study of ice formation in plants by measurement of heat evolution. *Fiziol. Rast.* **16,** 907-916.

Tumanov, I. I., Khvalin, N. N. Ogolevets, I. V., and Smetyuk, V. V. (1976a). Hardening of winter wheat breeding material, storage of such material in winter, and culling of forms lacking frost resistance from it by means of freezing. *Fiziol. Rast.* **23,** 348-354.

Tumanov, I. I., Khvalin, N. N., Ogolevets, I. V., and Smetyuk, V. V. (1976b). Hardening and storage in winter selection material of winter wheat and freezing of frost-susceptible forms from it. *Fiziol. Rast.* **23,** 405–412.

Tumanov, I. I., Kuzina, G. V., and Karnikova, L. D. (1972). Effect of the duration of vegetation in trees on the accumulation of reserve carbohydrates and the character of the photoperiodic reaction. *Fiziol. Rast.* **19,** 1122–1131.

Tumanov, I. I., Kuzina, G. V., and Karnikova, L. D. (1973a). Dormancy period and the ability of trees to harden at low temperatures. *Fiziol. Rast.* **20,** 5–16.

Tumanov, I. I., Kuzina, G. V., and Karnikova, L. D. (1973b). Growth regulators, vegetation length and first phase of hardening in frost-resistant arboreal plants. *Fiziol. Rast.* **20,** 1158–1169.

Tumanov, I. I., Kuzina, G. V., and Karnikova, L. D. (1974). Increasing the ability of arboreal plants for frost hardening by treatment with extracts of natural growth inhibitors. *Fiziol. Rast.* **21,** 380–390.

Tumanov, I. I., Trunova, T. I., Smirnova, N. A., and Zvereva, G. N. (1976c). Role of light in the development of frost-resistance in plants. *Fiziol. Rast.* **23,** 132–138.

Turner, N. C., and Jarvis, P. G. (1975). Photosynthesis in Sitka spruce (*Picea sitchensis* Bong). Carr: IV. Response to soil temperature. *J. Appl. Ecol.* **12,** 561–576.

Tysdal, H. M. (1933). Influence of light, temperature, and soil moisture on the hardening process in alfalfa. *J. Agr. Res.* **46,** 483–515.

Tysdal, H. M. (1934). Determination of hardiness in alfalfa varieties by their enzymatic responses. *J. Agr. Res.* **48,** 219–240.

Tysdal, H. M., and Pieters, A. J. (1934). Cold resistance of three species of lespedeza compared to that of alfalfa, red clover, and crown vetch. *J. Amer. Soc. Agron.* **26,** 923–928.

Turner, N. C. (1976). Use of the pressure chamber in membrane damage studies. *J. Exp. Bot.* **27,** 1085–1092.

Uemitsu, N., Ohashi, H., and Matsumiya, H. (1975). Nuclear magnetic resonance investigation of the state of water in protein solution. *J. Biochem.* **78,** 229–234.

Ullrich, H., and Heber, U. (1957). Uber die Schutzwirkung der Zucker bei der Frostresistenz von Winterweizen. *Planta* **48,** 724–728.

Ullrich, H., and Heber, U. (1958). Über das Denaturieren pflanzlicher Eiweisse durch Ausfrieren und seine Verbinderung. Ein Beitrag zur Klarung der Frostresistenz bei Pflanzen. *Planta* **51,** 399–413.

Ullrich, H., and Heber, U. (1961). Frostresistenz bei Winterweizen IV. Das Verhalten von Fermenten und Fermentsystemen gegenüber tiefen Temperaturen. *Planta* **57,** 370–390.

Ulmer, W. (1937). Über den Jahresgang der Frosthärte einiger immergruner Arten der alpinen Stufe, sowie der Zirbe und Fichte. Unter Berucksichtigung von osmotischen Wert, Zuckerspiegel und Wassergehalt. *Jahrb. Wiss. Bot.* **84,** 553–592.

Ungerson, J., and Scherdin, G. (1965). Untersuchungen über Photosynthese und Atmung unter naturlichen Bedingungen während des Winterhalbjahres bei *Pinus sylvestris.* L. *Picea excelsa* Link und *Juniperus communis* L. *Planta* **67,** 136–137.

Ungerson, J., and Scherdin, G. (1967). Jahresgang von Photosynthese und Atmung unter naturlichen Bedingungen bei *Pinus silvestris.* L. an ihrer Nordgrenze in der Subarktis. *Flora (Jena)* **157,** 391–434.

Uribe, E. G., and Jagendorf, A. T. (1968). Membrane permeability and internal volume as factors in ATP synthesis by spinach chloroplasts. *Arch. Biochem. Biophys.* **128,** 351–359.

Vajsablova, A., and Benko, P. (1971). Proline content in fruit trees on sandy soils of Zahorie (Western Slovakia). *Ved. Pr. Vysk. Ustav. Rastl. Innej. Vyroby. Piestanoch.* **9,** 233–238.

Van den Driessche, R. (1969b). Influence of moisture supply, temperature, and light on frost hardiness changes in Douglas Fir seedlings. *Can. J. Bot.* **47,** 1765–1772.

Van den Driessche, R. (1973). Prediction of frost hardiness in Douglas fir seedlings by measuring electrical impedance in stems at different frequencies. *Can. J. For. Res.* **3,** 256-264.

Van Fleet, D. S. (1954). The significance of the histochemical localization of quinones in the differentiation of plant tissues. *Phytomorphology* **4,** 300-310.

Van Huystee, R. B., Weiser, C. J., and Li, P. H. (1967). Cold acclimation in *Cornus stolonifera* under natural and controlled photoperiod and temperature. *Bot. Gaz. (Chicago)* **128,** 200-205.

Varenitsa, E. T., Zimina, T. K., and Zakharov, A. I. (1974). Increase of the depth of the tillering node and increase of the winter hardiness of F_1 winter wheat hybrids under the effect of chlorocholine chloride. *Dokl. Vses. (Ordena Lenina) Akad. S-Kh. Nauk. Im. V. I. Lenina* **2,** 2-3.

Varenitsa, E. T., Zimina T. K., and Zakharov, A. I. (1975). Effect of presowing treatment of seeds with chlorocholine chloride in winter hardiness of winter wheat. *S-Kh. Biol.* **10,** 304-306.

Vassiliev, I. M. (1939). Winter wheats as lagging behind the spring varieties in growth intensity when subjected to low temperatures. *C. R. Acad. Sci. URSS* **24,** 85-87.

Vasil'eva, I. M., Ishmukhametova, N. N., and Khisamutdinova, V. I. (1974). Effect of chlorocholine chloride on the state of water in winter wheat organs. *Fiziol. Biokhim. Kul't Rast.* **6,** 484-487.

Vasil'eva, I. M., and Rafikova, F. M., Khisamutdinova, V. I., Estrina, R. I., Smol'yaninov, S. N., and Galiev, N. A. (1973). The effect of chlorocholinechloride on winter hardiness and yield of winter wheat. *S-Kh. Biol.* **8,** 532-537.

Venkatesh, C. G., Rice, S. A., and Narten, A. H. (1974). Amorphous solid water: an x-ray diffraction study. *Science.* **186,** 927-928.

Vetuhova, A. (1936). Winter hardiness of winter wheat during winter in relation to phasic development of plants. *Zbirnik Prac. Agrofiziol.* **2,** 83-102. (*Herb. Abstr.* **8,** 1861, 1938).

Vetuhova, A. (1938). On the internal factors of resistance to frost in winter plants. *Z. Inst. Bot. Akad. Nauk. URSS* No. 18-19 (26-27): 57-59. (*Herb. Abstr.* **9,** 633, 1939).

Vetuhova, A. (1939). Colloidal changes in plants of winter wheat in relation to the dynamics of frost resistance. *Kolloid-Z.* **4,** 511-521. (*Herb. Abstr.* **10,** 173. 1940.)

Voinikov, V. K., Fedotova, V. D., and Usova, T. K. (1974). Study of the electrical resistance of winter-crop seedling tissue affected by low temperatures. *Izv. Sib. Otd. Akad. Nauk. SSSR. Ser. Biol. Nauk.* **3,** 103-107.

Volger, H. G., and Heber, U. (1975). Cryoprotective leaf proteins. *Biochim. Biophys. Acta.* **412,** 335-349.

Volger, H., Heber, U., and Berzborn, R. J. (1978). Loss of function of biomembranes and solubilization of membrane proteins during freezing. *Biochim. Biophys. Acta.* **511,** 455-469.

Volkova, R. I., Prusakova, L. D., Drozdov, S. N., and Ivanova, R. P. (1974). Effect of chlorocholine chloride on growth, tuber formation and frost resistance of potato plants. *Fiziol. Rast.* **21,** 1287-1292.

Waisel, Y., Borger, G. A., and Kozlowski, T. T. (1969). Effects of phenylmercuric acetate on stomatal movement and transpirations of excised *Betula papyrifera* marsh leaves. *Plant Physiol.* **44,** 685-690.

Waisel, Y., Kohn, H., and Levitt, J. (1962). Sulfhydryls-a new factor in frost resistance. IV. Relation of GSH-oxidizing activity to flower induction and hardiness. *Plant Physiol.* **37,** 272-276.

Waldman, M., Rikin, A., Dovrat, A., and Richmond, A. E. (1975). Hormonal regulators of morphogenesis and cold-resistance. II. Effect of cold-acclimation and of exogenous abscisic acid on gibberellic acid and abscisic acid activities in alfalfa *Medicago sativa* L. seedlings. *J. Exp. Bot.* **26,** 853-859.

Walter, C. A., Knight, S. C., and Farrant, J. (1975). Ultrastructural appearance of freeze-substituted lymphocytes frozen by interrupting rapid cooling with a period at −26°C. *Cryobiology,* **12,** 103-109.
Wlater, H. (1929). Die osmotischen Werte und die Kälteschaden unserer wintergrünen Pflanzen während der Winterperiode 1929. *Ber. Deut. Bot. Ges.* **47,** 338-348.
Walter, H. (1931). "Hydratur der Pflanze und ihre physiologischökologische Bedeutung." Fischer, Jena.
Walter, H. (1949). Über die Assimilation und Atmung der Pflanzen in Winter bei tiefen Temperaturen. *Ber. Deut. Bot. Ges.* **62,** 47-50.
Walter, H., and Weismann, O. (1935). Über die Gefrierpunkte und osmotischen Werte lebender und tote pflanzlicher Gewebe. *Jahrb. Wiss. Bot.* **82,** 273-310.
Walton, P. D. (1973a). Factors affecting the measurement of electrical impedance of alfalfa tissue. *Can. J. Plant Sci.* **53,** 119-123.
Walton, P. D. (1973b). Electrical impedance of the tissue of grasses receiving cold treatments. *Can. J. Plant Sci.* **53,** 125-127.
Walton, P. D. (1974). A quantitative evaluation of one aspect of frost hardiness in alfalfa. *Can. J. Plant Sci.* **54,** 343-348.
Walton, P. D. (1975). Methods for evaluating frost hardiness in alfalfa. *Can. J. Plant Sci.* **55,** 823-826.
Wasserman, A. R., and Fleischer, S. (1968). The stabilization of spinach chloroplast function. *Biochim. Biophys. Acta* **153,** 154-169.
Weaver, G. M., and Jackson, H. O. (1969). Assessment of winter hardiness in peach by a liquid nitrogen system. *Can. J. Plant Sci.* **49,** 459-463.
Weaver, G. M., Jackson, H. O., and Stroud, F. D. (1968). Assessment of winter hardiness in peach cultivars by electric impedance, scion diameter and artificial freezing studies. *Can. J. Plant Sci.* **48,** 37-47.
Webb, J. L. (1966). "Enzyme and Metabolic Inhibitors," Vol. III, p. 892. Academic Press. New York.
Weber, F. (1909). Untersuchungen über die Wandlungen des Starke-und Fettgehaltes der Pflanzen, insbesondere der Bäume. *Sitzber. Akad. Wiss. Wien Math. Naturwiss. Kl. Abt. I* **118,** 967-1031.
Weidner, M., and Salisbury, F. B. (1974). The temperature characteristics of ribulose-1, 5-diphosphate carboxylase, nitrate reductase, and pyruvate kinase from seedlings of two spring wheat varieties. *Z. Pflanzenphysiol.* **71,** 398-412.
Weimer, J. L. (1929). Some factors involved in the winterkilling of alfalfa. *J. Agr. Res.* **39,** 263-283.
Weinmann, R., and Kreeb, K. (1975). CO_2-exchange of sclerophylls in the northern Lake Garda district. *Ber. Dtsch. Bot. Ges.* **88,** 205-210.
Weise, G., and Polster, H. (1962). Investigations on the influence of cold hardiness on the physiological activity of forest growth. II. Metabolic-physiological investigation on the question of frost resistance of spruce and Douglas fir origin. *Biol. Zentralbl.* **81,** 129-243.
Weiser, C. J. (1970). Cold resistance and injury in woody plants. *Science* **169,** 1269-1278.
West, F. L., and Edlefsen, N. E. (1917). The freezing of fruit buds. *Utah Agr. Expt. Sta. Bull.* **151,** 1-24.
West. F. L., and Edlefsen, N. E. (1921). Freezing of fruit buds. *J. Agr. Res.* **20,** 655-662.
Whiteman, T. M. (1957). Freezing points of fruits, vegetables, and florist stocks. *U. S. Dept. Agr. Marketing Res. Rep.,* No. 196.
Whittam, J. H., and Rosano, H. L. (1973). Effects of the freeze-thaw process on α amylase. *Cryobiology* **10,** 240-243.

Wiegand, K. M. (1906a). Some studies regarding the biology of buds and twigs in winter. *Bot. Gaz. (Chicago)* **41,** 373–424.
Wiegand, K. M. (1906b). The occurrence of ice in plant tissue. *Plant World* **9,** 25–39.
Wilding, M. D., Stahmann, M. A., and Smith, D. (1960a). Free amino acids in alfalfa as related to cold hardiness. *Plant Physiol.* **35,** 726–732.
Wilding, M. D., Stahmann, M. A., and Smith, D. (1960b). Free amino acids in red clover as related to flowering and winter survival. *Plant Physiol.* **35,** 733–735.
Wildner, G. F., and Henkel, J. (1977). Temperature dependent conformation changes of ribulose-1, 5-bisphosphate carboxylase studied by the use of 1-aniline-8-naphthalene sulfonate. *Z. Naturforsch.* **32c,** 226–228.
Wildung, D. K., Weiser, C. J., and Pellett, H. M. (1973). Temperature and moisture effects on hardening of apple roots. *Hortscience.* **8,** 53–55.
Willemot, C. (1975). Stimulation of phospholipid biosynthesis during frost hardening of winter wheat. *Plant Physiol.* **55,** 356–359.
Willemot, C. (1977). Simultaneous inhibition of linolenic acid synthesis in winter wheat roots and frost hardening by BASF 13–338, a derivative of pyridazinone. *Plant Physiol.* **60,** 1–4.
Willemot, C., Hope, H. J., Williams, R. J., and Michaud, R. (1977). Changes in fatty acid composition of winter wheat during frost hardening. *Cryobiology* **14,** 87–93.
Williams, B. J., Pellett, N. E., and Klein, R. M. (1972). Phytochrome control of growth cessation and initiation of cold acclimation in selected woody plants. *Plant Physiol.* **50,** 262–265.
Williams, J. M., and Williams, R. J. (1976). Osmotic factors of dehardening in *Cornus florida L. Plant Physiol.* **58,** 243–247.
Williams, R. J. (1973). Osmotic properties of glycopeptides from hardy *Cornus spp. Cryobiology* **10,** 530.
Williams, R. J., and Harris, D. (1977). The distribution of cryoprotective agents into lipid interfaces. *Cryobiology* **14,** 670–680.
Williams, R. J., and Meryman, H. T. (1970). Freezing injury and resistance in spinach chloroplast grana. *Plant Physiol.* **45,** 752–755.
Williams, R. J., and Ramasastry, P. (1973). A possible role of cell membranes in cryoprotection of *Cornus* spp. *Cryobiology* **10,** 531.
Wilner, J. (1952). A study of desiccation in relation to winter injury. *Sci. Agr.* **32,** 651–658.
Wilner, J. (1955). Results of laboratory tests for winter hardiness of woody plants by electrolytic methods. *Proc. Amer. Soc. Hort. Sci.* **66,** 93–99.
Wilner, J. (1960). Relative and absolute electrolytic conductance tests for frost hardiness of apple varieties. *Can. J. Plant Sci.* **40,** 630–637.
Wilner, J., and Brach, E. J. (1974). Hardiness of roots in relation to shoot of container-grown plants by an electric method. *Can. J. Plant Sci.* **54,** 281–289.
Wilson, R. F., and Rinne, R. W. (1976). Effect of freezing and cold storage on phospholipids in developing soybean cotyledons. *Plant Physiol.* **57,** 270–273.
Winkler, A. (1913). Über den Einfluss der Aussenbedingungen auf die Kälteresistenz ausdauernder Gewächse. *Jahrb. Wiss. Bot.* **52,** 467–506.
Wirth, H., Daniel, E. E., and Carroll, P. M. (1970). The effect of adenosine deaminase inhibitor and 2-mercaptoethanol on frozen-thawed uterine horns. *Cryobiology,* **6,** 395–400.
Withers, L. A. (1978). The freeze-preservation of synchronously dividing cultured cells of *Acer pseudoplatanus* L. *Cryobiology* **15,** 87–92.
Wolosiuk, R. A., and Buchanan, B. B. (1977). Thioredoxin and glutathione regulate photosynthesis in chloroplasts. *Nature,* **266,** 565–567.
Wood, H., and Rosenberg, A. M. (1957). Freezing in yeast cells. *Biochim. Biophys. Acta* **25,** 78–87.

Woodroof, J. G. (1938). Microscopic studies of frozen fruits and vegetables. *Ga. Agr. Exp. Sta.*, No. 201.

Worzella, W. W. (1932). Root development in hardy and non-hardy winter wheat varieties. *J. Amer. Soc. Agron.* **24,** 626-637.

Wright, R. C. (1932). Some physiological studies of potatoes in storage. *J. Agr. Res.* **45,** 543-555.

Yablonskii, E. A. (1975). Dynamics of phosphorus-containing substances and winterhardiness of stone fruit crops. *Fiziol. Rast.* **22,** 1007-1012.

Yamasato, K., Okuno, D., and Ohtomo, T. (1973). Preservation of bacteria by freezing at moderately low temperatures. *Cryobiology,* **10,** 453-463.

Yariv, J., Kalb, A. J., Katchalski, E., Goldman, R., and Thomas, E. W. (1969). Two locations of the lac permease sulphydryl in the membrane of *E. coli. FEBS Lett.* **5,** 173-176.

Yasnikova, E. A. (1975). Pigments of hibernating plant buds and their participation of photochemical reactions. *Fiziol. Biokhim. Kul't Rast.* **7,** 603-606.

Yasmykova, O. O., and Tolmachov, I. M. (1967). The conversion of stored substances in fruit and nut trees during the fall and winter. *Ref. Zh. Biol.* No. 9G139.

Yelenosky, G. (1975a). Cold hardening in citrus stems. *Plant Physiol.* **56,** 540-543.

Yelenosky, G. (1975b). Tetrazolium reduction in citrus cold hardening. *Hortscience,* **10,** 384.

Yelenosky, G. (1978). Freeze survival of citrus trees in Florida. *In* "Plant Cold Hardiness and Freezing Stress" (P. H. Li and A. Sakai, eds.), pp. 297–312. Academic Press, New York.

Yelenosky, G., and Gilbert, W. (1974). Levels of hydroxyproline in citrus leaves. *Hortscience,* **9,** 375-376.

Yelenosky, G., and Guy, C. L. (1977). Carbohydrate accumulation in leaves and stems of "Valencia" orange at progressively colder temperatures. *Bot. Gaz.* **138,** 13-17.

Yoshida, S. (1969a). Studies on the freezing resistance in plants. I. Seasonal changes of glycolipids in the bark tissue of black locust tree. *Low Temp. Sci. Ser.* **B27,** 109-117.

Yoshida, S. (1969b). Studies on the freezing resistance in plants. II. Seasonal changes in phospholipids in the bark tissues of the black locust tree. *Low Temp. Sci. Ser.* **B27,** 119-124.

Yoshida, S. (1971). Lipid metabolism in developmental stages of poplar trees. *Low Temp. Sci. Ser. B. Biol. Sci.* **29,** 53-63.

Yoshida, S. (1973a). Seasonal changes in lipids and freezing resistance in poplar trees. *Low Temp. Sci. Ser. B. Biol. Sci.* **31,** 9-20.

Yoshida, S. (1973b). The intracellular localization of phospholipids and their behaviour upon hardening and dehardening in the cortical tissues of woody plants. *Low Temp. Sci. Ser. B. Biol. Sci.* **31,** 21-30.

Yoshida, S. (1978). Phospholipid degradation and its control during freezing of plant cells. *In* "Plant Cold Hardiness and Freezing Stress" (P. H. Li and A. Sakai, eds.), pp. 117–135. Academic Press, New York.

Yoshida, S., and Sakai, A. (1967). The frost hardening process of woody plants. XII. Relation between frost resistance and various substances in stem bark of black locust trees. *Low Temp. Sci. Ser. Biol. Sci. B* **25,** 29–44.

Yoshida, S., and Sakai, A. (1968). The effect of thawing rate on freezing injury in plants. II. The change in the amount of ice in leaves as produced by the change in temperature. *Low Temp. Sci. Ser. Biol. Sci. B* **26,** 23–31.

Yoshida, S., and Sakai, A. (1973). Phospholipid changes associated with the cold hardiness of cortical cells from poplar stem. *Plant Cell Physiol.* **14,** 353-359.

Yoshida, S., and Sakai, A. (1974). Phospholipid degradation in frozen plant cells associated with freezing injury. *Plant Physiol.* **53,** 509-511.

Young, R. (1969a). Cold hardening in "Redblush" grapefruit as related to sugars and water soluble proteins. *J. Amer. Soc. Hort. Sci.* **94,** 252-254.

Young, R. (1969b). Cold hardening in citrus seedlings as related to artificial hardening conditions. *J. Amer. Soc. Hort. Sci.* **94,** 612-614.

Young, R., and Yelenosky, G. (1973). Effect of freezing on cell ultrastructure in Citrus tissues. *Cryobiology* **10,** 531-532.

Younis, M. E. (1969). Studies in the respiratory and carbohydrate metabolism of plant tissues. XXVII. The effect of iodoacetate on the utilization of (^{14}C) glucose in strawberry leaves and on permeability barriers. *New Phytol.* **68,** 1059-1067.

Yu, N.-T., and Jo, B. H. (1973). Comparison of protein structure in crystals and in solution by laser Raman scattering: I. Lysozyme. *Arch. Biochem. Biophys.* **156,** 469-474.

Zacharowa, T. M. (1926). Über den Einfluss neidriger Temperaturen auf die Pflanzen. *Jahrb. Wiss. Bot.* **65,** 61-87.

Zalasky, H. (1975a). Chimeras, hyperplasia, and hypoplasia in frost burls induced by low temperature. *Can. J. Bot.* **53,** 1888-1898.

Zalasky, H. (1975b). Low-temperature-induced cankers and buds in test conifers and hardwoods. *Can. J. Bot.* **53,** 2526-2535.

Zech, A. C., and Pauli, A. W. (1960). Cold resistance in three varieties of winter wheat as related to nitrogen fractions and total sugar. *Agron. J.* **52,** 334-337.

Zeller, O. (1951). Über Assimilation und Atmung der Pflanzen im Winter bei tiefen Temperaturen. *Planta* **39,** 500-526.

Zhuravlev, Y. N., and Popova, L. I. (1968). Early spring photosynthesis in cryophilic plants. *Ref. Zh. Biol.* No. 5G136.

Ziganigirov, A. M. (1968). The morphological periodicity and winter hardiness of the dog rose. *Tr. Inst. Ekol. Rast. Zhivotn. Ural. Fil. Akad. Nauk. SSSR* **62,** 83-88.

Zimmer, G. (1970). Isolation and characterization of *in vitro* radioactively labelled SH-proteins from rat liver mitochondrial membranes. *FEBS Lett.* **9,** 108-112.

Zotochknia, T. V. (1962). A study of the redox processes in pears in relation to their winter-hardiness. *Tr. Tsentr. Genet. Lab. Im. I. V. Michurina* **8,** 160-170. *Ref. Zh. Biol.* 1963, No. 15G64.

IV

HIGH-TEMPERATURE STRESS

11. High-Temperature or Heat Stress

A. QUANTITATIVE EVALUATION OF STRESS

As in the case of chilling stress, a high-temperature stress is difficult to evaluate on an absolute basis. It is, nevertheless, possible to classify organisms on the basis of their response to the temperature stress (Table 11.1). (1) Psychrophiles (lovers of cold) grow and develop in a temperature range that includes chilling temperatures (0°-20°C). Any temperature above 15° to 20°C may be a heat stress for them. The term has been used mainly for microorganisms. Algae belonging to this group may actually grow on snow, and, therefore, have been called cryobionts or cryosestonic algae (Hindak and Komarek, 1968). (2) Mesophiles (lovers of middle temperatures), grow and develop at temperatures of about 10° to 30°C. Any temperatures above this range may be a heat stress for them. (3) Thermophiles (heat lovers) may grow and develop at temperatures between 30° and 100°C. Only temperatures above 45°C (moderate thermophiles) or much higher (extreme thermophiles) are heat stresses for them. Thus a quantitative evaluation of high-

TABLE 11.1

Classification of Organisms according to the Temperature Range Inducing Heat Stress

Classification of organisms	Threshold of high temperature stress (°C)	Organisms included
a. Normally hydrated for growth		
Psychrophiles	15-20	Algae, bacteria, fungi
Mesophiles	35-45	Aquatic and shade higher plants lichens, and mosses
Moderate thermophiles	45-65	Higher land plants, some cryptogams
Extreme thermophiles	65-100	Blue-green algae, fungi, bacteria
b. Air dry cells or tissues	70-140	Pollen grains, seeds, spores, lichens, and mosses

temperature stress might be defined as the number of degrees above 15°C, since this is the approximate threshold of heat injury for the least heat-resistant group—the psychrophiles. Such a choice, however, is arbitrary; a species may be found with a threshold below 15°C. The alga *Koliella tatrae,* for instance, grows optimally at 4°C and any long-lasting temperatures above 10°C are lethal (Hindak and Komarek, 1968). No quantitative definition for heat stress is, therefore, possible. It can only be said qualitatively that the specific heat stress for any organism increases with the temperature above the lowest one that imposes a stress.

B. LIMIT OF HIGH-TEMPERATURE SURVIVAL

The high temperature limit is generally lower for growing than for resting organisms (see below). Yet it may be astonishingly high for some growing organisms (Table 11.2). There are records of growth of blue-green algae at temperatures as high as 93°-98°C, although 80°-85°C is usually accepted as the upper limit for their growth (Vouk, 1923; Robertson, 1927), and it is more commonly lower, e.g., 63°-64°C (Castenholz, 1969). According to Brock (1967), and Brock and Darland (1970) there are some extreme thermophilic bacteria for which no upper limit can be determined. This is because they were able to grow and develop right up to the boiling point (92°-100°C), at which point the absence of liquid water brought growth to a stop. Nevertheless, they are still subject to heat stress and will be killed at some temperatures above 100°C. Thus, unlike low temperature stress, the most extreme of which can be tolerated by living cells under certain conditions, there is a limiting high-temperature stress for all organisms. Exactly what this limit is for plants, in general, cannot be stated with certainty. By analogy with low temperatures, the heat-killing temperature for a plant may be defined as the temperature at which 50% of the plant is killed. As indicated above, this varies markedly from plant to plant (Table 11.2). Probably the highest recorded temperatures for a growing higher plant are 60°-65°C (Table 11.7; Biebl, 1962). However, these temperatures were maintained for very short periods of time, usually in mid-afternoon, when growth probably had temporarily ceased.

In agreement with these older results, the heat-killing temperatures for some thirty-nine species of plants from August to September on the coast of Spain ranged from 44° to 55°C (Lange and Lange, 1963). The low limit for the higher, multicellular algae has been corroborated by Giraud (1958). At optimal illumination (5000-10,000 lux) the optimal temperature for growth of *Rhodosorus marinus* (a red alga) is 20°C. The cells die in a few hours above 29°C and in 2-3 days at 25°C.

TABLE 11.2

Heat Killing Temperature for Different Plants and Plant Parts

Plant	Heat-killing temperature (°C)	Exposure time	Reference
(a) Lower plants			
Cryptogams	42–47.5	15–30 min	de Vries, 1870
Ulothrix	24		Klebs, 1896 (see Belehradek, 1935)
Mastigocladus	52		Lowenstein, 1903
Blue-green algae	70–75	Few hr	Bünning and Herdtle, 1946
Thermoidium sulfureum	53		Miehe, 1907
Thermophilic fungi	55–62		Noack, 1920; Tansey and Brock, 1972
Hydrurus foetidus	16–20	Few hr	Molisch, 1926
Sea algae	27–42	12 hr	Biebl, 1939
Ceramium tenuissimum	38	8.5 min	Ayres, 1916
Gymnodinium Pascheri	18	10 min	Diskus, 1958
Mosses (turgid)	42–51		Noerr, 1974
(dry)	85–110		
(b) Higher Plants			
Herbaceous plants			
Nicotiana rustica			
Cucurbita pepo			
Zea mays	49–51	10 min	Sachs, 1864
Mimosa pudica			
Tropaeolum majus			
Brassica napus			
Aquatics	45–46	10 min	Sachs, 1864
Nineteen species	47–47.5	15–30 min	de Vries, 1870
Citrus aurantium	50.5	15–30 min	de Vries, 1970
Opuntia	>65		Huber, 1932
Shoots of iris	55		Rouschal, 1938b
Sempervivum arachnoideum	57–61		Huber, 1935
Succulents	>55	1–2 hr	Huber, 1935
Succulents	53–54	10 hr	Huber, 1935
Potato leaves	42.5	1 hr	Lundegårdh, 1949
Trees			
Pine and spruce seedlings	54–55	5 min	Münch, 1914
Cortical cells of trees	57–59	30 min	Lorenz, 1939
Seeds			
Barley grains (soaked 1 hr)	65	6–8 min	Goodspeed, 1911
Medicago seeds	120	30 min	Schneider-Orelli, 1910
Wheat grains (9% H_2O)	90.8	8 min	Groves, 1917
Wheat (soaked for 24 hr)	60	45–75 sec	Porodko, 1926b
Trifolium pratense seeds	70	Short time	Buchinger, 1929
Fruit			
Grapes (ripe)	63		Müller-Thurgau (see Huber, 1935)
Tomatoes	45		Huber, 1935
Apples	49–52		Huber, 1935
Pollen			
Red pine pollen	50	4 hr	Watanabe, 1953
Black pine pollen	70	1 hr	Watanabe, 1953

Resting tissues in the dehydrated state have long been known to tolerate much more severe treatment than when active and fully hydrated (Just, 1877). Dry seeds are able to survive as high as 120°C, in contrast to highly hydrated tissues that are killed by temperatures below 50°–60°C (Table 11.2). When dried and heated for 16 min in a vacuum, they can survive 122°–138°C (35°C higher than without a vacuum; Ben-Zeev and Zamenhof, 1962). Of course, not all seeds survive such high temperatures. Some may be killed by 50°–60°C (Crosier, 1956), while others may survive boiling in water for several hours, provided that they do not swell during the boiling (Just, 1877). Dry barley and oat grains can be made to survive high temperatures for even longer times if dried further. In the ordinary, air-dry state they survived 100°C for only 1 hr without injury. However, if they were carefully dried for 9 days at 50°C, 2 days at 60°C, 2 days at 80°C, and finally transferred to 100°C for 3 days, more than 58% were still able to germinate (Just, 1877). Even hard-coated seeds that survive autoclaving at 120°C for one-half hour, are killed by boiling for 10 min if their seed coats are first filed (Schneider-Orelli, 1910). Presumably the filing permits the living cells to take up enough water during this 10-min period to lower their heat tolerance. The relationship between the heat-killing point of seeds and their moisture content and the analogous relation to low-temperature tolerance are clearly shown in Fig. 11.1. The seeds (maize) with less than 10% moisture had killing temperatures above 80°C, but when their moisture content rose to 75%, their killing temperature dropped to about 40°C. It is not surprising then that some apparently air-dry seeds are killed at 60°C, due to a high internal water content (Crosier, 1956).

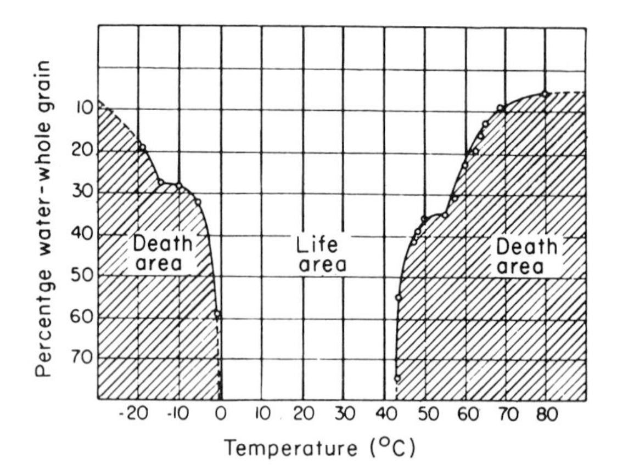

Figure 11.1. Relation of killing temperature to water content of corn grains. *Left:* Hogue's yellow Dent exposed for 24 hr to temperatures below zero (Kiesselbach and Ratcliff). *Right:* Reid's yellow Dent exposed for 2 hr to high temperatures. (From Robbins and Petsch 1932.)

Even in the case of nonresting tissues, the same relationship holds. Dallinger (see Lowenstein, 1903) showed that infusoria can survive 70°C if their water content is first reduced. Sea algae that succumb to seawater if kept at 35°C for 12 hr, survive 42°C for the same length of time if they are first dried on microscope slides (Biebl, 1939), *Nostoc muscorum* and *Chlorella* sp. in the lyophilized state survive 100°C for 10 min (Holm-Hansen, 1967).

C. THE TIME FACTOR

In contrast to the relatively minor role of exposure time in the case of freezing, the time subjected to high temperatures is of fundamental importance. This is true not only of vegetative plants (Table 11.3) but also of seeds, which are killed within minutes (or even seconds) by heat shocks at 60–120°C (Table 11.2). Longer periods (10–48hr) at considerably lower temperatures (40–70°C) may produce some degree of damage, retarding or decreasing germination (Onwueme, 1975; Onwueme and Lawanson, 1975; Onwueme and Adegoroye, 1975; Khan *et al.*, 1973; Makherjee *et al.*, 1973).

Not only does the heat-killing temperature vary inversely with the exposure time, but the relationship to time is actually exponential, so that Arrhenius plots give the expected straight-line relation between the log of the heat-killing rate and the reciprocal of the absolute temperature (Fig. 11.2). This has been thoroughly confirmed by Aleksandrov (1964). At intermediate temperatures, however, there is a break in the curve (Fig. 11.2), which is interpreted below. According to Lepeschkin (1912):

$$T = a - b \log Z$$

TABLE 11.3

Relationship between Heat-Killing Time and Temperature[a]

Temp. (°C)	Heat-killing time (min)				
	Tradescantia discolor	*Beta vulgaris*	*Brassica oleracea*	*Draparnaldia glomerata*	*Pisum sativum*
35				480	300–400
40	1300 (app)	>1500–2500	1100	80	32
45	725	420	577	7	2.2
50	243	90	45	1.2	0.27
55	44	4.3	3.8	0.32	0.095
60	7	0.7	0.8		
65	1.8				

[a] From Collander, 1924.

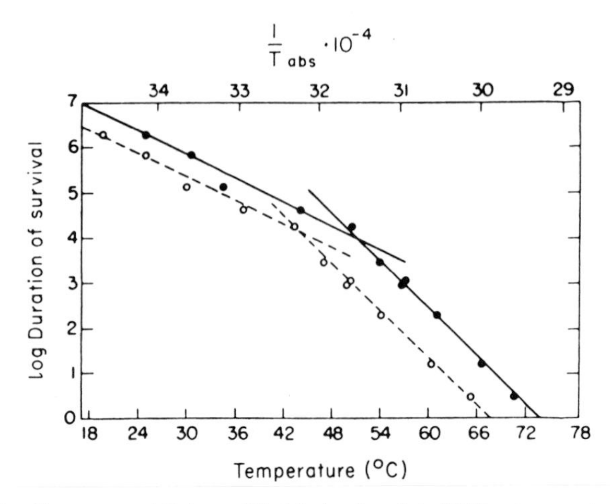

Figure 11.2. *Upper curve*(-●-): modified Arrhenius plot of killing temperature in relation to log of time. *Lower curve* (-O-): analogous Q_{10} relation. Break in curve indicates two distinct processes in *Helodea canadensis*. (From Belehradek and Melichar, 1930.)

where T = heat-killing temperature; Z = heating time; a and b are constants.

The agreement between the values calculated from this equation and the measured values is very good (Table 11.4). Other empirical equations have also been used successfully (Porodko, 1926b; Belehradek, 1935). S-Shaped curves relating heating time to injury may sometimes be obtained (Porodko, 1926b). According to Belehradek (1935), these are probability curves and simply mean that the organisms of a single type possess dissimilar heat-killing temperatures, in accord with the laws of variability.

TABLE 11.4

Heating Time and Coagulation Temperature *Tradescantia discolor*.[a]

Heating time (min)	Coagulation temperature (°C)	
	Determined	Calculated[b]
4	72.1	
10	69.6	67
25	63.2	62
60	57.0	57.1
80	55.7	55.5
100	54.1	54.2
150	52.0	

[a] From Lepeschkin, 1912.
[b] a = 79.8 and b = 12.8 in above equation.

D. OCCURRENCE OF HIGH-TEMPERATURE STRESS AND INJURY IN NATURE

1. High Plant Temperatures

All the above results were obtained with artificially induced high temperatures. The question is whether the plant is ever exposed to high enough temperatures to be injured under natural conditions. It is not enough to know the air temperatures exposed to, for it has long been known that the plant's temperature may rise above that of its environment. Dutrochet (1839) showed that this occurs if the plant is kept in saturated air to prevent the cooling effect of transpiration. The higher the external temperature, the greater was the elevation of the plant's temperature. In the case of fleshy organs (Table 11.5) with high metabolic activity, the elevation may be as high as 11°C (Vrolik and de Vriese, 1839), or even 14°R in the spadix of *Arum* (Goeppert—see Dutrochet, 1840). Due to their small specific surface such fleshy organs are unable to transfer all the excess heat to their environment, and their temperature rises. Thin leaves, on the other hand, may actually be cooled below the air temperature due to transpiration (see Chapter 12). More commonly, however, the temperatures of leaves exposed to sunlight may be well above that of the surrounding air (Table 11.6), due to absorption of 44 to 88% of the total radiation received (Raschke, 1960). According to Dörr (1941), the leaf temperature increases with its color in the following order: yellow, green, orange, and red. The strong absorption by red leaves caused both a rapid uptake and rapid loss of heat. Wilted leaves were always a few degrees warmer than turgid leaves. Herbaceous and especially woody stems reached temperatures as much as 12.2°C above those of the leaves. The temperature of the roots showed the closest agreement with the surrounding temperature. Different parts of the same leaf (or other plant part)

TABLE 11.5

Elevation of Temperature above That of the Air, in the Spadix of *Colocasia odora*[a]

Air temperature (°C)	Spadix temperature (°C)
20.0	27.8
16.7	26.1
15.6	26.5
20.7	28.9

[a] Vrolik and de Vriese, 1839.

TABLE 11.6

Plant Temperatures in Relationship to Temperatures of Surrounding Air

Plant	Plant part	Temperature	Air temperature	Reference
Sempervivum spp.	Succulent leaves	48-51°C	31°C	Askenasy, 1875
Tomato	Ripe fruit	100-106°F	80°-83°F	Hopp, 1947
Various	Leaves	44.25°C	36.5°C	Harder, 1930
Various	Leaves	37.6°C	24.7	Fritzsche, 1933
Conifer	Needles, twigs	9-11.8°C above air		Michaelis, 1935
	Stems (herbaceous and woody)	12.2°C above leaves		Dörr, 1941
Various	Thin leaves	6-10°C above air	20°-30°C	Ansari and Loomis, 1959
Herbaceous plants	Thick leaves	20°C above air	20°-30°C	Ansari and Loomis, 1959

may have different temperatures (Fig. 11.3), and those parts exposed to the most intense radiation reach the highest temperatures (Konis, 1950; Waggoner and Shaw, 1953). For this reason, position is also important. McGee (1916) found that *Opuntia* joints in the meridional position heat up more during a day in the sun than those in the equatorial position, but the heating above air temperature is great only in still air (Harder, 1930).

That dangerously high plant temperatures may occur under natural conditions seems obvious from all these observations. How high they actually do rise is shown in Table 11.7. In many cases they reach and may even exceed

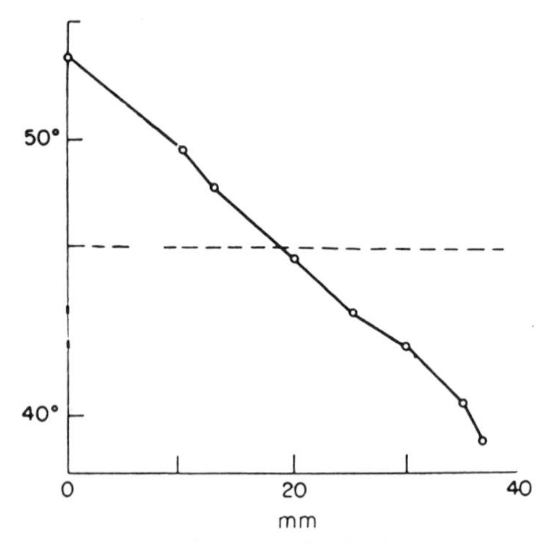

Figure 11.3. Temperature (°C) gradient with depth (abscissa) in a joint of *Opuntia*. (From Huber, 1932.)

TABLE 11.7

High Plant Temperatures Recorded under Natural Conditions

Plant	Highest temperature (°C)	Reference
Pine tree (south side cambium)	55	Hartig (Sorauer, 1924)
Opuntia	65	Huber, 1932
Sempervivum hirtum	50.2	Dörr, 1941
Globularia	48.7	Dörr, 1941
Tortula (dried out pulvinus)	54.8	Rouschal, 1938
Arum italicum (fruit)	50.3	Rouschal, 1938
Viburnum (leaves)	43.8	Rouschal, 1938
Iris (rhizomes)	42.5	Rouschal, 1938
Rhamnus alaternum	52.5	Konis, 1949
Fleshy fruit (various)	35–46	Huber, 1935

the 45°-55°C range that is usually accepted as the normal temperature limit for most plants (Huber, 1935).

2. Kinds of Injury

Heat injury has been described by many workers, especially in the case of bulky organs (see Sorauer, 1924; Huber, 1935). A well known type is the sun or bark burn that occurs on the south and southwest sides of thin-barked trees. This may be followed by drying and stripping of the bark, leading to defoliation and wood injury. The worst affected are the older, stronger stems; pole timber and branches are seldom injured in this way. The injury is thought to be due to overheating of the cambium. Hartig actually recorded a cambium temperature of 55°C on the southwest side of a spruce tree in the open at an air temperature of 37°C (Sorauer, 1924).

Burns have frequently been described in fleshy fruit—more commonly in grapes, cherries, and tomatoes, and less commonly in pears and gooseberries (Huber, 1935). As in the case of bark injury, they are usually confined to the most strongly heated southwest side (Huber, 1935). The burns may later dry up and become separated from the uninjured portion by a cork layer. Less often (e.g., in grapes) the whole fruit is killed. That true heat injury is involved follows from Müller-Thurgau's production of the same kind of injury by raising the fruit temperature artificially above 40°C (Huber, 1935). Szirmai (1938) was able to produce the "drought fleck disease" of paprika (which damaged 4-12% of the crop in Hungary) artificially by exposures to temperatures of 50° to 52°C when the surface was wet. In the case of dry fruit, 55°C was required to produce the injury. Direct exposure to sunlight produced the symptoms at air temperatures of 49°C.

Heat injury has been less often reported in leaves and other thin organs than in the above bulky plant parts (Sorauer, 1924; Huber, 1935). In practice, it has received the most attention when greenhouse-grown or shade plants are placed outdoors directly in the sun. Huber states, however, that it occurs most commonly on inland plains, but it is difficult to be sure that these are true cases of direct heat injury, since there is usually also a water deficiency. Thus, in spite of the many records of high temperatures in plants under natural conditions, Rouschal (1938) concluded that the maximum temperature the leaves can reach in the Mediterranean region during summer cannot produce any heat injury. By enclosing shoots in blackened tubes, he exposed them to temperatures higher than any he was able to observe under natural conditions; 55°C was reached for a short time, and 47°C was maintained for 10 to 15 hr. Even after several days of such treatment, no injury occurred. Konis (1949) used the same method and came to the same conclusion in the case of maquis plants under natural conditions in Israel. The lethal temperature was several degrees higher than the highest temperature recorded in the field. On the other hand, when sorghum plants were exposed at head emergence to simulated heat waves for 5 days (107°–108°F by day, 90°F by night) most of the enclosed flowers were killed (Pasternak and Wilson, 1969). High temperatures were apparently responsible since humidity (41 versus 70% by day) had little effect.

The greatest danger of heat injury occurs when the soil is exposed to insolation, reaching temperatures as high as 55° to 75°C (Lundegårdh, 1949). One of the most serious seedling "diseases," according to Münch (1913, 1914) is the killing of a narrow strip of bark around the stem of young woody plants at soil level (Fig. 11.4) when soil temperatures exceed 46°C. Since the seedlings usually die, he calls this "strangulation sickness." In laboratory tests, pear seedlings were found to succumb within 3 hr at 45°C or within 30 to 60 min at 50°C. Baker (1929) also concluded that fatal temperatures are reached in nature only at the base of the stem of 1- to 3-month-old conifers. He points out that surface soil temperatures of 130° to 160°F have been detected in temperate climates and that injury may occur at as low as 120°F. Henrici (1955) showed that temperatures of prostrate plants may exceed 55°C, and for a short time 60°C during summer in South Africa, although trees and other plants seldom reach 36°C. Young coffee plants suffer from collar injury, which Franco (1961) was able to induce by a temperature of 45°C. At 50°C most of the plants died, at 51°–55°C all died. Soil surface temperatures in the field reached 45°–51°C and sometimes above. Therefore, high-temperature injury must occur in nature, but this depends on the bulk of the structures. Although Rouschal (1938) measured maximum soil temperatures as high as 64°C, no injury occurred to the bulky iris rhizomes that he investigated, for their temperature did not rise above

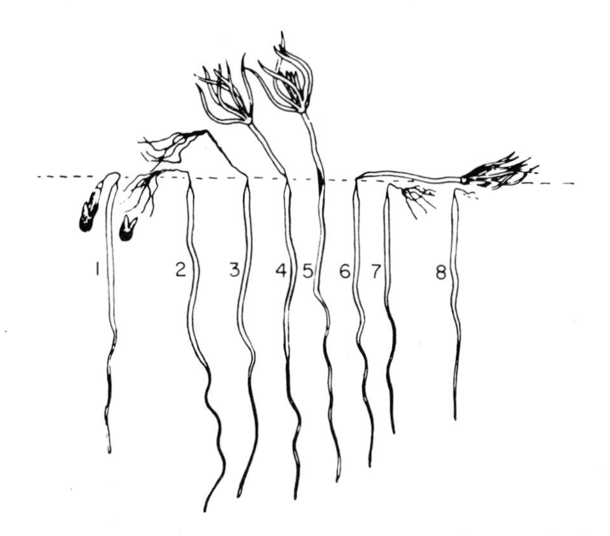

Figure 11.4. Killing of pine seedlings at soil level, due to overheating of soil surface. (From Münch, 1913.)

42.5°C. According to Julander's (1945) observations, however, the much thinner stolons of range grasses are in definite danger of injury. He observed a soil temperature of 51.5°C when the air temperature was 36°C. Since he was able to produce definite injury to the stolons at 48°C, and since air temperatures as high as 43°C are not uncommon under severe drought conditions, the possibility of heat injury under natural conditions seems obvious.

Lange (1953) showed that lichen temperatures may be well above that of the surrounding atmosphere (Fig. 11.5) and as high as 69.6°C under natural conditions. He suggested that they may rise to 75°–80°C, since these values have been observed for surface temperatures of soils that are their natural habitats. Since one-half hour periods at 70°–100°C were sufficient to kill a large number of different kinds of lichens in the dry state, and since the high temperatures he observed were maintained for much longer than one-half hour (Table 11.8), he concluded that they must be injured sometimes by heat under natural conditions.

Most of the above investigations were not conducted in the hottest climates. Researches by Lange (1958) in the Sahara desert have clearly demonstrated the importance of heat injury in this region. Even the most successful plants commonly attained leaf temperatures within 2°–6°C of their heat-killing points and some natural heat injury occurred similar to that produced under artificial conditions. Some of the plants (e.g., species of *Citrullus*), in fact, survived only if sufficient water was available for high transpiration

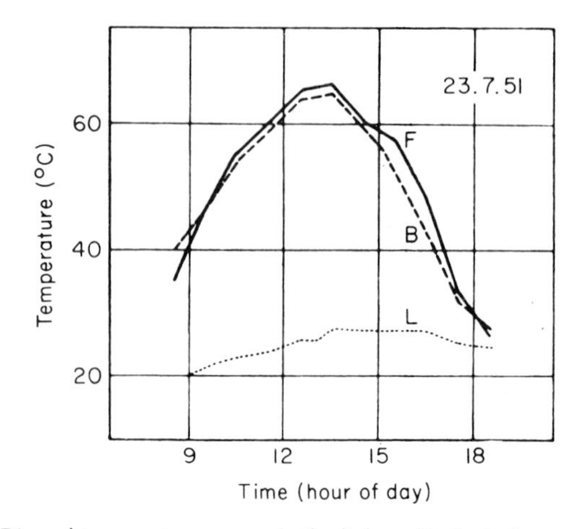

Figure 11.5. Diurnal temperature course in the lichen *Cladonia furcata* var. *palomoea* (F), the soil (B), and the air (L). (From Lange, 1953.)

rates (see Chapter 12). Attempts to grow cultivated plants introduced in an oasis during summer failed due to heat injury, even with plentiful watering. Even in temperate climates, Lange (1961) found that *Erica tetralix* owed its survival of summer heat to a marked summer rise in heat tolerance, without which its leaf temperatures would have been above the killing point (see Chapter 12). According to Khan and Laude (1969), heat stress during maturation of barley seed may aid in explaining differences in % germination at harvest of seed produced in successive years, or in different locations in the same year. The germination of the freshly harvested seed was depressed following a heat stress 7–10 days after awn emergence, but was enhanced by the same stress applied 3 weeks after emergence.

In view of all these observations, it cannot be denied that heat injury does occur in nature, though perhaps only on relatively rare occasions. By analogy with freezing injury, it may be expected only during relatively rare "test

TABLE 11.8

Time (min) during Which the Temperature of the Lichen Remained above the Listed Temperatures (°C)[a]

Species	62.5°	65.0°	67.5°	Max. temp.
Cladonia pyxidata	270	225	150	69.6°
Cladonia subrangiformis	195	125	—	67.0°
Lecidia decipiens	50	—	—	64.3°

[a] From Lange, 1953.

summers." It is most likely to occur when crop plants are grown in regions to which they are not adapted.

E. NATURE OF THE INJURY

A complicating factor in low-temperature injury is the change of the cell's water from the liquid to the solid state. At high temperatures, the analogous sudden change to vapor occurs only at temperatures that are not found under natural conditions. The slower vaporization at normal temperatures is, however, a possible cause of injury, but greater complexity is to be expected at high than at low temperatures, since all the reactions in the plant are already taking place rapidly, and a further rise in temperature might easily disturb the balance. Because of these complicating factors, the heat stress may produce direct or indirect stress injury as well as secondary stress injury.

1. Secondary Heat-Induced Drought Injury

Gäumann and Jaag (1936) measured the cuticular transpiration at temperatures of 20° to 50°C (Fig. 11.6). Their results show how pronounced the increase is at the higher temperatures. There are two reasons for the sharp rise in transpiration with the rise in environmental temperature: (a) the direct effect of temperature on the diffusion constant of water, and (b) the steepening of the vapor pressure gradient between the leaf and the external atmosphere. Curtis (1936b) points out that if the leaf temperature is 5°C above the atmospheric temperature, this is equivalent to a steepening of the gradient by a 30% lowering of the atmospheric r.h. (relative humidity). In other words, if the external atmospheric r.h. is 70%, the 5°C rise in leaf temperature would double the gradient and, therefore, the evaporation rate, aside from the increase due to the increased molecular velocity at the higher temperature. A 10°C rise would have a proportionately even greater effect than expected from the doubling of the temperature gradient. In addition to this direct effect of temperature on evaporation, it may further increase transpiration by maintaining the stomata open (Barabal'chuck and Chernyavskaya, 1974) and due to increased root temperature (Gur *et al.*, 1972).

The danger of drought injury under such conditions is obviously great, even without a deficiency in soil moisture. It is not surprising, therefore, that prolonged high temperature stress often results in injury due to desiccation, for instance, in the case of turfgrass (Krans and Johnson, 1974). A suggested method of distinguishing between primary and secondary heat injury is to

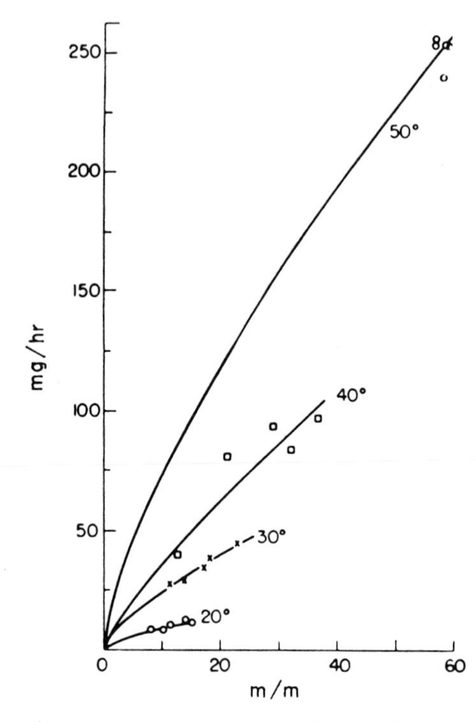

Figure 11.6. Cuticular transpiration of *Quercus robur* at different air temperatures. (From Gäumann and Jaag, 1936.) m/m = physiological saturation deficit (mm Hg).

observe for cell contraction (Balina, 1976)—a characteristic of cell drought (see Chapter 13) and therefore of secondary heat injury.

The above described secondary drought injury will be considered together with the primary drought stress. It must be eliminated in any attempt to study the primary heat injury, which in its turn may be either direct or indirect. The existence of two kinds of primary heat injury is demonstrated by the break in the Arrhenius plot, giving two straight-line relations between the log of the heat-killing rate and the reciprocal of the absolute temperature (Fig. 11.2). This break has been confirmed in a number of plants by Lorenz (1939) and by Aleksandrov (1964). The logical explanation is that the straight line in the lower temperature range is due to indirect heat injury, and in the upper range to direct heat injury.

2. Primary Indirect Heat Injury

a. GROWTH INHIBITION

The same kinds of indirect injury can occur at high temperatures that are not directly injurious, as at chilling temperatures. Growth inhibition, however, is of little importance at chilling temperatures, in contrast to its well-established importance as a kind of indirect heat injury. The reason for this can be seen from the cardinal temperatures for growth (Fig. 11.7). The decrease in growth rate below the T_{opt} is simply due to the known effect of temperature on the rates of any chemical reactions, and therefore on the metabolic processes basic to growth. Growth inhibition below the T_{opt} therefore does not involve injury. Above the T_{opt}, on the other hand, the curve does not continue upward but bends downward. This decrease in rate cannot be explained by the direct effect of temperature on ordinary chemical reactions, but must be due to injury leading to inhibition of some physical or chemical reaction.

Hilbrig (1900) showed that the injury at slightly supramaximal temperatures is gradual. The longer the plants are exposed to the high temperatures, the longer it takes them to recommence growth (Table 11.9). If the high temperature stress is maintained long enough, the growth inhibition is replaced by cell injury and death (Table 11.9). Similarly, fungus spores finally die after 52 days if kept at temperatures too high for growth. Temperatures

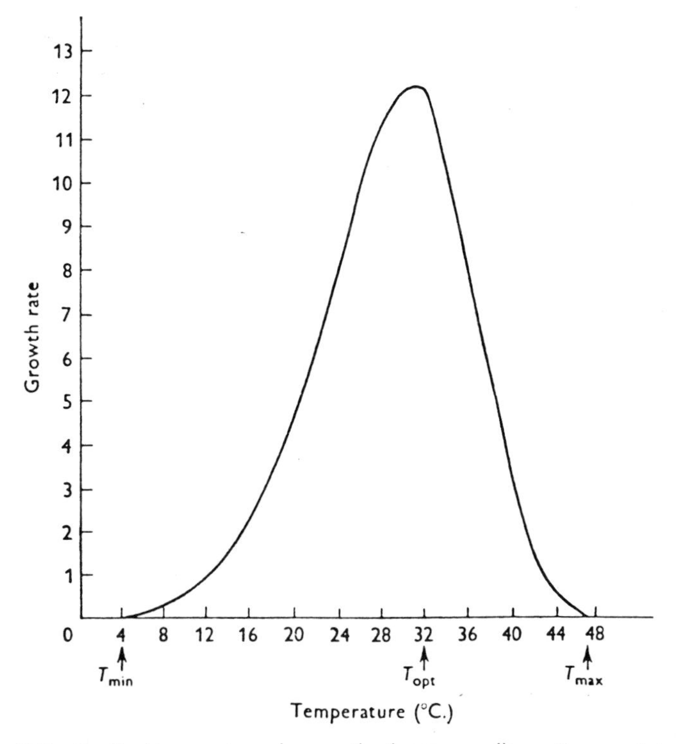

Figure 11.7. Cardinal temperatures for growth of maize seedlings. T_{min} = minimum temperature for growth. T_{opt} = optimum temperature for growth. T_{max} = maximum temperature for growth. From Lundegårdh (1957).

TABLE 11.9

Effect of Different Times of Exposure of *Penicillium* Cultures (1-Day Old) to 35°C on Subsequent Growth[a]

Exposure time (days)	Time for growth to recommence (days)
2	2
4	3
8	4
12	5
19	6
25	6
30	8
1 month	Dead

[a] From Hilbrig, 1900.

that are not quite high enough to stop growth completely, eventually may also be injurious (Hilbrig, 1900). Results with higher plants (bean, pea, and cucumber) were similar, though the temperature zone in which growth was stopped without immediate injury was smaller. Growth stoppage (at 45°C) could never be maintained for more than 1 hr and 45 min without killing the seedlings. Dangeard (1951c) found that cessation of cell elongation at 40–60°C may be reversed after several days at room temperature.

In contrast to this apparently general connection between growth inhibition and injury, moderate high temperatures may inhibit growth reversibly without any sign of injury. Alfalfa showed a significant reduction in number and length of its shoots within 7 days and persisting up to 6 wk. following a heat stress which produced no visible evidence of tissue injury (Pulgar and Laude, 1974). During such extended heat treatments of whole plants, however, it is not possible to eliminate water stress effects. Nevertheless, growth inhibition at high temperature may, in fact, be a protective device that prevents more serious high temperature injury. Species native to Death Valley, for instance, cease growth during the hottest time of the year, and survive due to their growth during the colder periods. That this is not simply a response to water stress is indicated by the exceptional thermophile native to the same climate—*Tidestromia oblongifolia*. It grows during the hottest season and may double its growth in 3 days at temperatures lethal to coastal species of plants (Björkman *et al.*, 1975). In the case of some seeds (e.g. lettuce), a moderately high temperature (27–36°C) induces a true *thermodormancy*—a noninjurious, reversible growth inhibition which is then maintained at lower, normal growing temperatures. The timing and duration of the high temperature period are critical (Heydecker and Joshua,

1977), and the thermodormancy may be prevented by a preceding cool period. It was also prevented by soaking the seeds for 15 min in a solution of kinetin in dichloromethane in confirmation of earlier results (Porto and Siegel, 1960; Ben-Zeev and Zamenhof, 1962). This agrees with other evidence of a fall in kinetin content as a result of heat treatment (Itai *et al.*, 1973; Gur *et al.*, 1972). Other hormones may also be involved in thermodormancy, and it has been concluded that the supraoptimal temperature raises the threshold concentration of ethylene required for germination (Dunlap and Morgan, 1977). Exogenous ethylene, therefore, partially restored the germination at 32°C. Others have associated an excess of ethylene with high temperature stress. The internal atmosphere of sweet potato plants held for 1 min at 48°C was higher in ethylene and O_2, lower in CO_2 than the controls. This was accompanied by an increased respiration rate (Patterson, 1974). Two soybean varieties showed a similar effect at an intermediate temperature. Their growth was abnormal at 25°C and this was accompanied by high levels of ethylene (Samimy and Lamotte, 1976). Growth was normal at both 20 and 30°C. Possibly because of such hormone effects, the growth character of *Agrostis palustris* was distinctly different at 40–30°C, compared to growth at 20–10°C (Duff and Bear, 1974a).

Growth inhibition of seeds may also be produced by more severe heat shocks. A period of 30 min at 50–85°C applied to air dry wheat seeds inhibited growth for 10 days (Musaelyan, 1975). This inhibition, however, may perhaps be due to a secondary desiccating effect of this severe heat shock.

In some cases, the maximum temperature for growth and the lethal temperature appear to be identical or nearly so—about 40°C in the case of the germination of cowpea and sorghum seeds (Kailasanathan *et al.*, 1976). In fact, the distinction between growth inhibition and other forms of indirect heat injury disappears with increase in length of time at the high temperature.

There are four possible kinds of indirect high temperature injury, any one of which may conceivably inhibit growth in mild doses and actually injure the plant in large doses: starvation, toxicity, biochemical lesion, and protein breakdown.

b. STARVATION

In the case of 25 species of plants, the temperature maximum for assimilation (36°-48°C) was from 3 to 12 degrees below the heat-killing temperature (44°-55°C; Pisek *et al.*, 1968). Similarly, when measured by the beginning of rapid loss of ions, heat killing may require 11° higher temperature than that at which photosynthesis is destroyed (Berry *et al.*, 1975). Starvation, however, occurs before this high temperature limit (the T_{max}) for photosynthesis is reached. This is because of the higher temperature optimum for respiration than for photosynthesis (50° and 30°C, respectively in potato leaves; Lundegårdh, 1949).

The temperature at which respiration and photosynthesis are equally rapid is called the *temperature compensation point*. Obviously, if the plant's temperature rises above the compensation point, the plant's reserves will begin to be depleted. A sufficiently long time at such temperatures would ultimately lead to starvation and death. Since the temperature compensation point drops with light intensity, this kind of injury can occur at relatively low temperatures in shade plants and may account for the low heat-killing points of some algae. This, however, would be the slowest kind of heat injury if the temperatures are moderate, and would normally occur only after one or more days during which the daylight temperature is maintained uninterruptedly at the supracompensation point. In the case of the higher compensation points of terrestrial sun plants, the process would be more rapid. As the temperature rises above the compensation point, respiration rate continues to increase and photosynthesis to decrease and, therefore, the starvation rate would increase exponentially.

The deficit increases particularly rapidly in plants with an active photorespiration (C_3 plants) in addition to the normal dark respiration. The pronounced increase in rate of photorespiration at high temperatures is apparently due to a higher Q_{10} for glycolate oxidase activity relative to that of catalase. As a result, there is an increased availability of H_2O_2 and a marked increase of glycolate oxidation to CO_2 by the peroxisomes of the leaf (3 times as rapid at 35°C as at 25°C; Grodzinski and Butt, 1977).

It is interesting that a thermophilic blue-green alga has no apparent compensation point, at least up to 65°C (Fig. 11.8) and is therefore in no danger of this kind of injury.

The existence of a compensation temperature is, of course, due to the lesser resistance to high temperature of the photosynthetic system than the respiratory apparatus (Hammer, 1972). Why is the photosynthetic system so sensitive? In none of the four species of thermophiles analyzed by Björkman (1975) was it due to decreased stomatal conductance of CO_2 above the optimum for photosynthesis. On the contrary, conductance was maximum at the highest temperature investigated, and increased concentrations of CO_2 failed to prevent the inhibition of photosynthesis by high temperature. Quantum yields, however, showed sharp reductions above T_{opt}, whether it was 47°C (in *Tidestromia oblongifolia*) or 40° (*Atriplex sabulosa*). The specific damage to chloroplasts has been investigated after heating the leaves by submerging them for 4 minutes in water at the test temperature (Berry *et al.*, 1975). The capacity of e^- transport by photosystem II was more susceptible to heat damage than were other activities. Thus,

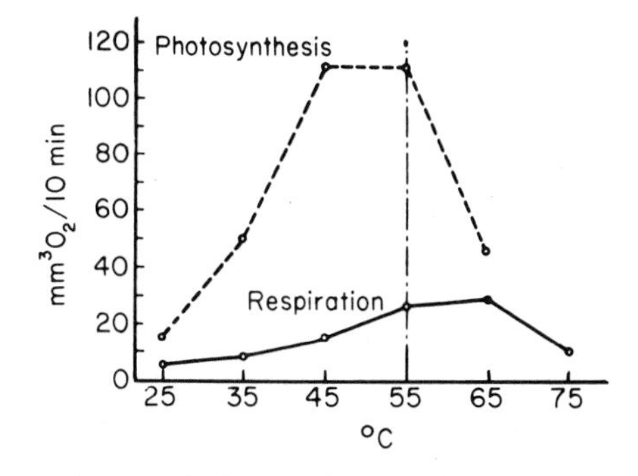

Figure 11.8. Change in rate of photosynthesis and respiration with temperature. Alga acclimatized at 55°C. (From Marré and Servettaz, 1956.)

heating to 44°C did not affect e^- transport by photosystem I, though reducing that of photosystem II by 25%. Chloroplast ribosome formation in grains is even more sensitive to heat. It was selectively prevented in developing rye leaves at 32–34°C and the leaves were chlorotic (Feierabend and Schrader-Reichhardt, 1976; Feierabend and Mikus, 1977). Between 28 and 34°C chlorophyll and carotenoid accumulation are strongly inhibited compared to 22°C. Feierabend (1977) concluded that inhibition of D-aminolevulinate synthase and photooxidation are responsible for chlorosis at 32°C. Chlorophyll accumulation is retarded in watermelon (*Calocynthis citrullus*) seedlings after exposure to 35–45°C for 4–24hr in darkness (Onwueme and Lawanson, 1973). This was due to a retardation in both the formation of protochlorophyll and its conversion to chlorophyll (Lawanson and Onwueme, 1973). RuDPC activity was low, as was the quantity of Fraction I protein. Photosynthesis was very low in these leaves. In *Atriplex lentiformis,* differences in photosynthetic activities were also correlated with RuDPC activities (Pearcy, 1977). In spinach. however, the activity of RuDPC decreased by only 10–20% at a temperature resulting in complete loss of photosynthetic activity (Yordanov and Vasil'eva, 1976). Furthermore, there was no significant change in composition of the structural proteins of the chloroplasts at a temperature strongly inhibiting photosynthesis (Yordanov *et al.*, 1975). In chlorella, (Tischner and Lorenzen, 1975) no specific inhibition of steps in phosphate metabolism were detected after exposure to 45°C although the partial reactions of photosynthesis were completely inhibited. The main effect of the heat shock was believed to be a decrease in the content of ADP and ATP.

Starvation injury is not necessarily due to a *net* decrease in assimilate, but may be due to effects on translocation. When japonica rice was grown at 35°D/30°N, the rate of ripening was more rapid than at normal growing temperatures, but the inflow of assimilate into the grain ended earlier. This resulted in a lower 1000-kernel weight, although there was no deficiency in the assimilate content of the plant (Sato and Inaba, 1976). Similarly, in the case of *Agrostis palustris* grown at 40° D/30°N the carbohydrate content of the leaves was higher than that of plants grown at lower temperatures. The growth reduction was, therefore, not due to carbohydrate starvation of the leaves (Duff and Beard, 1974b). In this case, however, some of the plants were killed.

The inhibitory effect of high temperature on photosynthesis is completely reversible if not too extreme—for instance a 2-min shock at 46°–51°C reduced $^{14}CO_2$ fixation in detached leaves of *Nicotiana rustica* (Ben Zioni and Itai, 1972). This decrease continued for 2½ h but recovery was complete 45h after treatment. Similarly, 5 min at 40°C was strongly inhibitory but 24h later the rate was restored almost to the control (Yordanov *et al.*, 1975). Even when photosynthesis was completely inhibited by a short-term heating (10–15 min at 37.5° to 45°C), all the plants were able to renew photosynthesis after 17hr in the dark at room temperature (Egorova, 1975, 1976).

c. TOXICITY

In contrast to chilling injury, there is little evidence that toxicity is associated with high temperature injury. Nevertheless, just as starvation injury is basically due to an inhibition of photosynthesis by the high temperature, toxicity injury, if it occurs, must be mainly a result of respiratory disturbances. Such disturbances occur gradually at moderately high temperatures (Table 11.10).

In *Crepis biennis,* respiration decreased with time even at 30°C (Kuijper, 1910). The marked decrease in reserves as a result of the high respiration rate affected both the starch and the proteins. The production of CO_2 by

TABLE 11.10

CO_2 Evolution by Pea Seedlings at High Temperatures[a]

Temperature (°C)	CO_2 evolution					
	1st hr	2nd	3rd	4th	5th	6th
30	51.7	50.9	52.2	53.6	53.5	53.5
35	68.7	62.8	60.1	61.7	60.9	60.9
40	73.3	55.2	49.0	45.3	43.0	41.2
45	73.5	48.4	41.9	35.9	31.9	28.6
50	74.0	38.8	17.8	12.0	8.0	5.9
55	35.7	12.8	9.7	5.4		

[a] From Kuijper, 1910. Similar results were obtained with wheat.

yeast shows no depression up to 45°C, but at 46° and higher it decreases with exposure time (van Amstel and Iterson, 1911; see Belehradek, 1935). The greater heat injury when submerged in water has actually been ascribed to an oxygen deficiency (Just, 1877). Other metabolic processes may be even more heat sensitive than respiration. A 1-hr exposure to 47°C in *Saccharomyces* and to 4°C in *Torula,* reduced oxygen consumption by 40 to 60% and nitrogen assimilation to zero (Van Halteren, 1950). After 2½ hours at room temperature, nitrogen assimilation recommenced, though at a lower rate than the control. The longer the exposure time, the slower the recovery rate and the rate of nitrogen assimilation when it does recommence (Table 11.11). Uptake of phosphate was also stopped for 2 to 3 hr after the high temperature exposure.

Respiratory disturbances may explain the typical heat injury symptoms induced in French prunes by oxygen concentrations of 1 and 2.5% at 86°F, and the protection against such injury at 86°–100°F by oxygen concentrations of 60–100% (Maxie, 1957). Possibly the heat injury is, in this case, due to toxic products of anaerobic respiration.

Direct evidence of heat-induced formation of such products has been obtained in the case of apple trees (Gur *et al.,* 1972). Both root and shoot growth were reduced at 30°C and above; and serious leaf damage occurred at 35°C. Differences in susceptibility of varieties were apparently due to damage by anaerobic respiration, for the more susceptible varieties had higher root contents of ethanol and a severe reduction in malic acid. Acetaldehyde was also found in the leaves and roots.

Petinov and Molotkovsky (1957) suggest that heat injury is due to the toxic effect of NH_3 produced at high temperatures and that this effect is counteracted by the respiratory production of organic acids. This may help to

TABLE 11.11

Duration of High Temperature, Recovery Time and Rate (on Recovery) of Nitrogen Assimilation in Yeast[a]

Duration of high temperature (hr)	Time for recovery of N assimilation (hr)	Rate of N assimilation after recovery (γ/cm³/2 hr)
0	0	6.3
¼	½	5.3
½	1	4.2
1	2½	3.8
1½	2½	3.3
2	5	2.5

[a] From van Halteren, 1950.

explain the heat tolerance of succulents, since they possess a highly acid metabolism. On the other hand, their acidity is at a maximum when danger of heat is minimal (at night) and at a minimum when the danger is maximal (in the afternoon). In seedlings of *Pennisetum typhoides,* after exposure to 48°C for 12–24 hr, ammonia N was found in detectable quantities (Lahiri and Singh, 1969), injuring the plants.

d. BIOCHEMICAL LESIONS

If the accumulation of an intermediate substance necessary for growth (a vitamin, cofactor, etc.) is inhibited at high temperatures, growth inhibition and eventually injury may occur (Kurtz, 1958). Evidence of this kind of injury was offered in the case of *Neurospora crassa.* A mutant that was able to grow in standard medium only up to 25°–28°C was induced to grow at as high temperatures as the wild type (35°–40°C) by supplying it with riboflavin. Adenine similarly induced growth at higher temperatures in the case of other plants. This agrees with Galston and Hand's (1949) evidence of adenine destruction at high temperature. Langridge and Griffing (1959) have supported this concept in the case of some races of *Arabidopsis thaliana.* Eight races showed marked decreases in growth at 31.5°C, five of them showing morphological symptoms of high temperature damage. Three of these showed increased growth at this temperature when supplied with vitamins, yeast extract, or nucleic acid. In the case of two of these races, biotin was specifically effective and completely prevented heat lesions. The third race showed a partial alleviation on the addition of cytidine. Not only biotin, but even 1% sucrose may increase the growth of *Arabidopsis thaliana* at supraoptimal temperatures, although the former had a much smaller effect on plants grown at optimal temperatures, and the latter actually decreased their growth (Shiralipour and Anthony, 1970). Sherman (1959) also found a greater nutrient requirement for yeast growth at elevated temperatures. In fact, the occurrence of such biochemical lesions has been suggested as the cause of the slow growth of even the thermotolerant yeasts (Loginova and Verkhovtseva, 1963), since these tend to be more heterotrophic than the mesophiles.

Neales (1968) showed that the vitamin requirement of the roots of one strain of *Arabidopsis thaliana* was greater at 31.5° than at 27°C. On the other hand, in the case of *Neurospora crassa,* riboflavin accumulated to a higher degree at a high temperature (37° and 40°C) than at 25°C (Ojha and Turian, 1968), even though injury occurred at 40°C (mitochondria shrunken, endoplasmic reticulum broken).

Just how these biochemical lesions arise is not known. One possibility has been suggested by Al'tergot (1963). He found that a temporary oxygen deficit decreased heat injury, while excess oxygen increased it. This is due to the thermostability of oxidases which, therefore, are able to destroy the readily oxidized substances. If these substances (e.g., ascorbic acid, glutathione), are essential, this destruction would produce a biochemical lesion.

Lipids have also been found to overcome heat-induced growth inhibition. Loginova *et al.* (1962) were able to accelerate the growth of a thermotolerant yeast by addition of Tween 80, ergosterol, or oleic acid. Similarly, Starr and Parks (1962) found that sterol synthesis by yeast was inhibited increasingly above 30°C, becoming critical at 40°C. Death at 40°C was averted by adding oleic acid. In order for cell growth to occur, however, both oleic acid and ergosterol had to be added.

The synthesis of lipid and B glucan in the primary leaves of bean plants was decreased by dipping in water at 46.5°–47.5° for 2 min. (Ordin *et al.*, 1974). This was accompanied by an inhibition of leaf growth within the first l2hr, recovery in the subsequent 24hr period. The most evident effect of the heat shock was a severe inhibition of the synthesis of β-1,4-glucosyl glycosidic linkages and stimulation of β-1,3-linkage synthesis (Musolan *et al.*, 1975).

The synthesis of several other substances is also inhibited by high temperature stress. A selective decrease in production of 23 S rRNA occurred in the thermophilic *Mycobacterium phlei* when exposed to sublethal temperatures (Rieber and Weinstein-Schonfeld, 1974). This was accompanied by the appearance of an inhibitor of RNA polymerizing activity, which is found at lower temperatures. At 40°C, synthesis of squalene and sterols in aerated yeast is only a third of that at 20°–30°C (Shimizu and Katsuki, 1975). This is attributed to the suppression of an enzyme involved in the synthesis of mevalonate from acetyl-CoA. A thermophilic fungus required the addition of a C_4 compound (succinate) for growth at 50°C (Asundi *et al.*, 1974).

Two distinct effects are indicated by the above evidence: (1) a biochemical lesion injuring the plant during its growth at supraoptimal temperatures, and (2) a biochemical lesion as a result of a heat shock, inducing injury during the post-shock growth at optimal temperatures. As an example of the latter, air-dry lettuce seed, were injured by a 1-hr exposure to 75°C as shown by a reduced subsequent germination. A concentration of 10^{-5} kinetin in the germination medium after the heat treatment prevented this injury (Porto and Siegel, 1960). This has been corroborated by Ben-Zeev and Zamenhof (1962).

e. PROTEIN BREAKDOWN

Lepeschkin (1935) was probably the first to produce indirect evidence of this process and of the importance of synthesis or repair of heat injury. He found that interrupting the exposure to high temperature at the mid-point by

2 min at 20°C had no effect on the total time needed to kill *Spirogyra*. When the interruption was for 2⅔ hr at 20°C, however, a longer total time at the high temperatures was needed for heat killing. From the brief heat break, he concludes that the protein denaturation is physically and chemically irreversible, but from the long break he concludes that it can be repaired by physiological activity. The same conclusion was arrived at by Allen (1950), who showed that, in the absence of nutrients, thermophilic bacteria die at 55°C just as rapidly as the mesophilic bacteria. The enzyme systems of the thermophiles were rapidly inactivated at this temperature. She concluded, therefore, that the thermophiles can synthesize enzymes and other cell constituents far faster than they are destroyed by heat, and that they have higher coefficients of enzyme synthesis than in the mesophiles. According to this concept, heat killing occurs when the speed of resynthesis of an indispensable component (e.g., an enzyme or an intermediate substance) is unable to compensate for its degradation. In support of this concept, some strains of *Neurospora crassa* are able to produce the hydrolytic enzyme cellulase in the presence of cellulose at 35°C (Hirsch, 1954). At 25°C they fail to produce the enzyme. It is perhaps possible that injury is due to a similar self-digesting proteolytic enzyme produced at high temperatures, perhaps from lysosomes. This would account for the liberation of amino acids and oligopeptides into the culture broth, by yeast cells grown at supraoptimal temperature (Chigaleichik *et al.*, 1975). Similarly, exposure of mycelium of *Papulaspora thermophila* to 60°C for 2hr resulted in a decreased variety of protein species, due presumably to a degradation to oligopeptides (Chapman and Ostrovsky, 1976).

On the other hand, the net loss may be due to a decreased synthesis rather than an increased breakdown. Several cases of uncoupling at high temperatures have been reported. In corn mitochondria, coupling between oxidation and phosphorylation began to decrease at 30°–35°C and at 40°C uncoupling was complete (Kurkova and Andreeva, 1966). The decreased phosphorylation would certainly lead to decreased synthesis of proteins as well as other substances. In *Chlorella* sp. K, when grown at 37°C and 43°C, accumulation of biological mass was uncoupled from cell multiplication, leading to hypertrophy (Semenenko *et al.*, 1969). Carbohydrates were synthesized at a high rate, and after 7 hr comprised up to 45% of the dry weight, while protein decreased to 18%. During the first hour the protein increased. These results have since been corroborated by Tischner and Lorenzen (1974).

Thermal uncoupling in chloroplasts may also occur at temperatures lower than those normally leading to heat inactivation (Emmett and Walker, 1969). Similarly, in the case of the psychrophile *Sclerotinia borealis* grown at 0°C, the maximum temperature for growth is 15°C, but the optimum for respiration is 25°C. It has, therefore, been postulated that growth becomes uncoupled from respiration above 15°C (Ward, 1968). In support of this explanation, the uncoupling agents, 2,4-DNP and dicoumarol stimulated oxygen uptake relatively more at 5° than at 25°C. In the case of peas, however, no change in the efficiency of respiration could be detected at 41°–43°C compared to 18°–20°C (Nikulina, 1969). In the case of wheat roots, DNP induced the same injury as a 45° heat shock (Skogqvist and Fries, 1970).

Uncoupling is not the only possible cause of decreased protein synthesis at high temperatures. When *Physarum polycephalum* was subjected to heat shocks at 40° for periods of 10 or 30 min (Schiebel *et al.*, 1969), the incorporation of amino acids into protein was decreased by approximately 40 or 70%, respectively. There was also a decrease in polyribosomes of more than 50%, which was postulated to be the cause of the decreased protein synthesis. In support of this explanation, an increase in rate of RNA degradation and a significantly higher RNase activity was observed in this myxomycete at supraoptimal temperatures (Bernstam, 1974). In the case of fission yeast, a 15 min. heat shock at 41°C produces the same effects as does cycloheximide—an inhibition of both RNA and protein synthesis and of growth (Polanshek, 1977). In young rye (*Secale cereale*) leaves, the formation of 70 S chloroplast ribosomes and of their rRNA were selectively prevented during growth at 32°C. Other organelles were not significantly affected, and at least some normal chloroplasts were formed on return to normal growing temperatures (Feierabend and Schrader-Reichhardt, 1976; Schaefers and Feierabend, 1976). In the psychrophile *Pseudomonas* sp., RNA synthesis was not detected at supramaximal temperatures (Harder and Veldkamp, 1968).

The occurrence of metabolic heat injury due to proteolysis, has now been established in higher plants by Engelbrecht and Mothes (1960, 1964), but it occurs in a different manner from that proposed by Allen for microorganisms. A 1 to 2-min exposure of a leaf of *Nicotiana rustica* to 49°–50°C produced a reversible, sublethal "heat-weakening". The leaf remained turgid after the heat treatment, but the normal yellowing (aging) occurred more rapidly than in the control. Amino acids were also translocated to the unheated half of the leaf if only the other half had been exposed to the heat treatment. When half the leaf was sprayed with kinetin, this prevented the above described (heat-induced) yellowing and the kinetin-treated leaf half accumulated amino acids, maintaining or increasing the protein content. This sublethal heat weakening could even be reversed by a kinetin treatment after the heating (confirming Porto and Siegel, 1960; see above), as well as by other treatments that checked the efflux of metabolites. A similar proteolysis occurred in 3-week-old plants of *Pennisetum typhoides* when exposed to 48° (± 1°C) for durations up to 24 hr (Lahiri and Singh, 1969). Soil water stress was negligible, so this was solely due to the heat stress. Wheat roots exposed to a heat shock of 45°C were uninjured if this was followed by 25°C, but were injured if it was followed by 35°C (Skogqvist and Fries, 1970). Kinetin overcame or prevented the injury, and so did chloramphenicol. In the case of a thermal blue-green alga (*Aphanocapsa thermalis*), the optimal temperature for growth and photosynthesis was 40°C (Moyse and Guyon, 1963). At higher temperatures, which had a strongly inhibiting effect, both amino acid synthesis and their condensation to proteins were slowed down to a greater extent than was carbohydrate synthesis.

Aleksandrov (1964) has produced evidence of a repair mechanism even during the heating. He observed cessation of cytoplasmic streaming after 90 min at 42.0°C, but after 200 min at the same temperature, some streaming recommenced, and streaming was practically normal at 6 hr. After a longer period, however, the cells eventually died.

This fourth kind of metabolic heat injury may therefore be due to a net loss of protoplasmic proteins, but it may occur after the heat stress has been removed. It may be suggested that a slightly more severe heat stress could perhaps produce the same kind of net protein loss during the heating. Engelbrecht and Mothes (1964) showed that this does not occur. The slightest increase in heat stress (1–2 min exposure to 50°–52°C) resulted in an "irreversible" heat injury. The normal degradation of proteins was actually inhibited, the heat damage occurred during the stress, and kinetin treatment failed to prevent the injury. Early evidence (Illert, 1924) had indicated that this kind of injury is not metabolic in nature for it was unaffected by CO_2 or oxygen supply. Yet Engelbrecht and Mothes (1964) were able to prevent it by preheating of the leaves at 46°C, i.e., by a hardening treatment. Heat tolerance (see Chapter 12), therefore, is induced by these hardening treatments against this metabolically "irreversible" rapid heat injury which is, therefore, a direct injury.

3. Primary Direct Heat Injury

a. OBSERVATIONS OF CELL INJURY

The above described kinds of injury, even if induced by heat shock, are produced relatively slowly (hours or days) as is characteristic of indirect stress injury. Primary direct stress injury, on the contrary appears rapidly, within seconds to about 30 min, although it may continue to progress for some time after cooling. In contrast to the indirect injury, which leads to a translocation of raw materials for protein synthesis to unheated parts, direct heat injury may actually result in damage to unheated parts. Yarwood (1961b) demonstrated that heat injury could be translocated from one primary bean leaf killed by heating for 10 sec at 65°C to the opposite unheated one. If the heated leaf was removed quickly, the unheated leaf was uninjured. Many attempts have been made to determine the nature of this direct injury by observation with the microscope.

i. Optical Microscopy. Sachs (1864) describes a heat solidification of protoplasm that may be reversible on cooling. As the small epidermal strips from young leaves or flower buds of *Cucurbita pepo* were warmed, protoplasmic streaming in the hair cells speeded up, until it became very violent. At higher temperatures, strands were pulled vigorously into one larger protoplasmic mass. Finally, the protoplasm all lay at rest against the cell wall. Five to ten minutes after cooling, protuberances gradually began to form, and the network of strands was slowly regenerated. This heat solidification of the protoplasm occurred when the strips were plunged into water at 46° to 47°C for 2 min. Even after exposure to 47°–48°C, streaming recommenced within 2 hr of cooling. In air, higher temperatures had to be used, e.g., 25 min at 50°–51°C. The solidification was then reversed only after a 4-hr cooling. Lepeschkin (1912) states that all layers of the protoplasm including the plasma membrane coagulate simultaneously at the heat-killing temperature. Therefore, he determined protoplasmic coagulation by the time semipermeability of the plasma membrane was lost, i.e., when the pigments of the cell sap were observed to diffuse out under the microscope. He later (1935) states that protoplasmic coagulation of plasmolyzed cells was recognized first by a rapid decrease in protoplast volume, and that the injury worked inward from the outer protoplasm layers (Lepeschkin, 1937). It required a higher temperature for the color to begin to leave the cell. Recovery was not possible after the chloroplasts had begun to coagulate, but the coagulation of the superficial protoplasm layer was reversible. In agreement with Lepeschkin's observations, soybean and *Elodea* exposed to sublethal temperatures showed a loss of chlorophyll and swollen chloroplasts (Daniell *et al.*, 1969). At the thermal death point, disorganization of the tonoplast, plasmalemma, and chloroplast membranes occurred. It was, therefore, concluded that the primary cause of the injury is disintegration of the cell membranes.

Other observations, however, indicated that injury could occur in the absence of observable coagulation. In opposition to Lepeschkin's earlier (1912) description, Bogen (1948) observed loss of color from *Rhoeo discolor* cells before any protoplasmic change could be detected. Lepeschkin's later (1935) observations seem to agree with Bogen. He describes four stages of heat coagulation in *Spirogyra* cells: (1) An imperceptible change in dispersion is detected by an increased permeability to water. The starch grains show just detectible swelling. (2) Starch swells significantly due to a greater increase in permeability, and coagulation of the protoplasmic surface begins. There is often a movement of the chloroplast ribbon toward the middle of the cell. (3) Complete heat swelling of the starch follows the complete coagulation of the chloroplast. (4) The proteins coagulate completely. Therefore, both investigators seem to agree that the first sign of injury is an increase in permeability.

Döring (1932) was able to detect heat swelling of chloroplast starch in living cells and suggests that this may actually injure the protoplasm. He concluded that the tonoplast is more heat tolerant than the rest of the protoplasm and that the changes in heat tolerance under different conditions may not be the same in these two protoplasmic components. Scheibmair (1937) observed the first signs of heat injury in the chloroplasts of mosses. They enlarged and became pale and irregular in contour. At the instant of death, the elaioplasts also changed observably and the whole protoplast contracted in apparent plasmolysis, which was easily distinguished from true plasmolysis by the angular form. Perhaps due to the chloroplast injury, the leaves of some plants (e.g., *Oxalis*) change from green to yellow on heat killing, though others (e.g., Polygonaceae) fail to show this color change (Illert, 1924).

Dangeard (1951a,b,c) attempted to find out how far disorganization by heat can proceed without causing death. He examined sections of fixed radicles after exposing them to heat-killing temperatures for various lengths of time. The chondriosomes (mitochondria) were rapidly destroyed, often by less than 1 min at 55° to 60°C. At lower temperatures, an hour or more was needed. They were rarely destroyed at 42°C or lower. In some cases their destruction was accompanied by the survival of a small percentage of the cells, judging survival by their appearance when fixed. Belehradek (1935) reviews many other changes observed by different workers in cells undergoing heat injury, e.g., the formation of granules, vacuolization, and protoplasmic contraction. Liberation of lipids has been recorded, even from the cell walls. The nucleus has been found the most heat-sensitive part of the cell, though more heat tolerant in the resting than the dividing state.

ii. Electron Microscopy. Essentially all of the organelles have been observed to show injury as a result of heat shock. A 5-min heating at 50°C decreased the nuclear volume about 15% in leaf cells adjoining stomata of *Tradescantia fluminensis*. Treatment at 59°C reduced them to a minimum volume less than half the original size (Barbal'chuk and Chernyavskaya, 1975). Longer heat treatments—growth at 38°C—gave rise to chromatin degeneration and chromosome lesions in sensitive varieties of wheat (Das, 1973). A shock at 44°C resulted in nucleolar segregation in meristematic cells of *Allium cepa*, but the granular ribonucleoproteins showed considerable stability (Risueno *et al.*, 1973). In *Physarum polycephalum*, polysomes were disaggregated by heat shock during either the S or G_2 period in the mitotic cycle (Brewer, 1972).

In nongreen cells, mitochondria show a sensitivity similar to that of chloroplasts. In animal tissues, the ultrastructure of the mitochondria is heat damaged at temperatures 8°–14° lower than the temperature that damages its isolated enzymes (Yakovleva *et al.*, 1974). Similarly, spores of *Dictyostelium discoideum* showed disrupted cristae in mitochondria as a result of heat damage (30 min at 45°C; Cotter and George, 1975). In the case of wheat roots, 45 minutes after heat shock (2 min at 45°C), the cisternae of the ER had expanded, giving rise to irregular vacuoles (Skogqvist, 1974a). The Golgi results were also irregular. Both recovered after 4hr, but the mitochondria remained irregular with fewer tubules and adhering membrane curls containing lipids.

b. MEMBRANE DAMAGE

Both the above optical and electron microscope observations point to membranes as the locus of injury—plasma, nuclear, chloroplast, mitochondrial, ER, and Golgi membranes. This conclusion is supported, not only by the visual observations, but also by the leakage of many substances from the heat-shocked cells. Thus, the following effects of 10 min. heat shocks on the plasmodium of *Physarum polycephalum* have been observed (Bernstam and Arndt, 1973a,b):

32°C	Slower cytoplasmic streaming, leakage of plasmodial pigments.
37°–38°C	Cessation of cytoplasmic streaming, respiration reduced, leakage of NA
41°C	Leakage of protein metabolites
47°–50°C	Highest level of leakage of substances and cessation of respiration.

Since decreased cytoplasmic streaming occurs also in the absence of injury, leakage was the first true sign of injury. Since heat-induced loss of ions requires as much as an 11°C higher temperature than loss of photosynthetic ability, the plasma membranes must be more stable to heat than the chloroplast membranes (Berry *et al.*, 1975). Similarly, the chloroplast envelope is apparently much less sensitive to heat stress than the photochemical reactions of the thylakoids (Krause and Santarius, 1975). 2 min. at 47.5°C increased the leakiness of tobacco leaves (Ben Zioni and Itai, 1973), although this was reversible, indicating that leakiness is an early step in heat injury. The loss of membrane integrity was reflected in the rate of incorporation of ^{32}P into the phospholipids. In the case of yeast cells, a temperature of

50°–60°C was required to induce escape of ions (Rudzyanok and Konyew, 1972). The rate of leakage increased with temperature, reaching a maximum rate close to the T_{max} for growth (Hagler and Lewis, 1974). Above the T_{max} they lost their ability to reabsorb the released substances. Leakage of amino acids was observed in eight of ten kinds of seeds when the imbibed seeds were exposed to 30–35°C (Hendricks and Taylorson, 1976). This was explained by a permeability change. However, amino acids are actively absorbed, and therefore the damage is more likely to be due to this active absorption system in the membranes than to its passive permeability. In favor of this explanation, the (Na^+, K^+, Mg^{2+}) ATPase of rat brain microsomes is increasingly inactivated at 25–49°C (Kirschmann *et al.*, 1973). K^+, Na^+, and Mg^{2+} protect against this heat inactivation and they suggest that a membrane conformation is stabilized through cation binding.

c. PROTEIN DENATURATION

On the basis of thermodynamic parameters, it has been suggested that heat injury involves only 1–3 catastrophic molecular events (Johnson, 1974). From the above observations, these should be sought in the membranes. Since all membranes consist of two substances—proteins and lipids—the molecular membrane changes must occur in these kinds of substances.

Protein denaturation was the earliest explanation of heat injury (Belehradek, 1935) and is the commonly accepted one to this day. Brock (1967) has recently rejected this explanation because heat injury follows first-order kinetics. However, first-order kinetics are characteristic of monomolecular reactions, and protein denaturation is simply the unfolding of a protein molecule. Christophersen and Precht (1952) even suggest that the death rate of cells follows the curve for a monomolecular reaction and therefore may be due to the breakdown of a single molecule. They admit that in the case of bacteria, deviations from the curve for monomolecular reactions are frequent. Brock's objection is, therefore, actually a point in favor of denaturation as the cause of heat injury. Unfortunately, the kinetics of the heat denaturation process are difficult to measure, because of the difficulty in separating it from the subsequent aggregation (see below). The early evidence in favor of the denaturation concept was the above-described coagulation of protoplasm in cells heated under the microscope.

Since many proteins are denatured at the heat-killing temperatures for many plants, this has long been accepted as the explanation of the observed cell coagulation, and of heat killing in general. Direct evidence in its favor is obtained from the reaction times. Lepeschkin (1912) points out that the same logarithmic relation to temperature has been found for the heat-killing time as for the protein coagulation time in solutions, although the former is more rapid.

Against the explanation of heat killing by protein denaturation is the long recognized fact that many organisms are killed at temperatures too low for the denaturation of known proteins. From such evidence, early investigators (Sachs, 1864; Just, 1877) concluded that protein coagulation cannot be involved. Collander (1924) points out that the temperature for denaturation of protoplasmic proteins is unknown. Furthermore, not all the cell proteins need be denatured in order to cause cell injury. There is always the possibility that a single sensitive protein may be responsible for the injury. Recent investigations lend support to this possibility. Aleksandrov (1964) found that cytoplasmic streaming is halted at a lower temperature than other cell processes. He, therefore, concluded that the proteins responsible for cytoplasmic streaming are more sensitive to heat denaturation than are other cytoplasmic proteins. Other results from his laboratory showed specific differences between proteins. The thermostability of urease in leaf homogenates was lower in the mesophilic *Leucojum vernum* than in the thermophilic species, *L. aestivum* (Fel'dman and Kamentseva, 1967). This was correlated with the thermotolerance of protoplasmic movement and respiration. The inactivation temperature for urease required much higher temperatures than the highest heat-killing temperature for higher plants. It required 40 min at 78°C to produce 50% inactivation in the less tolerant species, and the same time at 81°C in the more tolerant species. On the other hand, acid phosphatase (a much more heat-labile enzyme) was 50% inactivated at 48°–55°C (Fel'dman *et al.*, 1966). Similarly, a thermotolerant yeast showed a decrease in ATPase activity within the supraoptimal to submaximal temperature range for growth of the organism (Pozmogova and Mal'yan, 1976).

Lepeschkin (1935) points to the effects of various factors on heat injury as proof of the protein denaturation theory. Small amounts of acid (e.g., nitric) lowered the heat-killing temperature of *Spirogyra,* and also the heat-coagulation temperature of proteins. Brock and Darland (1970) have obtained a similar effect of pH on the maximum temperature for the most extreme thermophilic bacteria found growing at boiling temperatures (92°–100°C). At pH 2–3, the upper limit was lowered to 75°–80°C. Conversely, Lepeschkin raised both the heat-killing temperature and the heat coagulation temperature by use of very dilute alkalis. Narcotics, such as alcohol, ether, chloroform, and benzol, lowered both temperatures. Even the protective effect of plasmolyzing solutions against heat injury (see below) is in accord with the fact that denaturation of proteins occurs at higher temperatures in concentrated than in dilute solutions. He also points to the effect of salts in lowering the temperature for both processes.

Lepeschkin (1912, 1935) found that mechanical agents lower the heat-killing temperature. Cutting the tissues produces a sensitization that may last up to 15 hr in the case of beets. Bending of *Spirogyra* filaments has a similar effect. The farther from the cut or bend a cell is, the less its heat-killing temperature is affected. In contrast to Lepeschkin, Scheibmair (1937) found that centrifuging actually raised the heat-killing temperature of the moss *Plagiochila*. In a few cases, however, it was lowered.

Lepeschkin also found that light increased the speed of heat killing. Low concentrations of narcotics raised the heat-killing temperature though not affecting protein denaturation. Scheibmair (1937) was unable to confirm this last result and was able to obtain only a lowering of the heat-killing temperature.

It is obvious that these early investigators failed to differentiate between denaturation and coagulation of proteins. Modern investigations of pure proteins have, however, provided the basis for a reevaluation of the protein denaturation concept of heat injury. Protein denaturation is due to an unfolding of the molecule, with consequent loss of the activity possessed in the native state. According to Haurowitz (1959), such unfolding occurs when

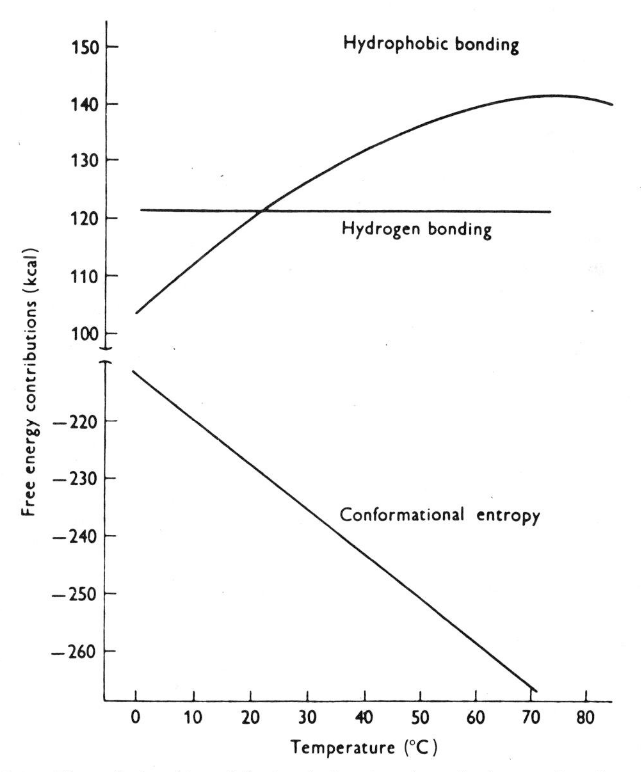

Figure 11.9. The relationship of hydrophobic bonding, hydrogen bonding, and conformational entropy to temperature. (From Brandts, 1967.)

the temperature is high enough to break the relatively weak H bonds (bond strength usually 2–3 kcal) which help to hold the folds in place in the native proteins. More recent evidence, however, has shown that the folds in the tertiary structure of the protein molecule are mainly held by hydrophobic bonds. As mentioned earlier, it is the weakening of these hydrophobic bonds that results in the reversible denaturation which occurs at low temperature. Evidence (Brandts, 1967) indicates that a similar reversible denaturation occurs at high temperatures. As the temperature rises, the conformational entropy favoring the denatured state increases more rapidly than the increase in strength of the hydrophobic bonds (Fig. 11.9) and a temperature is finally reached at which unfolding begins. The first effect of the high temperature on the proteins is, therefore, denaturation. This remains reversible, unless followed by aggregation:

$$N \underset{\text{normal } T}{\overset{\text{high } T}{\rightleftharpoons}} D \xrightarrow{\text{high } T} A$$

where N = native; D = denatured; A = aggregated proteins. Many proteins are converted so rapidly to the irreversible aggregated state that the reversible denaturation is difficult to detect. This is, no doubt, why it has been overlooked until recently. It also explains the observation of coagulation (an aggregation of proteins to form a gel structure) in heat-injured cells. Aggregation is irreversible only in the thermodynamic sense. Sachs' observation of reversal of heat coagulation may, perhaps, have been due to an enzymatic solubilization of the irreversible aggregate. On the other hand, it was reversible only for a short time, resembling the heat denaturation of proteins. It is, therefore, possible that the initial reversible coagulation was actually due to reversible denaturation, and that only the later irreversible coagulation was due to aggregation.

In agreement with this concept of denaturation, the thermophily of an alga (*Cyanidium caldarium*) was lost when treated with detergents, which are known to unfold (denature) proteins (Enami and Fukuda, 1977a).

Although Brandts' concept is an all-or-none conversion from $N \rightarrow D$, more recent evidence (Ananthanaryanan, 1973), indicates that heat denaturation of β-lactoglobuilin-A produces a partially unfolded state, retaining some degree of ordered structure which vanishes completely on addition of guanidine hydrochloride. It is, perhaps, this partial unfolding which is more readily repaired on cooling. The specific regions of the molecule thermally induced to unfold have been identified in bovine pancreatic RNase. Three sections of the chain are quite stable to thermal denaturation (Burgess and Scheraga, 1975).

In the case of the larger protein molecules, aggregation may be preceded by disaggregation on heating; for the breaking of bonds may occur not only within a protein subunit, but also between the subunits of a large multimer. For instance, a highly purified enzyme complex from bakers' yeast possessed two different enzymatic activities and a regulating site for both activities. After heating at 50°C for 5 min, the complex was disaggregated into subunits ¼ the size of the original complex, and these possessed only one of the enzymatic activities (Lue and Kaplan, 1970). Similarly, inactivation of peroxidase occurred at temperatures above 60°C and involved three processes: (1) dissociation of protohemin from the haloperoxidase, (2) a conformational change in the apoperoxidase, and (3) modification or degradation of the protohemin (Tamura and Morita, 1975). Some enzymes may actually be activated by a heat treatment (Eriksson and Vallentin, 1973) and this may also injure the cell. Peroxidase, in this way, may increase lipid oxidation and therefore membrane damage.

The hydrophobic bond strength increases with temperature up to and beyond the heat-killing temperature of most cells; at heat-killing temperatures their strength is greater than that of the hydrogen bonds (Fig. 11.9). Therefore, unlike low-temperature denaturation, heat denaturation may be expected to involve the breaking of H bonds even before the hydrophobic bonds. This conclusion has been supported by analysis of heat denaturation with infrared spectrometry (Boyarchuk and Vol'kenshtein, 1967). The number of CO and NH groups increased in all the six proteins tested. This indicates the disruption of secondary and tertiary protein structures by the abolition of many hydrogen bonds.

These recent determinations of heat denaturation of proteins at high temperatures have demonstrated that the less stable enzymes denature reversibly at temperatures considerably lower than previously suspected (Fig. 11.10). In the case of chymotrypsinogen A, for instance, the measured rates of H exchange at pH 5 and 37°C were higher and the activation energies lower than expected from thermal cooperative unfolding (Rosenberg and Enberg, 1969). The heat stability of enzymes, in fact, covers a wide range of temperatures (see Tables 12.9 and 12.10). It is, therefore, not surprising that denaturation of some proteins can occur at low enough temperatures to fall within the heat-killing range.

Unfortunately, it is difficult to interpret results obtained with pure proteins or cell extracts in terms of the normal, living cell. The stability of proteins may be greater when in the cell due to the presence of protective substances. Evidence of this is the 10°C higher inactivation point for crude amylase than for the crystalline product (Roy, 1956). Nakayama (1963), in fact, has isolated and partially purified a small-moleculed substance from Japanaese radish leaf which protects β-amylase from sweet potato against heat inactivation. The half-life of trehalase and invertase at 60° and 65°C was much greater when intact ascospores were heated than in the case of extracts of *Neuros-*

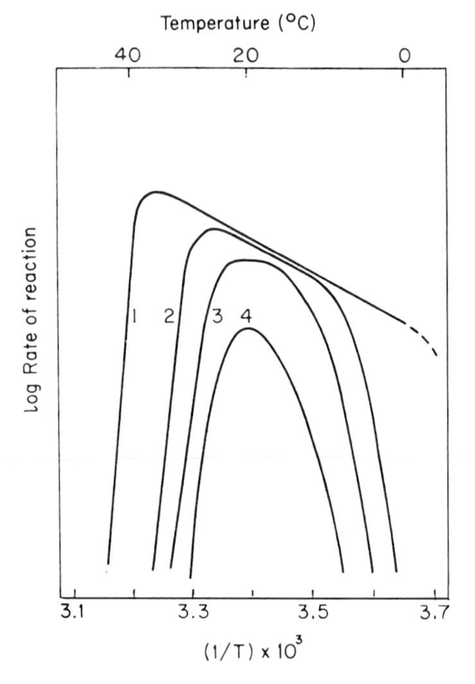

Figure 11.10. Arrhenius plots for four enzymatically controlled reactions. (From Brandts, 1967.)

pora tetrasperma (Yu *et al.*, 1967). Dialysis of the ascospores removed the protection. Similarly, complex formation with DNA (and also with CM-cellulose) prevented heat coagulation of serum albumin, ovalbumin, and chymotrypsinogen A (Hofstee and Bobb, 1968). Cystathionase and cysteine sulfinic acid decarboxylase were protected against heat denaturation by the cofactor pyridoxal phosphate (Chatagner *et al.*, 1968). Even the artificial attachment of a dye molecule (methyl orange) to a protein (serum albumin) increased its thermostability by 2.5°C (Vitvitskii, 1969). Many other examples are to be found in the literature.

It is even possible to distinguish large differences between the thermostability of proteins in the pure state, which are not detectible under natural conditions. The protein of tobacco mosaic virus (TMV) is denatured at 40.5° in the vulgare form and at 27°C in the mutant (Jockush, 1968). However, this difference between the two proteins was difficult to detect in the whole virus particle, due to stabilization by coaggregation with RNA. Similarly, quaternary aggregation (at pH 5.0) shifts the denaturation temperature upward by 20°-25°C.

On the other hand, substrate decreased the heat stability of succinate cytochrome *c* reductase but increased the heat stability of succinate oxidase (Luzikov *et al.*, 1967). The thermal inactivation of yeast cytochrome b_2 is increased in the presence of flavin mononucleotide (FMN) (Capeillare-Blandin, 1969). Similarly, though the presence of other proteins may have a protective effect, the opposite is also possible. The ovomucin-lysozyme complex is heat denatured at a lower temperature than that observed for either protein separately (Garibaldi *et al.*, 1968).

It must also be realized that activity and conformation are not always changed simultaneously by heat, so that injury to the cell might be expected at a lower or a higher temperature than the denaturation temperature. Lysozyme, for instance, is stable up to 75°C, yet the activity reached a maximum at 50°C, decreasing markedly above this temperature (Hayashi *et al.*, 1968). The loss of activity at 60°—70°C was believed due to the formation of an enzyme-substrate complex difficult to hydrolyze.

In the case of psychrophiles, the relationship of heat injury to protein denaturation appears to be clear-cut. Pyruvate carboxylase was 50% inactivated after it was heated for 10 min at 35°C (Grant *et al.*, 1968). This agreed with the cessation of glucose fermentation by resting cells (and cell-free extracts) after 30 min at 35°C. Similarly, the amino acid incorporating system of the psychrophilic yeast *Candida gelida* was completely inhibited after incubation at 35°C for 30 min (Nash *et al.*, 1969). This was due in part to unusually temperature-sensitive aminoacyl-sRNA synthetases. Of the thirteen tested, seven retained less than 50% of their activity after 30 min at 35°C. Leucyl-sRNA synthetase was inactivated after only 7 min at 35°C. *Candida gelida* also possessed thermolabile soluble enzymes involved in the formation of ribosomal-bound polypeptide chains. Similar results were obtained with another psychrophile, *Micrococcus cryophilus* (Malcolm, 1968, 1969). It cannot grow above 25°C due to an inhibition of protein synthesis. Heating for 10 min at 30°C inactivates three species of tRNA synthetase of this psychrophile, but has no effect on those of a mesophile or thermophile. It is primarily the glutamyl-tRNA synthetase and the prolyl-tRNA synthetase that are inactivated above 25°C. The tRNA is still functional. In the case of the marine psychrophile *Pseudomonas* sp., RNA synthesis could not be detected at supramaximal temperatures (Harder and Veldkamp, 1968). In contrast to *C. gelida*, the isoleucyl- and leucyl-tRNA synthetases from *Bacillus stearothermophilus* (a thermophile) and *Escherichia coli* (a mesophile) showed no striking difference in thermostability. They were able to aminoacylate the tRNA *in vitro* at temperatures above 70°C (Charlier *et al.*, 1969). It is, therefore, apparently only in the psychrophiles that this mechanism of heat injury functions. In these organisms, the mechanisms of indirect and direct injury are apparently identical; the maximum temperature for growth is due to cessation of protein synthesis (indirect heat injury) which in its turn is due to the direct heat inactivation of enzymes (synthetases).

From all these results, it must be concluded that protein denaturation or inactivation does, indeed, occur in the living cell at a low enough temperature to account for heat killing, at least in some cases. It must be remembered, however, that protein denaturation at high temperatures is a reversible process and does not become irreversible until followed by aggregation. It, therefore, does not seem likely that the rapid heat killing can occur unless the denatured proteins are also aggregated. What kind of bond formation may lead to this aggregation? Some evidence now points to a possible explanation. Just as in the case of freezing injury, heat injury results in an increased SS content of the proteins at the expense of SH groups (Table 11.13). These results have been confirmed by Kotlyar *et al.* (1969). When germinated wheat was heated to 40°–80°C, the level of SH groups in the water-soluble proteins decreased, and the SS: SH ratio increased. Similarly, the proteins of soybean hypocotyl show maximum thermostability after a 4-hour incubation of the tissue with 10^{-5} *M* 2,4-D and this was correlated with a decrease in SH content (Morré, 1970) and, therefore, a decrease in ability to form intermolecular SS bonds on heating. The oxidation of SH to SS groups may possibly explain the toxic action of O_2 on pea seedlings at supraoptimal temperatures (Aver'yanov and Veselovskii, 1976). Similarly, light decreases the thermostability of the Hill reaction in pea and spinach chloroplasts and suppresses other photosynthetic reactions whose irreversible inhibition is enhanced by the SH reagent NEM (Ageeva, 1977). Tempera-

TABLE 11.13

Effect of Heat Killing (15 Min at 58°C) on SH Content of Supernate from Kharkov Wheat after Absorbing 60 gm H_2O per 100 gm Grain[a]

	SH	SH + 2SS	%SH (of SH + 2SS)
A. Unvernalized (3 days at 20°C)			
Control	0.25	0.85	30
Heat killed (1)	0.16	0.80	20
(2)	0.12	0.90	13
B. Vernalized (40 days at +3°C)			
Control	0.52	1.36	38
Heat killed	0.20	0.96	21

[a] From Levitt, 1962.

tures of 40°C and higher decrease or stop cytoplasmic streaming in tobacco cells and simultaneously increase the number of SH groups (Bilanovskii and Oleinikova, 1974). A SH group is apparently required for the conformational stability of an enzyme from a thermophilic organism due to its participation in the tight organization of the active site (Mizusawa and Yoshida, 1973). The SH $\rightleftharpoons$ SS hypothesis has, indeed, been proposed to explain the heat injury in microorganisms (Harris, 1976; see Chapter 12). On the other hand, during the thermal denaturation of hen white lysozyme at 76°C (and perhaps even at 100°C) the four SS bonds remain intact (Chen *et al.*, 1973). In the case of some microorganisms, SH and SS groups can play no role in the thermal denaturation of certain proteins. For instance, the cysteine-cystine content of the flagellar proteins (as well as of the ribosomes; Friedman, 1968) of bacteria is very low, and does not differ significantly in thermophiles and mesophiles (Mallett and Koffler, 1957). Furthermore, the addition of 0.2 *M* sodium thioglycollate at pH 8 in 5 *M* urea had no effect on the thermostability. Therefore, SS bridges do not seem to form in these proteins. Similarly, an extracellular proteinase of *Escherichia coli* is completely inactivated by a 10 min exposure to 55°C, yet it contains no SH or SS groups (Nakajima *et al.*, 1974). Enzymes of thermophilic bacteria, in fact, owe their stability mostly to additional salt bridges (Perutz, 1978).

The inverse relationship between killing temperature and moisture content is also explainable by protein denaturation; there is a quantitative relation between moisture content and protein denaturation. This is because the unfolding can take place only in the presence of adequate water, which permits the necessary freedom of movement of the protein molecules. Thus, the rate of inactivation of a protein (e.g., sweet potato β-amylase at 63°C; Nakayama and Kono, 1957) increases with decreasing protein (i.e., enzyme) concentration. Consequently, dehydrated protoplasm can survive high temperatures without injury. In the case of some seeds, in fact, there is no injury until the temperature is high enough to break the valence bonds in the proteins and other protoplasmic substances, as shown by the charring after prolonged periods at these temperatures. The above two types of heat injury can therefore be called denaturation injury and decomposition injury, respectively.

Between the two extremes, a combination of the two types of injury may conceivably occur; because some purely chemical reactions (e.g., caramelization of sucrose) can also take place. That there are two types of injury is indicated by the break in the Arrhenius plot (Fig. 11.2) of the reciprocal of the killing temperature versus log of time (the reciprocal of the reaction rate). When this break occurs at relatively low temperatures (e.g., at 42°C; Fig. 11.2) it is explainable by indirect heat injury in the lower range and by direct heat injury in the upper range. When it occurs at a relatively high temperature (e.g., 60°C for white spruce; Lorenz, 1939) it is explainable by denaturation injury in the lower range and by decomposition injury in the upper range.

As mentioned above, heat denaturation of proteins has a temperature coefficient as high as that of heat killing. This is the strongest evidence in favor of the protein denaturation theory of heat injury in the intermediate heat-killing zone. In agreement with this, heat (decomposition) injury in the highest heat-killing zone does not have such high temperature coefficients (Belehradek, 1935). This conclusion has been corroborated for white spruce (Table 11.12), the Q_{10} dropping from 73 for temperatures below 60°C to 3.6 for temperatures above it (Lorenz, 1939). In terms of activation energy, the values were 94 and 28 kcal/mole, respectively. Siegel's (1969) more recent results are in good agreement with these—93 and 19 kcal/mole. He found that beet root tissue differs in stability toward oxygen at moderately high (45–60°C) and very high (60°–100°C) temperatures.

On the basis of the above evidence, protein denaturation may be the cause of heat killing in microorganisms, but may be insufficient in itself to account for heat killing of more complex organisms. Some investigators, for instance, (e.g., Rosenberg *et al.*, 1971—see Van Uden and Vidal-Leiria, 1976) have obtained nearly identical isokinetic temperatures for denaturation of proteins and thermal death in microorganisms suggesting a cause and effect relation. Others (Evans and Bowler, 1972; see Van Uden and Vidal-Leiria, 1976) obtained lower values for thermal death in multicellular animals and protozoans, suggesting that other macromolecules (phospholipids, nucleic acids) should be considered.

d. LIPID LIQUEFACTION

The observed liberation of lipids at high temperatures led Heilbrunn to suggest that heat killing may be due to liquefaction of protoplasmic lipids (see Belehradek, 1935). Many attempts have been made to support this theory, but without success (Belehradek, 1935; Campbell and Pace, 1968). Lepeschkin (1935) concludes that since the lipids occur in such thin layers in the protoplasm, they would liquefy instantly at a definite temperature and therefore could not account for the high Q_{10} values for heat killing. This objection is no longer valid, since (1) the lipid layer consists of several lipids, each with a different phase transition temperature, and (2) it is now known that the phase transition (from liquid crystalline to solid gel state) is not a true melting point and occurs over a temperature range of 10°–20° (see Chapter 4). Nevertheless, there is considerable evidence against the concept of lipid liquefaction as the cause of heat injury. Lepeschkin believed that lipids are freed as the result, rather than the cause, of death.

The strongest argument against the lipid liquefaction hypothesis is the now well-established fact that the phase transition of the membrane lipids occurs at temperatures below, not above the optimum growing temperature. Thus, in the case of *Mycoplasma laidlawi* (Reinert and Steim, 1970), the phase transition occurred at the same temperature whether determined (calorimetrically) in viable organisms, in isolated membranes, or in isolated membrane lipids. Furthermore, it occurred at a much lower temperature than the denaturation of the membrane protein—a transition temperature of 20°—45°C, compared to denaturation beginning slightly above 50°C. Since the lipid phase transition began at a temperature well below the temperature at which the organism grew, and the normal growing temperature (37°C) was well into the phase transition zone, the change in phase of the lipid seems to have no relation to heat injury in this organism. It was, in fact, possible by enriching the membranes in oleate to lower the phase transition temperature of the membrane lipids to −20°C. In the case of a protozoan, when the temperature rose above the optimum for growth, there was a significant reduction in protein synthesis, but lipid biosynthesis was slightly stimulated (Byfield and Scherbaum, 1967). This would certainly oppose the lipid concept even as an explanation of indirect heat injury, since lipid loss would be more readily replaced than protein loss. It also indicates that the energy source was still adequate.

In spite of the early evidence against it, the lipid liquefaction theory is supported by the more recent evidence. Thus even though the phase transition occurs largely at temperatures *below* the T opt for growth, nevertheless the mobility of the lipids must increase with the rise in temperature until a temperature is presumably reached at which lipid vesicle formation with minimum surface would be more stable thermodynamically than the maximum surface of the membrane bimolecular leaflet. At this point, membrane destruction and cell death would occur. Perhaps the best evidence of a role for lipids in heat injury is their well-documented increase in saturation with increase in heat tolerance (see Chapter 12). Lipid peroxidation may also conceivably be involved. The lipid oxidation may be produced by peroxidase, and therefore, paradoxically may be prevented by thermal destruction of the heme group of the peroxidase (Eriksson and Valentin, 1973). Therefore, lipid peroxidation may perhaps be a factor in heat injury only at moderately high temperature that is not high enough to inactivate the oxidases.

Heat injury may, therefore, involve both lipid and protein changes. This would seem to point to membrane damage as the cause of direct heat injury. A loss of semipermeability or inactivation of the active uptake system at high temperatures could then be due to either (1) excessive fluidity of the lipids, leading to disruption of the lipid layer, or (2) denaturation and aggregation of the membrane proteins.

e. NUCLEIC ACIDS

Like proteins, nucleic acids can also be denatured by heat, and the reaction is again first order (Peacocke and Walker, 1962). It is, therefore, not surprising that attempts have been made to implicate them in heat injury.

If, however, heat injury occurs via membrane damage, no direct role for NA can be visualized, since these substances are not essential components of cell membranes. They are, nevertheless, required for protein and lipid synthesis and therefore may be expected to play an indirect role in heat injury. They should, at least, be important in membrane repair, which was shown above to play a role in thermotolerance.

No relationship exists between the heat stability of DNA and thermophily, since both the base composition and the melting temperature are identical in thermophiles and mesophiles (Campbell and Pace, 1968). Similarly, the thermal denaturation of the sRNA's from a mesophile (*E. coli*) and a thermophile (*Bacillus stearothermophilus*) are virtually identical. Yet the ribosomes of the thermophiles are much more heat stable than those of the mesophiles. With few exceptions, the guanine and cytosine of the rRNA tended to increase, the adenine and uracil to decrease with increasing growth temperature. As in the case of DNA and sRNA, however, the rRNA is not significantly different in mesophiles and thermophiles, either as to its thermal denaturation or its gross base composition. Similar results have been obained with a psychrophile (*Micrococcus cryophilus*). It is unable to grow above 25°C due to an inhibition of protein synthesis (Malcolm, 1968). This was not due either to an inability to synthesize mRNA or to a degradation of existing RNA. On the other hand, three enzymes (tRNA synthetases) were found to be temperature-sensitive. Similarly, none of the ten tested sRNA species from the psychrophile *Candida gelida* was temperature-sensitive (Nash *et al.*, 1969).

Evidence in favor of a role for RNA was obtained in the case of the plasmodia of *Physarum polycephalum* at supraoptimal temperatures which inhibited protein synthesis (Bernstam, 1974). In the case of the extreme thermophile, (*Thermus thermophilus*) (former name *Flavobacterium thermophilum*), which can grow at 85°C, the results strongly suggest that most of the ribothymidine normally present is replaced, enabling the tRNA to synthesize protein at high temperatures (Watanabe and Oshima, 1974). In *Anacystis nidulans*, no direct relation was found between the thermotolerance of rRNA and degradation of ribosomes in the cell. In contrast to this organism, the rRNA of *Escherichia coli* is thermostable (Wollgiehn and Munsche, 1974). Fragmentation of one subunit of the rRNA of tobacco chloroplasts begins at 45–50°C and is completed at 60°C. The thermal instability of this component is the consequence of "hidden breaks" in the polynucleotide chain, originating by nuclease action in the course of aging of the ribosomes (Munsche and Wollgiehn, 1974).

4. Mechanisms of Heat Injury

On the basis of all the above evidence, what is the mechanism of heat injury? According to Crisan (1973), some 25 hypotheses have been proposed to explain it. In view of the many kinds of heat injury described above, any attempt to identify a single mechanism seems futile. The secondary, heat-induced, water stress which injures via desiccation is discussed in Vol. 2, Chapter 5, and will not be considered here. The following are the remaining possibilities.

(a) The five kinds of primary, indirect heat injury, occurring during continued exposure (hours or days) to moderately high temperature are all metabolic in nature and, therefore, may conceivably have a single basic mechanism. The first strain produced by high temperature is kinetic—an increase in reaction rates. As in the case of chilling injury (see Chapter 4), differences in slope of Arrhenius plots for different metabolic reactions may reverse relative reaction rates and lead to (1) an increase in net breakdown and therefore may produce a decrease in concentration or complete absence of an essential metabolite at heat-injuring temperatures. The resulting injury may be starvation, biochemical lesions, protein hydrolysis, and growth reduction or cessation. (2) Conversely, it may produce an increase in concentration of a toxic substance normally not detectibly present, or in too low a concentration to be injurious. In the case of soluble enzymes, the differences in slope may simply be due to differences in activation energy for different reactions, or if bends occur, to protein denaturation and therefore enzyme inactivation. In the case of membrane enzymes, breaks in the curve may be due to a phase transition or increased mobility of the membrane lipids associated with the enzyme. Membrane proteins are embedded in the membrane lipids, and due to their higher hydrophobicity than that of the soluble proteins, are not so likely to be denatured by high temperatures at which hydrophobic bonds are strengthened. Involvement of nucleic acids has also been suggested, but the evidence to date is negative.

(b) The above kinds of metabolic damage, occurring during hours or days at moderately high temperatures are unable to explain the primary, direct injury due to heat shocks for seconds or minutes at somewhat higher temperatures. A different mechanism must, therefore, be involved. The direct observations of heat shock injury, as well as the leakage of ions and amino acids, and the effects of salts, narcotics, etc., all point to membrane damage as the cause of primary, direct heat injury. Even membrane damage, however, may be due to more than one heat-induced strain. Protein denaturation and lipid phase transition or increased mobility are the logical explanations.

How can the same direct strains (protein denaturation and aggregation, increased mobility of lipid) produce both direct and indirect injury? Perhaps this is due to the differences in time and temperature. At the more moderate high temperatures, the strains are presumably insufficient to induce direct (heat shock) injury. During longer exposure, however, indirect injury develops. At the more extreme high temperatures, the rapid heat shock injury occurs before the indirect injury can develop. The time factor, however, is not the only explanation of indirect injury. In fact, indirect injury may be induced very rapidly—e.g. chloroplast damage, leading to disruption of photosynthesis. The simplest explanation, therefore, of a single cause (membrane damage) of both indirect and direct injury is that the chloroplast membrane is most sensitive to high temperature, the mitochondrial less so, and the plasma membrane least sensitive.

Therefore, indirect injury due to starvation would occur at temperatures just high enough to damage the chloroplast membranes. Biochemical lesions, toxicity, and protein net breakdown would occur at slightly higher temperatures due to mitochondrial membrane damage. Finally, direct injury would occur at still higher temperatures due to destruction of the integrity of the plasma membrane.

Post-heating recovery from heat shock, if not too severe, is possible by repair of the injury. Heat shock injury must, therefore, include damage to the heat repair capacity of the plant—its ability to return the altered proteins or lipids to their normal state. This prevention of repair would require a specific denaturation of the healing enzymes, during the heating. But denaturation is not enough, since it would be reversible on return to the lower, healing temperatures. Therefore, the healing enzymes must be converted from the denatured to the irreversibly aggregated state during the heating, in order to prevent repair after the heat shock.

A final kind of heat injury occurs at still higher temperatures in the case of dehydrated cells of seeds, pollen, or micoorganisms. The above kinds of heat injury cannot be expected to occur in such dehydrated cells. Denaturation cannot occur in the absence of water, nor can a phase transition of membrane lipids injure via disruption of active uptake or loss of semipermeability, since no uptake or diffusion can occur in the absence of water. In the absence of experimental evidence, it can only be postulated that the temperatures are high enough to induce direct chemical decomposition or oxidation—e.g. of membrane lipids.

In summary, then, there are four possible strains leading to heat injury: (1) reversal of relative rates of reactions, due to differences in activation ener-

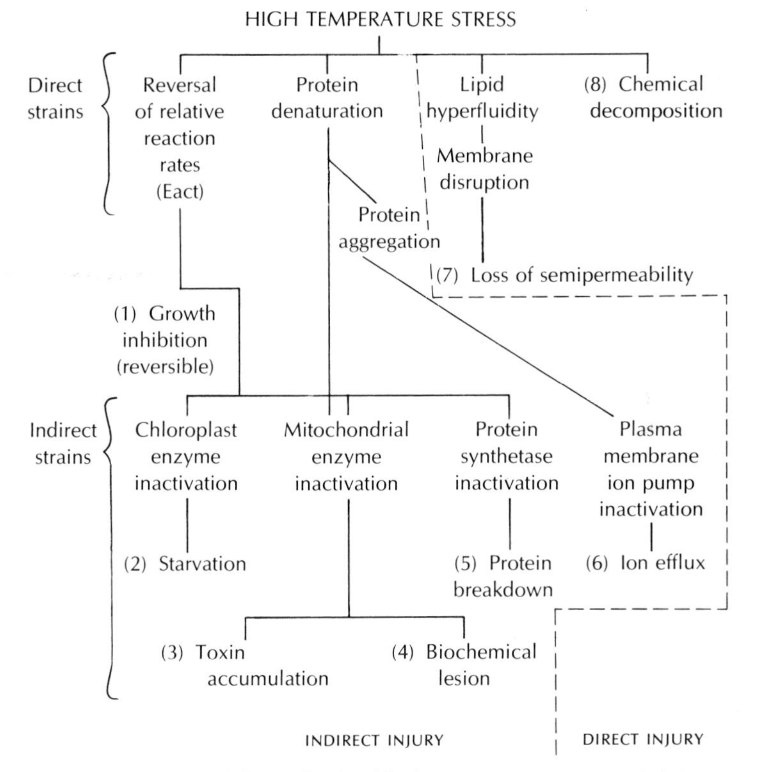

Diag. 11.1 Scheme for 8 different kinds of high temperature injury and their proposed relations to the 4 direct strains. Numbers in approximate order of development with rise in temperature.

gies, (2) protein denaturation and aggregation, (3) hyperfluidity of membrane lipids, and (4) direct chemical decomposition (Diag. 11.1). Other possible strains, such as nucleic acid denaturation, have not so far been implicated in heat injury.

F. PROTECTIVE SUBSTANCES

The first substances tested as possible protective agents against heat injury were salts and sugars. Several investigators obtained marked protection with them (Table 11.14). In contrast to these results, Kaho (1921, 1924, 1926) and Bogen (1948) found that salts *lowered* the heat killing temperatures of cells (observed microscopically) in proportion to their penetration into the cells. The reason for the discrepancy is perhaps the high temperature range used by Kaho—67.5 to 76.5°C—which permitted rapid penetration by the salts, especially since they were used in single salt solutions. The lower temperature range in Table 11.14 prevented appreciable penetration and permitted dehydrative protection.

Some positive evidence (see above) has been obtained of the utility of protective substances such as kinetin in the case of the proteolytic kind of metabolic heat injury. The protection by the kinetin was due to an increased protein synthesis relative to hydrolysis according to Engelbrecht and Mothes (1960). Porto and Siegel (1960) ascribe it to antioxidant activity of the kinetin. In agreement with Mothes, germination of lettuce seed in the light is inhibited above 27°C, but this cut-off point is shifted upward by kinetin solutions (0.1–1mg/liter) or downward by applying ABA (→1mg/liter; Reynolds and Thompson, 1971). Similarly, heat treatment (2 min at 46–47°C) of

TABLE 11.14

Protection by Dehydrating Salt or Sugar Solutions against Heat Killing or Shock

Plant	Solution	Killing or shocking temperature (°C)	Time exposed (min)	Reference
Agave americana	Water	57	10	de Vries, 1871
	10% NaCl	58.2	10	
Hyacinthus orientalis	Water	47.5	10	de Vries, 1871
	10% NaCl	55.4	10	de Vries, 1871
Saxifraga sarmentosa	Water	44.0	10	de Vries, 1871
	10% NaCl	58.2	10	de Vries, 1871
Fucus eggs	Seawater	45	1	Döring, 1932
	Seawater + 1*M* sucrose	50	1	
Plagiochila asplenioides	Water	48	4½	Scheibmair, 1937
	2½ *M* sucrose	>52	9	
Hookeria luscens	Water	52	3	Scheibmair, 1937
	1 *M* sucrose	>52	8	
Rhoeo discolor	Water	60	20	Bogen, 1948
	⅛ *M* sucrose (hypotonic)	60	17	
	½ *M* sucrose (hypertonic)	60	36	
Torula utilis	Water	44	60	Van Halteren, 1950
	1-3 *M* NaCl, glucose, sucrose	>44	60	
Torulopsis kefyr	Water	54		Christophersen and Precht, 1952
	⅓ - ⅔ *M* maltose	60		

TABLE 11.15

Protective Effects of Various Substances against Heat Injury or Heat Inactivation

Substance	Organism or system	Protection (+) or sensitization (−)	Reference
1. Salts			
0.25 *M* $CaCl_2$	Millet and wheat seed (presoaked 18–24h)	+	Genkel and Tsvetkova, 1955
Sea water (dil.)	*Chaetomorpha cannabina*	−	Biebl, 1969
Sea water (conc.)		+	Biebl, 1969
Chlorides	Bacteria	+	Ljunger, 1962
$ZnSO_4$ (0.05–0.08%)	Sugar beets, potatoes, sunflowers	+	Petinov and Molotkovsky, 1961, 1962
$ZnSO_4$, $CaCl_2Ca(NO_3)_2$	Barley, wheat	−	Onwueme and Laude, 1972
2. Organic acids	Sugar beets, potatoes, sunflowers	+	Petinov and Molotkovsky, 1961, 1962
3. Sugars		+	Feldman, 1962
Sucrose, glucose, lactose	5 species		
Rhamnose, maltose, and above	Heat coagulation of proteins	+	Tekman and Oztekin, 1972
4. Nucleotides	Inactivation of enzyme	+	Yon, 1973
UMP			
Adenine	Tomato plants	+	Emmerikh, 1973
5. Growth regulators			
Kinetin, ABA (see text)			
2,4-D(10^{-5}*M*)	Soybean hypocotyl protein	+	Morré, 1970
Maleic hydrazide	Growing leaves	+	Aleksandrov, 1964
IAA, maleic hydrazide	Wheat coleoptiles	−	Gorban, 1968
IAA, GA	Barley and wheat coleoptiles	−	Onwueme and Laude, 1972
6. Miscellaneous			
Polymers	Glucose oxidase activity	+	O'Malley and Ulmer, 1973
Chloroethanol	Wheat coleoptile	+	Miyamoto, 1963
Dipicolinic acid	Glucose dehydrogenase thermostability	+	Cross, 1968
Casein hydrolyzate or histidine	Wheat roots	+	Skogqvist and Fries, 1970

root systems of tobacco and beans reduced the CK levels and increased the ABA levels in the xylem exudate (Itai *et al.*, 1973).

Protection against heat injury has been reported for a wide variety of substances, although in the case of salts, some of the results are contradictory (Table 11.15). Some of the contradictions may be due to different conditions. In opposition to his positive results (Table 11.15) Morré (1970), for instance, also obtained no effect of 2,4-D when used at a concentration that stimulated growth.

12. Heat Resistance

A. HEAT AVOIDANCE

1. Occurrence

Since the term thermotolerance has come into more or less general use in recent years, as synonymous with heat tolerance, the parallel terms, thermoresistance and thermoavoidance, may also be used for heat resistance and heat avoidance, respectively.

As has already been mentioned, higher plants are poikilotherms. At first sight, this might seem to eliminate avoidance in the case of high temperature just as in the case of low-temperature injury. The problem is more complicated in the former. The intense absorption of radiant energy by leaves and other plant parts usually results in a much higher leaf temperature than the temperature of the transparent and practically nonabsorbing air (see Chapter 11). Even floating algae may possess a temperature in the middle of a mat up to 6.4°C higher than that of the surrounding water (Schanderl, 1955). Therefore, avoidance in the case of the heat-resistant plant does not necessarily imply a plant temperature lower than that of the air, but simply a temperature lower than that of a control, less avoiding plant under the same environmental conditions. Plants A and B may both have the same heat-killing temperature (and, therefore, tolerance), which is, for instance, 7°C above that of the surrounding air. If the temperature of plant A's leaves rises to 10°C above that of the air, and the temperature of plant B's leaves rises only to 5°C above that of the air, A's leaves will be killed, but B's will not. Since there is no difference in tolerance, the greater heat resistance of plant B is obviously due to avoidance.

That this kind of avoidance does, indeed, occur, has been shown by Lange and Lange (1963). Hard-leaved woody plants on the south coast of Spain attained leaf temperatures as much as 18.4°C above the air temperatures during August to September. The soft-leaved plants had leaf temperatures 10°–15°C below that of the hard-leaved plants, though still slightly above the air temperature. Since the hard-leaved plants attained leaf temperatures as high as 47.7°C and the killing temperature of the soft-leaved plants was as low as 44°C, the latter survived because of heat avoidance.

2. Possible Mechanisms

Heat avoidance may conceivably be developed in several ways.

a. INSULATION

Insulation cannot protect the plant whose temperature (due to absorption of radiant energy) is higher than that of its immediate environment. The insulation could only aggravate matters by preventing conductive loss of heat from the plant to the cooler environment. It can protect, however, if the plant part is in direct contact with a warmer environment above its killing temperature. This explains why mature tree seedlings with a good protective layer of bark are more heat resistant than immature seedlings with thin bark. The latter may be killed at soil level due to the high temperature of the soil surface. Of course, the soil temperature does not rise as high under the canopy formed by a mature seedling, so that protection of the stem by insulation may be superfluous.

b. DECREASED RESPIRATION

2. Respiration might conceivably have a harmful effect by contributing to the rise in temperature; a lowered respiration rate might conceivably induce avoidance. In the case of leaves, the quantity of heat released in this way is so much smaller than the radiant heat absorbed as to be insignificant (less than 10^{-5} and 0.7 kcal/cm^2 leaf/min, respectively). In the case of fleshy, insulated organs, however, such as the aroid inflorescence, respiration may be an important contributor to the temperature (Table 11.5) though this normally occurs at a time of year when there is no danger of heat injury. The major heat avoidance must, therefore, be due to the remaining two causes.

c. DECREASED ABSORPTION OF RADIANT ENERGY

A decreased absorption of radiant energy may be brought about in three ways.

1. Reflectance of radiant energy by leaves varies with the wavelength and the type of leaf (Fig. 12.1). Measurements (Billings and Morris, 1951) indicate a greater reflectance by species of hot habitats. In general, the reflectance curves of green leaves showed values of about 5% at 440 nm, rising to a peak of about 15% at 550 nm, with a gradual slope down to 5 or 6% at 675 nm. Above this, the curves rose steeply to a plateau of about 50% in the infrared region of 775–1100 nm. On the average, desert species reflected the greatest amount of visible radiation, followed by subalpine, west-facing pine forest, north-facing pine forest, and shaded campus species, respectively (Fig. 12.1). In the infrared, the differences between groups were not so marked, but the greatest reflectance here also was shown by the desert

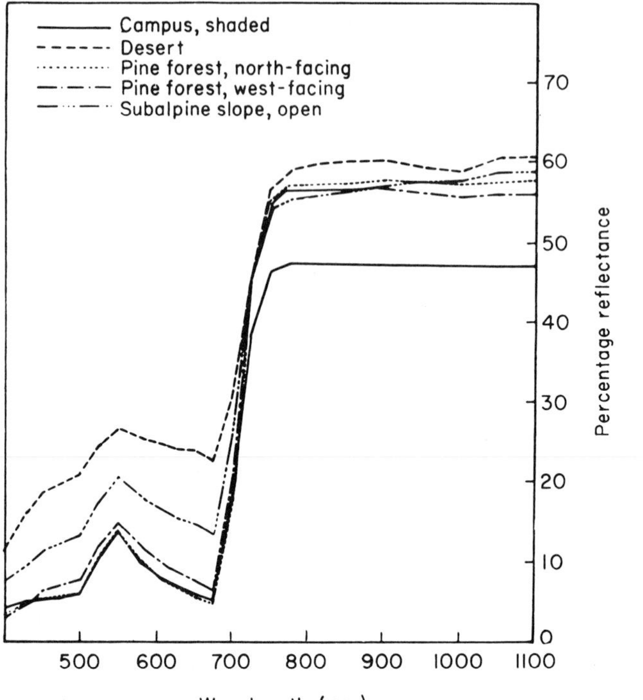

Figure 12.1. Percentage reflectance from leaves of various species. (From Billings and Morris, 1951.)

species, with an average value of about 60%. The greatest infrared reflectance (almost 70%), was measured from the glabrous leaves of the desert peach, *Prunus andersonii*. The hairs or scales on desert and subalpine plants were correlated with higher leaf reflectance (by as much as 56%-Ehleringer *et al.*, 1976) in the visible but not necessarily in the infrared region. The specific relationship depends on the particular range of infrared. By removing hairs from the upper surface of leaves of *Gynura aurantiaca,* Gausman and Cardenas (1969) showed that pubescence on young leaves increased the total and diffuse reflectance in the 750–1000 nm region but decreased both in the 1000–2500 nm region. Yet, according to Taleinshik and Usol'tseva (1967), the hairy coverings are responsible for the heat resistance of Manchu cherry. On the basis of the above results, this is possible only in the visible and the short infrared. However, results with several species indicate that refractive index discontinuities in leaves cause the reflectance of near IR light (Gausman, 1977). If this is correct, reflectance differences in the near IR between plants would depend on cell size, shape, and number.

On a clear day, considerably more than 50% (often as much as 65%) of the solar radiation incident at the earth's surface is in the infrared region (Gates and Tantraporn, 1952). The major effect on the plant's temperature must therefore be produced by reflectance in the infrared rather than in the visible region, the more so since the plant reflects so little of the latter. On the basis of Billings and Morris' results, shade leaves absorb about one-third more in the infrared than do leaves of desert plants. Their temperatures might therefore be expected to rise proportionately higher.

This assumes that the reflectance found by Billings and Morris for the short infrared (up to 1100 nm) holds as well for the long infrared, which includes most of the incident infrared in the radiation from the sun (Gates, 1965). Gates and Tantraporn (1952) have measured the reflectance by leaves in the long infrared region (1–20 μm). They find that it is generally small—less than 10% for an angle of incidence of 65°, and less than 5% for 20°. Since the transmissivity of the leaves is zero in the infrared region beyond 1.0 μ, this means that the absorption varies from above 90 to nearly 100% of the incident infrared. Such small differences can have little effect on leaf temperatures. Furthermore, *Opuntia* joints showed nearly the lowest reflectivity of the 27 species tested, and the hairy leaves of *Verbascum thapsus* failed to reflect any. On the other hand, *Citrus limonia* reflected the most. Shade leaves, in fact, reflected more than sun leaves of the same plant.

Therefore, in spite of the promising difference found in the short infrared, the above results with the long infrared (including most of the infrared radiation), lead to the conclusion that plants have not succeeded in developing heat avoidance by increasing their reflectance of the incident radiation. Unfortunately, however, the major portion of the infrared used by Gates and Tantraporn is of negligible importance in normal solar radiation since the solar energy above 2000 nm (or 2 μm) accounts for only 5–10% of the total (see Gates, 1965). It would, therefore, be desirable to confine measurements of leaf infrared reflectance to the range of 750–2000 nm. It is conceivable, in view of Billings and Morris's positive results for the lower portion of this range (which accounts for a little less than half of the solar energy in the infrared), that significant differences may exist between some heat resistant and nonresistant plants. This conclusion seems to be supported by Lange's third group of leaf temperature behavior (see below), and by an artificial increase in reflectance which has been achieved by spraying a white kaolin suspension on plots of sorghum (Fuchs *et al.*, 1976; Stanhill *et al.*, 1976). Reflectivity increased 24% immediately after spraying.

2. Transmissivity has long been known to vary markedly in leaves. It is obvious that a pale green leaf transmits much more visible radiant energy than a dark green leaf. Even the same leaf may show an increased transmissivity due to a change in orientation. Many plants are, in fact, known to turn

their leaf edges to the sun, thus decreasing the absorption of radiant energy and the consequent rise in temperature. This adaptation might conceivably lead to avoidance of heat injury. The plastids have also been observed to turn edge-on toward the surface of the leaf when intensely illuminated. This characteristic has no apparent relationship to heat resistance (Biebl, 1955). This empirical result is to be expected from the above mentioned fact that most of the absorbed radiant energy is due to the infrared radiations, which are mainly absorbed by the water.

3. Absorption by protective layers is possible since the external water layer protects the internal cells by filtering out most of the heat-producing infrared radiation. This would, however, require an external layer of cells with higher heat tolerance, in order to protect internal (chlorenchyma) cells of lower tolerance. In the case of succulents, this is a distinct possibility, not only because of the high water content of the cells, but also because measurements have revealed a marked negative temperature gradient from the surface of the organ to the center (Huber, 1932; Fig. 11.3). It is perhaps possible that a thick enough cuticle may protect in the same way.

d. TRANSPIRATIONAL COOLING

Due to the difficulties inherent in the measurement of leaf temperatures, earlier records of the cooling effect of transpiration were frequently exaggerated (Curtis, 1936b). For instance, the elimination of transpiration by vaselining of leaves resulted in a warming of only 1°–3°C above the temperatures of the freely transpiring leaves at air temperatures of 30°C or less (Ansari and Loomis, 1959). Carefully performed measurements have even led to the conclusion that the maximum cooling is 2°–5°C, and calculations confirmed this conclusion (Curtis, 1936a). Even this amount, of course, could be a deciding factor in heat survival. More recently, however, fully reliable methods of measuring leaf temperatures have been developed, using the older thermocouples as well as thermistors (Lange, 1965a), and an infrared radiometer which does not touch the leaf (Gates, 1963). Gates found some sunlit leaves as much as 20°C above the air temperature (a leaf temperature of 48°C in *Quercus macrocarpa* at an air temperature of 28°C), while shade leaves averaged 1.5°C below the air temperature. He concluded that transpiration must play a relatively strong role in reducing leaf temperatures, and that convection is a relatively inefficient process. The earlier estimates did not agree with this conclusion; according to the above calculations (Curtis and Clark, 1950), the heat absorbed by the evaporation of water from a rapidly transpiring leaf at 40°–50°C could account for only about 15% of the radiant energy absorbed in full sunlight. More recent calculations (Wolpert, 1962) indicate that transpiration may be expected to remove 23% of the incoming heat during the midday hours. These and other recalculations

(Raschke, 1956, 1960) have indicated that under extreme enough conditions of temperature and vapor pressure, transpiration can produce greater lowerings of leaf temperature than the above limit of 2°-5°C.

It is dangerous, however, to generalize from such calculations. Below 35°C, for instance, the average temperatures of *Xanthium* leaves were above the air temperature by an amount dependent on the wind velocity, an increase in wind velocity producing a decrease in transpiration (Drake *et al.*, 1970). At air temperatures above 35°C, on the other hand, the leaf temperatures were below the air temperatures and increasing wind markedly increased transpiration. A further complication not usually taken into account is the root temperature. Coffee plants transpired maximally at a root temperature of 33°C, with a drop in rate both above and below this temperature (Franco, 1958). Even in the case of plant organs with much smaller specific surface than leaves, for instance apples, artificial cooling by overhead irrigation lowered their temperature an average of 5.6°C (Unrath and Sneed, 1974).

In view of all this more recent information, it is no longer surprising that Lange (1958, 1959) was able to observe pronounced lowerings of leaf temperatures at air temperatures of 40°C and higher. That his measurements of leaf temperature were not in error was elegantly proved by parallel measurements of the heat-killing temperature of the leaf. The air temperatures were as much as 10°C above the heat-killing temperature. Since the leaves were alive, their temperatures must have been lower than the air temperatures by more than this 10°C difference. Lange found three types of leaf temperature behavior in the extreme heat of the desert and savannah regions of Mauretania:

Group a. The temperature of the horizontal leaves in full sunlight was

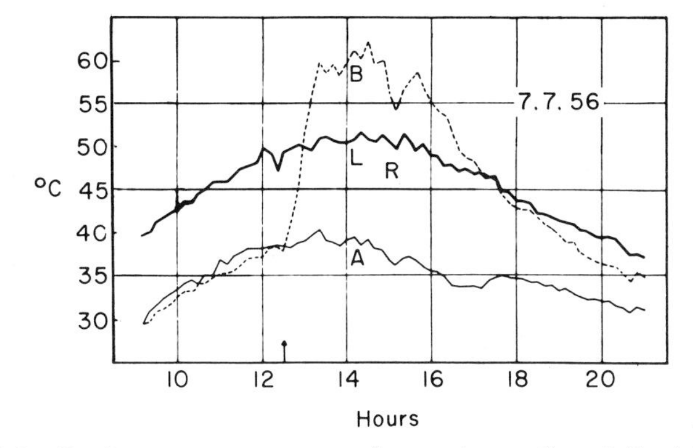

Figure 12.2. Daytime temperature curves for two leaves (A and B) of *Citrullus colocynthis*. Leaf B excised at 12:30 p.m. L, air temperature; R, heat-killing temperature of the leaf. (From Lange, 1959.)

markedly (as much as 15°C) below that of the surrounding air (Fig. 12.2), e.g., a leaf temperature of 42.5°C when the air temperature was 55°C. Similarly, when air temperatures ranged from 43–46°C in Death Valley, California, leaf temperatures of *Phragmites communis* were 8°C lower (Pearcy *et al.*, 1974). The importance of this decrease is evident from the optimum temperature for photosynthesis of this species (30°C). Such species that maintain leaf temperatures below the air temperature when in full sunlight, were called *undertemperature plants.* They accomplish this by maintaining their stomata open when heat is excessive (Sen *et al.*, 1972). Many plants open their stomata when subject to such high temperatures even in the dark (Babushkin and Barabal'chuk, 1974). This enables them to survive air temperatures that are otherwise lethal within 25–30 min if their stomata are closed.

Group b. The leaf temperature in full sun was above that of the surrounding air (Fig. 12.3), by as much as 13°C. Such species he calls *overtemperature plants.* If the temperatures were sufficiently extreme (under artificial conditions), even these overtemperature plants were able to make use of transpirational cooling. Thus, at temperatures of 30°–50°C, the undertemperature plant (*Cucumis prophetarum*) transpired at increasing rates with rise in temperature, nearly paralleling evaporation rates from an evaporimeter (Fig. 12.4). In the same temperature range, the overtemperature plant (*Phoenix dactylifera*) actually decreased its transpiration slightly with rise in temperature, due to closure of its stomata. Above 50°C, however, the stomata opened again and the overtemperature plant showed an even steeper rise in transpiration rate than the undertemperature plant, and a greater cooling effect per unit of water transpired (Figs. 12.4 and 12.5). They are, therefore, converted to undertemperature plants at the higher temperature.

Group c. Transpiration had no effect on the leaf temperature. Yet even when strongly irradiated, the leaf temperature rose very little above the air temperature.

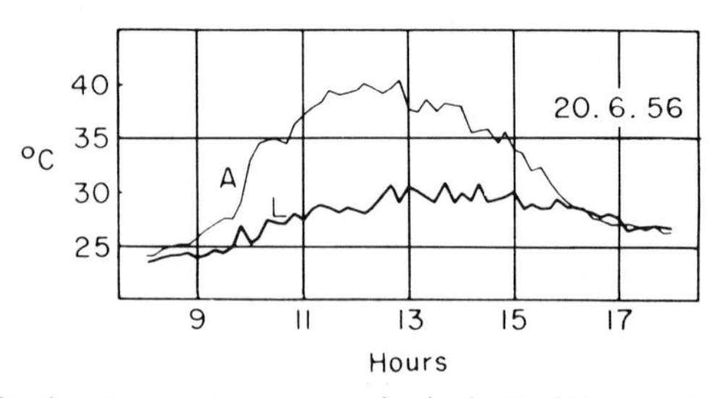

Figure 12.3. Daytime temperature curve of a leaf (A) of *Zygophyllum fontanesii*. L, air temperature. (From Lange, 1959.)

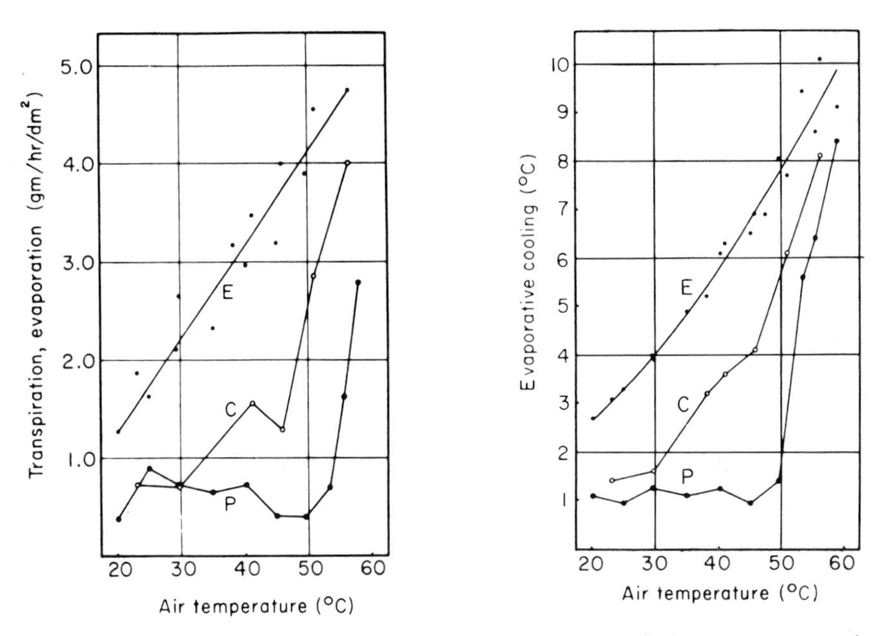

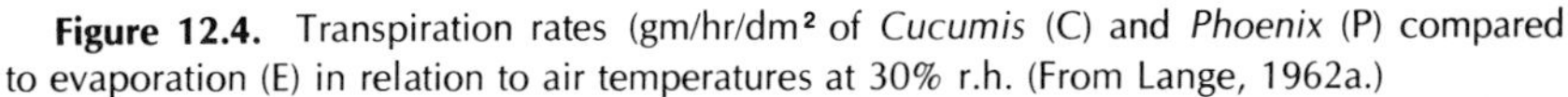

Figure 12.4. Transpiration rates (gm/hr/dm² of *Cucumis* (C) and *Phoenix* (P) compared to evaporation (E) in relation to air temperatures at 30% r.h. (From Lange, 1962a.)

Figure 12.5. Transpirational cooling (°C) of leaves of *Cucumis* (C) and *Phoenix* (P) and evaporation cooling of evaporimeter disc (E) in relation to air temperature at 30% r.h. (Evaporimeter cooling values reduced to half for purposes of comparison.) (From Lange, 1962a.)

That the first group owed their low temperatures to transpiration was proved by cutting off the shoot, thus severing it from its water supply. This resulted in a sudden rise in leaf temperature above that of the air (Fig. 12.2). On the other hand, the third group proved that other factors besides transpiration could control leaf temperature to some extent. These factors were not identified, but must have involved an increased reflectance and/or a decreased absorption of radiant energy (see above).

On the basis of measurements on 133 different species in Brazil, Coutinho (1969) observed the "De Saussure effect" (dark fixation of CO_2) in overtemperature plants, some of which developed leaf temperatures during daylight 10°–15°C above that of the surrounding air (e.g., a leaf temperature of 45°C at an air temperature of 30°C). It, therefore, seems likely that the most extreme overtemperature plants are to be found among those with an active dark CO_2 fixation system (or CAM metabolism), permitting them to keep their stomata closed during the day, thereby decreasing their transpiration rate, though permitting photosynthesis during daylight at the expense of CO_2 fixed at night.

The importance of transpirational cooling is well illustrated by the use of antitranspirants. These have resulted in a rise in leaf temperature of tobacco by as much as 9°F above the controls (Williamson, 1963). Similarly, when heat injury was determined in a temperature-controlled chamber, a doubling of the injury was obtained by raising the relative humidity from 50 to 75% or from 75 to 100%, at an air temperature of 43°C for 8 hr (Kinbacher, 1969).

3. Measurement

Although heat avoidance has never been measured, it should, at least be theoretically possible to do so, with the aid of modern control chambers. The temperature of the plant part would have to be measured under standard, steady-state conditions (e.g., an air temperature of 45°C and a soil temperature of 35°C, a light source of standard intensity and wave length distribution, a standard rate of air movement, and adequate soil moisture). The lower this temperature the greater the avoidance. A numerical value for avoidance might perhaps be obtained from a ratio of air temperature to leaf temperature at the steady state. It is obvious, however, that such measurements would give rise to many difficulties (e.g., due to the position of the leaf and the presence of others near it). Here again, however, a standardization of conditions could be developed. The heat avoidance should, of course, be measured at a temperature just below the heat-killing temperature.

B. HEAT TOLERANCE

The terms thermotolerance and thermostability have sometimes been used in the literature interchangeably for heat tolerance of organisms. In conformity with the general stress terminology defined earlier, the term thermotolerance will be used here interchangeably with heat tolerance—the ability of an organism to survive a heat stress within its tissues. The term thermostability will be used only for the ability of *substances*, e.g., proteins or nucleic acids, to remain stable, native, or active (e.g., as a catalytic enzyme), after exposure to heat stress. Lange and Lange (1963) found that avoidance was the main basis of the heat resistance in the woody plants characterized by mesomorphic leaf structure and a high transpiration rate. Their thermotolerance was low. On the other hand, the hard-leaved woody plants had a low transpiration rate and thermoavoidance, but a high thermotolerance. Unlike thermoavoidance, thermotolerance can be measured simply and with a high degree of precision in both of these types of plants. The main methods have remained essentially unchanged for over a century.

1. Methods of Measurement

No method has been developed for measuring the plant's overall tolerance of the indirect effects of heat stress. The only accepted method is to measure the T_{opt} and T_{max} for the specific process affected by the high temperature—growth, photosynthesis, respiration, etc. In the future, other methods may receive more attention, based, for instance, on Arrhenius plots of specific reactions.

Quantitative methods have long been used for measuring the plant's tolerance of the direct effects of heat stress. As in the case of low-temperature tolerance, direct heat tolerance is measured by determining the temperature at which 50% of the plant or plant part is killed—in this case called the heat-killing temperature. The two methods originally used by Sachs (1864) are still the standard procedures for measuring the heat-killing temperature of plants, though more elaborate equipment may now be employed. With the first, potted plants are placed in a heat chamber (whose temperature can be controlled) and maintained at a constant temperature for a standard length of time. They are then transferred to the greenhouse for a period of a week or two, following which the degree of injury is observed. Heyne and Laude (1940), for instance, exposed corn seedlings to 130°F at a relative humidity of 25 to 20% for 5 hr. With the second method, potted plants are overturned (Sachs, 1864), or small pieces or sections of the plant are plunged into water at a known temperature for a standard length of time, and then observed for growth or microscopically examined for injury. Sometimes the plant is not allowed to come in contact with the water. Julander (1945), for instance, cut 1½ inch pieces from stolons of range grasses, transferred them in lots of eight into stoppered glass tubes, which were then immersed in a constant-temperature bath at 48° ± 0.1°C for periods of 0, ½, 1, 2, 4, 8, and 16 hr. The stolon pieces were then planted and recovery was estimated after 4 weeks. Lange (1958) used this method in the field without removing the plant parts, by simply plunging them into the heated water in a thermos flask.

The first method has the advantage of using the whole plant, but there are definite objections to it. The actual temperatures attained by the tissues are not known, and may be far below those of the surrounding air due to the rapid transpiration at the high temperatures and low relative humidities usually employed. As evidence for this conclusion, Sapper (1935) found that the plants were able to survive air temperatures as much as 5°C higher in dry than in moist air; though wilted plants withstood higher air temperatures than turgid plants when heated in saturated air, in dry air the relationship was reversed. It is possible, therefore, to subject prairie grasses to hot winds at 135° to 145°F without injury, as long as soil moisture is available (Mueller and Weaver, 1942). Even if the plants are heated in still air, some time is needed for temperature equilibrium to be reached, and during this time the tissues are in danger of drought injury, especially at the low relative humidities that usually prevail (e.g., 25 to 30% in Heyne and Laude's 1940 experiments or 30–35% in Laude and Chaugule's 1953 experiments). Kinbacher (1963) demonstrated this in the case of eight varieties of winter oats. When subjected to 112°F (43°C) for 8 hr, only a heat stress resulted if the r.h. was maintained at 100%, but both a heat and a drought stress occurred if the r.h. was 50 or 75%. It is not surprising, therefore, that the "heat hardiness" of the plants tested by Heyne and Laude paralleled their field drought resistance. Consequently, in order to prevent drought injury during tests by the first method and to obtain plant temperatures identical with the measured air temperatures, Sachs' (1864) original precaution of maintaining 100% relative humidity should be adopted. Any injury produced at the high temperature will then be purely heat injury, as is always true of the second method. The two kinds of tolerance may, of course, be correlated, in which case a measurement of drought tolerance may yield a relative measurement of heat tolerance. This is not true in the case of some organisms, e.g., the green alga *Spongiochloris typica* (McLean, 1967). Even if the air is not saturated, the heat-killing point can be determined, provided that the actual temperature of the plant is used and not that of the surrounding air (Lange, 1967). Results obtained in this way do not differ by more than a degree or two from those obtained by the immersion method. Of course, as mentioned above, such tests should not be continued for long enough periods to permit drought injury.

It must be realized, however, that even when the above precautions are taken, the two methods may not always yield exactly the same results. Sachs (1864) found that with the first method (in air), 51°C was the killing temperature, though the plants withstood 49°–51°C for 10 min or more without injury. The same plants, however, were killed if plunged in water at 49° to 51°C for 10 min. According to de Vries (1870), killing occurs at about 2°C lower in water than in air. Yet, in spite of these results, agreement is often surprisingly good. Sapper (1935), for instance, gives a heat-killing temperature of 40.5°C for *Oxalis acetosella* exposed for one-half hour by the first method; Illert (1924) gives a value of 40°C for 20 min using the second (immersion) method on the same species.

The use of tissue pieces enables determinations on individual tissues or even cells. It has the disadvantage of judging survival by cellular methods instead of by the ability of the plant to continue normal metabolism and growth. In the hands of an untrained observer, cellular methods are dangerous to judge by, particularly if only one criterion is used (Weber, 1926a; de

Vries, 1871; Döring 1932). Scheibmair (1937) concluded that this danger can be avoided by combining vital staining, plasmolysis, observation for abnormalities, and finally deplasmolysis and replasmolysis. Even with perfectly reliable cellular methods of distinguishing living from dead cells, the second (immersion) method may be expected to yield slightly higher heat-killing temperatures than the first, since plants that are doomed to die after heating (by the first method) may retain full turgor and appear fully healthy for 24 hr (Sachs, 1864). To avoid overlooking such postheating changes, Scheibmair recommends leaving the sections for some time before testing. The actual differences obtained have been in the opposite direction (see above). This result must be at least partly due to the instantaneous rise of the cell temperature to that of the immersion water, the slower rise to the air temperature because of the far greater specific heat of water than of air, and the larger plant bulk.

A modification of the first method, used in the open, is to enclose a shoot in a blackened box, permitting its temperature to rise due to the insolation (Rouschal, 1938; Konis, 1949), but neither the temperature nor the time exposed to it can be controlled in this way. Besides the above mentioned methods of evaluating injury (plasmolysis, vital staining, observation of plants for recovery) several other criteria of survival have been used. Lange (1953) determined the temperature that reduced the subsequent equilibrium respiration of lichens to 50% of normal. The actual growth of the lichens after the heat stress gave similar results.

Aleksandrov (1964) developed a technique for observing cytoplasmic streaming (of the spherosomes) in the epidermal cells of intact leaves, as a criterion of heat survival. He covers the leaf surface with silicone oil, thus eliminating surface reflections when observed under the microscope. The leaves are heated at a series of temperatures for 5 min, and the temperature that is just sufficient to stop cytoplasmic streaming is taken as the heat-killing temperature. This method has been extended to the leaf parenchyma, in which the phototaxis of the chloroplasts was used as a criterion of heat injury (Lomagin *et al.*, 1966). The method is precise to 0.1°C, and it is even possible to distinguish between the thermostabilities of two kinds of streaming within the same cell (Lomagin, 1975).

One objection, pointed out by Aleksandrov, is that the cell can recover from this effect of heat if cooled rapidly enough, as shown by Sachs (1864). This objection may be valid for the absolute killing temperature; however, Aleksandrov obtained the same relative results by this method and by several others (e.g., plasmolysis, vital staining, respiration, photosynthesis).

Kappen and Lange (1968) object to Aleksandrov's method because it is confined to the epidermal cells, which may not respond to stresses in the same way as the leaf as a whole, and because injury is estimated immediately after heating instead of one or more days later. They point out that Till (1956) observed a winter rise in freezing tolerance of whole plants or leaves of *Hepatica nobilis,* but Aleksandrov's (1964) group failed to detect any rise by their method, perhaps due to the reversible effect of the temperature on cytoplasmic streaming in the epidermal cells. Similarly, Aleksandrov's method failed to reveal the changes in heat tolerance detected by Lange's modification of Sachs' first method, and which were produced by climatic or developmental changes. Again, when they observed cytoplasmic streaming immediately after the heat treatment (Aleksandrov's method), they were unable to detect the increase in heat tolerance following moderate drying of *Commelina africana,* which was established by their method. When, however, the cells were observed over a period of 200 hr, a faster recovery of protoplasmic streaming could be detected in the dried leaves than in those heated in the saturated state. In agreement with Lange, Biebl (1969) concluded that valid measurements of heat injury cannot be made earlier than 24 hr after heating *Chaetomorpha cannabina,* an intertidal alga. Aleksandrov's *et al.* (1970) evaluation of these objections will be considered below.

Oleinikova (1965) used the conductivity method to estimate heat injury in wheat and millet varieties, in the same way as it has been used for estimating freezing injury. One possible complication is a reported difference in electrical conductivity between varieties differing in tolerance, even in the absence of heat stress (Ivakin, 1975). Fel'dman and Lutova (1963) estimated the heat injury in algae by vital staining and by the effect on photosynthetic rate. A transpiration method has been used for determining the heat resistance of seven kinds of crop plants (Babushkin, 1975). The temperature at which the transpiration rate of the lower surface of the leaf rises sharply is taken as the lower limit, that for the upper surface gives the upper limit. The plants were killed if incubated in a chamber at the temperature of the upper limit for 30–35 min. The rate of photosynthesis decreased reversibly at the temperature of the lower limit of heat resistance. Thermostability of the luminescence of leaves can also be used as an indicator of heat injury, due presumably to thermal denaturation of the chloroplast lipoprotein complex (Aleksandrov and Dzhanumov, 1972).

As in the case of freezing tolerance, the relative heat tolerance of different species or varieties should be compared only when they are in the hardened condition, though few investigators take this precaution. Julander (1945) found little difference between range grass species when they were "unhardened," i.e., well watered and clipped and grazed. In the hardened condition, however, there were definite differences. Bermuda and Buffalo grass were the most tolerant; they were not killed by even 16 hr at 48°C. Bluestem was intermediate. Slender wheat, smooth brome, and Kentucky bluegrass were the least tolerant.

2. Effects of Environmental Factors

a. TEMPERATURE

As in the case of freezing tolerance, a relationship between the environmental temperature and heat tolerance is indicated by seasonal cycles in tolerance, although this was not clearly demonstrated until relatively re-

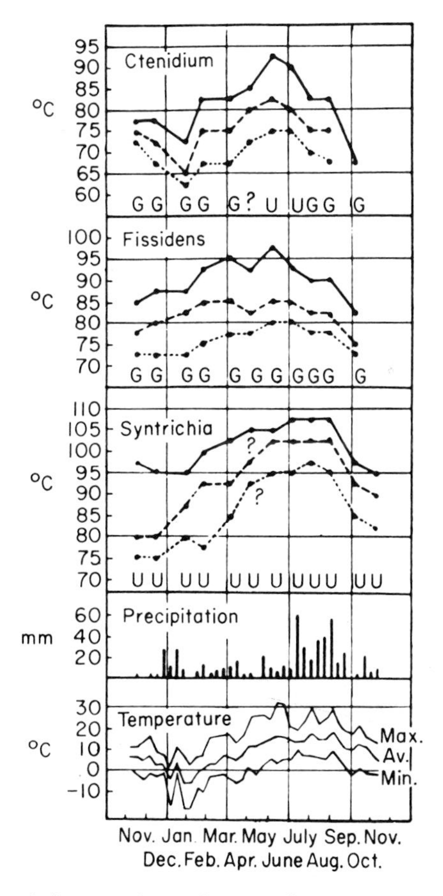

Figure 12.6. Seasonal change in heat tolerance of mosses, measured by exposure in the dry state for one-half hour to the high temperatures. Complete killing (top curve), severe injury (middle curve), and slight injury (bottom curve). G, injured by 50 hr, exposure over P_2O_5; U, uninjured. (From Lange, 1955.)

cently. The heat tolerance of mosses in the dry state shows an annual rise to a maximum during summer and a drop to a minimum during winter (Fig. 12.6). Wintergreen and evergreen plants in their normally moist state also show a seasonal cycle (Lange, 1961), but there may be two maxima in heat tolerance, one during the heat of summer, and the other during the cold of winter (Fig. 12.7). The cyclic change occurred not only in the new leaves but also in the previous year's leaves and, therefore, could not be explained simply by a change in growth and development. Furthermore, a higher heat tolerance was developed during the hot and dry summer of 1959 than during the cool summer of 1958. Although the differences are not as great as in the annual cycle of freezing tolerance, *Erica tetralix* showed a rise in heat tolerance of as much as 6.5°C during summer. The two winter and summer maxima also occur in evergreens of the subtropics on the Crimean south coast (Fal'kova, 1973). These maxima are primarily due to thermal hardening, both at high and low temperatures (Fal'kova and Galushko, 1974; Fel'dman and Kamentseva, 1974), though partial dehydration may be a factor in xeromesophytes (Fal'kova, 1973).

In other plants (e.g., grasses), the same winter maximum in heat tolerance was found, paralleling its freezing tolerance, but there was no summer maximum (Aleksandrov *et al.*, 1959; Aleksandrov, 1964). The third combination has also been reported. Fel'dman and Lutova (1963) observed a rise in heat tolerance and a drop in freezing tolerance during summer in the case of *Fucus* sp. In winter, the reverse occurred—freezing tolerance rose and

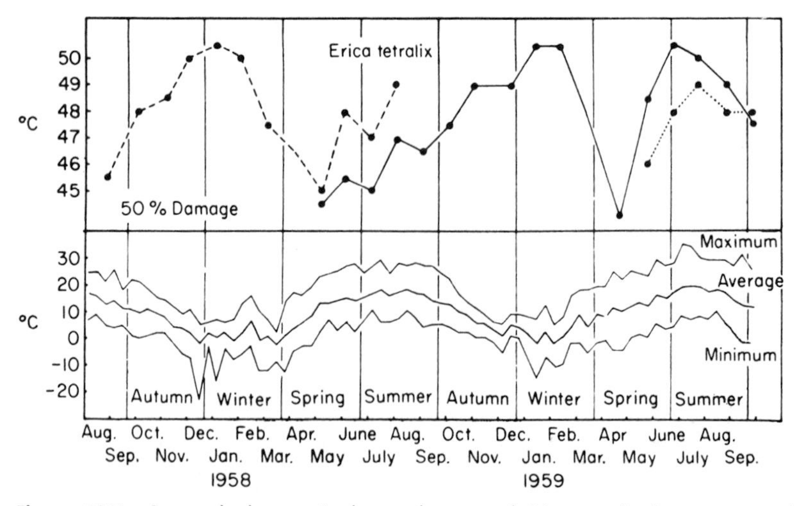

Figure 12.7. Seasonal changes in heat tolerance of *Erica tetralix* leaves exposed to high temperatures for 30 min (50% killing). Lower curves: the corresponding air temperatures. (From Lange, 1965b.)

heat tolerance fell. It is difficult to evaluate their measurements of freezing tolerance since these were obtained by plunging test tubes of the thalli into ice-salt mixtures at −30° to −20°C, and determining the length of time survived. This method may have measured freezing avoidance, or intracellular freezing avoidance, rather than freezing (i.e., freeze-dehydration) tolerance. Nevertheless, the results do agree with the seasonal changes in freezing tolerance found by others (Chapter 7).

The grasses that failed to show a summer maximum have a high degree of heat tolerance under any conditions, and this tolerance remains constant at growing temperatures of 22°-36°C (Aleksandrov, 1964). This may explain the lack of a summer maximum; when they are exposed artificially to 38°–44°C, their heat tolerance increases markedly. Presumably, the summer temperatures to which they are exposed in nature do not attain such high values and, therefore, no hardening occurs. Similarly, some mosses (e.g. *Atrichum undulatum*) may show no change in heat tolerance throughout the year, or may show an increase of 3°C only during winter in response to frosts (e.g., *Mnium affine;* Antropova, 1974). This lack of a summer hardening is apparently due to an insufficiently high temperature during the summer, for *M. affine* shows a rise in thermotolerance of 2.8°C as a result of artificial exposure to 42°C (Antropova, 1971).

Maximum high temperature hardening requires a temperature of 35–40°C in the case of 6 species of shrubs of the semiarid subtropics (Fal'kova 1975), as well as in the case of peas (Fel'dman *et al.*, 1975). An 18 hr treatment at this temperature range not only increased the thermotolerance of the leaves, but also increased the thermostability of three enzymes (G6PD, AP, Fd), and their resistance to tryptic digestion, acid and alkaline pH. Tobacco and bean leaves were only slightly protected against injury by such heat hardening, but their recovery from the injury was markedly improved (Ben Zioni and Itai, 1975). The ability of leaves (barley, wheat, Zebrina, Tradescantia) to heat harden decreased with senescence (Gorban *et al.*, 1974). Maximum hardening of wheat seedlings requires a more extended period at higher temperature—a regime of 44°6hr/25° 16hr, increasing to 54° 6hr/ 25° 16hr (Ignat'ev, 1973).

The relationship between temperature and hardening is obviously not as straightforward as in the case of freezing tolerance. This is further shown by comparing the heat tolerance of higher plants in different habitats. Thus, plants of dry, hot environments are frequently more heat tolerant than those of moist, cool environments (Table 12.1). Yet, in contrast to these results, plants of the Sahara desert range in heat tolerance from 44°-59°C (Lange, 1959), many of them thus showing no higher tolerance than those of the native mid-European plants used by Sapper. Furthermore, though most of the species did show a general correlation between heat tolerance and

TABLE 12.1

Heat-Killing Temperatures of Plants from Different Habitats[a]

Habitat or ecological type	Species	Heat-killing temp. (°C), ½-hr exposure
Submerged aquatics	*Helodea callitrichoides*	38.5–41.5
	Vallisneria species	
Shade	*Oxalis acetosella*	40.5–42.5
	Impatiens parviflora, etc.	
Partially shaded	*Geum urbanum*	
	Chelidonium majus	45–46
Very dry and sunny	*Alyssum montanum*	
	Teucrium montanum	
	Dianthus species	48–50 or over
	Iris chamaeiris	
	Verbascum thapsus, etc.	
Succulents	*Sedum* species	48.5–50
	Other succulents	50–54

[a] From Sapper, 1935.

habitat, there were several pronounced exceptions (Table 12.2). All the exceptions that survived the heat of the desert in spite of low heat tolerance were undertemperature plants and, therefore, owed their survival to heat avoidance. Consequently, if heat-avoiding plants are eliminated, the generalization still holds that the heat tolerance of most (but not all) higher plants varies directly with the temperature of their natural habitats. The relationship, of course, does not prove a direct hardening effect of high temperature on plants since they may have been selected for a stable heat tolerance.

In the case of mosses and lichens there is also usually a correlation between habitat and heat tolerance (Lange, 1953, 1955), although exceptions do occur. An unexpectedly high tolerance was found in two lichens in spite of their habitats (one in the shade and the other a hydrophyte). Similar relationships exist among algae, though the heat killing range is low (3 hr at 22°–38.5°C; Montfort *et al.*, 1957). The correlation with temperature changes under natural conditions may not necessarily prove a cause and effect relationship. *Chaetomorpha cannabina*, an intertidal alga, when cultivated in dishes for 5 consecutive days showed conspicuous fluctuations in heat tolerance in a "tidal rhythm" coinciding with the 6-hr intervals of ebb and flood (Biebl, 1969), and in the absence of any corresponding temperature cycle.

The rise in heat tolerance found by Lange during the heat of summer, suggests the possibility of artificial hardening by exposure to moderately warm temperatures. Early results, however, were the opposite of this expectation. Exposures to high, sublethal temperatures for extended periods actually reduced heat tolerance (Sapper, 1935). This was accompanied by a drain on the carbohydrate reserves at the high temperature. Conversely, low-temperature hardening that leads to an accumulation of reserves actually increases heat tolerance (Sapper, 1935). A cold treatment of pea plants has produced a similar effect (Highkin, 1959). These results were later corroborated (see above) by the maximum heat tolerance found during winter (Fig. 12.7).

Sapper's original failure to harden plants to heat has received some confirmation, but many have succeeded where he failed. Coffman (1957) found that exposure to high temperature increased the heat tolerance of winter oats, but not of spring oats. Aleksandrov (1964) has not succeeded in hardening grass species by exposure for long periods to moderately high but not injurious temperatures (22°–36.5°C). Similarly, growth at 21° or 29°C had no effect on the T_{opt} for photosynthesis of pitch pine (*Pinus rigida*) selections from different latitudes (Ledig *et al.*, 1977).

TABLE 12.2

Relation between Habitat and Heat Tolerance[a,b]

Species	Habitat	50% killing temp. (°C)
Trichomanes erosum	rain forest	44
Piptadenia africana	rain forest	45
Acacia tortilis	dune	45–46
Acacia senegal	savannah	47
Heisteria parvifolia	rain forest	47
Tamarix senegalensis	dune	47
Commelina africana	rain forest	47
Geophila obvallata	rain forest	47.5
Cassia aschreck	desert	49
Ziziphus nummularia	desert	49
Adenium honghel	savannah	50
Palisota hirsuta	rain forest	52
Culcasia angolensis	rain forest	52
Capparis decidua	desert	53
Boscia senegalensis	desert	55
Aristida pungens	desert	55
Phoenix dactylifera	desert	58

[a] Adapted from Lange, 1959.

[b] All the undertemperature desert plants and the overtemperature nondesert plants have been omitted. Expected order of increasing tolerance: rain forest < dune < savannah < desert.

At 38°C and higher, however, a reversible hardening occurred (Aleksandrov and Yazkulyev, 1961). They also, succeeded in hardening *Tradescantia fluminensis* by exposures to temperatures of 28°-36°C for 16-18 hr. A similar long period at 37.5°C killed the plant. Algae showed a direct relation between heat tolerance and growing temperature throughout the range tested (e.g., 10°-30°C for *Porphyra*). On the other hand, the green alga *Chlamydomonas* showed no capacity for heat hardening (Luknitskaya, 1967).

Lange (1961) succeeded in hardening *Commelina africana* and *Phoenix dactylifera* by cultivating them at high but not injurious temperatures. He (1962b) obtained a 4°C increase in heat tolerance (from 47° to 51°C) in the leaves of *Commelina africana* by growing the plants at 28°C for 5 weeks, compared to control plants grown for the same length of time at 20°C. Even tissue cultures (of Jerusalem artichoke) increase in heat tolerance as a result of an elevated growing temperature (Guern and Gautheret, 1969). It is possible to explain these discrepant results on the basis of the compensation point. Perhaps it is only those plants whose compensation point is near or below that of the hardening temperature used, that show a decrease in heat tolerance due to the consequent loss of reserves observed by Sapper (see above). Similarly, the absence of hardening at high but not injurious growing temperatures may be characteristic of plants (e.g., Aleksandrov's grasses) that already possess so high a heat tolerance in the unhardened state that only a more severe heat shock (see below) can harden them.

The Russian team under Aleksandrov's direction has conclusively and repeatedly shown that brief (even a single second) but severe heat shocks can induce a pronounced heat hardening of higher plants (Table 12.3). They have also succeeded in hardening lower plants by exposure to supraoptimal temperatures for brief periods, e.g., in the case of *Physarum polycephalum* (Lomagin and Antropova, 1968). Similar results have been obtained independently by Yarwood (1961a, 1963). In general, the higher the heat tolerance of the unhardened plant, the higher the temperature required for hardening by heat shock (Aleksandrov, 1964). Yarwood (1967) has succeeded in hardening bean, cowpea, corn, cucumber, fig, soybean, sunflower, and tobacco, as well as some rusts and viruses by exposure to air temperatures of 32°-40°C or water temperatures of 45°-55°C. The optimum exposure time was 20 sec at 50°C, with times at other temperatures corresponding to a temperature coefficient of 50.

Even cells in tissue culture are capable of hardening by heat shock (Schroeder, 1963). Cultures grown from pericarp of avocado fruit when treated for 10 min at 50°C and then incubated at 25°C for 3 days, survived 10 min at 55°C and grew during a subsequent 4 weeks at 35°C. Nonshocked

TABLE 12.3

Influence of 1-Second Heat Hardening at 59°C on the Heat Resistance of Cells of *Campanula persicifolia*[a]

Number	Maximal $t°$ at which protoplasmic motion is retained after 5-min heating		Difference[b]
	Pretreated cells	Control cells	
1	46.6	45.4	+1.2
2	46.6	45.8	+0.8
3	45.8	44.6	+1.2
4	45.4	43.8	+1.6
5	46.2	44.2	+2.0
6	45.4	44.2	+1.2
7	46.6	45.0	+1.6
Average	46.1	44.7	+1.4

[a] Lomagin, 1961. From Aleksandrov, 1964.
[b] For significance of the average difference, $P < 0.001$.

controls failed to grow after a similar exposure to 55°C for 10 min. This persistence of the hardening has also been shown by Aleksandrov (1964), although dehardening was perceptible after 24 hr, and was complete within 6 days.

Yarwood (1962) has also shown that a brief heat shock may actually sensitize a plant so that it is more injured by a later heat shock. Similarly, 5 min at 53°C sensitizes tissues of Jerusalem artichoke to injury at 37–42°C (Guern, 1974).

According to Wagenbreth (1965), hardening of beech requires a heat shock of 55°C, and high temperatures lower than this (e.g., 50°C) reduce its heat tolerance by increasing protein hydrolysis, accelerating yellowing, and increasing respiration. This may conceivably explain Yarwood's sensitization to heat.

Conversely, a moderate heat shock may increase heat tolerance only if this is measured some time after the heat stress (Lutova and Zavadskaya, 1966; Ben Zioni and Itai, 1975). This was explained by an increased ability to repair the damage, since no difference in heat injury could be detected immediately after the heat stress. In favor of this explanation, the injury was repaired by lowering the post-heating temperature to 25°C, but not at 34°C unless pretreated with kinetin, perhaps because of a requirement for protein synthesis (Skogqvist, 1974b).

Intermediate hardening periods of 3 hr at temperatures of 28°–50°C also increased heat tolerance, as measured by cytoplasmic streaming, chloroplast phototaxis, selective permeability, and respiration in leaves of *Tradescantia fluminensis* (Barabal'chuk, 1969). Maximum hardening occurred at 38°C.

b. MOISTURE

Sapper showed that dry cultivation increases heat tolerance by as much as 2°C (Table 12.4). Wilting alone increased hardiness as much as dry cultivation. This is in agreement with other results on the effect of dehydration (Chapter 11). Similarly, watered bluestem grass was killed by 4 hr at 48°C, and droughted bluestem only by 16 hr at the same temperature (Julander, 1945). The effect of moisture is also seen in the frequently observed burn injury to fleshy fruits after long rains when these are followed by hot, sunny weather (Sorauer, 1924; Huber, 1935). Similarly, fungal spores in younger developmental stages with higher water content are more sensitive to heat injury than those in later stages with lower water content (Zobl, 1950). The former become more heat tolerant on stepwise, careful drying, in agreement with results using seeds. Some species may fail to differ from each other in heat tolerance unless first exposed to drought hardening (Julander, 1945). The most extreme example of a relation between moisture and heat tolerance is, of course, the case of seeds (see above). The same phenomenon occurs in the vegetative state among lower plants. In the turgid state, mosses and lichens are killed by temperatures around 40°C; in the dry state (occurring regularly at midday), the killing temperature rises to 70°–100°C (Lange, 1955). Yet in opposition to this relation, succulent leaves have shown the

TABLE 12.4

Effect of Dry Cultivation on Heat Hardiness[a]

	Heat-killing temperature (°C)		Osmotic value at incipient plasmolysis (*M* sucrose)	
Species	Sparingly watered	Kept saturated	Sparingly watered	Kept saturated
Melilotus officinalis	46.5	45	0.55–0.60	0.45–0.50
Taraxacum officinale	46	44.5	0.65	0.45
Avena sativa	44.5	<43.5	0.65	0.40–0.45
Hordeum distichum	46–46.5	<44	0.80	0.45–0.50
Hieracium pilosella	>50.5	48.5	0.65–0.75	0.35–0.55
Ceterach officinarum	100–120	47		

[a] From Sapper, 1935.

highest heat tolerance of higher plants (Table 12.1). Lange (1958), however, found that the succulents of the Sahara desert are not especially heat tolerant. Their range of heat-killing temperatures (48°–53°C) falls in the middle of the range for all these desert plants. In fact, the plants with the highest tolerance were not the fleshy-leaved succulents but the hard-leaved species. In the case of *Kalanchoe blossfeldiana,* the strongly succulent leaves of the flowering plants have a higher heat tolerance than the weakly succulent leaves of vegetative plants. The heat-killing points are 50° and 47°C, respectively (Lange and Schwemmle, 1960). The two properties may, however, be separated, for a 10-day exposure to a short photoperiod was not enough to induce the succulence, yet it produced a significant rise in heat tolerance. Similarly, the differences in heat tolerance between leaves of different levels only partially paralleled succulence. These two investigators were therefore led to the conclusion that succulence is not in itself an indication of heat tolerance, but just as in the case of freezing tolerance, some plants possess much greater tolerance than is to be expected from their high water contents.

Even brief (6 hr) periods of dehydration leading to water saturation deficits of 22–27% were able to raise the heat-killing temperatures by about 3°C in the case of leaves of *Commelina africana, Hedera helix,* and *Phoenix reclinata* (Hammonda and Lange, 1962). Kappen and Lange (1968) showed that loss of water induced an increase in heat tolerance not only in lower plants and in xerophytic higher plants, but also in mesophytic higher plants. Zavadskaya and De'nko (1968) investigated twelve species of plants and found that the heat tolerance was higher when they were subjected to a water deficiency than when growing in humid habitats. As the leaves became saturated with water, heat tolerance gradually decreased.

The degree and speed of the dehydration may be critical. For instance, a decrease in soil moisture to one-third induced a maximum rise in thermotolerance of wheat, oats, and lettuce, but a further increase in soil moisture deficit was accompanied by a sharp fall in thermotolerance (Muromtsev *et al.,* 1972). Similarly, a drought sensitive barley cultivar required a gradual and prolonged dehydration to enhance heat hardening. A rapid dehydration (loss of 35% of leaf water in 3hr) was ineffective (Zavadskaya and Shukhtina, 1974).

c. LIGHT

The effect of light on heat tolerance is not so easily determined and some of the results are contradictory. Five days in the dark increased the heat tolerance of *Oxalis acetosella* (Illert, 1924). In agreement with these results, Sapper (1935) was unable to reduce the heat tolerance of plants by keeping them in the dark for as long as 3 days. Longer times than this, however, did reduce tolerance. Nevertheless, etiolated young seedlings were more tolerant than green, assimilating ones of the same age. In *Hordeum distichum* the heat-killing temperatures were 47° and 45°C, respectively. Sun plants, on the contrary, were always hardier than shade plants of the same species, though both were kept thoroughly watered. However, watering alone is unable to maintain the same hydration in the leaves under such atmospheric conditions. According to Heyne and Laude (1940), even a 1-hr exposure to light increased the heat tolerance of plants previously kept in the dark for 12 to 18 hr; as mentioned above, they were probably determining drought rather than heat tolerance. Similarly, older bromegrass seedlings deprived of light showed greatly reduced resistance (Laude and Chaugule, 1953). Nevertheless, bromegrass seedlings showed their highest heat tolerance immediately after emergence, even in seedlings deprived of light. Older seedlings deprived of light showed greatly reduced tolerance. When grown in a 12-hr day, *Kalanchoe blossfeldiana* shows a maximum heat tolerance at night in the middle of the dark period and a minimum at midday (Schwemmle and Lange, 1959a). These differences are not simply due to the direct effect of light, since the plant continues to show this "endogenous rhythm" for at least 2 days after transfer to continuous dark under constant temperature and moisture conditions. Furthermore, only the middle, actively growing leaves show this rhythm.

The heat tolerance of the photosynthetic apparatus of cotton increased with illumination, the maximum tolerance occurring during light saturation (Veselovskii *et al.*, 1976). Similarly, Oleinikova (1965) hardened wheat and millet plants by exposure to 33°–35°C. Although hardening occurred both in the light and in the dark, light accelerated the process, especially at the beginning of hardening. Some of the contradictions may be due to this increased hardening in the light as opposed to increased injury if exposed to light after the heat stress. In the case of leaves of *Tradescantia fluminensis*, exposure to light after short-term (5–10 min) heating induced injury at 10°C lower heating temperatures than if kept in darkness (Lomagin and Antropova, 1966). This injury in the light was accompanied by a considerable destruction of chlorophyll. In variegated leaves of *Chlorophytum elatum*, the light injured only the green parts of the leaf blades. The minimal light intensity causing injury was 1000 lux. The injury was sharply decreased if the air was replaced by nitrogen. The explanation suggested was a photooxidation sensitized by chlorophyll and occurring at the expense of energy not used in photosynthesis. It is, therefore, conceivable that light enhances heat hardening, but that it increases photooxidative heat injury. In possible agreement with this explanation, light significantly decreased the thermostability of the Hill reaction in isolated pea and spinach chloroplasts (Ageeva, 1977).

d. MINERAL NUTRITION

The effects of mineral nutrition have received little attention. Nutrient deficiency raised the heat-killing temperature of *Oxalis acetosella* (Illert, 1924). In agreement with these results, Sapper (1935) found that excess nitrogen or potassium reduced heat tolerance while deficiency very effectively raised it (by about 2°C). High nitrogen also reduced the heat tolerance of turf grass (Carroll, 1943).

There are many reports of a protective effect of Ca^{2+} against heat injury. The first evidence of this (see Chap. 11) was obtained by Kaho (1926) and Scheibmair (1937). Similarly, Ca^{2+} inhibited the reversible, heat-induced efflux of betacyanin from beet root exposed to 45°C for 90 min (Toprover and Glinka, 1976). The evidence is most conclusive for thermophilic microorganisms. The heat tolerance of *Bacillus stearothermophilus* depends on the maintenance of a high intracellular concentration of free Ca^{2+} (Ljunger, 1973). In several cases, (see Sec. C1), specific enzymes in these microorganisms are protected by Ca^{2+} against high temperature inactivation—amylase and protease in the bacterium, thermomycolase in *Malbranchea pulchella* (Voordouw *et al.*, 1974; Voordouw and Roche, 1975; Ong and Gaucher, 1976), alkaline protease in *Torula thermophila* (Karavaeva and Mukhiddinova, 1975). The latter two are thermophilic fungi. Trace elements (especially Zn) have also been reported to favor increased heat tolerance of leaves (Vlasyuk *et al.*, 1974), supposedly due to increased enzyme activity (Khodzhaev and Abaeva, 1975). The presence of divalent cations is important in maintaining the integrity of thermally stressed cells of *Pseudomonas fluorescens* (Gray *et al.*, 1977). Crosslinking by Mg^{2+} is suggested to increase the mechanical strength of the peptidoglycan layer. In the case of heterotrophic organisms (e.g. yeast) survival of high temperature may depend on an adequate quantity of organic nutrient, for instance sugar (Van Uden and Madeira-Lopes, 1975).

3. Relationship to Plant Characteristics

Age may markedly affect heat tolerance, although there are many apparent contradictions in the literature. In general, Sachs (1864) found that the blades of young, fully grown leaves were killed first. Younger leaves that were not fully grown and bud parts were more tolerant. The most tolerant of all were the old, healthy leaves. De Vries (1870) obtained the same results. According to Illert (1924), older leaves of *Oxalis acetosella* are less heat tolerant than younger ones; the significance of this observation is doubtful, since the leaves were already moribund. Seeds that were allowed to swell in water for 24 hr showed a decreased heat tolerance with age (Porodko,

1926a), even though no decrease in percentage germination of the unheated seeds could be detected. In fruit, heat tolerance increases with ripeness, e.g., from 43°C in unripe grapes to 62°C in ripe grapes (Müller-Thurgau, see Sorauer, 1924). Tree seedlings show an increased tolerance with unfolding of the cotyledons (Huber, 1935). Young leaves of *Helodea* are less tolerant than older ones, but the oldest are the least tolerant (Esterak, 1935). On the other hand, Scheibmair (1937) showed that young moss leaves are more tolerant than older ones, and younger cells more tolerant than older basal ones. According to Heyne and Laude (1940), 10- to 14-day-old corn seedlings are more tolerant than older ones. They also noted an exhaustion of the food material in the endosperm by about the fourteenth day. Bromegrass seedlings, on the contrary, decreased in tolerance about 14 days after planting, and increased with age after 28 days (Laude and Chaugule, 1953). As mentioned above, however, Heyne and Laude may actually have been measuring drought tolerance rather than heat tolerance. Sapper (1935) was unable to detect any difference in heat tolerance between seedlings and older plants. Baker (1929) concluded that the apparent increase in tolerance of conifer seedlings with age was not protoplasmic (i.e., true tolerance), but simply due to the development of mechanical protection (i.e., avoidance). Similarly, the age of tree seedlings and the associated maturity of the cortex and other tissues had little effect on heat tolerance, according to Smith and Silen (1963). According to Bogen (1948), however, younger leaves of *Rhoeo discolor* survive heating for a longer time than older leaves. The same was true of younger basal cells versus older tip cells.

Many other investigators showed similar tolerance gradients in a single organ that may not be related to age differences. The base of *Iris* and *Anthericum* leaves had higher killing temperatures than the tip (de Vries, 1870). In apples, the center may be killed at 50° to 52°C, and the outer layers only by temperatures above 52°C (Huber, 1935). According to Belehradek and Melichar (1930), the basal cells of *Helodea canadensis* leaves succumb first, the apical cells showing the greatest heat tolerance. This order was reversed at lower temperatures, and they point out that the gradient may actually be due to the injurious effect of cutting, which is perhaps propagated from the base to the tip. The accuracy of this suspicion was later demonstrated by Esterak (1935), who obtained a gradient in the reverse direction on the same material. The result apparently depended on whether the leaves were removed with forceps before testing, or by excising with scissors as done by Belehradek and Melichar. A real reversal was found by de Visser Smits (1926). Heat tolerance of beet roots decreased from the base to the tip, except at inception of the second growth period, when the gradient was reversed.

In the case of evergreen and wintergreen plants, Lange (1961) found the younger leaves to be much less heat tolerant than the older ones. Yet young leaves were more heat tolerant than mature leaves in the many species growing on the south coast of Spain (Lange and Lange, 1963). In the case of *Ilex aquifolium,* leaf no. 5 (a younger leaf) had a heat-killing point of 46°, and leaf no. 10, a heat-killing point of 48°C (Lange, 1965b). In *Kalanchoe blossfeldiana,* however, leaf no. 4 showed maximum sensitivity, and tolerance increased progressively above and below this leaf.

There seems to be an inverse relationship between growth rate and heat tolerance and this may complicate the above relations to age. The vegetative cone, youngest leaves, and fully grown older leaves of *Kalanchoe blossfeldiana* are more heat tolerant than the middle, not yet fully grown leaves (Schwemmle and Lange, 1959b). In summer, when growth was most intense, the growing cells of the leaves of *Zebrina* and *Echeveria* showed less heat tolerance than the mature cells (Gorban', 1962). A decrease in tolerance occurs in some plants on flowering (Henckel and Margolin, 1948; Henckel, 1964), though as mentioned above, *Kalanchoe blossfeldiana* shows the opposite relationship. Similarly, vernalization of peas markedly increased their heat tolerance, although this was independent of any floral initiation (Highkin, 1959).

There is some evidence of a relation to cell sap concentration. The effects of nutrition, low temperature, and moisture found by Sapper (1935) also paralleled the osmotic value (Table 12.4). When different species were compared, or even different parts of the same plant, this relation did not hold. Both with respect to their high water contents and low cell sap concentration the very thermotolerant succulents are an exception. Julander (1945) found, however, that hardened grasses had about twice as high carbohydrate contents as the unhardened, though there was very little starch. Sucrose accumulated but reducing sugars did not. The substances that accumulated the most were the colloidal carbohydrates, especially levulosans. Since Julander heat-hardened his plants by exposure to drought, these results are not comparable to any obtained by heat hardening. Sapper (1935), on the other hand, found etiolated plants with lower carbohydrate contents more tolerant than the normal green ones. Albino sunflower leaves, however, had a lower heat tolerance than that of the green plant (Avilova, 1962).

The relation of cell sap concentration to thermotolerance is also indicated by the opposite changes in heat tolerance that occur in the guard cells and subsidiary cells (Weber, 1926b). When the stomata are open in the light, the guard cells are more thermotolerant than the subsidiary cells; when they are closed in the dark, the subsidiary cells are more tolerant. In each case, the more thermotolerant cells are free of starch, but presumably contain more solute; the less thermotolerant contain starch, but probably less solute.

The photosynthetic rate of cucumbers in a hothouse rose to a maximum at

40°C, dropping at 47°C to the same value as at 27°C in the open (Fedoseeva, 1966). Relatively more oligosaccharides were formed at the higher temperatures. It was, therefore, proposed that the oligosaccharides play a protective role, raising the thermotolerance of the chloroplasts.

Parija and Mallik (1941) incubated seeds at 40° to 60°C for 8 to 120 hr and recorded the percentage germination. Oily seeds survived the high temperature better than starchy seeds; the higher the oil content, the better was the survival. Linseed was an exception, since it showed greater heat tolerance than cottonseed, though the latter had a higher oil content. They ascribe this to the mucilaginous seed coat. According to Zobl (1950), protein content is a factor in the heat tolerance of spores. This is based on nitrogen analyses which showed that the protein content (6.25 × N content) was three times as great in the heat tolerant bacterial spores as in the less tolerant fungal spores.

Protoplasmal properties have received little attention with regard to heat tolerance. Scheibmair (1937) observed more rapid rounding up in the more thermotolerant, upper, marginal cells of mosses on plasmolysis than in the less thermotolerant basal cells. On the other hand, the tip cells, which were just as thermotolerant as the upper marginal ones, never rounded up at all. This latter result, however, must be due to something other than true viscosity—perhaps a firmer adhesion to the cell wall. Henckel and Margolin (1948) conclude that two important reasons for the heat tolerance of succulents are high cytoplasmic viscosity and high bound water. According to their measurements, both of these factors are developed to a higher degree in the succulents than in the other xerophytes, which are less heat tolerant. They state that the bound water in succulents is much higher than in mesophytes—as high as 70% in certain cacti. During flowering, viscosity decreased suddenly but bound water increased and to some extent compensated for the decreased thermotolerance resulting from viscosity change. These conclusions are obviously due to a misunderstanding of the meaning of bound water, since nearly all the water of succulents is free (see Vol. 2, Chapter 4).

Swelling of mitochondria isolated from plants differing in heat tolerance differed at optimal (25°C) as well as at elevated (30°–50°C) temperatures. It was greater in mitochondria from pea plants and from nondrought hardened maize plants than from bean plants and drought-hardened maize (Andreeva, 1969). In the case of the heat-resistant bean and maize plants, shrinkage occurred on addition of ATP even at extreme temperatures (45°–50°C), while in the nonresistant pea plants the shrinkage was much less, indicating irreversible heat injury. This kind of mitochondrial injury is usually ascribed to membrane damage.

4. Mechanism of Heat Tolerance

Tolerance of the heat stress requires either strain avoidance or strain tolerance, but the strain may be either a direct or an indirect effect of the heat stress (see Diag. 11.1). The mechanism of heat tolerance, therefore, depends on which of these strains the plant must avoid or tolerate.

a. AVOIDANCE OR TOLERANCE OF INDIRECT STRAINS

1. Growth The temperature optimum for plant growth depends on the natural habitat of the plant. Thus the maximum growth rates (24–28%/day) are the same for two species native to a cool (coastal) climate and to a hot (Death Valley) climate respectively (Björkman *et al.*, 1974). In the case of the first species (*Atriplex* sp.), however, this maximum was attained at a temperature regime of 16°D/11°N. In the second case (*Tidestromia oblongifolia*) it was attained at 45°D/31°N. This relation to habitat, however, does not always hold. Pitch pines, for instance, had the same optimum temperatures for photosynthesis, though originating in three different latitudes and therefore different temperature regimes (Ledig *et al.*, 1977). Similarly, sorghum is considered less heat tolerant than some millet or maize varieties; yet young plants are capable of growth rates up to 22%/day at 45°C with natural light levels and day-length (Troughton *et al.*, 1974).

This ability to grow at high temperatures depends on the plant's ability to prevent the other four kinds of indirect heat injury.

2. Starvation If indirect heat injury is due to starvation or any other metabolic abnormality, those plants that are capable of continuing a normal type of metabolsim at high temperatures would be thermotolerant. Harder *et al.* (1932) have shown that many plants adapted to high temperatures have a much higher compensation point than unadapted plants. In contrast to potato and similar plants with low compensation points, these adapted plants are able to assimilate even at temperatures of 45°–53°C.

Though the temperature optima for photosynthesis are the same in tomato and cucumber, the latter is more thermotolerant, due to the less steep drop of its photosynthesis curve to zero at the maximum temperature (Lundegårdh, 1949).

As a result of such differences in compensation point, plants from warm habitats may have rates of net photosynthesis at 40°C nearly double those of plants from cool habitats (Williams and Kemp, 1976). This kind of metabolic adaptation may be accomplished in two ways.

1. The adapted plants may have a higher *actual* rate of photosynthesis at the high temperature. Thus, the T_{opt} for growth of the thermophilic blue-

green alga *Synechococcus lividus* was 45°C. An upward or downward adaptation of the T_{opt} for photosynthesis occurred as a result of growth at supraoptimum (57°C) or suboptimum (35°C) temperatures. Associated with the increase in T_{opt} was an increase in chlorophyll *a,* plastoquinone A, and RuP_2 carboxylase activity. Reduced rates above 55°C, apparently resulted from damage to ferricyanide reducing systems and decreased RuP_2 carboxylase activity (Sheridan and Ulik, 1976). The desert shrub, *Larrea divaricata* is able to maintain a relatively high and constant photosynthetic activity, throughout the year, even though the mean daily maximum temperature may vary by nearly 30° from winter to summer. None of several factors tested (stomatal opening, respiration, O_2 inhibition of photosynthesis, limitation of CO_2 diffusion) accounted for the acclimation (Mooney *et al.,* 1978). Two adaptations were (a) an increased capacity for photosynthesis at low temperature and (b) an increased thermal stability of key components of the photosynthetic apparatus with increased temperature. The latter adaptation stabilized the interactions between the light-harvesting pigments and the photosynthetic reaction centers (Armond *et al.,* 1978). The temperature rise decreased the quantum yield of the electron transport at limiting light, followed by a loss of electron transport activity at rate saturating light. This was attributed to a block in the transfer of excitation energy from chlorophyll *b* to chlorophyll *a,* and to changes in distribution of excitation energy between photosystems II and I. There was a loss in effectiveness of 480 nm light (absorbed primarily by chlorophyll *b*) to drive the photochemical processes.

2. The adapted plant may have a lower respiration rate. The respiration rate of thermophilic blue green algae rises very slowly with temperature, and even at high temperatures, assimilation is in excess. According to Marré and Servettaz (1956) and to Prat and Kubin (1956), there is no compensation point in thermophilic blue-green algae, or at least it is above 55°C (Fig. 11.7).

In four desert shrubs, it is the dark respiration that adapts, for it drops in summer to about half the winter value when both are measured at 25°C (Strain, 1969). Desert plants, in general, have a low sensitivity to high temperature in comparison with cultivated plants (Fig. 12.8). For example, the apparent CO_2 absorption of apricot sinks to the compensation point below 35°C, though all the desert plants still exhibit a pronounced or even a maximum assimilation at this temperature, and do not attain the compensation point until the leaf temperature rises above 40°C, and in some cases 50°C (Lange *et al.,* 1969). In the case of *Atriplex polycarpa,* even intraspecific differences in CO_2 exchange were correlated with the climates of origin (Chatterton *et al.,* 1970). Furthermore, a daily temperature regime of 43°D/32°N for a week reduced the rate of dark CO_2 evolution; an equal exposure

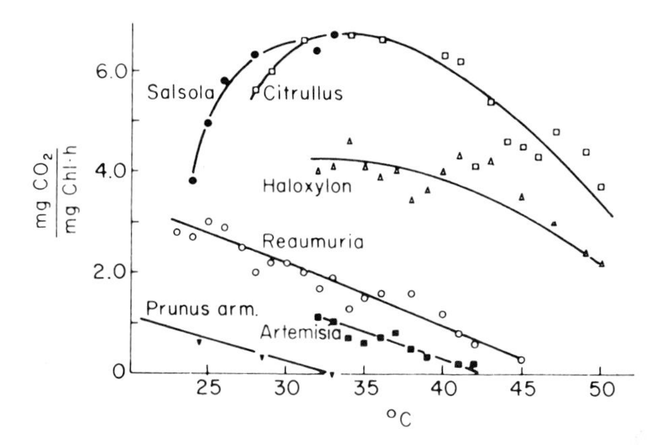

Figure 12.8. Effect of high temperatures on net assimilation of cultivated plants (e.g., apricot—*Prunus arm.*) and of desert plants (e.g., *Citrullus*). (From Lange *et al.*, 1969.)

to 16°D/5°N increased it. Cultivated grains adapted to high temperatures also show this respiratory adaptation (Razmaev, 1974).

The best adapted plants use both of these metabolic changes. It is now known that C_4 plants are generally adapted to higher temperatures than C_3 plants, for instance in the case of grasses. This is associated with a higher compensation point in the C_4 plants (e.g. *Bouteloua gracilis*) than in the C_3 plants (*Agropyron Smithii;* Williams, 1974). The higher compensation point is due to both an increased ability to photosynthesize at high temperatures, because of a more efficient CO_2 absorbing system and a decreased rate of respiration at high temperatures during daylight due to an almost complete absence of photorespiration. It is due to these adaptations that species of amaranth, and especially *Tidestromia oblongifolia* can thrive and continue to grow rapidly in the high temperatures of Death Valley (e.g. 52°C; Seagrave, 1976).

3. Biochemical Lesions, Toxins, and Protein Breakdown. There has been little or no attempt to investigate the plant's ability to avoid or tolerate these indirect strains. If, however, the high-temperature induced metabolic disturbances are due to differences between the slopes of the Arrhenius plots of two reactions, this should be recognized by differences in their energies of activation. Activation energies for some enzymes from thermophilic organisms are low: 35 kcal/mole for carbonic anhydrase from heat-tolerant cotton (Chang, 1975), only 17 and 8kcal/mole below and above 50°C for homoserine dehydrogenase from the still more thermophilic micoorganism *Thermus flavus* (Saiki *et al.*, 1973). In opposition to these results, it is higher in desert plants than in tomato (Weber *et al.*, 1977). These difference, however, are only apparently contradictory. A toxin-producing reaction, for instance, should have a lower activation energy in thermotolerant plants, a toxin-destroying enzyme a higher activation energy. No information is available as to these specific differences.

If the metabolic injury is due to a decreased rate of formation of essential intermediates at the high temperature leading to a biochemical lesion, tolerance would require a higher rate of production (see Chapter 11). Similarly, in the case of heat injury due to toxins, thermotolerance would require a decreased production of the toxin, or its detoxification. Neutralization by the synthesis of organic acids, has been suggested as the thermotolerance mechanism against NH_3 toxicity (Chapter 11). For the fourth kind of metabolic heat injury, protein breakdown, thermotolerance due to a repair mechanism has already been indicated (Chapter 11). This may perhaps be the basis of some differences in tolerance among four grass species. The two species that are more heat sensitive contained less RNA and DNA per unit dry weight than the two more resistant species (Baker and Jung, 1970). When subjected to a heat stress (35°C for up to 15 hr) the RNA content was reduced more in the two sensitive species. This could, of course, affect protein synthesis both during and after the heat stress.

Further evidence that repair after exposure to the heat stress may be a factor has been produced by Aleksandrov (1964). Epidermal cells of *Catabrosa aquatica* showed greater heat tolerance than those of *Campanula persicifolia* as judged by the temperature that stops protoplasmic streaming in 5 min (44.7° and 43.5°–44.0°C, respectively). The protoplasmic streaming can be restored if the heating is not excessive, and in contrast to the above result, *Campanula persicifolia* is better able to repair this injury, for it recovers its streaming even after heating to 51.5°C for 5 min, as compared with 49.0°C in the case of *Catabrosa aquatica*. The temperature difference between these two points varies from 3.8° to 9.0°C, depending on the species (Table 12.5). This repair of injury to cytoplasmic streaming may continue up to 18 days after the heat stress.

Metabolic adaptation leading to avoidance of another kind of injury has been suggested by Ljunger (1970). He found that thermophilic bacteria die at temperatures ordinarily employed for their cultivation, if the medium lacks Ca^{2+}. Heat tolerance also required K^+, phosphate ions, and glucose or another energy source. He, therefore, proposes that their heat tolerance is dependent on an active transport of Ca^{2+} from the environment into the cells.

These metabolic adaptations are not constitutive but must be developed during acclimation. Plants of *Atriplex lentiformis* from desert habitats acclimate when grown at high temperatures (43°D/30°N) by developing higher CO_2 uptake rates at these high temperatures (and lower rates at low tempera-

TABLE 12.5

Temperature Differences in Certain Plants between the Temperature That Stops Protoplasmic Streaming and the Temperature That Prevents Resumption of Protoplasmic Streaming[a]

Species	t° stopping motion after 5-min heating I	Maximum t° after which restoration is possible II	II–I	References
Tradescantia fluminensis	44.5	52.0	7.5	Aleksandrov, 1955
Zebrina pendula	46.7	53.0	6.3	I. S. Gorban', unpublished
Campanula persicifolia	44.0	52.0	8.0	Aleksandrov, 1955
Ficus radicans	45.8	54.6	8.8	Derteva, unpublished
Chlorophytum elatum	47.0	56.0	9.0	Liutova, unpublished
Catabrosa aquatica	44.7	48.5	3.8	E. I. Den'ko, unpublished
Oxilatoria tenuis	49.2	54.0	4.8	Mel'nikova, 1960
Phormidium autumnale	52.2	56.4	4.2	Mel'nikova, 1960

[a] From Aleksandrov, 1964.

tures) than those grown at low temperatures (23°D/18°N). Coastal plants were unable to acclimate, and in fact showed reduced photosynthesis at all temperatures when grown at high temperatures (Pearcy, 1977). Such acclimation of net photosynthesis also occurs in oak trees (*Quercus rubra var. borealis*), but is a response to the major portion of the 24hr period and is insensitive to the amplitude of the temperature variations (Chabot and Lewis, 1976). In plants of *Atriplex lentiformis* at low temperatures, the differences in photosynthetic activities were correlated with differences in RUDP carboxylase activities (Pearcy, 1977). At high temperatures, they appeared due to a combination of factors—decreased respiration rates, decreased temperature-dependence of respiration and increased thermostability of the photosynthetic apparatus, preventing the marked reduction in photosystem II activity which occurred in the unadapted plants when exposed to 46°C (Pearcy *et al.*, 1977). Respiration rate also adapts to the temperature of exposure in the case of the desert population of *Isomeris arborea*. They show complete temperature compensation, the dark respiration remaining unchanged with increase in growth temperature (Tobiessen, 1976). The coastal populations showed the normal increase in dark respiration with increasing temperatures. The thermally stable coastal climate apparently did not induce the thermal homeostasis which developed in plants of the thermally fluctuating desert climate. Wheat seedlings show a clear metabolic relation to their growing temperature (Weidner and Ziemens, 1975). When grown at a low temperature (4°C), the temperature optimum for protein synthesis ([^{14}C]leucine incorporation) was 27.5°C; when grown at a high temperature

(36°C) it was at 35°C. At 4°C, the low temperature plants had 2 times the rate of protein synthesis of the high temperature plants. At 35°C, the high temperature plants had 3 times the rate of the low temperature plants. The activation energy of the plants grown at medium or reduced temperatures was lower than that of the plants grown at the high temperature (11 versus 15 kcal/mole).

These results demonstrate the ability of the plant to adapt metabolically to high temperatures. This metabolic adaptation, however, may be distinct from the thermotolerance of the plant. Thus, the leaves of *Agrostis palustris* grown at 40°D/30°N had 51–53% more soluble carbohydrates than those grown at lower temperatures. This was not due to a blockage of translocation, and photosynthesis (O_2 evolution) increased with temperature from 20–40°C (Duff and Beard, 1974b). The carbohydrate synthesis was, therefore, not a limiting factor in the growth reduction and plant death at supraoptimal temperatures (Duff and Beard, 1974b). Starvation, therefore, was not involved.

b. AVOIDANCE OR TOLERANCE OF DIRECT STRAINS

Each of the above kinds of heat tolerance involves a different kind of indirect strain and therefore requires a different metabolic mechanism. Where the direct effects of the heat stress are involved, the mechanism must again be different. Furthermore, a heat-tolerant plant does not necessarily develop the different kinds of tolerance to the same degree. Pisek *et al.* (1968), for instance, measured the tolerance of the direct effects of the heat stress by the rapid (30 min) heat-killing temperature (using Lange's modification of Sachs' method), and tolerance of the indirect effects by the maximum temperature for net assimilation. Among twenty-two species of higher plants, the difference between the two ranged from 3°–12°C, and the difference had no relation to the heat-killing temperature, which ranged from 43°–55°C.

Many attempts have been made to explain the heat tolerance that is due to avoidance or tolerance of the direct strain. In most cases, the direct strain is assumed to be protein denaturation. Therefore, tolerance is assumed to be due to (1) thermostability of the proteins leading to avoidance of the denaturation strain, or (2) increased speed of resynthesis of the proteins (repair), leading to tolerance of the denaturation strain. Several mechanisms of one or other of these two kinds of thermotolerance have been suggested.

i. Thermostability of Proteins.

The earlier concepts were highly speculative by necessity, due to lack of knowledge of the true nature of denaturation and the detailed structure of the proteins. Molisch (1926) suggested the presence of protective substances which would prevent coagulation. Bünning and Herdtle (1946) and Christopherson and Precht (1952) proposed protection by bound water. Bogen (1948), however, was perhaps the first to suggest an increase in the structural strength of the proteins. Aleksandrov's (1960; Aleksandrov *et al.*, 1961) more recent concept includes (a) increased resynthesis, (b) increase in antidenaturing substances, and (c) structural changes in the protein molecules, leading to conformational flexibility.

At pointed out above, the earlier ideas were actually based on observations of coagulation rather than denaturation, and of whole cells rather than proteins. More direct evidence of the protein denaturation theory has since then accumulated by the numerous demonstrations of a relationship between the thermostability of proteins and the heat tolerance of the tissues or cells in which they occur.

This relationship has been demonstrated in bacteria, fungi, algae, and higher plants (Table 12.6). Even in the case of 26 kinds of mammals, the thermostability of a single enzyme (catalase) varied from 48.1° to 67.1°C (Feinstein *et al.*, 1967). The differences are not only constitutive, but also adaptive. Thus, when a microorganism adapts to high temperature, either during its natural development (Agrawal *et al.*, 1974), or as a result of artificial hardening (Haberstick and Zuber, 1974; Air and Harris, 1974) the thermostability of its enzymes increases correspondingly. In higher plants, similar results have been reported (Shcherbakova *et al.*, 1973; Shcherbakova, 1972), but the results are less clearcut.

That these results are due to the proteins themselves and not to the presence or absence of stabilizing factors was indicated by heat hardening of cucumber and wheat plants (Fel'dman *et al.*, 1966). The thermostability of their acid phosphatase increased by 0.5° to 3.0°C. Since this increase in thermostability persisted after dialysis, it does not depend on the accumulation of a dialyzable protective substance. Similarly, the thermostability of lipase from the thermophilic fungus *Humicola lanuginosa* was not influenced by protective cofactors but was due to the enzyme itself (Liu *et al.*, 1973), and a thermal transition in ATPase conformation occurred at 50°C both in the presence and absence of its substrate, ATP (Hachimori and Noson, 1973). In the case of pyruvate kinase, the increase in thermostability is due, apparently, to a conformational change when the fungus Neurospora is grown at 42°C compared to 28°C (Kapoor *et al.*, 1976).

Cations, however, may stabilize ATPase against heat inactivation, the cation concentration required for protection increasing with the inactivating temperature (Kirschmann *et al.*, 1973). Similarly, prolyl-tRNA synthetase *was* protected against thermal denaturation by various substrates (Norris and Fowden, 1973), and protection was observed in sucrose vs. buffer solutions

TABLE 12.6

Proteins (Enzymes) Possessing a Thermostability Correlated with the Thermotolerance of the Organisms in Which They Exist

Thermophilic organisms	Enzymes	References
Bacteria	Eleven	Amelunxen and Lins, 1968
	Pyruvate carboxylase	Sundaram *et al.*, 1969
	D-Alanine carboxy-peptidase	Yocum *et al.*, 1974
	6-Phosphogluconate dehydrogenase	Pearse and Harris, 1973
	Homoserine dehydrogenase	Cavari and Grossowicz, 1973
Fungi	Cellulase	Loginova and Tashpulatov, 1967
	Hexokinase	Christophersen, 1963
	Cellobiase, aryl- β-D-glucosidases	Lusis and Becker, 1973
	Acid protease	Hashimoto *et al.*, 1972
	Lipase	Liu *et al.*, 1973
	α-galactosidase	Arnaud *et al.*, 1976
Algae	TPNH-cytochrome *c* reductase	Marré and Servettaz, 1956
	Phosphorylase	Fredrick, 1973
	C-phycocyanin	Kao *et al.*, 1975; Maccoll *et al.*, 1974
Higher plants	Urease, acid phosphatase, ATPase	Feldman, 1968, 1973; Feldman and Kamentseva, 1967
	malate dehydrogenase	McNaughton, 1966, 1974; Kinbacher *et al.*, 1967
	Ferredoxin	Mukhin and Gins, 1974; Mukhin *et al.*, 1973
	Nitrate reductase	Pal *et al.*, 1976
	RuDPC	Weber *et al.*, 1977

(Neucere and St. Angelo, 1972). Surface effects have also been demonstrated. The enzymes trypsin and papain when covalently coupled to porous glass show a marked increase in thermostability (Weetall, 1969), and the thermostability of chymotrypsin and trypsin was increased by covalently binding them to a complementary surface of a polmyer support in a multipoint fashion (Martinek *et al.*, 1977a). The authors suggest that this mechanism may function in vivo and may be responsible for the stability of membrane proteins (Martinek *et al.*, 1977b). Similarly, the association of trypsin, with trypsin inhibitors increased its heat stability (Donovan and Beardslee, 1975).

In contrast to the increase in *thermostability* of the enzyme, its *activity* declines with adaptation to high temperature (Christophersen, 1963; Fel'dman, 1968). The activity of acid phosphatase was decreased by half due to the hardening treatment, although its thermostability increased 0.5° to 3.0°C (Fel'dman *et al.*, 1966). Urease, however, is an exception, since its activity does not decline with increase in thermostability. The concentration of protein also decreases as a result of heat hardening (Fel'dman, 1966).

Some enzymes have not shown a relation between their thermostability and the heat tolerance of the plant from which they are extracted. In the case of *Typha latifolia* (McNaughton, 1966), in contrast to malate dehydrogenase, glutamate-oxaloacetate transaminase was quite resistant regardless of origin, and aldolase was rapidly inactivated regardless of origin. This may be interpreted to mean that only certain specific enzymes are important in the heat tolerance of plants, or that some enzymes are not thermostable unless the plant is first hardened (see below). On the other hand, the thermostability of an enzyme in a living cell may be quite different from that in extracts of the cell, due to pH differences, presence of substrate, etc. Since, in the above experiments, the leaves were blended with 30 volumes of water and no buffer was used, only the most stable enzymes would be unaltered. Aldolase is an unstable enzyme (see above) and any differences between plants would therefore quickly disappear. According to Kurkova (1967), ATPase activity rose more with heating (40°C) in the unhardened than in hardened corn shoots. This difference explained the observed phosphorylation which continued in the hardened but not in the unhardened shoots at 40°C. This result agrees with Fel'dman's observation (see above) that although heat hardening increases the thermostability of enzymes, it decreases their activity.

Others, however, have also shown that not all an organism's enzymes may possess thermostabilities related to its tolerance (e.g., Yutani *et al.*, 1973). An enzyme may be thermostable even at a temperature producing irreversible injury to the cells (Leblova and Mares, 1975; Aleksandrov and Alekseeva, 1972). Within the same plant, different enzymes may have large differences in thermostability—e.g. inactivation temperatures of 36°C for NRase, 65°C for NADPH-glutamate dehydrogenase in soybean (Magalhaes *et al.*, 1976). Some plant enzymes may be even more thermostable than these (Chen *et al.*, 1975). There may even be isoenzymes with highly specific heat responses [e.g. esterases of Lilium and Nicotiana (Pandey, 1973)], some of the isoenzymes completely losing their activity, others retaining their activity. In 3 arctic grasses (C_3 plants), as well as in a few tested C_4 plants, heat injury could not be explained by inactivation of carboxylating enzymes, since the optima were 50°C for RUDPC, 40°C for PEPC (Tieszen and Sigurdson, 1973). The interpretation of such results, however, is not always clearcut. The *in vivo* inactivation temperature for urease in cucumber (by heating the cotyledon) is lower than *in vitro* (by heating the homogenate; Gorban', 1972).

It must be realized that protein thermostability is required for prevention of indirect proteolytic heat injury as well as for prevention of direct heat injury; in order to resynthesize its protoplasmic proteins more rapidly at higher temperatures, the enzymes involved in protein synthesis must be fully active at these temperatures. Since these enzymes are proteins, the heat-tolerant plant can prevent heat injury only if its protein synthesizing enzymes are thermostable (see above). The thermotolerant plant must, therefore, possess the following two kinds of protein thermostability. (1) Its protein synthesizing enzymes must be thermostable, for prevention of indirect injury due to proteolysis. (2) The protoplasmic proteins whose denaturation and aggregation would cause instant death must be thermostable, for prevention of direct injury. These must be the membrane proteins, for they (and probably not the soluble enzymes) can be expected to cause instant death if denatured and aggregated (see Chapter 10). This conclusion is supported by direct observation of cells during heat killing (see Chapter 11).

Lange (1961) has postulated that these two kinds of thermotolerance occur at two different times of the year. The greater rate of protein synthesis would account for the summer increase in thermotolerance that is not accompanied by an increase in freezing tolerance. He believes that the winter rise in freezing tolerance involves a general stabilization of the proteins, accounting for the accompanying rise in thermotolerance. Since such a general stabilization would also be expected to lead to increased rate of protein synthesis, Lange's hypothesis appears less likely than the above explanation—a winter increase in the thermostability only of the membrane proteins.

ii. Role of Lipids.

In the case of poikilothermic animals, Ushakov and Glushankova (1961) found no correlation between iodine number of protoplasmic lipids and cell thermotolerance. Ushakov (1964) was also unable to detect a significant relationship between the melting point of the lipids and the thermostability of the cells. On the other hand, he was able to establish a close relationship between cell thermotolerance and the denaturation temperature of the proteins. In 88.9% (57 species) of the pairs of allied, but distinct species tested, parallel differences were found in the thermotolerance of the homologous cells and thermostability of their proteins (Ushakov, 1966).

In contrast to Ushakov's results, a relationship between lipid properties and growing temperature has been found in a number of microorganisms (see Huang *et al.*, 1974); for instance, in the case of a thermophilic, unicellular alga! eucaryote (*Cyanidium caldarium*) that grows in acid hot springs (Kleinschmidt and McMahon, 1970a). It can grow at temperatures from below 20°-56°C. When grown at 55°C, total lipid content decreased to one-half. The ratio of unsaturated: saturated fatty acids also decreased three times at the higher temperature. At 20°C, fully 30% of the fatty acid was linolenic, though none was detected at 55°C. The cells grown at 55°C were more heat tolerant than those grown at 20°C by 10°-15°C (Kleinschmidt and McMahon, 1970b). This was attributed to the higher saturation of the membrane fatty acids. The decrease in unsaturation at the higher temperature was explained by the decreased solubility of oxygen, which was apparently required for desaturation. In the case of *Bacillus cereus,* however, it also involved enzyme activation, for a system catalyzing the biohydrogenation of oleate to stearate was induced at 37°C but not at 20°C (Kepler and Tove, 1973). The appearance of this system was blocked by chloramphenicol or rifampin.

A similar direct relationship between lipid saturation and heat tolerance was found in nine thermophilic and nine mesophilic species of seven genera of fungi (Mumma *et al.*, 1970). Total lipids varied between 8 and 54.1%, most falling between 8 and 18.3%. The predominant fatty acids were palmitic, oleic, and linolenic. The mesophiles contained 0-18.5% linolenic acid; the thermophiles inappreciable amounts (<0.5%). Fatty acids of the thermophiles were, therefore, more saturated than those of the mesophiles.

These results were supported by the low degree of unsaturation in four thermophilic fungi (Raju *et al.*, 1976). Similarly, a mutant of *Neurospora crassa,* grown at 34°C had approximately 2 times the saturated fatty acid content of the wild type grown at 22°C (Friedman, 1977). The phase transitions occurred at −11° and −31°C, respectively. Among seven psychrophilic, mesophilic, and thermophilic species of yeasts (Arthur and Watson, 1976), there was a direct correlation between the growth temperature and the degree of membrane lipid saturation. Even in the case of higher plants, Kuiper (1970) has shown that lipid saturation increases with the growing temperature (see Chapter 10).

In the higher plant (*Atriplex lentiformis*), the leaf lipids were more saturated when grown at 43°D/30°N than at 23°D/18°N (Pearcy, 1978). In the monogalactosyl diglyceride, the major change was the presence of hexadecatrienoic acid (16:3) at low but not at high *T*. In the other lipids a decrease in linolenic acid (18:3) occurred at the higher growing *T*. The increased lipid saturation was correlated with greater thermostability of the photosynthetic apparatus.

Besides the increased saturation, other lipid properties have been related to thermotolerance. A thermotolerant strain of the yeast *Candida tropicalis* when grown at a lower temperature (30°C) produced half the amount of triglycerides, more sterols, and a higher proportion of short-chain fatty acids than when grown at a higher temperature (40°C), but there were only minor changes in the phospholipid content (Thorpe and Ratledge, 1973). *Acholeplasma laidlowii* can regulate the fluidity of its membrane when its growing temperature is altered (Huang *et al.*, 1974). In the case of an extreme thermophile, *Flavobacterium thermophilum* growing in hot springs above 70°C, a novel glycolipid was purified (Oshima and Yamakawa, 1974). The authors suggest that the glycolipids might have some function in stabilizing the membrane at high temperature.

In five extremely thermophilic bacteria of the Caldariella group, the lipids were based on the same type of cyclic diether or tetraether, combining glycerol and one of a series of very unusual C_{40} isoprenoid diols (de Rosa *et al.*, 1976, 1977).

Treatment with Triton x-100 decreased the thermophily of the blue-green alga *Cyanidium caldarium* (Enami and Fukuda, 1977a). The T_{opt} for the Hill reaction was lowered from 45° to 30°C, and this was accompanied by a decrease in the photosystem I reaction. The thermophily was completely recovered by washing with distilled water. The thermotolerance was also destroyed by lipase (especially galactolipase (Enami and Fukuda, 1977b), in agreement with the proposal of Kleinschmidt and McMahon that lipids are involved in thermotolerance due to their high degree of saturation of the fatty acids.

Nucleic acids have failed to show a relation to thermotolerance in wheat species (Utkhede and Jain, 1976) or in a thermophilic alga (Kikuchi *et al.*, 1973). The ribosomes of a thermophilic fungus, however, were more thermostable than those from a mesophilic species (Miller and Shepherd, 1973).

iii. Repair of Heat Injury.

In the case of freezing, either post-thawing injury or repair may occur, depending on the conditions. Similarly, heat shock may also be followed by either a progress of the injury or by repair. Which occurs, will depend on the environment to which the cells are exposed after the heat shock. In the case of *E. coli,* and yeast, the normal temperature for growth (37°C) favors repair (Bhaumik and Bhattacharjee, 1975; Stevenson and Richards, 1976). The illumination of corn leaves after heat shock (10 min at 42–44°C) significantly accelerated the recovery of photosynthesis (Semikhatova and Egorova, 1976). In the case of microorganisms, the medium is also important. Heat-injured bacteria and yeast cells recovered better on media which did not support growth, although 5–7 days were required for normal recovery of yeast (Bhaumik and Bhattacharjee, 1975; Stevenson and Richards, 1976).

Heat hardening previous to the heat shock of tobacco and bean leaves had only a slight protective effect against the short term effects of the heat treatment, but had a pronounced effect on recovery (Ben Zioni and Itai, 1975; Gorban', 1974). Similarly, kinetin (and to a lesser extent ABA) pretreatment increased the heat damage but post-treatment enhanced recovery (Itai *et al.*, 1978).

The internal factors responsible for the repair are less understood. It has been suggested that mild heat damage activates the repair mechanism (whatever this is) via feedback (Aleksandrov and Barabal'chuk, 1972). Translocation of readily used metabolites may be involved (Al'tergot and Dzhekshenaliev, 1973). Perhaps it is the sparing of these metabolites for repair that accounts for the above-mentioned better recovery on media that do not support growth. Somewhat surprisingly, repair occurs not only in the absence of growth, but also in the absence of protein synthesis (Bernstam and Arndt, 1974a,b). Furthermore, wheat roots thermosensitized by a heat shock were completely protected by two inhibitors of protein synthesis (chloramphenicol and ethanol; Skogqvist, 1973). Yet in the case of thermophiles, this must not be true, since the rate of breakdown in them is greater than in mesophiles (Miller *et al.*, 1974), and would lead to injury in the absence of rapid synthesis.

C. MOLECULAR ASPECTS OF THERMOTOLERANCE

Most of the above concepts and experimental evidence involve protein denaturation as a major cause of direct heat injury, and protein thermostability as the consequent basis for plant thermotolerance. According to Brandts' concepts of native (N) proteins, their denaturation (D) and aggregation (A) (see Chapter 11):

$$N \underset{\text{normal } T}{\overset{\text{high } T}{\rightleftharpoons}} D \xrightarrow{\text{high } T} A$$

The thermostability of the proteins, on this basis, can only be due to the prevention of either the reversible denaturation ($N \rightleftharpoons D$) or of the irreversible aggregation ($D \rightarrow A$).

1. Prevention of Reversible Denaturation ($N \rightleftharpoons D$)

This could be due to a strengthening of the protein bonds. Thus Koffler's *et al.* (1957) analyses of the amino acid content of the proteins led him to conclude that the thermostability of the proteins from thermophilic bacteria is due to their higher hydrophobicities. Similar results were obtained by Ohta *et al.* (1966). This conclusion was supported by the disintegration of the flagella of the mesophile in 4–6 *M* urea and those of the thermophile only in 9 *M* urea (Mallett and Koffler, 1957). It is now known that these high concentrations of urea break the hydrophobic bonds. The higher concentration needed to disintegrate the thermophile's protein indicates a greater strength or number of hydrophobic bonds. They also demonstrated a greater resistance to sodium dodecylsulfate, an anionic detergent that breaks hydrophobic bonds. Similar results were obtained by Marré *et al.* (1958). Acetamide and urea inhibited the activity of cytochrome *c* reductase to a lesser extent when the enzyme was extracted from a thermophilic blue-green alga (*Aphanocapsa thermalis,* than when extracted from a thermosensitive alga *Anabaena cylindrica*). All the above results are explained by Brandts' concept. Since the strength of hydrophobic bonds increases with the rise in temperature up to about 75°C (Brandts, 1967), proteins with a higher proportion of hydrophobic bonds would remain in the folded, native state at temperatures high enough to denature proteins with a lower proportion of hydrophobic bonds. In agreement with this concept, the β form of polylysine is the more stable form at 50°C and appears to owe a large part of its stability to hydrophobic interactions between the lysyl residues. The α-helical polylysine is stabilized largely by interamide hydrogen bonds and is the more stable form at 4°C (Davidson and Fasman, 1967).

Bigelow (1967) has calculated the hydrophobicities of the amino acid side chains and has used these values to obtain a quantitative comparison between those proteins of thermophilic and mesophilic organisms for which amino acid analyses are available. The phycocyanins of five mesophilic species yielded average hydrophobicities of 980–1110 cal/residue, compared to 1190 cal/residue for one thermophilic species. The α-amylases of four mesophilic species had hydrophobicities of 1020–1130 cal/residue, compared to 1210 for the one thermophile. Friedman (1968), on the other hand, has compared the amino acid analyses for the ribosome protein of a mesophilic and a thermophilic bacterium and has concluded that there is no obvious difference between the two. This conclusion was based solely on visual comparison. Calculation of the hydrophobicities, with the use of Bigelow's values, supports his conclusion, yielding average values of 1062 cal/residue and 1095 cal/residue, respectively (Table 12.7). However, some of the hydrophobicity values for the side chains are so low (0.45–1.70 kcal/residue) that they may not be able to form bonds strong enough to hold the protein folds in place. Furthermore, these differences in strength are exaggerated at high temperatures (Brandts, 1967). The strength of the weaker hydrophobic bonds rises only slightly with temperature to a maximum at

TABLE 12.7

Relative Hydrophobicities of Proteins from a Thermophile (*Bacillus stearothermophilus*) and a Mesophile (*Escherichia coli*)[a,b]

Amino acids with hydrophobic side chains	Hydrophobicity (kcal/res)	*B. stearothermophilus*			*E. coli*		
		Mole (%)	kcal	Av. cal/res	Mole (%)	kcal	Av. cal/res
Trp	3.00	—	—	—	—	—	—
Ile	2.95	6.64	19.55	—	5.51	16.25	—
Tyr	2.85	2.16	6.15	—	1.78	5.07	—
Phe	2.65	3.55	9.40	—	3.03	8.03	—
Pro	2.60	4.44	11.52	—	3.67	9.55	—
Leu	2.40	8.34	20.00	—	7.40	17.75	—
Val	1.70	9.07	15.40	*666*	9.63	16.35	*566*
Lys	1.50	6.30	9.45	—	9.01	13.50	—
Met	1.30	2.33	3.02	—	2.40	3.12	—
Cys/2	1.00	0.82	0.82	—	0.53	0.53	—
Ala	0.74	10.51	7.90	—	10.98	8.23	—
Arg	0.75	5.01	3.76	—	7.30	5.46	—
Thr	0.45	5.53	2.49	—	5.22	2.35	—
Totals		64.70	109.46	*1095*	66.46	106.19	*1062*

[a] Data from Bigelow, 1967 and Friedman, 1968. Numbers in italics indicate averages.
[b] From Levitt, 1969.

50°C, only 200 cal above the value at 0°C. The strength of the stronger hydrophobic bonds rises much more steeply with temperature to a maximum at 80°C, which is 750 cal above the value at 0°C. If only side chains with values of 2 kcal/residue and above are compared, all five amino acid residues are found to be present in larger quantity in the thermophilic than in the mesophilic protein (Table 12.7). On the basis of these five amino acids, the average hydrophobicity of the protein of the thermophile is 666 cal/residue compared to 566 cal/residue in the mesophile. In terms of numbers of strong hydrophobic bonds, there would be a maximum of 1 bond/8 residues in the thermophile, and only 1/9.3 residues in the mesophile. It is even conceivable that some of the weaker hydrophobic side chains may be effective in the proteins of the thermophile because of proximity to the one strong SS bond, which apparently occurs in the protein of the thermophile but not in the protein of the mesophile. A similar analysis of Matsubara's (1967) data for the heat-resistant protein thermolysin again reveals no higher value for total hydrophobicity (1010 cal/residue), but a very high value for the six groups having high hydrophobicities (706 cal/residue).

This mechanism could conceivably also arise due to hardening. If the proteins are broken down and resynthesized during this exposure to moderately high temperature, the newly synthesized proteins, though having the same amino acid sequence, would fold in a somewhat different manner due to the increased affinity of the hydrophobic groups for each other at the higher temperature and the weakening of the H bonds. They would therefore have a higher proportion of hydrophobic to hydrogen bonds and would be more stable at the high temperatures. This would account for the slow natural hardening during summer (see above). Similarly, even if the proteins are not broken down and resynthesized, if some of the hydrogen bonds are broken at the high temperature, this may permit the formation of new and stronger hydrophobic bonds, between groups previously separated sterically, and which now have a stronger affinity for each other because of the high temperature. This more rapid hardening would occur due to heat shock.

As mentioned earlier, the change in conformation may or may not alter the activity of the enzyme, depending on whether or not the active site is altered. Thus, heat-hardening of leaves, which leads to increased thermotolerance of their cells, increases the thermostability of their enzymes by 1.5°–7°C in the case of ATPase, acid phosphatase, and urease (Fel'dman, 1968). In the case of the first two, this was accompanied by a decrease in activity, but the activity of urease was not affected by its increase in thermostability.

In agreement with this concept, a protein conversion during heat hardening has been observed by Jacobson (1968) in the case of *Drosophila*. An increase in thermostability accompanied the conversion of the electrophoretically slowest moving isoenzyme of alcohol dehydrogenase to the fastest moving form. This result is predicted by the above concept, since the increased strength of the hydrophobic bonds and the rupture of some of the hydrophilic bonds would lead to a greater proportion of internal hydrophobic bonds, and a greater proportion of external, free hydrophilic (polar) groups, leading to a greater surface charge and, therefore, greater electrophoretic mobility.

More recent results suggest that the thermostability of an enzyme may be greatly increased by an increase in hydrophobicity due to the substitution of only a few suitable acids (Yutani *et al.,* 1977). A marked effect, in fact, resulted from replacement of even a single Glu residue by Gln or Met at position 49, in the α-subunit of tryptophan synthetase from *Escherichia coli.* Similarly, cytochrome *c* from the thermophilic fungus *Humicola lanuginosa* is unique in having phenylalanine as residue 74 (Morgan *et al.,* 1972; Morgan and Riehm, 1973). Similar results have been obtained with β-galactosidase and glyceraldehyde-3-P dehydrogenase, and it must be concluded that the thermostability of proteins is often a much more subtle property than previously believed (Amelunxen and Singleton, 1976).

It does not necessarily follow that all protein thermostability is due to an increase in hydrophobicity alone. For instance, in comparison with the enzyme from mesophiles, the thermostability of glutamine synthetase from the thermophilic *Bacillus stearothermophilus* correlates with both an increased average hydrophobicity and with carboxylate side-chain residue contents, but not with α-helix, B-sheet, or H-bonding residues (Wedler and Hoffmann, 1974). Similarly, where an increase in thermostability appears to be due to substitution of a single amino acid residue, this amino acid may not be more hydrophobic than the one for which it substitutes. For instance, the thermostability of collagens was correlated with their amino acid contents, e.g., the proline, hydroxyproline, and threonine contents (Rigby, 1967). Jockush (1968), in fact, has suggested that any pair of homologous proteins will differ in thermostability if there is at least one difference in amino acid sequence. He investigated the coat protein of TMV, whose tertiary structure is not complicated by covalent (SS) cross-links. At low ionic strength (0.02) and a slightly alkaline reaction (pH 7.5) the protein is mildly heat denatured and stays in solution. This mild denaturation consists of a partial unfolding of protein subunits leading to exposure of hydrophobic regions, which in turn causes limited aggregation, and at higher ionic strength or lower pH results in precipitation. The temperatures of half denaturation were 40.5° and 27°C for the vulgare form and the mutant, respectively. The following lowerings of the denaturation temperature resulted from amino acid replacement:

tyrosine by cysteine	40°C → 35°C
asparagine by lysine	30
asparagine by glycine	32
asparagine by alanine	31
proline by threonine	29
proline by leucine	27
threonine by isoleucine	32
proline by serine	32
isoleucine by threonine	37
proline by leucine	37

Similarly, it has been found that glutamic acid tends to stabilize a protein against thermal denaturation, serine to destabilize it (Bull and Breese, 1973).

The strongest of all intramolecular bonds responsible for the conformation of the proteins is the covalent SS bond. The importance of this bond in the thermostability of proteins can be seen by comparing thermostable with thermolabile enzymes. The thermostable are characterized by absence of SH groups and presence of SS bonds (Table 12.8). Similarly, when different lysozymes are compared, hen's egg lysozyme has four SS bonds and is the most thermostable, human lysozyme has three SS bonds and is less ther-

TABLE 12.8

Properties of Heat-Stable Enzymes

Protein	Inactivation temperature (°C)	Time (min)	Inactivation (%)	Molecular weight	SH (groups/molecule)	SS (groups/molecule)
1. Pepsin	65 (in acid)	15	50	35,000	0	3
2. α-Amylase (*Bacillus subtilis*)	65	30		48,700	0	0
3. Arginase	70	177	50	140,000	0	0
4. α-Amylase (thermophile)	90	60	10	15,000	0	2
5. Inorganic pyrophosphatase	90-100	—	—	63,000	Trace	
6. Cytochrome *c*	>100	—	—	13,000	0	0
7. Muramidase (lysozyme)	>100	—	—	15,000	0	4
8. Ribonuclease	>100 (in acid)	—	—	12,700	0	4
9. Trypsin	>100	—	—	24,000	0	6
10. Myokinase (adenylate kinase)	>100	—	—	21,000	2	0

[a] From Levitt, 1966.

mostable, and goose egg lysozyme has two SS bonds and is the least stable (Jollès, 1967). Similar results have been obtained with bovine lens proteins (Mehta and Maisel, 1966).

Although analyses of pure plant proteins have not yet been made, indirect evidence points to a possible role of SS bonds in the thermotolerance of plants. Fraction I protein was found to be more thermostable when isolated from the leaves of heat-hardened bean plants than from unhardened plants (Sullivan and Kinbacher, 1967). Blocking the SH groups with PCMB (parachloromercuribenzoate) did not change the thermostability of the protein from hardened leaves. On the other hand, cleavage of the SS bonds with ME and sodium sulfite decreased the thermostability of the protein from hardened plants to that of the protein from unhardened plants. No significant difference was found, however, between the number of SS bonds in the hardened and unhardened plants. It was therefore suggested that the hardening process led to a repositioning of the SS bonds in such a way as to increase protein stability. This agrees with the above concept of a change in conformation leading to a larger number of hydrophobic bonds. Similarly, large effects on the thermal stability of lactate dehydrogenase resulted from

SH modification (Jacobson and Braun, 1977). In the case of an alkaline protease from a thermophilic actinomycete, it has been suggested (Mizusawa and Yoshida, 1976) that the SH group of the single cysteine residue is bound to another residue via an H bond, to tighten the active site conformation.

Some enzymes apparently owe their thermostability to structural factors other than hydrophobic or SS bonds. A thermostable acid protease of *Penicillium duponti* differs from other acid proteases by being a glycoprotein with 4.33% carbohydrate (Emi *et al.*, 1976). An ATPase from a thermophilic bacterium consists of 5 subunits and is thermostable if reconstituted from only 2 of the 3 different kinds of subunits, thermolabile if the third is added (Yoshida *et al.*, 1977). In both cases, the reconstituted enzyme regained its hydrolyzing activity.

At least two enzymes from thermophilic microorganisms (thermolysin and α-amylase) owe their stability to bound Ca^{2+} (Pangburn *et al.*, 1976; Weaver *et al.*, 1976; Fontana *et al.*, 1976; Heinen and Lauwers, 1976; Yutani, 1976). Four Ca^{2+} ions are apparently bound to each molecule of thermolysin (a metallic endopeptidase), and removal of 3 of these results in loss of thermostability but not activity. The Ca^{2+} may perhaps act in place of SS bonds, which are absent from some proteins of thermophiles—e.g. the above thermolysin. It is now believed that the enzymes of the thermophilic bacteria owe their extra stability mostly to the additional salt bridges (Perutz, 1978). Thus, three additional salt bridges made by each subunit to the other, in the tetramer of D-glyceraldehyde-3-phosphate dehydrogenase from *Bacillus stearothermophilus* apparently make a major contribution to its thermostability, compared to the corresponding enzyme from lobster muscle (Biesecker *et al.*, 1977).

From all the above evidence of the effects of small differences, it is not surprising that the structure of proteins from thermophiles is very similar to corresponding proteins of mesophiles. No differences in hydrophobicity, H-bonding, or average residual size could be detected between the thermostable phosphofructokinase from the extreme thermophile *Thermus* X-1 and the enzyme from the mesophiles *E. coli* and *Clostridium pasteurianum* (Cass and Stellwagen, 1975). Similarly, the enhanced thermostability of thermolysin relative to thermolabile enzymes, could not be attributed to a common determinant such as hydrophobic stabilization or a metal ion. In any given instance, the difference in thermostability may be due to a subtle difference in hydrophobicity, metal binding, H-bonding, ionic interactions, SS covalent bonding, etc., or a combination of two or more of these (Matthews *et al.*, 1974). Only the knowledge of the complete structure and composition of the molecule can be expected to reveal with certainty which of these factors is involved.

There is a simple, thermodynamic basis for the marked effect of such subtle differences. The net free energy difference between the folded and unfolded (denatured) protein is typically only 5–10 kcal/mole. This is equivalent to several H bonds, or to the hydrophobic interaction energy generated by a single leucine side-chain (Weaver *et al.*, 1976; Hocking and Harris, 1976).

2. Prevention of Irreversible Aggregation ($D \rightarrow A$)

Aggregation could be prevented only by eliminating or protecting the chemical groups capable of interacting with each other to form intermolecular bonds. It is still a question as to which chemical groups can form such bonds. As in the case of intramolecular bonds, the strongest *intermolecular* bond that can be formed is the covalent SS bond. In agreement with this, SH-containing enzymes are thermolabile (Table 12.9), presumably because the SH groups of adjacent molecules can combine to form intermolecular SS bonds. The thermostable enzymes, on the other hand, are mostly free of SH groups (Table 12.8). The most striking difference is between the same enzyme (α-amylase) from a thermotolerant and a thermosensitive species of bacterium. The former has two SS bonds per molecule, while the latter has none (Table 12.8). Similarly, the proteins of some thermophilic bacteria are free of any SH or SS groups (Koffler *et al.*, 1957; Ohta *et al.*, 1966; Matsubara, 1967).

Considerable evidence points to the conversion of SH groups of proteins to intermolecular SS bonds during freezing (see Chapter 10). Since, as mentioned above, an increase in freezing tolerance is accompanied by an increase in heat tolerance, the same chemical groups must be involved in protein aggregation at both temperatures. In favor of this conclusion, a decrease in protein SH and an apparent increase in SS groups has been found during heat killing (Levitt, 1962). In agreement with these results, heat injury was decreased when oxygen was deficient and was increased when oxygen was increased (Al'tergot, 1963). These results were explained by the thermostability of the oxidases and the consequent oxidation of ascorbic acid, GSH, and tannins at high temperature. Porto and Siegel (1960) also implicate oxidation in heat injury. They explain the kinetin-induced restoration of heat-injured germination on the basis of antioxidant activity. Similarly, the results of Levy and Ryan (1967) indicate that heat inactivation of the relaxing site of actomyosin occurs because certain labile SH groups are oxidized to the SS form. Dithiothreitol led to both prevention and reversal of this effect. Mishiro and Ochi (1966) found that human serum albumin becomes turbid at 60°–95°C. This turbidity was completely prevented in 0.05 *M* solution in the presence of 10^{-3} *M* sodium dipicolinate, which decreased the number of

TABLE 12.9

Properties of Heat-Labile Enzymes[a]

Enzyme	Inactivation temperature (°C)	Time at temperature (min)	Molecular weight	SH (groups/ molecule)	SS (groups/ molecule)
1. β-Galactosidase	55	1	750,000	12	2
2. L-Glutamate dehydrogenase	55	—	1,000,000	90–120	
3. α-Glycerophosphate dehydrogenase	55	1 (53% activity lost)	78,000	15–16	
4. Glyceraldehyde-3-phosphate dehydrogenase	Room temperature (stabilized by ethylene-diaminetetraacetate at 39)	—	120,000	11±2	0
5. Succinic dehydrogenase	Unstable even at 25 at 25	—	200,000	SH enzyme	
6. Xanthine oxidase	56	—	290,000	SH enzyme	
7. Glucose oxidase	Above 40	—	154,000	SH enzyme	

[a] From Levitt, 1966.

SH groups. Brandt and Andersson (1976) have also found SS bonding to be involved in heat aggregation of human serum albumin. A similar dependence on SH and SS groups is indicated in the case of an enzyme (uridine diphosphate-glucose 4-epimerase) from *Saccharomyces fragilis*. The enzyme is inactivated by 42°C for 7 minutes and is reactivated by ME (a SH reagent) and NAD (Ray and Bhaduri, 1976).

Harris (1976) has pointed out that the enzymes of thermophilic microorganisms tend to have fewer SH groups than their counterparts in mesophiles. He suggests that they may have been deleted as an adaptation against inactivation of the enzyme at high temperatures. He also suggests incorrect *intra*molecular (rather than *inter*molecular) SS bonding as one kind of heat inactivation of enzymes. GPDH, for instance, is heat inactivated by formation of an intrachain SS bond between Cys-149 and Cys-143. Formation of this bond seems to trigger off a conformational change that results in precipitation and irreversible inactivation of the enzyme. He also points out that the *number* of SH groups is not necessarily the deciding factor, since their exact location in the molecule is also important.

However, even SH-free proteins of thermophilic bacteria are inactivated if the temperature is high enough. Even gelatin, which is denatured and free of SH groups, forms covalent cross-links between adjacent molecules when the water content falls below 0.2 gm/100 gm protein (Yannas and Tobolsky, 1967). It therefore seems reasonable to conclude that as the temperature rises, and more and more bound water is converted to free water (Petrochenko and Privalov, 1973; Rueegg *et al.*, 1975), previously protected chemical groups become available for the intermolecular covalent bond formation. Consequently, at moderately high temperatures, aggregation may be primarily due to intermolecular SS bonding by oxidation of free SH groups (which do not bind water); at still higher temperatures, other groups may become available for intermolecular bonding. What these other groups are, is not known.

In the absence of SH groups (e.g., in the case of proteins from thermophiles), Awad and Deranleau (1968) suggest that intermolecular hydrophobic bonding is possible. Nearly all the hydrophobic groups are internal to the native molecule and therefore unavailable for intermolecular bond formation as long as the molecules remain native. When the proteins are denatured reversibly at the high temperature, this is due to the greater thermodynamic tendency to unfold than to remain bonded by the hydrophobic groups in the folded state (see Fig. 11.8). Although these hydrophobic bonds are too weak to hold the protein in the folded state, they may be strong enough to aggregate the unfolded protein molecules. On the other hand, the initial unfolding also involves the breaking of hydrophilic bonds (e.g., H bonds) which are weaker than the hydrophobic bonds at the high temperature. This may permit the formation of new hydrophobic bonds previously sterically impossible. It is, therefore, conceivable that as soon as the hydrophilic bonds are broken, the molecule quickly refolds to a conformation that is more stable at the high temperature. This would clearly explain the difference between rapid thermohardening, and thermoinjury. A brief exposure to a supraoptimal temperature would permit partial unfolding and immediate refolding to a more hydrophobically bonded, and therefore, more thermostable form. A longer exposure or a higher temperature would permit aggregation and, therefore, injury.

An apparently opposite relation has been suggested by Berns and Scott (1966). They conclude that the protein from a thermophilic member of the Cyanophyta (*Synechococcus lividus*) has a greater number of charged and polar groups than that of mesophiles. According to their interpretation, aggregation reversibly inactivates the enzymes at 25°C, and only by raising the temperature to 50°C would the smaller, active form of the enzyme occur. This explanation is difficult to accept, since the association of monomer protein subunits to form di-, tri- and multimers usually involves hydrophobic bonding which increases in strength from 5°–50°C (see Fig. 11.8). Association would, therefore, be more likely at the higher than at the lower temperature. Furthermore, it is usually the larger, associated protein molecule that is active, and the monomer that is inactive. Aggregation of a protein has also been reported for another thermophilic blue-green alga (*Oscillatoria*) from a hot spring (Castenholz, 1967). Since these organisms survive as much as 30° higher temperatures than the most thermotolerant higher plant (see above), the mechanism is perhaps different.

On the basis of the above concepts, thermostability of proteins depends on their possession of one or more of the following characteristics: (1) Specific, strong intramolecular bonds—hydrophobic, SS, salt, Ca^{2+}, and perhaps others, that oppose high temperature-induced denaturation, and (2) absence of SH groups which could lead to SS- induced aggregation at high temperature.

Thermotolerance of plants is apparently due mainly to protein thermostability and therefore presumably dependent on these same characteristics. Since soluble enzymes are less hydrophobic than membrane proteins (see Chapter 10), this concept also explains why the indirect metabolic injury due to protein loss occurs at a lower temperature than the direct (membrane) injury. The different kinds of thermotolerance are tabulated in Diag. 12.1. Although the evidence points to the proteins as the main factors in heat tolerance, this does not, of course, preclude a role for saturated lipids with high phase transition temperatures and, therefore, a fluidity suited to growth

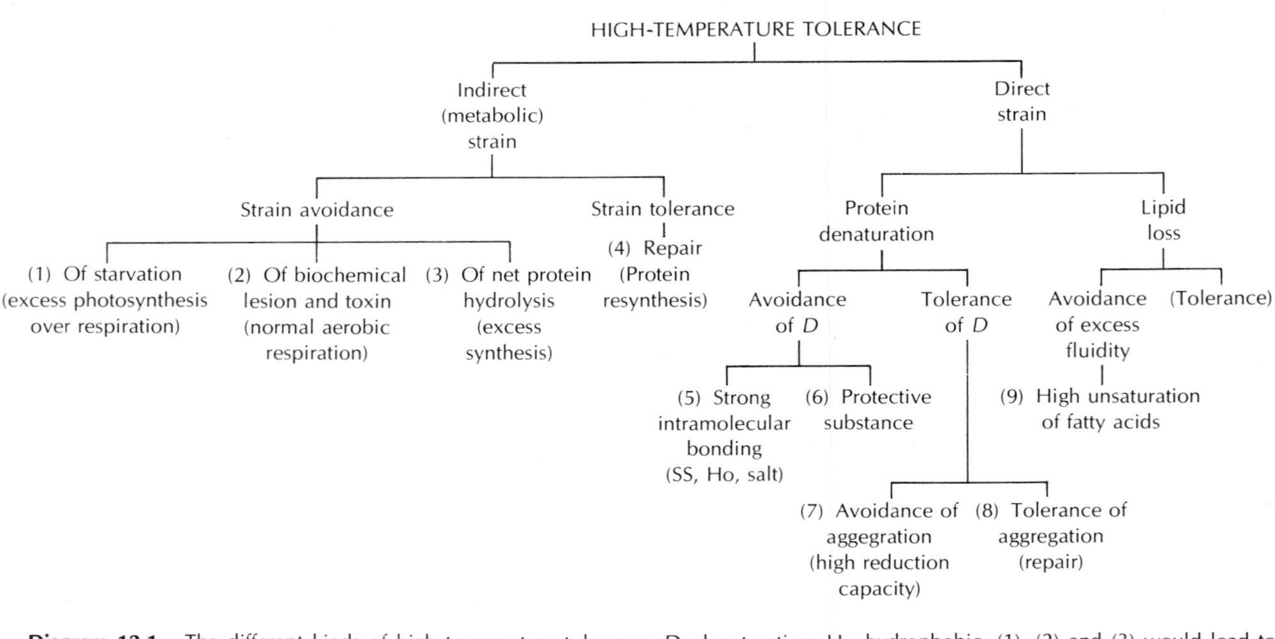

Diagram 12.1. The different kinds of high-temperature tolerance. D, denaturation. H_o, hydrophobic. (1), (2) and (3) would lead to avoidance of growth inhibition.

at high temperatures. In fact, it must be concluded that both the proteins and the membrane lipids must be involved in the adaptation of cells to high temperature. Neither alone can confer thermotolerance on the plant.

How can the above concepts of heat injury apply to organisms with different degrees of thermotolerance?

1. Psychrophilic microorganisms. In at least some cases the damage is due to denaturation of thermolabile, soluble enzymes at temperatures normal for the growth of mesophiles but supraoptimal for the growth of the psychrophiles (e.g. 15–20°C). This may be partially due to their unusually high enzyme concentration, which is apparently responsible for their ability to grow and develop at low temperatures (Berry, 1975). At the higher temperatures, normal for the growth of mesophiles, due to the higher reaction rates, there would be an excess of enzyme over substrate. Since attached substrate increases the stability of enzymes, these unattached enzymes would be denatured and aggregated at lower temperatures than similar enzymes in mesophiles.

2. Mesophiles. The surprisingly low T_{opt} for photosynthesis (20–30°C) may be due to denaturation and therefore, inactivation of a single enzyme among the many required for normal photosynthesis. Simple denaturation, however, is insufficient to account for the after effect—the continued inhibition of photosynthesis for hours or days after return to the T_{opt} for photosynthesis. In such cases, denaturation must be followed by aggregation at the supraoptimal temperature—for instance by intermolecular SS formation. Chloroplast enzymes are particularly susceptible to aggregation via intermolecular SS bonding, because of the light-induced SH groups and the metabolic systems capable of oxidizing them (see Chapter 10). The time for recovery would presumably be required for hydrolysis of the denatured and aggregated enzyme followed by its resynthesis.

Direct injury to mesophiles, due to membrane damage, is not so readily explained by protein denaturation. The mesophiles are, however, able to survive and grow over a wider temperature range than the psychrophiles (e.g. 5–45°C versus 0–15°C for psychrophiles). They must, therefore, possess a high enough fluidity of their membrane lipids for maintenance of both normal semi-permeability and active uptake at the low temperatures above their T_{min}. The membranes may, therefore, be too fluid at the high, injurious temperatures. This must lead to inactivation of membrane enzymes required for active absorption, and if severe enough, to a loss of semipermeability.

3. True thermophiles (able to survive temperatures of 70–100°C). These are found only among microorganisms. According to Tansey and Brock (1972), the failure of eukaryotes to evolve the ability to grow at as high

temperatures as the thermophilic prokaryotes may be due to their inability to form organellar membranes that are both thermostable and functional. Conversely, it may be due to the lesser development of membranes within the protoplasm of microorganisms than in higher organisms, for instance the absence of membranes of nuclei, plastids, etc. and also the absence of SH groups from their proteins. Thermophily is also associated with an inability to grow and develop normally at the lower temperatures normal for the growth of most organisms. At the high temperatures at which they grow, the proteins of psychrophiles and mesophiles would all be denatured, and the membrane lipids would be excessively fluid, permitting the biomolecular membrane leaflets with maximum surface, to form vesicles with minimum surface. If the thermophiles are heated to a high enough temperature for injury, these same two factors would presumably be involved.

D. RELATIONSHIP BETWEEN THERMOTOLERANCE AND LOW-TEMPERATURE TOLERANCE

1. Thermotolerance and Chilling Tolerance

The different kinds of indirect stress injury induced by the chilling (Chapter 3) and the heat (Chapter 11) stresses are apparently identical. It does not necessarily follow, of course, that the respective tolerances of these two stresses are correlated. If, for instance, the metabolic disturbance leading to the injury is due to a reversible denaturation and, therefore, an inactivation of certain enzymes, the proteins most readily denatured at a high temperature would be those with low hydrophobicities, and this would lead to a greater stability at low temperature. On this basis, thermotolerance and chilling tolerance of the indirect stress effects would be mutually exclusive. This prediction agrees with the known facts; chilling-sensitive plants are commonly those adapted to growth at high temperatures and therefore possessing thermotolerance. Similarly, conversion of mesophilic to psychrophilic bacteria lowered both the T_{min} and T_{max} for growth (Olsen and Metcalf, 1968).

Tolerance of the direct effects of the chilling stress is apparently related to the membrane lipids (see Chapter 3). The evidence indicates an increase in unsaturation of lipids in chilling tolerant and a decrease in thermotolerant plants. This again leads to the prediction that the two tolerances are mutually exclusive.

2. Thermotolerance and Freezing Tolerance

Two opposite relations between thermotolerance and freezing tolerance must be explained.

1. Thermotolerance increases in some heat resistant plants during the summer rise in temperature, at the same time as freezing resistant plants achieve their minimal freezing tolerance. This mutually exclusive hardening response is easily explained by (a) the lipid properties which change in opposite directions during acclimation to low and high temperature respectively—a high fatty acid unsaturation and low phase transition temperature at low temperatures, the reverse at high temperatures. This adaptation in opposite directions would also apply to avoidance of lipid peroxidation, for the higher unsaturation in the freezing tolerant plants would favor peroxidation at high temperatures, the low unsaturation in the high temperature tolerant plants would be an avoidance mechanism against peroxidation. (b) Another possible factor is a decrease in relative hydrophobicity of the proteins at low temperature, resulting in a prevention of $N \rightleftharpoons D$ in both.

2. The fall and winter increase in freezing tolerance may carry with it an increase in heat tolerance. Two factors may again be involved. (a) An increase in avoidance of protein aggregation ($D \rightarrow A$) in the plants acclimated to low temperature may also prevent the aggregation at high temperature—for instance if due to a higher reducing capacity. This high reduction would also protect both kinds of plants against lipid peroxidation, if the plants with high temperature tolerance are not already sufficiently protected by an absence of unsaturation. (b) An increase in ability to repair the damage after thawing of the previously frozen plant, may also conceivably be operative at normal growing temperatures following exposure to a heat stress. However, the kind of protein capable of retaining enzyme activity needed for repair at low temperatures may be inactivated at high temperature and vice versa.

It follows from the above that the only plausible factor common to both freezing and heat tolerance is a high reduction capacity. Since there *is* one kind of tolerance common to both stresses, this is evidence in favor of high reduction capacity as an important factor in tolerance against temperature stresses.

Bibliography

(Chapters 11 and 12)

Ageeva, O. G. (1977). Effects of light on thermostability of Hill reaction in pea and spinach chloroplasts. *Photosynthetica* **11,** 1–4.

Agrawal, P. K., Narayan, R., and Gollakota, K. G. (1974). Study of heat resistance in a bacterial nucleosidase. *Biochem. Biophys. Res. Commun.* **60,** 111–117.

Air, G. M., and Harris, J. I. (1974). DNA-dependent RNA polymerase from the thermophilic bacterium *Thermus aquaticus. FEBS Lett.* **38,** 277–281.

Aleksandrov (Alexandrov), V. Ya. (1964). Cytophysiological and cytoecological investigations of heat resistance of plant cells toward the action of high and low temperature. *Q. Rev. Biol.* **39,** 35–77.

Aleksandrov, V. Ya., and Alekseeva, N. N. (1972). Thermostability of oxidative phosphorylation and its role in the development of thermal injury of plant cells. *Tsitologiya* **14,** 591–597.

Aleksandrov, V. Ya., and Barabal'chuk, K. A. (1972). Repair of thermal injury of *Tradescantia* leaf cells after exposure to heat hardening. *Tsitologiya* **14,** 1328–1334.

Aleksandrov, V. Ya., and Dzhanumov, D. A. (1972). Effect of heat injury and heat-hardening on a photoinduced long-term after- luminescence of *Tradescantia fluminensis* Vell. leaves. *Tsitologiya* **14,** 713–720.

Aleksandrov, V. Ya., and Yazkulyev, A. (1961). Heat hardening of plant cells under natural conditions. *Tsitologiya* **3,** 702–707.

Aleksandrov, V. Ya., Lutova, M. Y., and Feldman, H. L. (1959). Seasonal changes in the resistance of plant cells to the action of various agents. *Cytologia* **1,** 672–691.

Aleksandrov, V. Ya., Ouchekov, B. P., and Poljansky, G. I. (1961). La mort thermique des cellules par rapport au problem de l'adaptation des organismes á la température du milieu. *Pathol. Biol.* **9,** 849–854.

Aleksandrov, V. Ya., Lomagin, A. G., and Feldman, N. L. (1970). The responsive increase in thermostability of plant cells. *Protoplasma* **69,** 417–458.

Allen, M. B. (1950). The dynamic nature of thermophily. *J. Gen. Physiol.* **33,** 205–214.

Al'tergot, V. F. (1963). The action of high temperatures on plants. *Izv. Akad. Nauk SSSR, Ser. Biol.* **28,** 57–73.

Al'tergot, V. F., and Dzhekshenaliev, K. D. (1973). Enhancement of root regeneration by heat injury metabolites. *Dokl. Akad. Nauk. SSR ser. Biol.* **212,** 510–512.

Amelunxen, R. E., and Lins, M. (1968). Comparative thermostability of enzymes from *Bacillus stearothermophilus* and *Bacillus cereus. Arch. Biochem. Biophys.* **125,** 765–769.

Amelunxen, R. E., and Singleton R., Jr. (1976). Thermophilic glyceraldehyde-3-phosphate dehydrogenase. *Experientia, Suppl.* **26,** 107–120.

Ananthanarayanan, V. S. (1973). Thermal denaturation of B-lactoglobulin-A in water. *Curr. Sci.* **42,** 845–846.

Andreeva, I. N. (1969). Effect of temperature on the swelling and shrinkage of isolated mitochondria. *Fiziol. Rast.* **16,** 221–227.

Ansari, A. Q., and Loomis, W. E. (1959). Leaf temperatures. *Am. J. Bot.* **46,** 713-717.

Antropova, T. A. (1971). Temperature adaptation of cells of the moss *Mnium affine* Bland. *Bot. Zh.* **56,** 1681-1686.

Antropova, T. A. (1974). Seasonal changes of cold and heat resistance of cells of two moss species. *Bot. Zh.* **59,** 117-122.

Armond, P. A. Schreiber, U., and Björkman, O. (1978). Photosynthetic acclimation to temperature in the desert shrub *Larrea divaricata.* II. Light harvesting efficiency and electron transport. *Plant Physiol.* **61,** 411-415.

Arnaud, N., Bush, D. A., and Horisberger, M. (1976). Study of an intracellular a-galactosidase from the thermophilic fungus *Penicillium duponti. Biotechnol. Bioeng.* **18,** 581-585.

Arthur, H., and Watson, K., (1976). Thermal adaptation in yeast: Growth temperatures, membrane lipid, and cytochrome composition of psychrophilic, mesophilic, and thermophilic yeasts. *J. Bacteriol.* **128,** 56-68.

Askenasy, E. (1875). Ueber die Temperatur, welche Pflanzen in Sonnenlicht annehmen. *Bot. Z.* **33,** 441-444.

Asundi, S., Mattoo, A. K., and Modi, V. V. (1974). Studies on a thermophilic *Talaromyces* sp.: Mitochondrial fatty acids and requirement for a C_4 compound. *Indian J. Microbiol.* **14,** 75-80.

Aver'yanov, A. A., and Veselovskii, V. A. (1976). Toxic action of oxygen on plants at superoptimal temperatures. *Fiziol. Rast.* **23,** 1248-1254.

Avilova, L. D. (1962). The thermostability of epidermal cells of albino, green, and etiolated sunflower plants. *Tsitologiya* **4,** 73-76.

Awad, E. S., and Deranleau, D. A. (1968). Thermal denaturation of myoglobin. I. Kinetic resolution of reaction mechansim. *Biochemistry* **7,** 1791-1795.

Ayres, A. A. (1916). The temperature coefficient of the duration of life of *Ceramium tenuissimum. Bot. Gaz. (Chicago)* **62,** 65-69.

Babushkin, L. N. (1975). Transpiration method for determining heat resistance of plants. *Fiziol. Rast.* **22,** 647-653.

Babushkin, L. N., and Barabal'chuk, K. A. (1974). Effect of rapid stomatal opening in darkness at a temperature lethal for plants. *Fiziol. Rast.* **21,** 644-646.

Baker, B. S., and Jung, G. A. (1970). Response of four perennial grasses to high temperature stress. *Proc. Int. Grass. Congr., 11th, 1969* pp. 499-502.

Baker, F. S. (1929). Effect of excessively high temperatures on coniferous reproduction. *J. For.* **27,** 949-975.

Balina, N. V. (1976). Effect of dry wind on microsporogenesis of drought-hardened beans. *Fiziol. Rast.* **23,** 146-151.

Barabal'chuk, K. A. (1969). Reaction of thermolabile and thermostable functions on the plant cell to heat-hardening action. *Tsitologiya* **11,** 1021-1032.

Barabal'chuk, K. A., and Chernyavskaya, V. N. (1974). Effect of higher temperature on stomatal movement in *Tradescantia* leaves. *Tsitologiya* **16,** 1481-1487.

Barabal'chuk, K. A., and Chernyavskaya, V. N. (1975). Effect of high temperature on the nuclear volume in spiderwort leaf cells. *Tistologiya* **17,** 1223-1226.

Belehradek, J. (1935). Temperature and living matter. *Protoplasma-Monogr.* (Berlin) **8,** 277 pp.

Belehradek, J., and Melichar, J. (1930). L'action différente des températures élevées et des températures normales sur la survie de la cellule végétale (*Helodea canadensis,* Rich.). *Biol. Gen.* **6,** 109-124.

Ben-Zeev, N., and Zamenhof, S. (1962). Effects of high temperatures on dry seeds. *Plant Physiol.* **37,** 696-699.

Ben Zioni, A., and Itai, C. (1972). Short-and long-term effects of high temperatures (47-49°C) on tobacco leaves. I. Photosynthesis. *Physiol. Plant.* **27,** 216-219.

Ben Zioni, A., and Itai, C. (1973). Short-and long-term effects of high temperatures (47-49°C) on tobacco leaves. III. Efflux and ^{32}P incorporation into phospholipids. *Physiol. Plant.* **28,** 493–497.

Ben Zioni, A., and Itai, C. (1975). Pre-conditioning of tobacco and bean leaves to heat shock by high temperature or NaCl. *Physiol. Plant.* **35,** 80–84.

Berns, D. S., and Scott, E. (1966). Protein aggregation in a thermophilic protein: Phycocyanin from *Synechoccus lividus. Biochemistry* **5,** 1528–1533.

Bernstam, V. A. (1974). Effects of supraoptimal temperatures on the myxomycete *Physarum polycephalum.* II. Effects on the rate of protein and ribonucleic acid synthesis. *Arch. Microbiol.* **95,** 347–356.

Bernstam, V. A., and Arndt, S. (1973a). Effects of supraoptimal temperatures on the myxomycete *Physarum polycephalum.* I. Protoplasmic streaming, respiration and leakage of protoplasmic substances. *Arch. Mikrobiol.* **92,** 251–261.

Bernstam, V. A., and Arndt, S. (1973b). The thermostability of some indices of vitality in *Physarum polycephalum* plasmodium. *Tsitologiya* **15,** 1091–1096.

Bernstam, V. A., and Arndt, S. (1974a). Effects of supraoptimal temperatures on the myxomycete *Physarum polycephalum.* III. Effects of starvation and cycloheximide on repair of thermal injury in plasmodia. *Arch. Microbiol.* **95,** 357–363.

Bernstam, V. A., and Arndt, S. (1974b). Effects of starvation and cycloheximide on the repair of heat injury of Myxomycete *Physarum polycephalum* plasmodium. *Tsitologiya* **16,** 381–385.

Berry, J. A. (1975). Adaptation of photosynthetic processes to stress. *Science* **188,** 644–650.

Berry, J. A., Fork, D. C., and Garrison, S. (1975). Mechanistic studies of thermal damage to leaves. *Carnegie Inst. Washington, Yearb.* **74,** 751–759.

Bhaumik, G., and Bhattacharjee, S. B. (1975). Repair of heat injury in thymine auxotrophs of *Escherichia coli. Can. J. Microbiol.* **21,** 1274–1277.

Biebl, R. (1939). Über die Temperaturresistenz von Meeresalgen verschiedener Klimazonen und verschieden tiefer Standorte. *Jahrb. Wiss. Bot.* **88,** 389–420.

Biebl, R. (1955). Tagesgange der Lichttransmission verschiedener Blätter. *Flora (Jena)* **142,** 280–294.

Biebl, R. (1962). Protoplasmatische Ökologie der Pflanzen. Wasser und Temperatur. *Protoplasmatologia* **1,** 1–344.

Biebl, R. (1969). Studien zur Hitzeresistenz der Gezeitenalge *Chaetomorpha cannabina* (Aresch.) Kjellm. *Protoplasma* **67,** 451–472.

Biesecker, G., Harris, J. I., Thierry, J. C., Walker, J. E., and Wonacott, A. J. (1977). Sequence and structure of D-glyceraldehyde 3-phosphate dehydrogenase from *Bacillus stearothermophilus. Nature (London)* **266,** 328–332.

Bigelow, C. C. (1967). On the average hydrophobicity of proteins and the relation between it and protein structure. *J. Theor. Biol.* **16,** 187–211.

Bilanovskii, N. F., and Oleinikova, T. N. (1974). Cytological study of tobacco cells in culture under the effect of high temperature. *Tsitol. Genet.* **8,** 237–240.

Billings, W. D., and Morris, R. J. (1951). Reflection of visible and infrared radiation from leaves of different ecological groups. *Am. J. Bot.* **38,** 327–331.

Björkman, O. (1975). Thermal stability of the photosynthetic apparatus in intact leaves. *Carnegie Inst. Washington, Yearb.* **74,** 748–751.

Björkman, O., Nobs, M., Mooney, H., Troughton, J., Berry, J., Nicholson, F., and Ward, W. (1974). Growth responses of plants from habitats with contrasting thermal environments. Transplant studies in the Death Valley and the Bodega Head experiment gardens. *Carnegie Inst. Washington, Yearb.* **73,** 748–767.

Björkman, O., Mooney, H. A., and Ehleringer, J. (1975). Photosynthetic responses of plants

from habitats with contrasting thermal environments. *Carnegie Inst. Washington, Yearb.* **74,** 743-748.
Bogen, H. J. (1948). Untersuchungen über Hitzetod und Hitzeresistenz pflanzlicher Protoplaste. *Planta* **36,** 298-340.
Boyarchuk, Yu. M., and Vol'kenshtein, M. V. (1967). The effects of protein conformation on hydrogen bonds between peptide groups. *Biofizika* **12,** 341-343.
Brandt, J., and Andersson, L. O. (1976). Heat denaturation of human serum albumin: Migration of bound fatty acids. *Int. J. Pept. Protein Res.* **8,** 33-37.
Brandts, J. F. (1967). Heat effects in proteins and enzymes. *In* "Thermobiology" (A. H. Rose, ed.), pp. 25-72. Academic Press, New York.
Brewer, E. N. (1972). Polysome profiles, amino acid incorporation in vitro, and polysome reaggregation following disaggregation by heat shock through the mitotic cycle in *Physarum polycephalum. Biochim. Biophys. Acta* **277,** 639-645.
Brock, T. D. (1967). Life at high temperatures. *Science* **158,** 1012-1019.
Brock, T. D., and Darland, G. K. (1970). Limits of microbial existence: Temperature and pH. *Science* **169,** 1316-1318.
Buchinger, A. (1929). Der Einfluss hoher Anfangstemperaturen auf die Keimung, dargestellt an *Trifolium pratense. Jahrb. Wiss. Bot.* **71,** 149-153.
Bull, H. E., and Breese, K. (1973). Thermal stability of proteins. *Arch. Biochem. Biophys.* **158,** 681-686.
Bünning, E., and Herdtle, H. (1946). Physiologische Untersuchungen an thermophilen Blaualgen. *Z Naturforsch.* **1,** 93-99.
Burgess, A. W., and Scheraga, H. A. (1975). A hypothesis for the pathway of the thermally-induced unfolding of bovine pancreatic ribonuclease. *J. Theor. Biol.* **53,** 403-420.
Byfield, J. E., and Scherbaum, O. H. (1967). Temperature effect on protein synthesis in a heat-synchronized protozoan treated with actinomycin D. *Science* **156,** 1504-1505.
Campbell, L. L., and Pace, B. (1968). Physiology of growth at high temperatures. *J. Appl. Bacteriol.* **31,** 24-35.
Capeillare-Blandin, C. (1969). Interactions substrat-flavine-hemoproteine et stabilité thermique du cytochrome b_2 (L-lactate deshydrogenase de la levure). *FEBS Lett.* **4,** 311-315.
Carroll, J. C. (1943). Effects of drought, temperature, and nitrogen on turf grasses. *Plant Physiol.* **18,** 19-36.
Cass, K. H., and Stellwagen, E. (1975). A thermostable phosphofructokinase from the extreme thermophile *Thermus* x-1. *Arch. Biochem. Biophys.* **171,** 682-694.
Castenholz, R. W. (1967). Aggregation in a thermophilic Oscillatoria. *Nature (London)* **215,** 1285-1286.
Castenholz, R. W. (1969). The thermophilic cyanophytes of Iceland and the upper temperature limit. *J. Phycol.* **5,** 360-368.
Cavari, B. Z., and Grossowicz, N. (1973). Properties of homoserine dehydrogenase in a thermophilic bacterium. *Biochim. Biophys. Acta* **302,** 183-190.
Chabot, B. F., and Lewis, A. R. (1976). Thermal acclimation of photosynthesis in northern red oak. *Photosynthetica* **10,** 130-135.
Chang, C. W. (1975). Activation energy for thermal inactivation and K_{m-} of carbonic anhydrase from the cotton plant. *Plant Sci. Lett.* **4,** 109-113.
Chapman, E. S., and Ostrovsky, D. S. (1976). Protein changes in response to temperature stress in *Papulaspora thermophila. Mycologia* **68,** 678-681.
Charlier, S., Grosjean, H., Lurquin, P., Vanhunbeck, J., and Werenne, J. (1969). Comparative studies on the isoleucyl- and luecyl-tRNA synthetases from *Bacillus stearothermophilus* and *Escherichia coli:* Thermal stability of the aminoacyl-adenylate-enzyme complexes. *FEBS Lett.* **4,** 239-242.

Chatagner, F., Durieu-Trautmann, O., and Rain, M. C. (1968). Influence of pyridoxal phosphate and some other derivatives of pyridoxine on the stability *in vitro* of cystathionase and of cysteine sulfinic acid decarboxylase. *Bull. Soc. Chim. Biol.* **50,** 129-141.

Chatterton, N. J., McKell, C. M., and Strain, B. R. (1970). Intraspecific differences in temperature-induced respiratory acclimation of desert saltbush. *Ecology* **51,** 545-547.

Chen, J. F., Yang, Y. C., and Hsieh, W. T. (1975). A thermostable nuclease in *Penicillium citrinum*. *J. Chin. Biochem. Soc.* **4,** 51-56.

Chen, M. C., Lord, R. C., and Mendelsohn, R. (1973). Laser-excited Raman spectroscopy of biomolecules. IV. Thermal denaturation of aqueous lysozyme. *Biochim. Biophys. Acta* **328,** 252-260.

Chigaleichik, A. G., Shul'ga, A. V., Termhitarova, N. G., and Rylkin, S. S. (1975). Physiological-biochemical characteristics of *Candida tropicalis* grown on n-alkanes at supraoptimal temperature. *Mikrobiologiya* **44,** 61-66.

Christophersen, J. (1963). The effect of the adaptation temperature on resistance to heat and activity of the hexokinase in yeast cells. *Arch. Mikrobiol.* **45,** 58-64.

Christophersen, J., and Precht, H. (1952). Untersuchungen zum Problem der Hitzeresistenz. II. Untersuchungen an Hefezellen. *Biol. Zentralbl.* **71,** 585-601.

Coffman, F. A. (1957). Cold resistant oat varieties also resistant to heat. *Science* **125,** 1298-1299.

Collander, R. (1924). Beobachtungen über die quantitativen Beziehungen zwischen Tötungsgeschwindigkeit und Temperatur beim Wärmetod pflanzlicher Zellen. *Commentat. Biol, Soc. Sci. Fenn* **1,** 1-12.

Cotter, D. A., and George, R. P. (1975). Germination and mitochondrial damage in spores of *Dictyostelium discoideum* following supraoptimal heating. *Arch. Microbiol.* **103,** 163-168.

Coutinho, L. M. (1969). Novas observações sobre a ocorrencia do "efeito de De Saussure" e suas relações com a suculencia a temperatura folhear e os movimentos estomaticos. *Bol. Fac. Filos. Cienc. Let., Univ. Sao. Paulo, Bot.* **24,** 77-102.

Crisan, E. V. (1973). Current concepts of thermophilism and the thermophilic fungi. *Mycologia* **65,** 1171-1198.

Crosier, W. (1956). Longevity of seeds exposed to dry heat. *Proc. Assoc. Off. Seed Anal.* **46,** 72-74.

Cross, T. (1968). Thermophilic actinomycetes. *J. Appl. Bacteriol.* **31,** 36-53.

Curtis, O. F. (1936a). Leaf temperatures and the cooling of leaves by radiation. *Plant Physiol.* **11,** 343-364.

Curtis, O. F. (1936b). Comparative effects of altering leaf temperatures and air humidities on vapor pressure gradients. *Plant Physiol.* **11,** 595-603.

Curtis, O. F., and Clark, D. G. (1950). "An Introduction to Plant Physiology." McGraw-Hill, New York.

Dangeard, P. (1951a). Observations sur la résistance des radicules de diverses plantes à des températures entre 40 et 60°. *C. R. Hebd. Seances Acad. Sci.* **232,** 913-915.

Dangeard, P. (1951b). Observations sur la destruction du chondriome par la chaleur. *C. R. Hebd. Seances Acad. Sci.* **232,** 1274-1276.

Dangeard, P. (1951c). Observations sur la résistance des radicules à des temperatures entre 40 et 60°. *Botaniste* **35,** 237-243.

Daniell, J. W., Chappell, W. E., and Couch, H. B. (1969). Effect of sublethal and lethal temperatures on plant cells. *Plant Physiol.* **44,** 1684-1689.

Davidson, B., and Fasman, G. (1967). The conformation transitions of unchanged poly-L-lysine: α-helix-random coil-B structure. *Biochemistry* **6,** 1616-1629.

Das, P. K. (1973). Developmental stability and thermosensitivity of different varieties of wheat. *Nucleus* **16,** 175-179.

de Rosa, M., Gambacorta, A., and Bu'lock J. D. (1976). The Caldariella group of extreme thermoacidophile bacteria: Direct comparison of lipids in *Sulfolobus, Thermoplasma* and the MT strains. *Phytochemistry* **15,** 143-145.

de Rosa, M., de Rosa, S., Gambacorta, A., Minale, L., and Bu'lock, J. D. (1977). Chemical structure of the ether lipids of thermophilic acidophilic bacteria of the Caldariella group. *Phytochemistry* **16,** 1961-1965.

de Visser Smits, D. (1926). Einfluss der Temperatur auf die Permeabilität des Protoplasmas bei *Beta vulgaris L. Recl. Trav. Bot. Neerl.* **23,** 104-199.

de Vries, H. (1870). Matériaux pour la connaissance de l'influence de la température sur les plantes. *Arch. Neerl. Sci. Exactes Nat.* **5,** 385-401.

de Vries, H. (1871). Sur la mort des cellules végétales par l'effet d'une température élevée. *Arch. Neerl. Sci. Exactes Nat.* **6,** 245-295.

Diskus, A. (1958). Das Osmoseverhalten einiger Peridineen des Susswassers. *Protoplasma* **49,** 187-196.

Donovan, J. W., and Beardslee, R. A. (1975). Heat stabilization produced by protein-protein association: A differential scanning calorimetric study of the heat denaturation of the trypsin-soybean trypsin inhibitor and trypsin-ovomucoid complexes, *J. Biol. Chem.* **250,** 1966-1971.

Döring, H. (1932). Beiträge zur Frage der Hitzeresistenz pflanzlicher Zellen. *Planta* **18,** 405-434.

Dörr, M. (1941). Temperaturmessungen an Pflanzen des Frauensteins bei Mödling. *Beih. Bot. Zentralbl.* **60,** 679-728.

Drake, B. G., Raschke, K., and Salisbury, F. B. (1970). Temperatures and transpiration resistances of Xanthium leaves as affected by air temperature, humidity, and wind speed. *Plant Physiol.* **46,** 324-330.

Duff, D. T., and Beard, J. B. (1974a). Supraoptimal temperature effects upon *Agrostis palustris.* I. Influence on shoot growth and density, leaf blade width and length, succulence and chlorophyll content. *Physiol. Plant.* **32,** 14-17.

Duff, D. T., and Beard, J. B. (1974b). Supraoptimal temperature effects upon *Agrostis palustris.* Part II. Influence on carbohydrate levels, photosynthetic rate, and respiration rate. *Physiol. Plant.* **32,** 18-22.

Dunlap, J. R., and Morgan, P. W. (1977). Characterization of ethylene/gibberellic acid control of germination in *Lactuca sativa* L. *Plant Cell Physiol.* **18,** 561-568.

Dutrochet, M. (1839). Recherches sur la température propre des végétaux. *Ann. Sci. Nat.* **12,** 77-84.

Dutrochet, M. (1840). Recherches sur la chaleur propre des êtres vivants à basse température. *Ann. Sci. Nat., Bot. Biol. Veg.* [2] **13,** 5-49 and 65-85.

Egorova, L. I. (1975). Aftereffect of short-term heating of leaves on photosynthesis. *Bot. Zh.* **60,** 1000-1004.

Egorova, L. I. (1976). The kinetics of photosynthesis after high temperature treatment. *Bot. Zh.* **61,** 945-950.

Ehleringer, J., Björkman, O., and Mooney, H. A. (1976). Leaf pubescence: Effects on absorptance and photosynthesis in a desert shrub. *Science* **192,** 376-377.

Emi, S., Myers, D. V., and Jacobucci, G. A. (1976). Purification and properties of the thermostable acid protease of *Penicillium duponti. Biochemistry* **15,** 842-848.

Emmerikh, F. D. (1973). Effect of adenine on heat resistance of tomato plants. *Fiziol. Rast.* **20,** 1288-1290.

Emmett, J. M., and Walker, D. A. (1969). Thermal uncoupling in chloroplasts. *Biochim. Biophys. Acta* **180,** 424–425.

Enami, I., and Fukuda, I. (1977a). Mechanisms of the acido-and thermophily of *Cyanidium caldarium* Geitler. III. Loss of these characteristics due to detergent treatment. *Plant Cell Physiol.* **18,** 671–680.

Enami, I., and Fukuda, I. (1977b). Mechanisms of the acido-and thermophily of *Cyanidium caldarium* Geitler. IV. Loss of these characteristics due to enzyme treatment. *Plant Cell Physiol.* **18,** 707–710.

Engelbrecht, L., and Mothes, K. (1960). Kinetin als Faktor der Hitzeresistenz. *Ber. Dtsch. Bot. Ges.* **73,** 246–257.

Engelbrecht, L., and Mothes, K. (1964). Weitere Untersuchungen zur experimentelle Beeinflussung der Hitzewirkung bei Blättern von *Nicotiana rustica. Flora (Jena)* **154,** 279–298.

Eriksson, C. E., and Vallentin, K. (1973). Thermal activation of peroxidase as a lipid oxidation catalyst. *J. Am. Oil Chem. Soc.* **50,** 264–268.

Esterak, K. B. (1935). Resistenz-Gradienten in Elodea Blättern. *Protoplasma* **23,** 367–383.

Fal'kova, T. V. (1973). Seasonal changes in thermoresistance of higher plants cells under conditions of Mediterranean type subtropics. *Bot. Zh.* **58,** 1424–1438.

Fal'kova, T. V. (1975). Thermal hardening of the cells of higher plants under semiarid subtropical conditions. *Ekologiya* **6,** 90–98.

Fal'kova, T. V., and Galushko, R. V. (1974). Seasonal changes of protoplasmic thermostability of *Daphne laureola* L. on the southern shore of the Crimea. *Ekologiya* **5,** 10–15.

Fedoseeva, G. P. (1966). The adaptation of photosynthesis in cucumbers to high temperatures in protected sites. *Uch. Zap. Ural. Gos. Univ.* **58,** 67–73.

Feierabend, J. (1977). Capacity for chlorophyll synthesis in heatbleached 70-S ribosome-deficient rye leaves. *Planta* **135,** 83–88.

Feierabend, J., and Mikus, M. (1977). Occurrence of a high temperature sensitivity of chloroplast ribosome formation in several higher plants. *Plant Physiol.* **59,** 863–867.

Feierabend, J., and Schrader-Reichhardt, U. (1976). Biochemical differentiation of plastids and other organelles in rye leaves with a high-temperature-induced deficiency of plastid ribosomes. *Planta* **129,** 133–145.

Feinstein, R. N., Sacher, G. A., Howard, J. B., and Braun, J. T. (1967). Comparative heat stability of blood catalase. *Arch. Biochem. Biophys.* **122,** 338–343.

Fel'dman, N. L. (1962). The influence of sugars on the cell stability of some higher plants to heating and high hydrostatic pressure. *Tsitologya* **4,** 633–643.

Fel'dman, N. L. (1966). Increase in the thermostability of urease in the thermal hardening of leaves. *Dokl. Akad. Nauk SSSR* **167,** 946–949.

Fel'dman, N. L. (1968). The effect of heat hardening on the heat resistance of some enzymes from plant leaves. *Planta* **78,** 213–225.

Fel'dman, N. L. (1973). Temperature dependence of enzymatic activity and Michaelis constant of acid phosphatase from the leaves of spring (*Leucojum vernum*) and summer (*L. aestivum*) snowflake species. *Tsitologiya* **15,** 170–176.

Fel'dman, N. L., and Lutova, M. I. (1963). Variations de la thermostabilité cellulaire des algues en fonction des changements de la température du millieu. *Cah. Biol. Mar.* **4,** 436–458.

Fel'dman, N. L., and Kamentseva, I. E. (1967). Urease thermostability in leaf extracts of two Leucojum species with different periods of vegetation. *Tsitologiya* **9,** 886–889.

Fel'dman, N. L., and Kamentseva, I. E. (1974). Heat resistance of cells of central Asian Allium species with ephemeroid and long vegetative cycles. *Ekologiya* **5,** 48–52.

Fel'dman, N. L., Kamentseva, I. E., and Yurashevskaya, K. N. (1966). Acid phosphatase thermostability in the extracts of cucumber and wheat seedling leaves after heat hardening. *Tsitologya* **8,** 755–759.

Fel'dman, N. L., Lutova, M. I., and Shcherbakova, A. M. (1975). Resistance of some proteins of *Pisum sativum* L. leaves after heat hardening to elevated temperature, proteolysis and shifts in pH. *J. Therm. Biol.* **1,** 47-51.

Fontana, A., Boccu, E., and Veronese, F. M. (1976). Effect of EDTA on the conformational stability of thermolysin. *Experientia, Suppl.* **26,** 55-59.

Franco, C. M. (1958). "Influence of Temperature on Growth of Coffee Plant." IBEC Res. Inst., Rockefeller Plaza, New York.

Franco, C. M. (1961). Lesao do colo do cafeeiro, causada pelo calor. *Bragantia* **20,** 645-652.

Frederick, J. F. (1973). Differences of the primer-independent phosphorylase isozyme in thermophilic and mesophilic algae. *Plant Cell Physiol.* **14,** 443-448.

Friedman, K. J. (1977). Role of lipids in the *Neurospora crassa* membrane I. Influence of fatty acid composition on membrane lipid phase transitions. *J. Membr. Biol.* **32,** 33-48.

Friedman, S. M. (1968). Protein-synthesizing machinery of thermophilic bacteria. *Bacteriol. Rev.* **32,** 27-38.

Fritzsche, G. (1933). Untersuchungen über die Gewebetemperaturen von Strandpflanzen unter dem Einfluss der Insolation. *Beih. Bot. Zentralbl.* **50,** 251-322.

Fuchs, M., Stanhill, G., and Moreshet, S. (1976). Effect of increasing foliage and soil reflectivity on the solar radiation balance of wide-row grain sorghum. *Agron. J.* **68,** 856-871.

Galston, A. W., and Hand, M. E. (1949). Adenine as a growth factor for etiolated peas and its relation to the thermal inactivation of growth. *Arch. Biochem. Biophys.* **22,** 434-443.

Garibaldi, J. A., Donovan, J. W., Davis, J. G., and Cimino, S. L. (1968). Heat denaturation of the ovomucin-lysozyme electrostatic complex—a source of damage to the whipping properties of pasteurized egg whites. *J. Food Sci.* **33,** 514-524.

Gates, D. M. (1963). Leaf temperature and energy exchange. *Arch. Meteorol., Geophys. Bioklimatol., Ser. B* **12,** 321-336.

Gates, D. M. (1965). Heat transfer in plants. *Sci. Am.* **213,** 76-84.

Gates, D. M., and Tantraporn, W. (1952). The reflectivity of deciduous trees and herbaceous plants in the infrared to 25 microns. *Science* **115,** 613-616.

Gäumann, E., and Jaag, O. (1936). Untersuchungen über die pflanzliche Transpiration. *Ber. Schweiz. Bot. Ges.* **45,** 411-518.

Gausman, H. W. (1977). Reflectance of leaf components. *Remote Sens. Environ.* **6,** 1-9.

Gausman, H. W., and Cardenas, R. (1969). Effect of leaf pubescence of *Gynura aurantiaca* on light reflectance. *Bot. Gaz. (Chicago)* **130,** 158-162.

Genkel, P. A., and Tsvetkova, I. V. (1955). Increasing the heat resistance of plants. *Dokl. Akad. Nauk SSSR* **102,** 383-386.

Giraud, G. (1958). Sur la vitesse de croissance d'une Rhodophycee monocellulaire marine, le *Rhodosorus marinus* Geitler, cultivée en milieu synthétique. *C. R. Hebd. Seances Acad. Sci.* **246,** 3501-3504.

Goodspeed, T. H. (1911). The temperature coefficient of the duration of life of barley grains. *Bot. Gaz. (Chicago)* **51,** 220-224.

Gorban', I. S. (1962). On the correlation between the growth and thermostability of plant cells. *Tsitologya* **4,** 182-192.

Gorban', I. S. (1968). Effect of beta-indoleacetic acid and maleic hydrazide on the elongation and thermostability of wheat coleoptile. *Tsitologiya* **10,** 76-87.

Gorban', I. S. (1972). Thermostability of urease from cucumber cotyledon seedlings assayed in intact organs and in homogenates during heating. *Tsitologiya* **14,** 1504-1512.

Gorban', I. S. (1974). Change in reparative properties of plant cells under the effect of heating at superoptimal temperatures. *Tsitologiya* **16,** 1111-1116.

Gorban', I. S., Zavadskaya, I. G., Shikhtina, G. G., and Shcherbakova, A. M. (1974). The ability of cells for the heat hardening in coleoptiles and leaves of different ages. *Tsitologiya* **16,** 1036-1040.

Grant, D. W., Sinclair, N. A., and Nash, C. H. (1968). Temperature-sensitive glucose fermentation in the obligatory psychrophilic yeast *Candida gelida. Can. J. Microbiol.* **14,** 1105-1110.

Gray, R. J. H., Ordal, Z. J., and Witter, L. D. (1977). Diluent sensitivity in thermally stressed cells of *Pseudomonas fluorescens. Appl. Environ. Microbiol.* **33,** 1074-1078.

Grodzinski, B., and Butt, V. S. (1977). The effect of temperature on glycollate decarboxylation in leaf peroxisomes. *Planta* **133,** 261-266.

Groves. J. F. (1917). Temperature and life duration of seeds. *Bot. Gaz.* (*Chicago*) **63,** 169-189.

Guern, N. (1974). Augmentation de la thermosensibilité des tissus de Topinambour cultivés *in vitro* par un traitement bref á 53°. *C. R. Hebd. Seances Acad. Sci.* **279,** 149-152.

Guern, N., and Gautheret, R. (1969). The hardening of the tissues of the Jerusalem artichoke at high temperatures. *C.R. Hebd. Seances Acad. Sci., Ser.* **269,** 332-334.

Gur, A., Bravdo, B., and Mizrahi, Y. (1972). Physiological responses of apple trees to supraoptimal root temperature. *Physiol. Plant.* **27,** 130-138.

Haberstick, H.-U., and Zuber, H. (1974). Thermoadaptation of enzymes in thermophilic and mesophilic cultures of *Bacillus stearothermophilus* and *Bacillus caldotenax. Arch. Microbiol.* **98,** 275-287.

Hachimori, A., and Noson, Y. (1973). Conformational change with temperature of ATPase from *Bacillus stearothermophilus. Biochim. Biophys. Acta* **315,** 481-484.

Hagler, A. N., and Lewis, M. J. (1974). Effect of glucose on thermal injury of yeast that may define the maximum temperature of growth. *J. Gen. Microbiol.* **80,** 101-109.

Hammer, L. (1972). Temperature tolerance of tropical marine algae and phanerogams. *Mitt. Inst. Colombo-Aleman Invest. Cient. "Punta de Betin"* **6,** 53-64.

Hammouda, M., and Lange, O. L. (1962). Zur Hitzeresistenz der Blätter hoherer Pflanzen in Abhängigkeit von ihrem Wassergehalt. *Naturwissenschaften* **49,** 500-501.

Harder, R. (1930). Beobachtungen über die Temperatur der Assimilations-organe sommergrüner Pflanzen der algerischen Wüste. *Z Bot.* **23,** 703-744.

Harder, R., Filzer, P., and Lorenz, A. (1932). Über versuche zur Bestimmung der Kohlensäureassimilation immergrüner Wüstenpflanzen während der Trockenzeit in Beni Unif (algerische Sahara). *Jahrb. Wiss. Bot.* **75,** 45-194.

Harder, W., and Veldkamp, H. (1968). Physiology of an obligately psychrophilic marine Pseudomonas sp. *J. Appl. Bacteriol.* **31,** 12-23.

Harris, J. I. (1976). *Experientia, Suppl.* **26,** 413-414.

Hashimoto, H., Iwaasa, T., and Yokotsuka, T. M. (1972). Thermostable acid protease produced by *Pencillium duponti* K1014, a true thermophilic fungus newly isolated from compost. *Appl. Microbiol.* **24,** 986-992.

Haurowitz, F. (1959). "Progress in Biochemistry Since 1949." Wiley (Interscience), New York.

Hayashi, K., Kugimiya, M., and Fanatsu, M. (1968). Heat stability of lysozyme-substrate complex. *J. Biochem.* (*Tokyo*) **64,** 93-97.

Heinen, W., and Lauwers, A. M. (1976). Amylase activity and stability at high and low temperature depending on calcium and other divalent cations. *Experientia, Suppl.* **26,** 77-89.

Henckel, P. A. (1964). Physiology of plants under drought. *Annu. Rev. Plant Physiol.* **15,** 363-386.

Henckel, P. A., and Margolin, K. P. (1948). Reasons of resistance of succulents to high temperatures. *Bot. Zh.* (*Leningrad*) **33,** 55-62.

Hendricks, S. B., and Taylorson, R. B. (1976). Variation in germination and amino acid leakage of seeds with temperature related to membrane phase change. *Plant Physiol.* **58,** 7-11.

Henrici, M. (1955). Temperatures of karroo plants. *S. Afr. J. Sci.* **51,** 245-248.

Heydecker, W., and Joshua, A. (1977). Alleviation of the thermodormancy of lettuce seeds. *J. Hortic. Sci.* **52,** 87-98.

Heyne, E. G., and Laude, H. H. (1940). Resistance of corn seedlings to high temperatures in laboratory tests. *J. Am. Soc. Agron.* **32,** 116-126.

Highkin, H. R. (1959). Effect of vernalization on heat resistance in two varieties of peas. *Plant Physiol.* **34,** 643-644.

Hilbrig. H. (1900). Ueber den Einfluss supramaximaler Temperatur auf das Wachstum der Pflanzen. Inaugural Dissertation, Universität Leipzig, Leipzig.

Hindak, F., and Komarek, J. (1968). Cultivation of the cryosestonic alga *Koliella tatrae* (Kol) Hind. *Biol. Plant.* **10,** 95-97.

Hirsch, H. M. (1954). Temperature-dependent cellulose production by *Neurospora crassa* and its ecological implications. *Experientia* **10,** 180-182.

Hocking, J. D., and Harris, J. I. (1976). Glyceraldehyde-3-phosphate dehydrogenase from an extreme thermophile, *Thermus aquaticus. Experientia, Suppl.* **26,** 121-133.

Hofstee, B. H. J., and Bobb, D. (1968). Heat denaturation of chymotrypsinogen A in the presence of polyanions. *Biochim. Biophys. Acta* **168,** 564-566.

Holm-Hansen, O. (1967). Factors affecting the viability of lyophilized algae. *Cryobiology* **4,** 17-23.

Hopp, R. (1947). Internal temperatures of plants. *Proc. Am. Soc. Hortic. Sci.* **50,** 103-108.

Huang, L., Lorch, S. K., Smith, G. G., and Haug, A. (1974). Control of membrane lipid fluidity in *Acholeplasma laidlawii. FEBS Lett.* **43,** 1-5.

Huber, H. (1932). Einige Grundfragen des Wämehaushaltes der Pflanzen I. Die Ursache der hohen Sukkulenten-Temperaturen. *Ber. Dtsch. Bot. Ges.* **50,** 68-76.

Huber, H. (1935). Der Wärmehaushalt der Pflanzen. *Naturwiss. Landwirtsch.* **17,** 148.

Ignat'ev, L. A. (1973). Role of physiological plant activity in adaptive increase of the heat resistance. II. Heat regime effect on wheat seedlings. *Izv. Sib. Otd. Akad. Nauk SSSR, Ser. Biol. Nauk* **2,** 24-31.

Illert, H. (1924). Botanische Untersuchungen über Hitzetod und Stoffwechselgifte. *Bot. Arch.* **7,** 133-141.

Itai, C., and Ben Zioni, A. (1973). Short- and long-term effects of high temperatures (47-49°C) on tobacco leaves. II. O_2 Uptake and amylolytic activity. *Physiol. Plant.* **28,** 490-492.

Itai, C., Ben Zioni, A., and Ordin, L. (1973). Correlative changes in endogenous hormone levels and shoot growth induced by short heat treatments to the root. *Physiol. Plant.* **28,** 355-360.

Itai, C., Ben Zioni, A., and Munz, S. (1978). Heat stress: Effects of abscisic acid and kinetin on response and recovery of tobacco leaves. *Plant Cell Physiol.* **19,** 453-459.

Ivakin, A. P. (1975). Study of heat resistance of vegetable plants by the method of measuring the electrical resistance of leaf tissues. *Dokl. Vses. Ordena Lenina Akad. S-kh. Nauk im. V. I. Lenina* **1,** 11-12.

Jacobson, A. L., and Braun, H. (1977). Differential scanning calorimetry of the thermal denaturation of lactate dehydrogenase. *Biochim. Biophys. Acta* **493,** 142-153.

Jacobson, K. B. (1968). Alcohol dehydrogenase of Drosophila: Interconversion of isoenzymes. *Science* **159,** 324-325.

Jockush, H. (1968). Stability and genetic variation of a structural protein. *Naturwissenschaften* **55,** 514-518.

Johnson, H. A. (1974). On the thermodynamics of cell injury: Some insights into the molecular mechanisms. *Am. J. Pathol.* **75,** 13-25.

Jollès, P. (1967). Rapports entre la structure et l'activité de quelques lysozymes. *Bull. Soc. Chim. Biol.* **49,** 1001-1012.

Julander, O. (1945). Drought resistance in range and pasture grasses. *Plant Physiol.* **20,** 573-599.

Just, L. (1877). Ueber die Einwirkung hoher Temperaturen auf die Erhaltung der Keimfähigeit der Samen. *Beitr. Biol. Pflanz.* **2,** 311-348.

Kaho, H. (1921). Über die Beeinflussung der Hitzekoagulation des Pflanzenprotoplasmas durch Neutralsalze. I. *Biochem. Z.* **117,** 87-95.

Kaho, H. (1924). Über die Beeinflussung der Hitzekoagulation des Pflanzenplasmas durch die Salze der Erdalkalien. VI. *Biochem. Z.* **151,** 102-111.

Kaho, H. (1926). Über den Einfluss der Temperatur auf die koagulierende Wirkung einiger Alkalisalze auf das Pflanzenplasma. VIII. *Biochem. Z.* **167,** 182-194.

Kailasanathan, K., Rao, G.G.S.N., and Sinha, S. K. (1976). Effect of temperature on the partitioning of seed reserves in cowpea and sorghum. *Indian J. Plant Physiol.* **19,** 171-177.

Kao, O. H. W., Mercedes, R. E., and Berns, D. S. (1975). Physical-chemical properties of C-phycocyanin isolated from an acidothermophilic eukaryote, *Cyanidium caldarium. Biochem. J.* **147,** 63-70.

Kapoor, M., O'Brien, M., and Braun, A. (1976). Modification of the regulatory properties of pyruvate kinase of *Neurospora* by growth at elevated temperatures. *Can. J. Biochem.* **54,** 398-407.

Kappen, L., and Lange, O. L. (1968). Heat resistance of half-dried leaves of *Commelina africana:* Two research methods compared. *Protoplasma* **65,** 119-132.

Karavaeva, N. N., and Mukhiddinova, N. G. (1975). Tolerance of alkaline protease of the fungus *Torula thermophila* strain UzPT-1 to heating and Ca^{2+} ions. *Prikl. Biokhim. Mikrobiol.* **11,** 704-710.

Kepler, C. R., and Tove, S. B. (1973). Induction of biohydrogenation of oleic acid in *Bacillus cereus* by increase in temperature. *Biochem. Biophys. Res. Commun.* **52,** 1434–1439.

Khan, R. A., and Laude, H. M. (1969). Influence of heat stress during seed maturation on germinability of barley seed at harvest. *Crop Sci.* **9,** 55-58.

Khan, R. A., Ahmad, S., and Hussain, S. (1973). Effects of pre-sowing high temperature stress on seedling emergence and yield of seed cotton. *Exp. Agric.* **9,** 9-14.

Khodzhaev, D. K., and Abaeva, S. S. (1975). Change of enzymatic activity in cotton seeds and seedlings under the effect of trace elements and thermal hardening. *Biol. Nauki (Moscow)* **18,** 84-88.

Kiesselbach, T. A., and Ratcliff, J. A. (1918). Freezing injury of seed corn. *Nebr., Agric. Exp. Stn., Bull.* **163,** 1-16.

Kikuchi, T., Ishida, M. R., Matsubara, T., Tsushimoto, G., and Mizuma, N. (1973). Some features of RNA and protein systheses in thermophilic alga *Cyanidium caldarium:* The effect of antibiotics on in vivo incorporation of ^{14}C-uracil and ^{14}C-leucine. *Ann. Rep. Res. React. Inst., Kyoto Univ.* **6,** 29-37.

Kinbacher, E. J. (1963). Relative high-temperature resistance of winter oats at different relative humidities. *Crop Sci.* **2,** 466-468.

Kinbacher, E. J. (1969). The physiology and genetics of heat tolerance. *In* "Physiological Limitations on Crop Production under Temperature and Moisture Stress" (E. R. Lemon et al., eds.), pp. 000-000. Natl. Acad. Sci., Washington, D.C.

Kinbacher, E. J., Sullivan, C. Y., and Knull, H. R. (1967). Thermal stability of malic dehydrogenase from heat hardened *Phaseolus acutifolius.* Tepary. Buff. *Crop Sci.* **7,** 148-151.

Kirschmann, C., Levy, I., and de Vries, A. (1973). Stabilization by cations of microsomal ATPase against heat inactivation. *Biochim. Biophys. Acta* **330,** 167-172.

Kleinschmidt, M. G., and McMahon, V. A. (1970a). Effect of growth temperature on the lipid composition of *Cyanidium caldorium.* I. Class separation of lipids. *Plant Physiol.* **46,** 286-289.

Kleinschmidt, M. G., and McMahon, V. A. (1970b). Effect of growth temperature on the lipid composition of *Cyanidium caldorium*. II. Glycolipid and phospholipid components. *Plant Physiol.* **46,** 290-293.

Koffler, H., Mallett, G. E., and Adye, J. (1957). Molecular basis of biological stability to high temperatures. *Proc. Natl. Acad. Sci. U.S.A.* **43,** 464-477.

Konis, E. (1949). The resistance of maquis plants to supramaximal temperatures. *Ecology* **30,** 425-429.

Konis, E. (1950). On the temperature of Opuntia joints. *Palest. J. Bot., Jerusalem Ser.* **5,** 46-55.

Kotlyar, G. I., Lebedeva, G. Ya., and Zhuravskaya, T. G. (1969). Sulfhydryl groups and disulfide bonds in water-soluble proteins of heated germinated grain. *Prikl. Biokhim. Mikrobiol.* **5,** 366-368.

Krans, J. V., and Johnson, G. V. (1974). Some effects of subirrigation on bentgrass during heat stress in the field. *Agron. J.* **66,** 526-530.

Krause, G. H., and Santarius, K. A. (1975). Relative thermostability of the chloroplast envelope. *Planta* **127,** 285-299.

Kuijper, J. (1910). Ueber den Einfluss der Temperatur auf die Atmung höhere Pflanzen. *Recl. Trav. Bot. Neerl.* **7,** 131-240.

Kuiper, P. J. C. (1970). Lipids in alfalfa leaves in relation to cold hardiness. *Plant Physiol.* **45,** 684-686.

Kurkova, E. B. (1967). The effect of elevated temperatures on the oxidative phosphorylation of the mitochondria of corn roots. *Uch. Zap. Mosk. Obl. Pedagog. Inst.* **169,** 185-191.

Kurkova, E. B., and Andreeva, I. N. (1966). Changes in the morphology and biochemical activity of mitochondria due to temporary injury (in 3-day-old corn shoots). *Fiziol. Rast.* **13,** 1019-1023.

Kurtz, E. B., Jr. (1958). Chemical basis for adaptation in plants. Understanding of heat tolerance in plants may permit improved yields in arid and semiarid regions. *Science* **128,** 1115-1117.

Lahiri, A. N., and Singh, S. (1969). Effect of hyperthermia on the nitrogen metabolism of *Pennisetum typhoides*. *Proc. Natl. Inst. Sci. India, Part B* **35,** 131-138.

Lange, O. L. (1953). Hitze und Trockenresistenz der Flechten in Beziehung zu ihrer Verbreitung. *Flora (Jena)* **140,** 39-97.

Lange, O. L. (1955). Untersuchungen über die Hitzeresistenz der Moose in Beziehung zu ihrer Verbreitung. I. Die Resistenz stark ausgetrockneter Moose. *Flora (Jena)* **142,** 381-399.

Lange, O. L. (1958). Hitzeresistenz in Blattemperaturen mauretanisher Wüstenpflanzen. *Ber. Dtsch. Bot. Ges.* **70,** 31-32.

Lange, O. L. (1959). Untersuchungen über Warmehaushalt und Hitzeresistenz mauretanischer Wusten-und Savannenpflanzen. *Flora (Jena)* **147,** 595-651.

Lange, O. L. (1961). Die Hitzeresistenz einheimischer immer-und wintergrüner Pflanzen im Jahreslauf. *Planta* **56,** 666-683.

Lange, O. L. (1962a). Über die Beziehung zwischen Wasser-und Warmehaushalt von Wüsten pflanzen. *Veroeff. Geobot. Inst. Ruebel* **37,** 155-168.

Lange, O. L. (1962b). Versuche zur Hitzeresistenz—Adaptation bei hoheren Pflanzen. *Naturwissenschaften* **49,** 20-21.

Lange, O. L. (1965a). Leaf temperatures and methods of measurement. *In* "Methodology of Plant Ecophysiology" (F. E. Eckardt, ed.), pp. 203-209. UNESCO, Paris.

Lange, O. L. (1965b). The heat resistance of plants, its determination and variability. *In* "Methodology of Plant Eco-physiology" (F. Eckhardt, ed.), pp. 399-405. UNESCO, Paris.

Lange, O. L. (1967). Investigations on the variability of heat-resistance in plants. *In* "The Cell and Environmental Temperatures" (A. S. Troshin, ed.), pp. 131-141. Pergamon, Oxford.

Lange, O. L., and Lange, R. (1963). Untersuchungen über Blattemperaturen, Transpiration und Hitzeresistenz an Pflanzen mediterraner Standorte (Costa brava, Spanien). *Flora (Jena)* **153,** 387–425.

Lange, O. L., and Schwemmle, B. (1960). Untersuchungen zur Hitzeresistenz Vegetativer und Blühender Pflanzen von *Kalanchoe blossfeldiana. Planta* **55,** 208–225.

Lange, O. L., Koch, W., and Schulze, E. D. (1969). CO_2-gas exchange and water relationships of plants in the Negev desert at the end of the dry period. *Ber. Dtsch. Bot. Ges.* **82,** 39–61.

Langridge, J., and Griffing, B. (1959). A study of high temperature lesions in *Arabidopsis thaliana. Aust. J. Biol. Sci.* **12,** 117–135.

Laude, H. M., and Chaugule, B. A. (1953). Effect of stage of seedling development upon heat tolerance in bromegrasses. *J. Range Manage.* **6,** 320–324.

Lawanson, A. O., and Onwueme, I. C. (1973). Effect of prior heat stress on protochlorophyll and chlorophyll formation in seedlings *Colocynthis citrullus. Z. Pflanzenphysiol.* **69,** 461–463.

Leblova, S., and Mares, J. (1975). Thermally stable phosphoenolpyruvate carboxylase from pea, tobacco and maize green leaves. *Photosynthetica* **9,** 177–184.

Ledig, F. T., Clark, J. G., and Drew, A. P. (1977). The effects of temperature treatment on photosynthsis of pitch pine from northern and southern latitudes. *Bot. Gaz. (Chicago)* **138,** 7–12.

Lepeschkin, W. W. (1912). Zur Kenntnis der Einwirkung supramaximaler Temperaturen auf die Pflanze. *Ber. Dtsch. Bot. Ges.* **30,** 703–714.

Lepeschkin, W. W. (1935). Zur Kenntnis des Hitzetodes des Protoplasmas. *Protoplasma* **23,** 349–366.

Lepeschkin, W. W. (1937). Zell-Nekrobiose und Protoplasma-Tod. *Protoplasma-Monogr. (Berlin)* No. 12.

Levitt, J. (1962). A sulfhydryl-disulfide hypothesis of frost injury and resistance in plants. *J. Theor. Biol.* **3,** 355–391.

Levitt, J. (1966). Cryochemistry of plant tissue: Protein interactions. *Cryobiology* **3,** 243–251.

Levitt, J. (1969). Growth and survival of plants at extremes of temperature—a unified concept. *Proc. Soc. Exp. Biol. Med.* **23,** 395–448.

Levy, H. M., and Ryan, E. M. (1967). Heat inactivation of the relaxing site of actomyosin. *Science* **156,** 73.

Liu, W.-H., Beppu, T., and Arima, K. (1973). Physical and chemical properties of the lipase of thermophilic fungus *Humicola lanuginosa* S-38. *Agric. Biol. Chem.* **37,** 2493–2499.

Ljunger, C. (1962). Introductory investigations of ions and thermal resistance. *Physiol. Plant.* **15,** 148–160.

Ljunger, C. (1970). On the nature of the heat resistance of thermophilic bacteria. *Physiol. Plant.* **23,** 351–364.

Ljunger, C. (1973). Further investigations on the nature of the heat resistance of thermophilic bacteria. *Physiol. Plant.* **28,** 415–418.

Loginova, L. G., and Tashpulatov, Zh. (1967). Multicomponent celluloylytic enzymes of thermotolerant and mesophilic fungi related to *Aspergillus fumigatus. Mikrobiologiya* **36,** 988–992.

Loginova, L. G., and Verkhovtseva, M. I. (1963). Amino acid requirements of thermotolerant yeasts. *Mikrobiologiya* **32,** 216–222.

Loginova, L. G., Gerasimova, N. F., and Seregina, L. M. (1962). Requirement of thermotolerant yeasts for supplementary growth factors. *Mikrobiologiya* **31,** 21–25.

Lomagin, A. G. (1975). The thermostability of two types of protoplasmic streaming in *Physarum polycephalum* plasmodia. *Tsitologiya* **17,** 1273-1277.

Lomagin, A. G., and Antropova, T. A. (1966). Photodynamic injury to heated leaves. *Planta* **68,** 297-309.

Lomagin, A. G., and Antropova, T. A. (1968). A study of the capacity of *Physarum polycephalum* to adapt to different temperatures. *Tsitologiya* **10,** 1094-1104.

Lomagin, A. G., Antropova, T. A., and Semenichina, L. V. (1966). Phototaxis of chloroplasts as a criterion of viability of leaf parenchyma. *Planta* **71,** 119-124.

Lorenz, R. W. (1939). High temperature tolerance of forest trees. Minn. *Agric. Exp. Stn., Tech. Bull.* **141,** 1-25.

Lowenstein, A. (1903). Über die Temperaturgrenzen des Lebens bei der Thermalalge *Mastigocladus laminosus* Cohn. *Ber. Dtsch. Bot. Ges.* **21,** 317-323.

Lue, P. F., and Kaplan, J. G. (1970). Heat-induced disaggregation of a multifunctional enzyme complex catalyzing the first steps in pyrimidine biosynthesis in baker's yeast. *Can. J. Biochem.* **48,** 155-159.

Luknitskaya, A. F. (1967). Do chlamydomonads have a heat hardening capacity? *Tsitologiya* **9,** 800-803.

Lundegårdh, H. (1949). "Klima und Boden," 3rd ed. Fischer, Jena.

Lundegårdh, H. (1957). "Klima und Boden," 4th ed. Fischer, Jena.

Lusis, A. J., and Becker, R. R. (1973). The B-glucosidase system of the thermophilic fungus *Chaetomium thermophile* var. *coprophile* n. var. *Biochim. Biophys. Acta* **329,** 5-16.

Lutova, M. I., and Zavadskaya, I. G. (1966). Effects of the plant keeping duration at different temperatures on the cell heat resistance. *Tsitologiya* **8,** 484-493.

Luzikov, V. N., Rakhimov, M. M., Saks, V. A., and Berezin, I. V. (1967). Heat-inactivation of succinate oxidase and its fragments. *Biokhimiya* **32,** 1032-1035.

Maccoll, R., Edwards, M. R., Mulks, M. H., and Berns, D. S. (1974). Comparison of the biliproteins from two strains of the thermophilic cyanophyte *Synechococcus lividus*. *Biochem. J.* **141,** 419-425.

McGee, J. M. (1916). The effect of position upon the temperature and dry weight of joints of Opuntia. *Carnegie Inst. Washington, Yearb.* **15,** 73-74.

McLean, R. J. (1967). Desiccation and heat resistance of the green alga *Spongichloris typica*. *Can. J. Bot.* **45,** 1933-1938.

McNaughton, S. J. (1966). Thermal inactivation properties of enzymes from *Typha latifolia* L. ecotypes. *Plant Physiol.* **41,** 1736-1738.

McNaughton, S. J. (1974). Natural selection at the enzyme level. *Am. Nat.* **108,** 616-624.

Magalhaes, A. C., Peters, D. B., and Hageman, R. H. (1976). Influence of temperature on nitrate metabolism and leaf expansion in soybean (*Glycine max* L. Merr.) seedlings. *Plant Physiol.* **58,** 12-16.

Makherjee, J., Mukherji, S., and Sircar, S. M. (1973). High temperature-induced changes in germination, seedling vigor and the metabolic activities in rice seeds. *Biol. Plant.* **15,** 65-71.

Malcolm, N. L. (1968). A temperature-induced lesion in amino acid-transfer ribonucleic acid attachment in a psychrophile. *Biochim. Biophys. Acta* **157,** 493-503.

Malcolm, N. L. (1969). Molecular determinants of obligate psychrophily. *Nature* (*London*) **221,** 1031-1033.

Mallett, G. E., and Koffler, H. (1957). Hypothesis concerning the relative stability of flagella from thermophilic bacteria. *Arch. Biochem. Biophys.* **67,** 254-256.

Marré, E., and Servettaz, O. (1956). Richerche sull'adattamento proteico in organismi termoresistent. I. Sul limite de resistenza all'inattivazione termica dei sistemi fotosintetico a respiratorio di alghe di acque termali. *Atti. Accad. Naz. Lincei, Cl. Sci. Fis., Mat. Nat., Rend.* [8] **20,** 72-77.

Marré, E., Albertario, M., and Vaccari, E. (1958). Richerche sull' adattamento proteico in organismi termoresistenti. III. Relativa insensibilita di enzimi di Cianoficee a denaturanti che agiscomo rompendo i legami di idrogeno. *Atti Accad. Naz. Lincei, Cl. Sci. Fis., Mat. Nat., Rend.* [8] **24,** 351-353.

Martinek, K., Klibanoc, A. M., Goldmacher, V. S., and Berezin, I. V. (1977a). The principles of enzyme stabilization. I. Increase in thermostability of enzymes covalently bound to a complementary surface of a polymer support in a multipoint fashion. *Biochim. Biophys. Acta* **485,** 1-12.

Martinek, K., Klibanov, A. M., Goldmacher, V. S., Tchernysheva, A. V., Mozhaev, V. V., Berezin, I. V., and Glotov, B. O. (1977b). The principles of enzyme stabilization. II. Increase in the thermostability of enzymes as a result of multipoint noncovalent interaction with a polymeric support. *Biochim. Biophys. Acta* **485,** 13-28.

Matsubara, H. (1967). Some properties of thermolysin. *Publ. Am. Assoc. Adv. Sci.* **84,** 283-294.

Matthews, B. W., Weaver, L. H., and Kester, W. R. (1974). The conformation of thermolysin. *J. Biol. Chem.* **249,** 8030-8044.

Maxie, E. C. (1957). Heat injury in prunes. *Proc. Am. Soc. Hortic. Sci.* **69,** 116-121.

Mehta, P. D., and Maisel, H. (1966). The effect of heat on bovine lens proteins. *Experientia* **22,** 818-820.

Michaelis, G. P. (1935). Okologische Studien an der alpinen Baumgrenze. *Beih Bot. Zentralbl.* **52b,** 333-377.

Miehe, H. (1907). *Thermoidium sulfureum* n.g.n.sp., ein neuer Wärmepilz. *Ber. Dtsch. Bot. Ges.* **25,** 510-515.

Miller, H. M., and Shepherd, M. G. (1973). Thermal stability of ribosomes from a thermophilic and a mesophilic fungus. *Can. J. Microbiol.* **19,** 761-763.

Miller, H. M., Sullivan, P. A., and Shepherd, M. G. (1974). Intracellular protein breakdown in the thermophilic and mesophilic fungi. *Biochem. J.* **144,** 209-214.

Mishiro, Y., and Ochi, M. (1966). Effect of dipicolinate on the heat denaturation of proteins. *Nature (London)* **211,** 1190.

Miyamoto, T. (1963). The effect of seed treatment with the extracts of organisms and the solutions of some chemical substances on the resistance to salt concentration in wheat seedlings. *Physiol. Plant.* **16,** 333-336.

Mizusawa, K., and Yoshida, F. (1973). Thermophilic *Streptomyces* alkaline proteinase. II. The role of a sulfhydryl group and the conformational stability. *J. Biol. Chem.* **248,** 4417-4423.

Mizusawa, K., and Yoshida, F. (1976). Role of sulfhydryl group in the structure and function of alkaline proteases from a thermophilic actinomycete, *Streptomyces rectus* var. *proteolyticus. Experientia, Suppl.* **26,** 61-66.

Molisch, H. (1926). "Pflanzenbiologie in Japan." Fischer, Jena.

Montfort, C., Reid, A., and Reid, I. (1957). Gradation of functional heat resistance in marine algae and its relation to the environment and hereditary advantage. *Biol. Zentralbl.* **76,** 257-289.

Mooney, H. A., Björkman, O., and Collatz, G. J. (1978). Photosynthetic acclimation to temperature in the Desert Shrub, *Larrea divaricata.* 1. Carbon dioxide exchange characteristics of intact leaves. *Plant Physiol.* **61,** 406-410.

Morgan, W. T., and Riehm, J. P. (1973). Proteins of the thermophilic fungus *Humicola lanuginosa.* II. Some physicochemical properties of a cytochrome c. *Arch. Biochem. Biophys.* **154,** 415-421.

Morgan, W. T., Hensley, C. P., Jr., and Riehm, J. P. (1972). Proteins of the thermophilic fungus

Humicola lanuginosa. I. Isolation and amino acid sequence of a cytochrome c. *J. Biol. Chem.* **247,** 6555-6565.

Morré, D. J. (1970). Auxin effects on the aggregation and heat coagulability of cytoplasmic proteins and lipoproteins. *Physiol. Plant.* **23,** 38-50.

Moyse, A., and Guyon, D. (1963). Effet de la température sur l'efficacité de la phycocyanine et de la chlorophylle chez *Aphanocapsa* (*Cyanophyceae*). *In* "Studies on Microalgae and Photosynthetic Bacteria" (S. Miyachi, ed.), pp. 253-270. Univ. of Tokyo Press, Tokyo.

Mueller, I. M., and Weaver, J. E. (1942). Relative drought resistance of seedlings of dormant prairie grasses. *Ecology* **23,** 387-398.

Mukhin, E. N., and Gins, V. K. (1974). Differential effects of temperature on ferredoxin from pea and maize leaves. *Plant Sci. Lett.* **2,** 115-118.

Mukhin, E. N., Gins, V. K., Kulikov, A. V., and Likhtenstein, G. I. (1973). Heat resistance of ferredoxins of higher plants in relation to their activity in NADP photoreduction. *Fiziol. Rast.* **20,** 1007-1012.

Mumma, R. O., Fergus, C. L., and Sekura, R. D. (1970). The lipids of thermophilic fungi: Lipid composition comparisons between thermophilic and mesophilic fungi. *Lipids* **5,** 100-103.

Münch, E. (1913). Hitzeschaden an Waldpflanzen. *Naturwiss. Z. Forst.- Landwirtsch.* **11,** 557-562.

Münch, E. (1914). Nochmals Hitzeschaden an Waldpflanzen. *Naturwiss. Z. Forst.- Landwirtsch.* **12,** 169-188.

Munsche, D., and Wollgiehn, R. (1974). Age-dependent lability of chloroplast ribosomal RNA from *Nicotiana rustica*. *Biochim. Biophys. Acta* **340,** 437-445.

Muromtsev, N. A., Magomedov, Z. G., and Doskoch, Y. E. (1972). Dependence of plant thermoresistance on thermodynamic properties of soil moisture. *S-kh. Biol.* **7,** 559-564.

Musaelyan, M. S. (1975). Aftereffect of heating seeds by superoptimal temperatures on initial growth of wheat sprouts. *Biol. Zh. Arm.* **28,** 98-101.

Musolan, C., Ordin, L., and Kindinger, J. I. (1975). Effects of heat shock on growth and on lipid and B-glucan synthetases in leaves of *Phaseolus vulgaris* and *Vigna sinensis*. *Plant Physiol.* **55,** 328-332.

Nakajima, M., Mizusawa, K., and Yoshida, F. (1974). Purification and properties of an extracellular proteinase of psychrophilic *Escherichia freundii*. *Eur. J. Biochem.* **44,** 87-90.

Nakayama, S. (1963). Properties of a substance in Japanese radish leaf (*Raphanus sativus* L. var. *acanthiformis* Makino) which protects sweet potato B-amylase *Agric. Biol. Chem.* **27,** 326-331.

Nakayama, S., and Kono, Y. (1957). Studies on the denaturation of enzymes. I. Effect of concentration on the rate of heat-inactivation of enzymes. *J. Biochem.* (*Tokyo*) **44,** 25-31.

Nash, C. H., Grant, D. W., and Sinclair, N. A. (1969). Thermolability of protein synthesis in a cell-free system from the obligately psychrophilic yeast *Candida gelida*. *Can. J. Microbiol.* **15,** 339-343.

Neales, T. F. (1968). Effects of high temperature and genotype on the growth of excised roots of *Arabidopsis thaliana*. *Aust. J. Biol. Sci.* **21,** 217-223.

Neucere, N. J., and St. Angelo, A. J. (1972). Physiochemical properties of peanut proteins in sucrose. *Anal. Biochem.* **47,** 80-89.

Nikulina, G. N. (1969). Relative quantitative evaluation of the energy efficiency of respiration at high temperatures. *Bot. Zh.* **54,** 1242-1253.

Noack, K. (1920). Der Betriebstoffwechsel der thermophilen Pilze. *Jahrb. Wiss. Bot.* **59,** 413-466.

Noerr, V. M. (1974). Heat resistance of mosses. *Flora* (*Jena*) **163,** 388-397.

Norris, R. D., and Fowden, L. (1973). A comparison of the thermal stability and substrate binding constants of prolyl-tRNA synthetase from *Phaseolus aureus* and *Delonix regia. Phytochemistry* **12,** 2109-2121.

Ohta, Y., Ogura, Y., and Wada, A. (1966). Thermostable protease from thermophilic bacteria. I. Thermostability, physicochemical properties, and amino acid composition. *J. Biol. Chem.* **241,** 5919-5925.

Ojha, M. N., and Turian, G. (1968). Thermostimulation of conidiation and succinic oxidative metabolism of *Neurospora crassa. Arch. Mikrobiol.* **62,** 232-241.

Oleinikova, T. V. (1965). High temperature and light effects on the permeability of cells of spring cereal leaves. *Sci. Counc. Cytol. Probl. Akad. Nauk SSSR* pp. 70-81.

Olsen, R. H., and Metcalf, E. S. (1968). Conversion of mesophilic to psychrophilic bacteria. *Science* **162,** 1288-1289.

O'Malley, J. J., and Ulmer, R. W. (1973). Thermal stability of glucose oxidase and its admixtures with synthetic polymers. *Biotechnol. Bioeng.* **15,** 917-925.

Ong, P. S., and Gaucher, G. M. (1976). Production purification and characterization of thermomycolase, the extracellular serine protease of the thermophilic fungus *Malbranchea pulchella* var. *sulfurea. Can. J. Microbiol.* **22,** 165-176.

Onwueme, I. C. (1975). Changes in imbibition rate of okra (Hibiscus esculentus) seeds following high temperature treatments. Phyton (*Buenos Aires*) **33,** 139-142.

Onwueme, I. C., and Adegoroye, S. A. (1975). Emergence of seedlings from different depths following high temperature stress. *J. Agric. Sci.* **84,** 525-528.

Onwueme, I. C., and Laude, H. M. (1972). Heat-induced growth retardation and attempts at its prevention in barley and wheat coleoptiles. *J. Agric. Sci.* **79,** 331-333.

Onwueme, I. C., and Lawanson, A. O. (1973). Effect of heat stress on subsequent chlorophyll accumulation in seedlings of *Colocynthis citrullus. Planta* **110,** 81-84.

Onwuene, I. C., and Lawanson, A. O. (1975). Chlorophyll accumulation in cowpea (Vigna) leaves and melon (Colocynthis) cotyledons as influenced by prior heat stress and seedling age. *Phyton* (*Buenos Aires*) **33,** 69-73.

Ordin, L., Itai, C., Ben Zioni, A., Musolan, C., and Kindinger, J. I. (1974). Effect of heat shock on plant growth and on lipid and B-glucan syntheses. *Plant Physiol.* **53,** 118-121.

Oshima, M., and Yamakawa, T. (1974). Chemical structure of a novel glycolipid from an extreme thermophile, *Flavobacterium thermophilum. Biochemistry* **13,** 1140–1146.

Pal, U. R., Johnson, R. R., and Hageman, R. H. (1976). Nitrate reductase activity in heat (drought) tolerant and intolerant maize genotypes. *Crop Sci.* **16,** 775-779.

Pandey, K. K. (1973). Heat sensitivity of esterase isozymes in the styles of *Lilium* and *Nicotiana. New Phytol.* **72,** 839-850.

Pangburn, M. K., Levy, P. L., Walsh, K. A., and Neurath, H. (1976). Thermal stability of homologous neutral metalloendopeptidases in thermophilic and mesophilic bacteria: structural considerations. *Experientia, Suppl.* **26,** 19-30.

Parija, P., and Mallik, P. (1941). Nature of the reserve food in seeds and their resistance to high temperature. *J. Indian Bot. Soc.* **19,** 223-230.

Pasternak, D., and Wilson, G. L. (1969). Effects of heat waves on grain sorghum at the stage of head emergence. *Aust. J. Exp. Agric. Anim. Husb.* **9,** 636-638.

Patterson, D. R. (1974). Stress ethylene production and the respiration rate, internal atmosphere, early growth, and yield of *Ipomoea batatas* plants. *J. Am. Soc. Hortic. Sci.* **99,** 481-483.

Peacocke, A. R., and Walker, I. O. (1962). The thermal denaturation of sodium deoxyribonuclease. II. Kinetics. *J. Mol. Biol.* **5,** 560-563.

Pearcy, R. W. (1977). Acclimation of photosynthetic and respiratory carbon dioxide exchange to growth temperature in *Atriplex lentiformis* (Torr.) Wats. *Plant Physiol.* **59,** 795-799.

Pearcy, R. W. (1978). Effect of growth temperature on the fatty acid composition of the leaf lipids in *Atriplex lentiformis* (Torr.) Wats. *Plant Physiol.* **61,** 484-486.

Pearcy, R. W., Berry, J. A., and Bartholomew, B. (1974). Field photosynthetic performance and leaf temperatures of *Phragmites communis* under summer conditions in Death Valley, California. *Photosynthetica* **8,** 104-108.

Pearcy, R. W., Berry, J. A., and Fork, D. C. (1977). Effects of growth temperature on the thermal stability of the photosynthetic apparatus of *Atriplex lentiformis* (Torr.) Wats. *Plant Physiol.* **59,** 873-878.

Pearse, B. M. F., and Harris, J. I. (1973). 6-phosphogluconate dehydrogenase from *Bacillus stearothermophilus*. *FEBS Lett.* **38,** 49-52.

Perutz, M. F. (1978). Electrostatic effects in proteins. *Science* **201,** 1187-1191.

Petinov, N. S., and Molotkovsky, U. G. (1957). Protective reactions in heat-resistant plants induced by high temperatures. *Fiziol. Rast.* **4,** 221-228.

Petinov, N. S., and Molotkovsky, U. G. (1961). The protective processes of heat-resistant plants. *Arid Zone Res.* **16,** 275-283.

Petinov, N. S., and Molotkovsky, Yu, G. (1962). Heat stability of plants and ways of increasing it. *Vestn. Akad. Nauk SSSR* **8,** 62-64.

Petrochenko, S. I., and Privalov, P. L. (1973). Change of hydration of globular proteins during their heat denaturation. *Biofizika* **18,** 555-557.

Pisek, A., Larcher, W., Pack, I., and Unterholzner, R. (1968). Kardinale Temperaturbereiche der Photosynthese und Grenztemperaturen des Lebens der Blätter verschiedener Spermatophyten. II. Temperaturmaximum der Netto-Photosynthese und Hitzeresistenz der Blätter. *Flora (Jena)* **158,** 110-128.

Polanshek, M. M. (1977). Effects of heat shock and cycloheximide on growth and division of the fission yeast. *Schizosaccharomyces pombe. J. Cell Sci.* **23,** 1-23.

Porodko, T. M. (1926a). Über die Absterbegeschwindigkeit der erhitzten Samen. *Ber. Dtsch. Bot. Ges.* **44,** 71-80.

Porodko, T. M. (1926b). Einfluss der Temperatur auf die Absterbegeschwindigkeit der Samen. *Ber. Dtsch. Bot. Ges.* **44,** 80-84.

Porto, F., and Siegel, S. M. (1960). Effects of exposures of seeds to various physical agents. III. Kinetin-reversible heat damage in lettuce seed. *Bot. Gaz. (Chicago)* **122,** 70-71.

Pozmogova, I. N., and Mal'yan, A. N. (1976). ATP content in *Candida tropicalis* cells growing at different temperatures. *Mikrobiologiya* **45,** 413-416.

Prat, S., and Kubin, S. (1956). Photosynthesis and respiration of thermophilic blue-green algae. *Fiziol. Rast.* **3,** 508-515.

Pulgar, C. E., and Laude, H. M. (1974). Regrowth of alfalfa after heat stress. *Crop Sci.* **14,** 28-30.

Raju, K. S., Maheshwari, R., and Sastry, P. S. (1976). Lipids of some thermophilic fungi. *Lipids* **11,** 741-746.

Raschke, K. (1956). Über die physikalischen Beziehungen zwischen Wärmeübergangszahl. Strahlungsaustausch, Temperatur und Transpiration eines Blattes. *Planta* **48,** 200-238.

Raschke, K. (1960). Heat transfer between the plant and the environment. *Annu. Rev. Plant Physiol.* **11,** 111-126.

Ray, M., and Bhaduri, A. (1976). Uridine-diphosphate-glucose 4- epimerase from *Saccharomyces fragilis:* Inactivation by heat and reconstitution of the inactive enzyme. *Eur. J. Biochem.* **70,** 319-323.

Razmaev, I. I. (1974). Comparative characterization of heat resistance of plants according to their respiration rate. *Uzb. Biol. Zh.* **18,** 14-17.

Reinert, J. C., and Steim, J. M. (1970). Calorimetric detection of a membrane-lipid phase transition in living cells. *Science* **168,** 1580-1582.

Reynolds, T., and Thompson, P. A. (1971). Characterisation of the high temperature inhibition of germination of lettuce (*Lactuca sativa*). *Physiol. Plant.* **24,** 544-547.

Rieber, M., and Weinstein-Schonfeld, F. (1974). Appearance of an inhibitor of RNA polymerase activity in thermophilic *Mycobacterium phlei* exposed to sublethal temperatures. *Biochim. Biophys. Acta* **361,** 236-240.

Rigby, B. J. (1967). Correlation between serine and thermal stability of collagen. *Nature (London)* **214,** 87-88.

Risueno, M. C., Stockert, J. C., Gimenez-Martin, G., and Diez, J. L. (1973). Effect of supraoptimal temperatures on meristematic cells nucleoli. *J. Microsc. (Paris)* **16,** 87-94.

Robbins. W. J., and Petsch, K. F. (1932). Moisture content and high temperature in relation to the germination of corn and wheat grains. *Bot. Gaz. (Chicago)* **93,** 85-92.

Robertson, A. H. (1927). Thermophile and thermoduric microorganisms, with special reference to species isolated from milk. *N.Y. Agric. Exp. Stn., Geneva, Tech. Bull.* **130**.

Rosenberg, A., and Enberg, J. (1969). Studies of hydrogen exchange in protein. II. The reversible thermal unfolding of chymotrypsinogen A as studied by by exchange kinetics. *J. Biol. Chem.* **244,** 6153-6159.

Rouschal, E. (1938). Zum Wärmehaushalt der Macchienpflanzen. *Oesterr. Bot. Z.* **87,** 42-50.

Roy, D. K. (1956). Heat stability of fungal alpha-amylase. *Ann. Biochem. Exp. Med.* **16,** 111-112.

Rudzyanok, A. M., and Konyew, S. V. (1972). Use of various methods of differential staining of yeast cells for studying the characteristics of heat damage of their membranes. *Vyestsi. Akad. Navuk. B. SSR Syer. Biyal. Navuk.* **6,** 66-70.

Rueegg, M., Moor, U., and Blanc, B. (1975). Hydration and thermal denaturation of B-lactoglobulin: A calorimetric study. *Biochim. Biophys. Acta* **400,** 334-342.

Sachs, J. (1864). Ueber die obere Temperatur-Grenze der Vegetation. *Flora (Jena)* **47,** 5-12, 24-29, 33-39, and 65-75.

Saiki, T., Shinshi, H., and Arima, K. (1973). Studies on homoserine dehydrogenase from an extreme thermophile, *Thermus flavus* HT-62. Partial purification and properties. *J. Biochem. (Tokyo)* **74,** 1239-1248.

Samimy, C., and Lamotte, C. E. (1976). Anomalous temperature dependence on seedling development in some soybean (*Glycine max* (L.) Merr.) cultivars: Role of ethylene. *Plant Physiol.* **58,** 786-789.

Sapper, I. (1935). Versuche zur Hitzresistenz der Pflanzen. *Planta* **23,** 518-556.

Sato, K., and Inaba, K. (1976). High temperature injury of ripening in rice plant. V. On the early decline of assimilate storing ability of grains at high temperature. *Proc. Crop Sci. Soc. Jpn.* **45,** 156-161.

Schaefers, H.-A., and Feierabend, J. (1976). Ultrastructural differentiation of plastids and other organelles in rye leaves with high-temperature-induced deficiency of plastid ribosomes. *Cytobiologie* **14,** 75-90.

Schanderl, H. (1955). Studies on the internal temperature of submerged water plants. *Ber. Dtsch. Bot. Ges.* **68,** 28-34.

Scheibmair, G. (1937). Hitzresistenz-Studien an Moos-Zellen. *Protoplasma* **29,** 394-424.

Schiebel, W., Chayka, T. G., de Vries, A., and Rusch, H. P. (1969). Decrease of protein synthesis and breakdown of polyribosomes by elevated temperature in *Physarum polycephalum*. *Biochem. Biophys. Res. Commun.* **35,** 338-345.

Schneider-Orelli, O. (1910). Versuche über die Widerstandsfähigket gewisser Medicago-Samen (Wollkletten) gegen hohe Temperaturen. *Flora (Jena)* **100,** 305-311.

Schroeder, C. A. (1963). Induced temperature tolerance of plant tissue in vitro. *Nature* (*London*) **200,** 1301-1302.

Schwemmle, B., and Lange, O. L. (1959a). Endogen-Tagesperiodische Schwankungen der Hitzeresistenz bei *Kalanchoe blossfeldiana*. *Planta* **53,** 134–144.

Schwemmle, B., and Lange, O. L. (1959b). Neue Beobachtungen über die Endogene Tagesrhythmik. *Nachr. Akad. Wiss. Goettingen, II. Math.-Phys. Kl., 2* No. 3, pp. 31-35.

Seagrave, S. (1976). Scientists learn from wild plants. *BioScience* **26,** 153-156.

Semenenko, V. E., Vladimirova, M. G., Orleanskaya, O. B., Raikov, N. I., and Kovanova, E. S. (1969). Physiological characteristics of Chlorella during disturbance of cell functions by high temperatures. *Fiziol. Rast.* **16,** 210-220.

Semikhatova, O. A., and Egorova, L. I. (1976). Effect of light on the reactivation of photosynthesis in corn leaves after heat stress. *Bot. Zh.* (*Leningrad*) **61,** 313-323.

Sen, D. N., Chawan, D. D., and Sharma, K. D. (1972). Ecology of Indian desert. V. On the water relations of Salvadora species. *Flora* (*Jena*) **161,** 463-471.

Shcherbakova, A. M. (1972). Thermostability of glucose-6-phosphate dehydrogenase from leaves of winter wheat hardened by heat and cold. *Dokl. Akad. Nauk SSSR, Ser. Biol.* **205,** 993-996.

Shcherbakova, A. M., Fel'dman, N. L., and Shukhtina, G. G. (1973). Changes in the thermostability of some proteins after the heat hardening of wheat leaves of different ages. *Tsitologiya* **15,** 391-398.

Sheridan, R. P., and Ulik, T. (1976). Adaptive photosynthesis responses to temperature extremes by the thermophilic cyanophyte *Synechococcus lividus*. *J. Phycol.* **12,** 255-261.

Sherman, F. (1959). The effects of elevated temperatures on yeast. I. Nutrient requirements for growth at elevated temperatures. *J. Cell. Comp. Physiol.* **54,** 29-36.

Shimizu, I., and Katsuki, H. (1975). Effect of temperature on ergosterol biosynthesis in yeast. *J. Biochem.* (*Tokyo*) **77,** 1023-1028.

Shiralipour, A., and Anthony, D. S. (1970). Chemical prevention of growth reduction caused by supraoptimal temperatures in *Arabidopsis thaliana*. *Phytochemistry* **9,** 463-469.

Siegel, S. M. (1969). Further studies on factors affecting the efflux of betacyanin from beet root: A note on the thermal effects. *Physiol. Plant.* **22,** 327-331.

Skogqvist, I. (1973). Induction of thermosensitivity in wheat roots: Salt sensitivity and effects of chloramphenicol and ethanol. *Physiol. Plant.* **28,** 77-80.

Skogqvist, I. (1974a). Induction of heat sensitivity of wheat roots and its effects on mitochondria, adenosine triphosphate, triglyceride, and total lipid content. *Exp. Cell Res.* **86,** 285-294.

Skogqvist, I. (1974b). Induced heat sensitivity of wheat roots and protecting effect of ethanol and kinetin. *Physiol. Plant.* **32,** 166-169.

Skogqvist, I., and Fries, N. (1970). Induction of thermosensitivity and salt sensitivity in wheat roots and the effect of kinetin. *Experientia* **26,** 1160-1162.

Smith, F. H., and Silen, R. R. (1963). Anatomy of heat-damaged Douglas-fir seedlings. *For. Sci.* **9,** 15-32.

Sorauer, P. (1924). "Handbuch der Pflanzenkrankheiten," Vol. 1. Parey, Berlin.

Stanhill, G., Moreshet, S., and Fuchs, M. (1976). Effect of increasing foliage and soil reflectivity on the yield and water use efficiency of grain sorghum. *Agron. J.* **68,** 329-332.

Starr, P. R., and Parks, L. W. (1962). The effect of temperature on sterol metabolism in yeast. *J. Cell. Comp. Physiol.* **59,** 107-110.

Stevenson, K. E., and Richards, L. J. (1976). Thermal injury and recovery of *Saccharomyces cerevisiae*. *J. Food Sci.* **41,** 136-137.

Strain, B. R. (1969). Seasonal adaptations in photosynthesis and respiration in four desert shrubs growing in situ. *Ecology* **50,** 511-513.

Sullivan, C. Y., and Kinbacher, E. J. (1967). Thermal stability of Fraction I protein from heat-hardened *Phaseolus acutifolius* Gray, "Tepary Buff." *Crop Sci.* **7,** 241-244.

Sundaram, T. K., Cazzulo, J. J., and Kornberg, H. L. (1969). Anaplerotic CO_2 fixation in mesophilic and thermophilic bacilli. *Biochim. Biophys. Acta* **192,** 355-357.

Szirmai, J. (1938). Die Dörrfleckenkrankheit (Hitzeschaden) des Paprikas. *Phytopahtol. Z.* **11,** 1-13.

Taleinshik, E. D., and Usol'tseva, S. M. (1967). Some data on the water balance and heat resistance of Manchu, sour, ground and Sakhalin cherry trees. *Fiziol. Rast.* **14,** 1065-1070.

Tamura, Y., and Morita, Y. (1975). Thermal denaturation and regeneration of Japanese radish peroxidase. *J. Biochem.* (*Tokyo*) **78,** 561-57i.

Tansey, M., and Brock, T. D. (1972). The upper temperature limit for eukaryotic organisms. *Proc. Natl. Acad. Sci. U.S.A.* **69,** 2426-2428.

Tekman, S. and Oztekin, S. (1972). The preventive effect of some sugars and mannitol on the heat coagulation of proteins. *Istanbul Univ. Eczacilik Fak. Mecm.* **8,** 41-46.

Thorpe, R. F., and Ratledge, C. (1973). Fatty acids of triglycerides and phospholipids from a thermotolerant strain of *Candida tropicalis* grown on n-alkanes at 30° and 40°C. *J. Gen. Microbiol.* **78,** 203-206.

Tieszen, L. L., and Sigurdson, D. C. (1973). Effect of temperature on carboxylase activity and stability in some Calvin cycle grasses from the arctic. *Arct. Alp. Res.* **5,** 59-66.

Till, O. (1956). Über die Frosthärte von Pflanzen sommergrüner Laubwälder. *Flora* (*Jena*) **143,** 499-542.

Tischner, R., and Lorenzen, H. (1974). Effect of high growth temperature on cell compounds and enzymes in synchronized Chlorella. *New Bot.* **1,** 72-88.

Tischner, R., and Lorenzen, H. (1975). Physiological effects of heat shocks in synchronous Chlorella during their most sensitive developmental stage. *Biochem. Physiol. Plfanz.* **168,** 233-245.

Tobiessen, P. (1976). Thermal acclimation of dark respiration in coastal and desert populations of *Isomeris arborea*. *Am. Midl. Nat.* **96,** 462-466.

Toprover, Y., and Glinka, Z. (1976). Calcium ions protect beet root cell membranes against thermally induced changes. *Physiol. Plant.* **37,** 131-134.

Troughton, J. H., Björkman, O., and Berry, J. A. (1974). Growth of sorghum at high temperatures. *Carnegie Inst. Washington, Yearb.* **73,** 835-838.

Unrath, C. R., and Sneed, R. E. (1974). Evaporative cooling of "Delicious" apples: The economic feasibility of reducing environmental heat stress. *J. Am. Hortic. Sci.* **99,** 372-375.

Ushakov, B. P. (1964). Thermostability of cells and proteins of poikilotherms and its significance in speciation. *Physiol. Rev.* **44,** 518-560.

Ushakov, B. P. (1966). The problem of associated changes in protein thermostability during the process of speciation. *Helgol. Wiss. Meeresunters.* **14,** 466-481.

Ushakov, B. P., and Glushankova, M. A. (1961). On the lack of a deficit correlation between the iodine number of protoplasmic lipids and the cell thermostability. *Tsitologiya* **3,** 707-710.

Utkhede, R. S., and Jain, H. K. (1976). Ribonucleic acid and high temperature sensitivity in wheat. *Cytologia* **41,** 1-4.

Van Halteren, P. (1950). Effets d'un choc thermique sur le métabolisme des levures. *Bull. Soc. Chim. Biol.* **32,** 458-463.

Van Uden, N., and Madeira-Lopes, A. (1975). Dependence of the maximum temperature for growth of *Saccharomyces cerevisiae* on nutrient concentration. *Arch. Microbiol.* **104,** 23-28.

Van Uden, N., and Vidal-Leiria, M. M. (1976). Thermodynamic compensation in microbial thermal death: Studies with yeasts. *Arch. Microbiol.* **108,** 293-298.

Veselovskii, V. A., Leshchinskaya, L. V., Markarova, E. N., Veselova, T. V., and Tarusov, B. N. (1976). Effect of illumination of cotton leaves on heat resistance of the photosynthetic apparatus. *Fiziol. Rast.* **23,** 467-472.

Vitvitskii, V. N. (1969). Effect of methyl orange on the tryptic hydrolysis and heat denaturation of serum albumin. *Mol. Biol.* **3,** 678-682.

Vlasyuk, P. A., Balaganskaya, V. E., and Goryacheva, L. O. (1974). Trace elements and corn heat tolerance. *Fiziol. Biokhim. Kul't. Rast.* **6,** 33-36.

Voordouw, G., and Roche, R. S. (1975). The role of bound calcium ions in thermostable, proteolytic enzymes. I. Studies on thermomycolase, the thermostable protease from the fungus *Malbranchea pulchella. Biochemistry* **14,** 4659-4666.

Voordouw, G., Gaucher, G. M., and Roche, R. S. (1974). Physiochemical properties of thermomycolase, the thermostable, extracellular, serine protease of the fungus *Malbranchea pulchella. Can. J. Biochem.* **52,** 981-990.

Vouk, V. (1923). Die Probleme der Biologie der Thermen. *Int. Rev. Hydrobiol.* **11,** 89-99.

Vrolik, G., and de Vriese, W. H. (1839). Nouvelles expériences sur l'élévation de température du spadice d'une Colocasia odora (*Caladium odorum*) faites au Jardin Botanique d'Amsterdam. *Ann. Sci. Nat. Bot. Biol. Veg.* [2] **11,** 65-85.

Wagenbreth, D. (1965). Durch Hitzeschocks induzierte Vitalitätsänderungen bei Laubholzblättern. *Flora (Jena)* **156,** 63-75.

Waggoner, P. E., and Shaw, R. H. (1953). Temperature of potato and tomato leaves. *Plant Physiol.* **27,** 710-724.

Ward, E. W. B. (1968). The low maximum temperature for growth of the psychrophile *Sclerotinia borealis.* Evidence for the uncoupling of growth from respiration. *Can. J. Bot.* **46,** 385-390.

Watanabe, K., and Oshima, T. (1974). Replacement of ribothymidine by 5-methyl-2-thiouridine in sequence GTΨC in tRNA of an extreme thermophile. *FEBS Lett.* **43,** 59-63.

Watanabe, M. (1953). Effect of heat application upon the pollen viability of Japanese black pine and Japanese red pine. *J. Jpn. For. Soc.* **35,** 248-251.

Weaver, L. H., Kester, W. R., Ten Eyck, L. F., and Matthews, B. W. (1976). The structure and stability of thermolysin. *Experientia, Suppl.* **26,** 31-36.

Weber, D. J., Andersen, W. R., Hess, S., Hansen, D. J., and Gunasekaran, M. (1977). Ribulose-1,5-bisphosphate carboxylase from plants adapted to extreme environments. *Plant Cell Physiol.* **18,** 693-699.

Weber, F. (1926a). Hitzeresistenz funktionierender Schliesszellen. *Planta* **1,** 553-557.

Weber, F. (1926b). Hitzeresistenz funktionierender Stomata-Nebenzellen. *Planta* **2,** 669-677.

Wedler, F. C., and Hoffmann, F. M. (1974). Glutamine synthetase of *Bacillus stearothermophilus.* II. Regulation and thermostability. *Biochemistry* **13,** 3215-3221.

Weetall, H. H. (1969). Trypsin and papain covalently coupled to porous glass: Preparation and characterization. *Science* **166,** 615-616.

Weidner, M., and Ziemens, C. (1975). Preadaptation of protein synthesis in wheat seedlings to high temperature. *Plant Physiol.* **56,** 590-594.

Williams, G. J., III (1974). Photosynthetic adaptation to temperature in C_3 an C_4 grasses. A possible ecological role in the shortgrass prairie. *Plant Physiol.* **54,** 709-711.

Williams, G. J., III, and Kemp, P. R. (1976). Temperature relations of photosynthetic response in populations of *Verbascum thapsus* L. *Oecologia* **25,** 47-54.

Williamson, R. E. (1963). The effect of a transpiration-suppressant on tobacco leaf temperature. *Soil Sci. Soc. Am., Proc.* **27,** 106.

Wollgiehn, R., and Munsche, D. (1974). Thermolability of the ribosomal RNA from *Anacystis nidulans*. *Biochem. Physiol. Pflanz.* **165,** 407–418.

Wolpert, A. (1962). Heat transfer analysis of factors affecting plant leaf temperature. Significance of leaf hair. *Plant Physiol.* **37,** 113–120.

Yakovleva, V. I., Komissarova, E. N., and Gubnitskii, L. S. (1974). The thermal stability of rat liver glutamate dehydrogenase in a system of isolated mitochondria. *Tsitologiya* **16,** 167–171.

Yannas, I. V., and Tobolsky, A. V. (1967). Cross linking of gelatin by dehydration. *Nature* (*London*) **215,** 509–510.

Yarwood, C. E. (1961a). Acquired tolerance of leaves to heat. *Science* **134,** 941–942.

Yarwood, C. E. (1961b). Translocated heat injury in plants. *Nature* (*London*) **192,** 887.

Yarwood, C. E. (1962). Acquired sensitivity of (*Phaseolus vulgaris* var. Pinto) leaves to heat. *Plant Physiol.* **37,** Suppl., 70.

Yarwood, C. E. (1963). Sensitization of leaves to heat. *Adv. Front. Plant Sci.* **7,** 195–203.

Yarwood, C. E. (1967). Adaptations of plants and plant pathogens to heat. *Publ. Am. Assoc. Adv. Sci.* **84,** 75–92.

Yocum, H. R., Blumberg, P. M., and Strominger, J. L. (1974). Purification and characterization of the thermophilic D-alanine carboxypeptidase from membranes of *Bacillus stearothermophilus*. *J. Biol. Chem.* **249,** 4863–4871.

Yon, R. J. (1973). Wheat-germ aspartate transcarbamoylase: The effects of ligands on the inactivation of the enzyme by trypsin and denaturing agents. *Biochem. J.* **131,** 699–706.

Yordanov, I. T., and Vasil'eva, V. S. (1976). Effect of elevated temperature on the rate of photosynthesis and the activity of ribulosediphosphate and phosphoenolpyruvate carboxylases. *Fiziol. Rast.* **23,** 812–817.

Yordanov, I. T., Zeinalov, Y., and Stamenova, M. (1975). Influence of post-action of high temperatures on proteins and spectral characteristics of lamellar proteins and spectral characteristics of pigment systems I and II. *Biochem. Physiol. Pflanz.* **168,** 567–673.

Yoshida, M., Sone, N., Hirata, H., and Kagawa, Y. (1977). Reconstitution of adenosine triphosphatase of thermophilic bacterium from purified individual subunits. *J. Biol. Chem.* **252,** 3480–3485.

Yu, S. A., Sussman, A. S., and Wooley, S. (1967). Mechanisms of protection of trehalase against heat inactivation in Neurospora. *Bacteriology* **94,** 1306–1312.

Yutani, K. (1976). Role of calcium ion in the thermostability of α-amylase produced from *Bacillus stearothermophilus*. *Experientia, Suppl.* **26,** 91–103.

Yutani, K., Sasaki, I., and Ogasahara, K. (1973). Comparison of thermostable α-amylases from *B. stearothermophilus* grown at different temperatures. *J. Biochem. (Tokyo)* **74,** 573–579.

Yutani, K., Ogasahara, K., Sugino, Y., and Matsushiro, A. (1977). Effect of a single amino acid substitution on stability of conformation of a protein. *Nature* (*London*) **267,** 274–275.

Zavadskaya, I. G., and Den'ko, E. I. (1968). Effect of an insufficient water supply on the stability of leaf cells of some plants of the Pamirs. *Bot. Zh.* **53,** 795–805.

Zavadskaya, I. G., and Shukhtina, G. G. (1974). Effect of dehydration and elevated temperature on leaf cell thermoresistance of a drought-sensitive barley cultivar. *Tsitologiya* **16,** 950–955.

Zobl, K. (1950). Ueber die Beziehungen zwischen chemischer Zusammensetzung von Pilzsporen und ihrem Verhalten gegen Erhitzen. *Sydowia* **4,** 175–184.

Index

C

G

I

N

X

Y, Z